# 2nd Edition of Integrated Human Exposure to Air Pollution

# 2nd Edition of Integrated Human Exposure to Air Pollution

Nuno Canha
Marta Almeida
Evangelia Diapouli

MDPI

Basel • Beijing • Wuhan • Barcelona • Belgrade • Novi Sad • Cluj • Manchester

*Editors*

Nuno Canha
Safety and Risk
Assessment Unit
HyLab–Green Hydrogen
Collaborative Laboratory
Sines
Portugal

Marta Almeida
Centro de Ciências e
Tecnologias Nucleares
Universidade de Lisboa
Lisbon
Portugal

Evangelia Diapouli
National Centre for Scientific
Research "DEMOKRITOS"
Athens
Greece

*Editorial Office*
MDPI AG
Grosspeteranlage 5
4052 Basel, Switzerland

This is a reprint of articles from the Special Issue published online in the open access journal *International Journal of Environmental Research and Public Health* (ISSN 1660-4601) (available at: www.mdpi.com/journal/ijerph/special_issues/2nd_Air_Pollution_Exposure).

For citation purposes, cite each article independently as indicated on the article page online and as indicated below:

Lastname, A.A.; Lastname, B.B. Article Title. *Journal Name* **Year**, *Volume Number*, Page Range.

**ISBN 978-3-7258-2334-5 (Hbk)**
**ISBN 978-3-7258-2333-8 (PDF)**
**doi.org/10.3390/books978-3-7258-2333-8**

# Contents

# About the Editors

**Nuno Canha**

Nuno Canha holds an MSc in Chemistry from the Instituto Superior Técnico (IST), University of Lisbon (Portugal), and completed his PhD in Environmental Sciences at the Delft University of Technology (The Netherlands) in 2014. His doctoral research, carried out at IST's Centre for Nuclear Sciences and Technologies (C2TN), focused on indoor air quality in primary schools, identifying and characterising sources of particulate matter (PM) in classrooms. For 15 years, Nuno Canha has conducted research on indoor and outdoor air quality, focusing on the health effects of PM exposure (based on its toxicity and oxidative potential) and on the human exposure to air pollutants in different micro-environments (including sleeping environments). He has co-authored 69 papers in peer reviewed scientific journals and has made 172 contributions to national and international scientific conferences.

Since January 2024, Nuno Canha has been a Lead Researcher at HyLab (Green Hydrogen Collaborative Laboratory), coordinating the Safety and Risk Assessment Unit. His main research topics include optimising methodologies to assess hydrogen leaks through simulation and experimentation, developing strategies to minimise risks in hydrogen storage, transport, and use, and defining best practices to ensure safety throughout the hydrogen value chain. Nuno Canha has always been committed to science communication, actively promoting and participating in outreach activities for different audiences.

**Marta Almeida**

Marta Almeida graduated in Environmental Engineering in 1998 from the NOVA School of Science and Technology. She completed her PhD in Environmental Sciences in 2004 at the University of Aveiro and obtained her Habilitation Degree in Environmental Engineering from Instituto Superior Técnico in 2023. For about 25 years, she has conducted research on air quality, climate change, and health in urban systems. She worked at Instituto de Soldadura e Qualidade, where, in addition to her research activities, she provided consultancy services on air quality management to national and international industries. She is currently a Principal Researcher at the Department of Nuclear Sciences and Engineering (DECN) and carries out research at the Centre for Nuclear Sciences and Technologies (C2TN), where she coordinates a team of young researchers with whom she shares her passion for research. In partnership with industries, companies, authorities, and non-governmental organisations, her team studies the characteristics and origin of air pollutants, assesses the health effects of exposure to air pollution, and defines strategies and tools to improve air quality and the well-being of citizens.

**Evangelia Diapouli**

Evangelia Diapouli is a Principal Researcher at the Institute of Nuclear & Radiological Sciences and Technology, Energy & Safety, of the National Centre for Scientific Research "Demokritos". She holds a Chemical Engineering Diploma from the National Technical University of Athens (2000), an MSc in Environmental Engineering from Johns Hopkins University, USA (2002), and a PhD focusing in Indoor Air Quality from the School of Chemical Engineering of the National Technical University of Athens (2008). She has more than 20 years of experience in the study of indoor and ambient air quality. Her research focuses on the physico-chemical characterization of airborne particulate matter, aerosol source apportionment, and population and occupational risk assessment.

Dr Diapouli has participated in several European and national research projects, in some of them acting as Principal Investigator either for the NCSR "Demokritos" team or for the whole project. She has also been actively involved in European initiatives for the development of standard methods for aerosol characterization, including the quantification of carbonaceous species in airborne particles and the adoption of a European harmonized methodology in the application of receptor models for PM source apportionment. She has co-authored 85 papers in peer reviewed journals and has presented more than 130 papers at national and international conferences. Since 2019, she has been a member of the Management Board of the Hellenic Association for Aerosol Research (HAAR).

# Preface

This reprint, entitled "Second Edition of Integrated Human Exposure to Air Pollution", presents new insights into key aspects of human exposure to air pollution. It compiles seventeen innovative studies, exploring five main areas: advanced methodologies for exposure assessment, citizen engagement in addressing pollution, outdoor air pollution, indoor air quality, and the health impacts of pollution. Notably, the reprint features two comprehensive reviews, one on the effects of climate change on indoor air and another on the health impacts of long- and short-term exposure to $NO_2$ and $PM_{2.5}$.

**Nuno Canha, Marta Almeida, and Evangelia Diapouli**
*Editors*

International Journal of
*Environmental Research and Public Health*

MDPI

*Editorial*

# Integrated Human Exposure to Air Pollution: A Step Further

**Nuno Canha [1,*], Evangelia Diapouli [2] and Susana Marta Almeida [1]**

1 Centro de Ciências e Tecnologias Nucleares ($C^2TN$), Instituto Superior Técnico, Universidade de Lisboa, Estrada Nacional 10, Km 139.7, 2695-066 Bobadela LRS, Portugal; smarta@ctn.tecnico.ulisboa.pt

2 National Centre for Scientific Research "Demokritos", Agia Paraskevi, 15341 Athens, Greece; ldiapouli@ipta.demokritos.gr

* Correspondence: nunocanha@ctn.tecnico.ulisboa.pt

Along with climate change, air pollution is one of the biggest environmental problems affecting everyone in the world today. According to the World Health Organization, air pollution causes an estimated 7 million premature deaths and reduces the healthy life years of millions more [1]. The burden of disease attributable to air pollution is now thought to be comparable to that of other major global health risks, such as unhealthy diets and tobacco smoking.

In order to reduce the negative impact of human exposure to air pollution on the health and well-being of citizens, it is essential to understand and develop strategies and mitigation measures to control it. However, the exposure of citizens to air pollutants is typically based only on the concentrations of pollutants measured at air quality monitoring stations operated by national environmental agencies. These monitoring stations focus on outdoor air quality and are usually located in urban centres.

This approach does not take all components of exposure into account, as people spend a large proportion of their time indoors and have different time–activity patterns, and there is also high variability in air pollutant concentrations within a given city.

Therefore, human exposure over a whole day cannot be reflected only by outdoor exposure, and should consider all micro-environments wherein people spend their time (e.g., home, workplace, transport, leisure, and others) and the time spent in them. The characterisation of indoor and outdoor environments is essential to assess integrated human exposure to air pollutants.

The need to increase knowledge in this area led us to create a Special Issue of IJERPH dedicated to integrated human exposure to air pollutants (https://www.mdpi.com/journal/ijerph/special_issues/IHETAP, accessed on 1 September 2023). Considering the great interest that this Special Issue has received from researchers all over the world in its first edition, we felt that a second edition would be useful to collect additional valuable information and research that has been carried on this important topic out since then.

With this second edition of the IHETAP Special Issue, we invited colleagues to contribute with research focusing on human exposure in different microenvironments, individual exposure under specific conditions and activities, and methodologies to understand pollution sources and their impact on indoor and ambient air quality, with the main aim of developing effective mitigation measures to reduce human exposure and protect public health.

In total, the second edition of this Special Issue brings together eighteen peer-reviewed open access articles that provide new insights into important topics in the field of human exposure to air pollution. Overall, five main areas have been discussed and explored in this Special Issue, namely (i) new methodologies for human exposure assessment (contribution 1); (ii) citizen empowerment with regard to air pollution (contributions 2, 3, and 4); (iii) outdoor air pollution (contributions 5, 6, 7, and 8); (iv) indoor air quality (contributions 9, 10, 11, 12, and 13); and (v) the health effects of human exposure to air pollution (contributions 14, 15, 16, and 17). We highlight two review articles in this edition of our



**Citation:** Canha, N.; Diapouli, E.; Almeida, S.M. Integrated Human Exposure to Air Pollution: A Step Further. *Int. J. Environ. Res. Public Health* **2023**, *20*, 7061. https://doi.org/10.3390/ijerph20227061

Received: 13 October 2023
Accepted: 30 October 2023
Published: 13 November 2023

Special Issue that provide comprehensive understanding of the impact of climate change on indoor air quality (contribution 12) and the health effects of long- and short-term exposure to ambient $NO_2$ and $PM_{2.5}$ (contribution 18). We believe that these studies will make a significant contribution to the advancement of knowledge in this field.

As editors of this Special Issue of IJERPH, we would like to acknowledge the great efforts of everyone involved in this process, from the dedication of all the reviewers who provided valuable feedback to the authors and the editorial team.

**Funding:** N.C. acknowledges the funding by national funds through FCT—Fundação para a Ciência e Tecnologia, I.P. (Portugal) for his contract (reference 2021.00088.CEECIND), and for the project HypnosAir (reference PTDC/CTA-AMB/3263/2021, https://doi.org/10.54499/PTDC/CTA-AMB/3263/2021). FCT support is also acknowledged by the $C^2TN$/IST authors (UIDB/04349/2020 + UIDP/04349/2020).

**Acknowledgments:** The guest editors of this Special Issue of the International Journal of Environmental Research and Public Health are grateful to all of the authors, reviewers, and MDPI staff.

**Conflicts of Interest:** The authors declare no conflict of interest.

**List of Contributions:**

1. Correia, C.; Martins, V.; Matroca, B.; Santana, P.; Mariano, P.; Almeida, A.; Almeida, S.M. A Low-Cost Sensor System Installed in Buses to Monitor Air Quality in Cities. *Int. J. Environ. Res. Public Health* **2023**, *20*, 4073. https://doi.org/10.3390/ijerph20054073.
2. Fernandes, J.; Brandão, T.; Almeida, S.M.; Santana, P. An Educational Game to Teach Children about Air Quality Using Augmented Reality and Tangible Interaction with Sensors. *Int. J. Environ. Res. Public Health* **2023**, *20*, 3814. https://doi.org/10.3390/ijerph20053814.
3. Canha, N.; Justino, A.R.; Gamelas, C.A.; Almeida, S.M. Citizens' Perception on Air Quality in Portugal—How Concern Motivates Awareness. *Int. J. Environ. Res. Public Health* **2022**, *19*, 12760. https://doi.org/10.3390/ijerph191912760.
4. Xue, W.; Li, X.; Yang, Z.; Wei, J. Are House Prices Affected by PM2.5 Pollution? Evidence from Beijing, China. *Int. J. Environ. Res. Public Health* **2022**, *19*, 8461. https://doi.org/10.3390/ijerph19148461.
5. Brągoszewska, E.; Mainka, A. Impact of Different Air Pollutants ($PM_{10}$, $PM_{2.5}$, $NO_2$, and Bacterial Aerosols) on COVID-19 Cases in Gliwice, Southern Poland. *Int. J. Environ. Res. Public Health* **2022**, *19*, 14181. https://doi.org/10.3390/ijerph192114181.
6. Pardo, N.; Sainz-Villegas, S.; Calvo, A.I.; Blanco-Alegre, C.; Fraile, R. Connection between Weather Types and Air Pollution Levels: A 19-Year Study in Nine EMEP Stations in Spain. *Int. J. Environ. Res. Public Health* **2023**, *20*, 2977. https://doi.org/10.3390/ijerph20042977.
7. Fatima, S.; Mishra, S.K.; Ahlawat, A.; Dimri, A.P. Physico-Chemical Properties and Deposition Potential of PM2.5 during Severe Smog Event in Delhi, India. *Int. J. Environ. Res. Public Health* **2022**, *19*, 15387. https://doi.org/10.3390/ijerph192215387.
8. Abecasis, L.; Gamelas, C.A.; Justino, A.R.; Dionísio, I.; Canha, N.; Kertesz, Z.; Almeida, S.M. Spatial Distribution of Air Pollution, Hotspots and Sources in an Urban-Industrial Area in the Lisbon Metropolitan Area, Portugal—A Biomonitoring Approach. *Int. J. Environ. Res. Public Health* **2022**, *19*, 1364. https://doi.org/10.3390/ijerph19031364.
9. Aldekheel, M.; Altuwayjiri, A.; Tohidi, R.; Jalali Farahani, V.; Sioutas, C. The Role of Portable Air Purifiers and Effective Ventilation in Improving Indoor Air Quality in University Classrooms. *Int. J. Environ. Res. Public Health* **2022**, *19*, 14558. https://doi.org/10.3390/ijerph192114558.
10. Wei, W.; Little, J.C.; Nicolas, M.; Ramalho, O.; Mandin, C. Modeling Primary Emissions of Chemicals from Liquid Products Applied on Indoor Surfaces. *Int. J. Environ. Res. Public Health* **2022**, *19*, 10122. https://doi.org/10.3390/ijerph191610122.
11. Tashakor, M.; Behrooz, R.D.; Asvad, S.R.; Kaskaoutis, D.G. Tracing of Heavy Metals Embedded in Indoor Dust Particles from the Industrial City of Asaluyeh, South of Iran. *Int. J. Environ. Res. Public Health* **2022**, *19*, 7905. https://doi.org/10.3390/ijerph19137905.
12. Mansouri, A.; Wei, W.; Alessandrini, J.-M.; Mandin, C.; Blondeau, P. Impact of Climate Change on Indoor Air Quality: A Review. *Int. J. Environ. Res. Public Health* **2022**, *19*, 15616. https://doi.org/10.3390/ijerph192315616.
13. Araja, A.; Bertins, M.; Celma, G.; Busa, L.; Viksna, A. Distribution of Minor and Major Metallic Elements in Residential Indoor Dust: A Case Study in Latvia. *Int. J. Environ. Res. Public Health* **2023**, *20*, 6207. https://doi.org/10.3390/ijerph20136207.

14. Wikuats, C.F.H.; Nogueira, T.; Squizzato, R.; de Freitas, E.D.; Andrade, M. de F. Health Risk Assessment of Exposure to Air Pollutants Exceeding the New WHO Air Quality Guidelines (AQGs) in São Paulo, Brazil. *Int. J. Environ. Res. Public Health* **2023**, *20*, 5707. https://doi.org/10.3390/ijerph20095707.
15. Chalvatzaki, E.; Chatoutsidou, S.E.; Almeida, S.M.; Morawska, L.; Lazaridis, M. The Representativeness of Outdoor Particulate Matter Concentrations for Estimating Personal Dose and Health Risk Assessment of School Children in Lisbon. *Int. J. Environ. Res. Public Health* **2023**, *20*, 5564. https://doi.org/10.3390/ijerph20085564.
16. Mohd Isa, K.N.; Jalaludin, J.; Mohd Elias, S.; Mohamed, N.; Hashim, J.H.; Hashim, Z. Evaluation of the Relationship between Fractional Exhaled Nitric Oxide (FeNO) with Indoor $PM_{10}$, $PM_{2.5}$ and $NO_2$ in Suburban and Urban Schools. *Int. J. Environ. Res. Public Health* **2022**, *19*, 4580. https://doi.org/10.3390/ijerph19084580.
17. Hisamuddin, N.H.; Jalaludin, J.; Abu Bakar, S.; Latif, M.T. The Influence of Environmental Polycyclic Aromatic Hydrocarbons (PAHs) Exposure on DNA Damage among School Children in Urban Traffic Area, Malaysia. *Int. J. Environ. Res. Public Health* **2022**, *19*, 2193. https://doi.org/10.3390/ijerph19042193.
18. Mainka, A.; Żak, M. Synergistic or Antagonistic Health Effects of Long- and Short-Term Exposure to Ambient $NO_2$ and $PM_{2.5}$: A Review. *Int. J. Environ. Res. Public Health* **2022**, *19*, 14079. https://doi.org/10.3390/ijerph192114079.

## Reference

1. WHO Global Air Quality Guidelines. Available online: https://apps.who.int/iris/handle/10665/345329 (accessed on 1 February 2022).

International Journal of
*Environmental Research and Public Health*

MDPI

*Article*

# A Low-Cost Sensor System Installed in Buses to Monitor Air Quality in Cities

**Carolina Correia [1,*], Vânia Martins [1], Bernardo Matroca [1], Pedro Santana [2,3], Pedro Mariano [1], Alexandre Almeida [2,4] and Susana Marta Almeida [1]**

1 Centro de Ciências e Tecnologias Nucleares, Instituto Superior Técnico, Universidade de Lisboa, Estrada Nacional 10, 2695-066 Bobadela, Portugal
2 ISCTE—Instituto Universitário de Lisboa (ISCTE-IUL), Av. das Forças Armadas, 1649-026 Lisboa, Portugal
3 ISTAR—Information Sciences and Technologies and Architecture Research Center, Av. das Forças Armadas, 1649-026 Lisboa, Portugal
4 Instituto de Telecomunicações, Av. Rovisco Pais, 1, 1049-001 Lisboa, Portugal
* Correspondence: ccorreia@ctn.tecnico.ulisboa.pt

**Abstract:** Air pollution is an important source of morbidity and mortality. It is essential to understand to what levels of air pollution citizens are exposed, especially in urban areas. Low-cost sensors are an easy-to-use option to obtain real-time air quality (AQ) data, provided that they go through specific quality control procedures. This paper evaluates the reliability of the ExpoLIS system. This system is composed of sensor nodes installed in buses, and a Health Optimal Routing Service App to inform the commuters about their exposure, dose, and the transport's emissions. A sensor node, including a particulate matter (PM) sensor (Alphasense OPC-N3), was evaluated in laboratory conditions and at an AQ monitoring station. In laboratory conditions (approximately constant temperature and humidity conditions), the PM sensor obtained excellent correlations ($R^2 \approx 1$) against the reference equipment. At the monitoring station, the OPC-N3 showed considerable data dispersion. After several corrections based on the k-Köhler theory and Multiple Regression Analysis, the deviation was reduced and the correlation with the reference improved. Finally, the ExpoLIS system was installed, leading to the production of AQ maps with high spatial and temporal resolution, and to the demonstration of the Health Optimal Routing Service App as a valuable tool.

**Keywords:** air quality; low-cost sensors; particulate matter; mobile experiments; calibration



**Citation:** Correia, C.; Martins, V.; Matroca, B.; Santana, P.; Mariano, P.; Almeida, A.; Almeida, S.M. A Low-Cost Sensor System Installed in Buses to Monitor Air Quality in Cities. *Int. J. Environ. Res. Public Health* **2023**, *20*, 4073. https://doi.org/10.3390/ijerph20054073

Academic Editors: Andrew S. Hursthouse and Chunrong Jia

Received: 16 December 2022
Revised: 26 January 2023
Accepted: 21 February 2023
Published: 24 February 2023



## 1. Introduction

Air pollution represents one of the biggest environmental risks to human health leading to increased morbidity and mortality due to cardiovascular and respiratory diseases [1–3]. While several measures have already been implemented in urban areas to tackle air quality (AQ) deterioration such as the improvement of public transportation [4–6], the levels of pollutants to which the citizens are exposed are still high and above the guidelines defined by the World Health Organization (WHO). The road transport sector is one of the major contributors to air pollution in urban areas [7]. In fact, in the transports microenvironment, commuters are particularly exposed to high levels of air pollution [8–10].

In urban areas, air pollutant monitoring and exposure assessment studies are traditionally performed through complex stationary equipment at fixed monitoring stations. Although very reliable, these stations are only able to characterize AQ in their vicinity and, due to the high acquisition and maintenance costs, they are installed in a small number of points within cities. Consequently, the spatial characterization of air pollutants' concentrations in urban areas is not fully achieved, compromising a correct assessment of the population's exposure to atmospheric pollutants [11], which can be accomplished through, for example, location-specific analysis [12].

Nowadays, low-cost sensors (LCS) are an increasingly attractive option to complement the data produced by the reference equipment from fixed monitoring stations [13]. These AQ sensors are low-cost, small, easy-to-use, and portable, which allows for the generation of massive quantities of information. Thus, they have been playing an important role in monitoring pollution trends, finding hot spots, identifying pollution sources, informing the population, and supporting environmental management [14–17].

However, although they have the capability to produce high temporally and spatially resolved AQ data, they tend to have more severe data quality problems compared to the conventional AQ monitoring instruments. To guarantee the quality of the data produced, LCS need to be calibrated and corrected against several factors. The complexity of particulate matter (PM) due to its emission sources, chemical composition, particle sizes, concentration ranges, and the environments in which the sensors are used led to the development of different calibration methods. The calibration of LCS may be developed at a laboratory or on-field. Laboratory caalibration is often seen as the most appropriate method since it makes it possible to exploit how different factors influence the sensors' response. On the other hand, on-field calibration can be the most appropriate method as it is carried out under the conditions in which the sensors will be used, although without the opportunity to analyze to what extent different factors influence the sensors' measurements. In both cases, calibration is usually performed by placing an LCS next to reference equipment.

PM LCS' readings are mostly influenced by meteorological factors, such as temperature (T) or relative humidity (RH) [15–20]. The most common methods to account for this effect in the calibration process include linear and multilinear regression analysis and specific RH correction techniques, which are further investigated in Section 2.3. In addition to this method, machine learning techniques [21–26] can also be employed to correct the sensors' data, having the advantage of including other factors in its analysis, such as the period of the day and characteristics of the area where the data is collected. While this may be seen as an advantage, it could transform the calibration process into a very location and condition-specific process, failing to generalize the calibration procedure [26]. In this paper, the LCS' calibration considers only the correlation with reference equipment and the influence of meteorological factors, as they are recognized as some of the main factors influencing the sensors' performance.

The ExpoLIS project developed a low-cost sensing system to be installed on the top of public buses which can cover an entire city generating a massive quantity of data with high spatial and temporal resolution, and a Health Optimal Routing Service App that aims to expand the citizens' knowledge about AQ and to reduce their exposure during commuting. This study aims to evaluate the performance of these ExpoLIS tools to measure PM concentrations in several conditions.

## 2. Methods

### 2.1. ExpoLIS LCS Sensor Node Prototype

The ExpoLIS sensor node prototype (for more details about the Sensor Node prototype, including the code and the model for 3d printing, see https://github.com/ExpoLIS-project—accessed on 23 February 2022) (Figure 1a) resulted from several trials that intended to improve its robustness and the accuracy of the installed sensors [27]. It is composed of an air inlet (in surface A in Figure 1b) to which can be connected a conductive tube to extend the air intake point. To ensure that an adequate quantity of air passes through the sensors without any delays, the exhaust point (in surface B in Figure 1b) is prepared to be connected to a conventional air pump to create an adequate airflow, preferably at a rate of 9 L $min^{-1}$ (a Leland Legacy air pump from SKC Inc. was used in this study). The sensor node's power supply can fluctuate between 12 volts of direct current (VDC) and 24 VDC, which can be provided either by a car lighter, a battery pack, or an AC/DC converter connected to an electrical outlet.

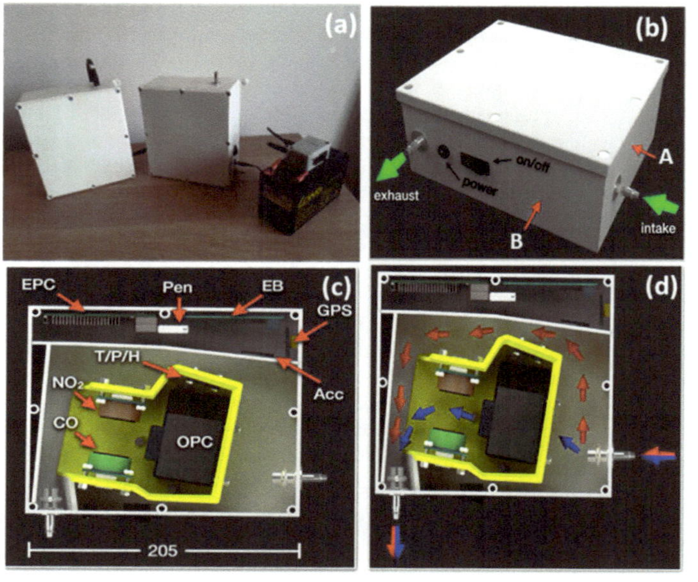


**Figure 1.** Sensor node details: (**a**) ExpoLIS sensor node prototype; (**b**) sensor node 3D model, including the representation of the air inlet (A) and the exhaust (B) surfaces; (**c**) LCSs and electronic devices included in the sensor box (EPC: Embedded Computer, Pen: USB Drive, EB: Custom Electronics Board, GPS: Global Positioning System device, Acc: Accelerometer, T/P/H: T, P and RH sensors, OPC: OPC-N3 sensor, $NO_2$: NO2-A43F sensor and CO: CO-B4 sensor) and (**d**) air flow in the interior of the sensor node.

In the interior of the sensor node, the allocation of each component reflects the need to ensure the smallest occupied space possible and an adequate airflow through the AQ LCS. The sensor node hardware is divided in two different categories: (1) the LCS and (2) the electronic devices, as represented in Figure 1c.

The sensor node is equipped with three AQ sensors and a meteorological sensor. A miniaturized optical particle counter (Model OPC-N3 from Alphasense) employs the principle of light scattering to count the numerical concentrations of PM that reach the detection chamber. These concentrations are converted on board to mass concentrations according to the European standard EN 481. Although the sensor provides concentrations of several size bins, for this study we only considered the concentrations of PM with aerodynamic diameters smaller than 2.5 μm (PM2.5) and 10 μm (PM10). Although this work only focuses on the measurement of PM, two electrochemical four-electrode gas sensors from Alphasense were also installed to monitor the concentrations of CO (CO-B4, Alphasense) and $NO_2$ (NO2-A43F, Alphasense). These sensors' working principle is based on the reaction of the given gas with a specific electrolyte that generates an electrical current. The sensors' output is composed of two different voltages, one from a working electrode and one from an auxiliary electrode, whose difference is proportional to the target gas concentrations. These voltages are periodically sampled by an analog-to-digital converter (ADC). Regarding the meteorological conditions, the sensor node monitors T, RH (SHTC3, Adafruit), and the barometric pressure (P) (LPS25, Adafruit). Each one of the measured concentrations is delivered at the rate of one second to accurately monitor concentrations while in motion.

As the electronic devices need to be protected against water droplets or moisture, they are placed in a section separated from the air flow (Figure 1c). Power distribution and sensor interfacing are handled by a custom electronics board. This board includes

16-bit analog-to-digital converters (ADS1115, Adafruit) for interfacing the gas sensors, and an accelerometer (MSA 301, Adafruit) for registering the sensor node's vibration and tilt angles. A GPS device (PA1010D, Adafruit) that geo references the data is fixed to one of the interior surfaces of the sensor node to ensure a proper sky view. Finally, the electronics board is connected to an embedded computer, RaspberryPI 3B+, responsible for processing and storing the gathered sensor data in an USB drive.

The sensor node's configuration allows the air that enters to circulate in the area allocated to the AQ LCS through the OPC-N3 that already has an integrated fan promoting airflow (blue arrows in Figure 1d). To avoid overflowing the OPC-N3 with any remaining air, a secondary path allows this air to easily surround the yellow 3D printed object and exit the sensor node at the exhaust point (red arrows in Figure 1d).

### 2.2. *Prototype Quality Control*

#### 2.2.1. Laboratory Tests

The prototype's quality control, and subsequent calibration, was performed through two laboratory tests to evaluate how specific factors can influence the data produced by the LCS. The first test was performed in a laboratory for 39 h and aimed at evaluating the performance of the sensors under controlled T and RH conditions. The LCS were exposed to an air pollution source (incense stick) to evaluate their response to several PM concentrations. Secondly, the sensors were placed in a garage for 14 h, where they were exposed to vehicle exhaust and higher variations of T and RH. In these tests, measurements were made in parallel with a light-scattering laser photometer (DustTrak, Model 8533, TSI Inc. Shoreview, MN, USA). This equipment is an optical instrument that measures simultaneously the mass concentrations of particles with several aerodynamic diameters, namely PM2.5 and PM10. The DustTrak measurements were corrected against reference PM concentration data obtained by gravimetry using a sampler (MVS6, Sven Leckel, Germany), which is certified according to CEN EN 12341. This sampler collected PM2.5 in Teflon filters and PM2.5-10 in nucleopore filters during 24 h periods at a constant flow rate of 2.3 $m^3h^{-1}$ (following the same approach as in [28]), which were weighted before and after sampling in a microbalance (R160P, Sartorius, Germany).

#### 2.2.2. Field Tests

Ambient environment is complex, so laboratory calibration may not be sufficient to correct for the outdoor conditions to which the LCS are exposed. Thus, it is necessary to perform field calibrations that take into account meteorological factors that can affect the LCS' performance [29,30]. The outdoor evaluation of the ExpoLIS sensor node system performance was accomplished in one of the Lisbon urban background AQ monitoring stations (Olivais station) from the Portuguese AQ Monitoring Network. The station is equipped with reference instruments that measure PM2.5 and PM10 concentrations using beta attenuation technology (Environment MP101M, Envea, France). At this location, PM concentrations measured by the ExpoLIS sensor node system were compared with concentrations provided by the station's reference equipment and the DustTrak 8533 for 14 days.

### 2.3. *Correction of the Factors Affecting PM Concentrations*

LCS' PM readings are known to be affected by meteorological conditions, thus requiring a correction over the original data. A straightforward correction of the data against reference equipment can be achieved through a simple linear regression (SLR). On the other hand, to correct the influence of T and RH, two approaches can be considered: the application of a multiple linear regression (MLR), and the use of the k-Köhler theory to correct for the effect of the particles' hygroscopicity.

In this study, SLR and MLR were used to correct the OPC-N3 data against the reference equipment without accounting for T and RH (Equation (1)), as well as to consider the effect of T (Equation (2)), RH (Equation (3)), and of both meteorological parameters (Equation (4)):

$$PM_{OPC-N3} = a \times PM_{reference} + b, \tag{1}$$

$$PM_{OPC-N3} = c \times PM_{reference} + d \times T + e, \tag{2}$$

$$PM_{OPC-N3} = f \times PM_{reference} + g \times RH + h, \tag{3}$$

$$PM_{OPC-N3} = i \times PM_{reference} + j \times T + k \times RH + l \tag{4}$$

where $PM_{OPC\text{-}N3}$ are the concentrations measured by the OPC-N3, and $PM_{reference}$ the concentrations measured by the reference equipment, both in μg $m^{-3}$.

During periods of high levels of RH, PM concentration readings may be incorrectly higher due to the ability of particles to absorb water [15,19]. The k-Köhler theory relates the particles hygroscopicity and their volume, as presented in Equation (5) [15,20]:

$$\frac{m}{m_0} = 1 + \frac{\frac{\rho_w}{\rho_p} k}{-1 + \frac{1}{a_w}}, \tag{5}$$

where $\frac{m}{m_0}$ represents the ratio between the wet $(m)$ and dry $(m_0)$ particles' mass, $\rho_w$ and $\rho_p$ are the water density (1 g $cm^{-3}$) and particle density (assumed to be 1.65 g $cm^{-3}$), respectively; $k$ represents the slope of the exponential line of the humidogram and, finally, $a_w$ is the water activity (determined as 1/RH).

In this study, the wet particle mass corresponds to the OPC-N3 readings and the dry particle mass to the Environment MP101M instrument, as it is equipped with an air dryer that removes humidity from the entering air.

### 2.4. Prototype Quality Control

Finally, four ExpoLIS sensor nodes were installed in four public buses to test their ability to measure AQ during movement. In each measurement route, concentrations were measured inside and outside of buses; being that these data was also used as an input to the ExpoLIS Health Optimal Routing Service App. This app was designed to show AQ maps produced by the sensors installed in the buses and to help citizens to reduce their exposure to air pollutants while commuting [31].

The Municipality of Lisbon defined a system of hierarchical levels to characterize the road networks that consists of four different levels [32]:

- Level 1: Structuring road network, which supports long distance routes;
- Level 2: Main distribution network, whose function is to distribute traffic between the different sectors of the city;
- Level 3: A secondary distribution network that supports the proximity distribution;
- Level 4: A network of proximity that works at the levels of collection and distribution within neighborhoods and as local accesses.

Following this approach, four routes were selected for the measurement campaign (Figure 2) considering the need to obtain diverse routes with different characteristics, as presented in Table 1.

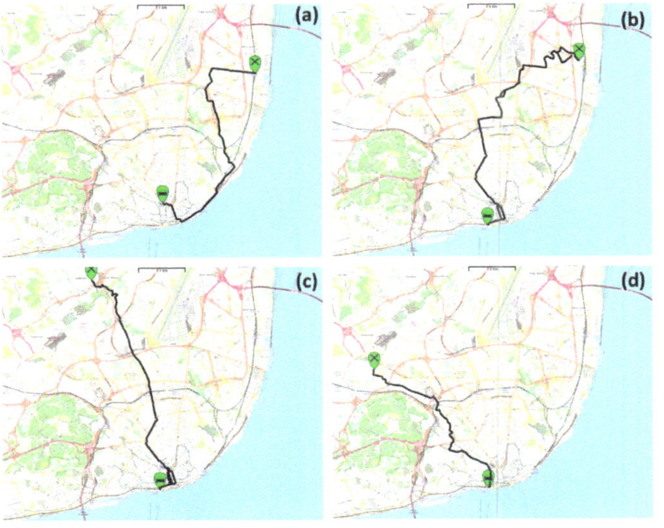


**Figure 2.** Bus routes selected for the measurements campaign: (**a**) Restauradores—Oriente; (**b**) Restauradores—Moscavide; (**c**) Odivelas—Cais do Sodré and (**d**) Portas de Benfica—Cais do Sodré.

**Table 1.** Bus Routes starting and ending points and hierarchical levels.

| Routes (Starting Point-Ending Point) | Bus Route | Hierarchical Level (%) | | | |
|---|---|---|---|---|---|
| | | Level 1 | Level 2 | Level 3 | Level 4 |
| Restauradores—Oriente | 759 | 0 | 10.3 | 63.6 | 26.1 |
| Restauradores—Moscavide | 744 | 0 | 54.8 | 30.4 | 14.8 |
| Odivelas—Cais do Sodré | 736 | 5.9 | 31.2 | 50.9 | 12.0 |
| Portas de Benfica—Cais do Sodré | 758 | 0 | 3.8 | 91.7 | 4.5 |

## 3. Results

### 3.1. Intercomparison between Reference Instruments

The DustTrak 8533 and the Environment MP101M were used as reference equipment because they provide good time resolution, fast response signal, excellent signal-to-noise ratio, and simplicity. The data provided by these instruments was firstly corrected against simultaneous gravimetric PM2.5 and PM10 measurements.

Figure 3 displays the comparison of PM2.5 and PM10 concentrations determined gravimetrically (by a Leckel MVS6) against those measured by the DustTrak 8533 and the Environment MP101M, which were averaged in 24 h periods. A low correlation between the DustTrak 8533 and Leckel measurements was obtained, and therefore PM2.5 and PM10 concentrations measured during this work were corrected based on the linear regression equations shown in Figure 3a. On the contrary, a good correlation (Figure 3b) was obtained in the intercomparison between the Environment MP101M and Leckel data ($R^2$ = 0.86 for PM2.5 and $R^2$ = 0.96 for PM10).

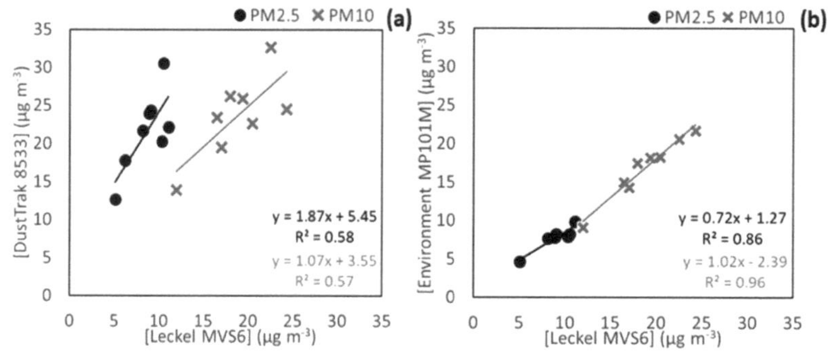


**Figure 3.** (**a**) Correlation between the gravimetric PM mass concentrations and the DustTrak 8533 and (**b**) the Environment MP101M measurements.

## 3.2. Quality Control of the ExpoLIS Sensor Node in Laboratory Tests

The ExpoLIS sensor nodes were firstly tested in laboratory with small variations of T (27.9–28.9 °C) and RH (45.4–48.4%) values against the DustTrak 8533. The LCS sensor nodes were tested under steadily increasing concentrations, which varied between 0.3 and 38.0 µg m$^{-3}$ for PM2.5, and between 0.3 and 51.7 µg m$^{-3}$ for PM10 (as measured by the OPC-N3).

The OPC-N3 and the DustTrak had an excellent correlation ($R^2$ equal to 0.99). However, the OPC-N3 underestimated the concentrations indicating the need to correct the measured concentrations based on the SLR equation presented in Figure 4a. As indicated in Figure 4b, the underestimation of the sensors' reading was solved with this approach, while keeping the correlation between the two pieces of equipment.

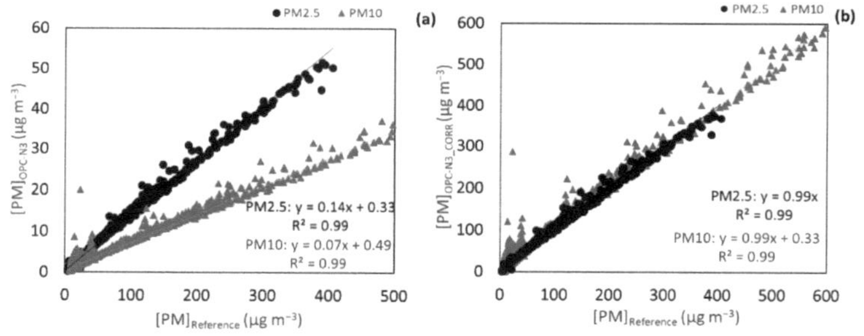


**Figure 4.** (**a**) SLR of the concentrations measured by the OPC-N3 and DustTrak 8533 for PM2.5 and PM10 and (**b**) consequent correction. Measurements done in a laboratory setting.

Furthermore, tests were performed in a garage where the ExpoLIS sensor nodes were exposed to vehicle exhaust and a higher variation of T (26.5–31.8 °C) and RH (28.2–41.0%). Figure 5 shows that the agreement between the OPC-N3 and the DustTrak got worse, possibly because of the lower range of concentrations inside the garage (from 0.7 to 9.1 µg m$^{-3}$ for PM2.5 and from 0.8 to 58.5 µg m$^{-3}$ for PM10 concentrations, as measured by the OPC-N3), considering that, as it has already been seen, the correlation tends to be higher for larger ranges of concentrations. In addition, the variation of T and RH was higher than in the previous test, and therefore, data was corrected with MLR considering RH and T according Equation (5). Figure 5b shows that with a correction considering the effect of T and RH, the agreement improved and the underestimation of concentrations by the OPC-N3 was corrected.

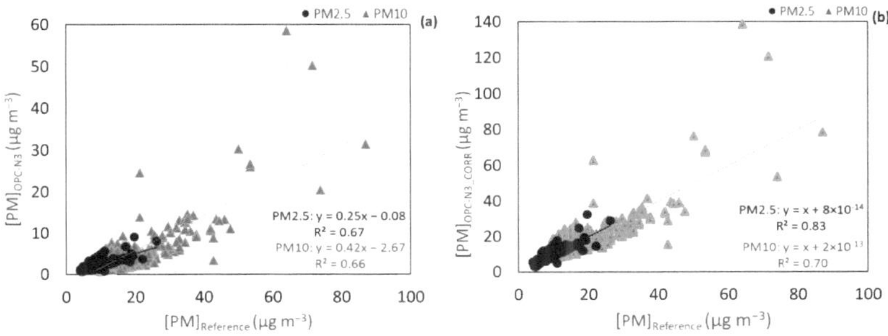


**Figure 5.** (**a**) Correlation between the OPC-N3 PM2.5 and PM10 concentrations and the correspondent concentrations measured by the DustTrak and (**b**) correlation between OPC-N3 PM2.5 and PM10 concentrations corrected with a MLR considering T and HR and concentrations measured by the DustTrak. Measurements done in an indoor environment with a vehicle exhaust as a pollution emission source.

### 3.3. Quality Control of the ExpoLIS Sensor Node in Field Tests

While the results from laboratory measurements are a good performance indicator, they are not representative of the actual conditions in which the sensors will be measuring air pollutants concentrations. There are several factors, including meteorological conditions (T and RH), which influence the sensors performance and their ability to measure PM concentrations. Therefore, the ExpoLIS sensor node system was installed at the Olivais AQ Monitoring Station where it measured concentrations in parallel with the equipment Environment MP101M and DustTrak 8533.

For the OPC-N3, the concentration range of PM2.5 was from 0.7 to 21.9 µg m$^{-3}$, and for PM10 it varied between 1.6 and 44.9 µg m$^{-3}$ during a test period of 14 days. The results from the comparison between the OPC-N3 and the Environment MP101M were not satisfactory. Figure 6a shows weak correlations ($R^2$ equal to 0.04 and 0.12 for PM2.5 and PM10, respectively) and a high dispersion of the values, which has already been reported in other studies that evaluated the performance of this sensor in real-world conditions [33].

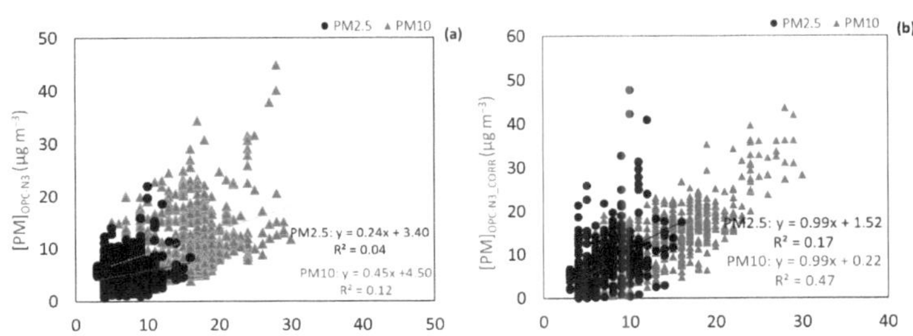


**Figure 6.** (**a**) Correlation of PM2.5 and PM10 OPC-N3 concentrations data against the Environment MP101M (**b**) corrected PM2.5 and PM10 concentrations. Measurements done at an AQ monitoring station.

During the tests, a high variation of RH (21.9–89.2%) was observed. In addition, it was possible to verify that high RHs were usually associated with high OPC-N3 readings, especially for PM10 (Figure S1). The k-Köhler theory was applied to correct the effect of RH. For its application, the data from the Environment MP101M was considered as the mass of dry particles, and the OPC-N3 concentrations data the mass of wet particles, through the principles of Equation (5). After the correction, the correlation coefficients of

PM10 improved with $R^2$ values increasing from 0.12 to 0.47. In addition, the slopes' values improved from 0.45 to 0.99, which indicates that the PM10 concentrations measured by the sensors were in more agreement with the values reported by the reference.

For PM2.5, although there was an effect of RH on the measured concentrations, there was not an evident exponential tendency (Figure S2), which is a criterion for the application of the k-Köhler theory. Therefore, the correction, which considered the effects of RH and T, was applied using a MLR analysis. As a result, there was an improvement in both the values of $R^2$ (from 0.04 to 0.17) and in the slopes of the correlation line (from 0.25 to 0.99). As it also happened for PM10, the corrections for PM2.5 and PM10 successfully improved the measured data, in spite of the dispersion of the data that was not totally solved.

### 3.4. Deployment of the ExpoLIS Exposure System in Lisbon

After the evaluation of the sensor node performance and the implementation of the correction obtained in the field tests, the ExpoLIS system was deployed in Lisbon to assess the behavior of a sensor node in the field, to produce concentration maps, and to test all the functionalities of the ExpoLIS App (for more details regarding the ExpoLIS App see https://github.com/ExpoLIS-project/expolis-mobile-app—accessed on 23 February 2022) [27]. In each bus route, two sensor nodes were installed to simultaneously measure the indoor and outdoor PM concentrations, and measurements were made twice a day, at 8 a.m. and at 8 p.m. Table 2 summarizes the average PM concentrations measured by the sensor node along the four selected bus routes and the corresponding meteorological conditions. The indoor average concentrations were 6.9 μg m$^{-3}$ for PM2.5 and 27.7 μg m$^{-3}$ for PM10. However, the concentrations to which the passengers were exposed (PM Indoor concentrations) varied according to the route and starting time. In general, PM2.5 indoor concentrations were higher than the concentrations measured outdoors (4.1 μg m$^{-3}$ for PM2.5 and 15.0 μg m$^{-3}$ for PM10) with I/O ratios varying between 0.9 and 3.3 for PM2.5, and between 1.2 and 5.0 for PM10.

**Table 2.** Indoor and outdoor PM concentrations (average ± standard deviation) derived I/O ratios for the bus routes analyzed and the range of variation of the meteorological conditions (namely T and RH).

| Routes | Starting Time | PM Indoor Concentrations | | PM Outdoor Concentrations | | I/O | | Meteorological Conditions | |
|---|---|---|---|---|---|---|---|---|---|
| | | PM2.5 (μg m$^{-3}$) | PM10 (μg m$^{-3}$) | PM2.5 (μg m$^{-3}$) | PM10 (μg m$^{-3}$) | PM2.5 | PM10 | T (°C) | RH (%) |
| Restauradores—Oriente | 8 a.m. | 7.9 ± 2.6 | 31.0 ± 14.3 | 5.8 ± 2.0 | 20.1 ± 10.0 | 1.4 | 1.5 | 23.2–31.2 | 39.2–57.8 |
| | 8 p.m. | (*) | (*) | (*) | (*) | (*) | (*) | | |
| Restauradores—Moscavide | 8 a.m. | 5.3 ± 2.5 | 25.1 ± 14.7 | 2.5 ± 1.2 | 13.4 ± 11.4 | 2.1 | 1.9 | 22.1–34.3 | 35.1–62.5 |
| | 8 p.m. | 5.8 ± 0.7 | 18.3 ± 7.9 | 3.3 ± 0.7 | 10.2 ± 4.6 | 1.8 | 1.8 | | |
| Odivelas—Cais do Sodré | 8 a.m. | 8.7 ± 2.6 | 29.9 ± 14.3 | 4.5 ± 2.1 | 15.2 ± 9.9 | 1.9 | 2.0 | 20.7–31.2 | 35.8–62.0 |
| | 8 p.m. | 8.8 ± 5.2 | 42.0 ± 39.3 | 9.5 ± 5.4 | 35.9 ± 26.9 | 0.9 | 1.2 | | |
| Portas de Benfica—Cais do Sodré | 8 a.m. | 4.9 ± 2.8 | 21.1 ± 15.6 | 1.4 ± 0.8 | 4.2 ± 3.7 | 3.3 | 5.0 | 19.0–28.9 | 28.4–47.9 |
| | 8 p.m. | 7.2 ± 2.1 | 26.5 ± 10.6 | 2.3 ± 1.0 | 6.2 ± 4.2 | 3.1 | 4.3 | | |

(*) Due to a systematic error in the sensor's reading the night period measurements were not considered.

The results show that the high PM concentrations to which the citizens are exposed during commuting can lead to a substantial contribution to their total daily exposure and inhalation of air pollutants, especially in high vehicle-density metropolitan areas. Figure 7 shows the ability of the ExpoLIS Health Optimal Routing Service App to provide to Lisbon's citizens not only the spatial distribution of outdoor PM concentrations along a selected route, but also the total PM inhaled dose and emission. Moreover, this app provides a routing service in which the users can select their starting and ending points, as well as the option "avoid pollution", receiving in turn the route that minimizes exposure to air pollutants. The App allows the user to compare (i) the dose of pollutants inhaled during commuting in different transport modes (i.e., Bus, Car, Motorcycle, Bicycle and Walking) and (ii) the pollutants' emission of the motorized modes, considering not only the type of fuel but also the Euro standard.

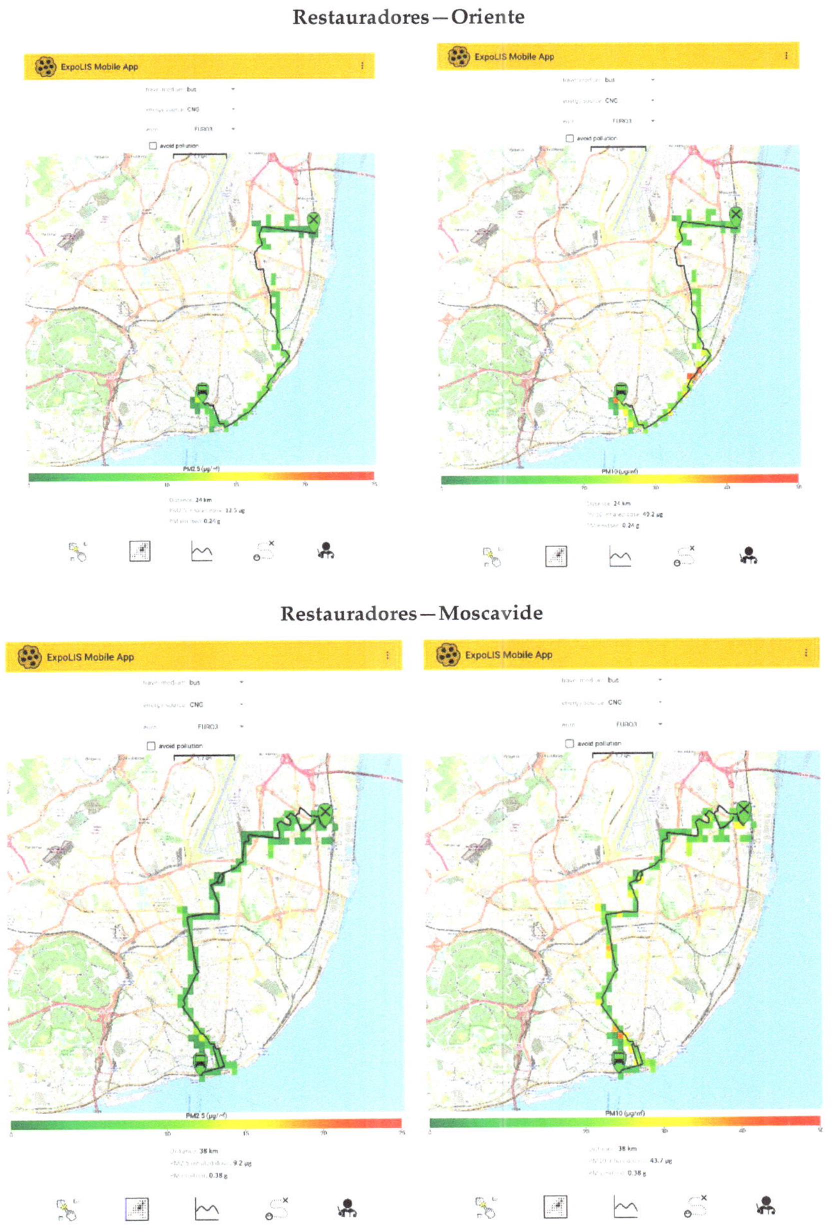


**Figure 7.** *Cont.*

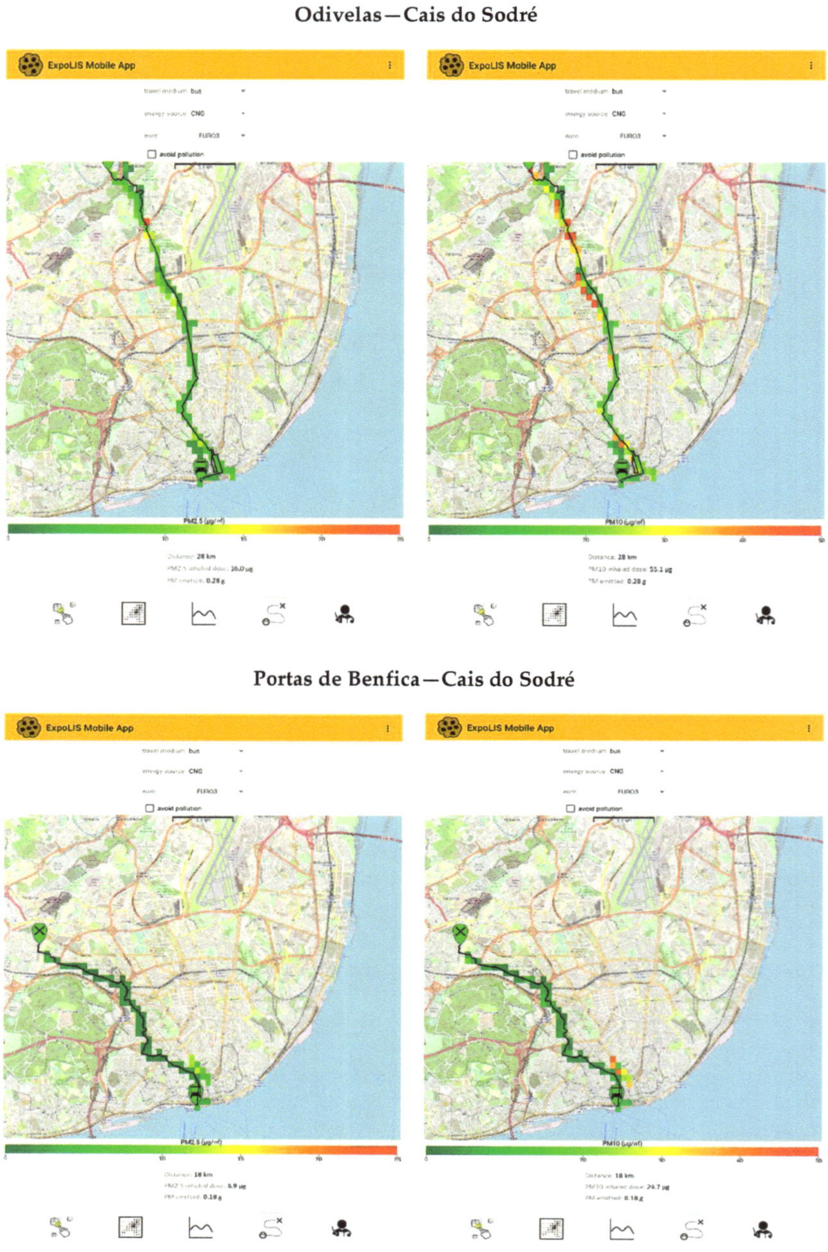


**Figure 7.** Screenshots of the ExpoLIS Health Optimal Routing Service App for the four selected routes when commuting by a Euro 3 bus powered by Compressed Natural Gas (CNG), with the representation of the spatial distribution of outdoor PM2.5 and PM10 concentrations, and the indication of the respective PM inhaled dose and emission.

#### 3.4.1. Reducing the PM Inhaled Dose through the Use of the ExpoLIS App

In the ExpoLIS App the PM inhaled dose ($D_{PM}$ in µg) along a selected route is estimated using Equation (6):

$$D_{PM} = PM_{out} \times \frac{I}{O} \times t \times IR, \quad (6)$$

where $PM_{out}$ is the average outdoor PM concentration (µg m$^{-3}$), I/O the indoor-to-outdoor PM ratio, t the travel time (h) and IR the inhalation rate (m$^3$ h$^{-1}$). Although in this work the indoor and outdoor concentrations in buses were measured simultaneously (Section 2.4), the deployment of the system in cities only included the installation of the sensors outdoors. Therefore, in the App, the concentrations to which the citizens are exposed during commuting are derived from the outdoor concentrations through the application of the I/O ratios. The PM I/O ratios generated in this work for buses, and in a previous study [8] for cars, were a valuable base information for the development of the ExpoLIS App. The I/O values considered were 2.10 and 2.50 in buses and 0.96 and 0.93 in cars, for PM2.5 and PM10, respectively. These values should be updated in the App whenever more scientific knowledge is available.

For motorcycle, bicycle, and walking, the I/O ratio used was equal to 1. The inhalation rates are based on two studies developed previously [34,35]. Table 3 displays the IR used for the different commuting modes.

**Table 3.** Inhalation rate (m$^3$ h$^{-1}$) applied in the ExpoLIS app for each transport mode.

| Transport Mode | IR (m$^3$ h$^{-1}$) |
|---|---|
| Bicycle | 1.41 |
| Car | 0.71 |
| Bus | 0.76 |
| Motorcycle | 0.94 |
| Walking | 1.15 |

Table 4 presents the PM2.5 and PM10 inhaled doses reported by the ExpoLIS App for each selected route, for different transport modes and considering a two-way trip. The highest PM2.5 and PM10 inhaled doses were obtained when travelling by bus and on bicycles, due to the high indoor PM concentrations and inhalation rate, respectively. In buses, the high indoor PM concentrations are mainly associated with the resuspension of particles, created by the passengers' movement and the air flowing in and out when the doors are opened [36]. Moreover, bus passengers are exposed to high pollution at stops when the doors are opened, often in places where queues of idling vehicles are releasing high levels of air pollutants.

**Table 4.** PM2.5 and PM10 inhaled doses (µg) for each selected route during a two-way trip.

| Routes | Starting Time | Travelling Time (min) | PM2.5 Inhaled Doses (µg) | | | | PM10 Inhaled Doses (µg) | | | |
|---|---|---|---|---|---|---|---|---|---|---|
| | | | Bus | Car | Motorcycle | Bicycle | Bus | Car | Motorcycle | Bicycle |
| Restauradores—Oriente | 8 a.m. | 125 | 12.5 | 3.9 | 4.9 | 13.1 | 49.2 | 13.0 | 16.9 | 45.3 |
| | 8 p.m. | - | - | - | - | - | - | - | - | - |
| Restauradores—Moscavide | 8 a.m. | 137 | 9.2 | 2.6 | 3.3 | 8.9 | 43.7 | 13.7 | 17.8 | 47.9 |
| | 8 p.m. | 111 | 8.2 | 3.5 | 4.4 | 11.9 | 25.8 | 10.4 | 13.5 | 36.4 |
| Odivelas—Cais do Sodré | 8 a.m. | 145 | 16.0 | 3.5 | 4.4 | 11.8 | 55.1 | 11.4 | 14.9 | 40.0 |
| | 8 p.m. | 121 | 13.5 | 7.4 | 9.3 | 25.0 | 64.5 | 27.0 | 35.1 | 94.5 |
| Portas de Benfica—Cais do Sodré | 8 a.m. | 111 | 6.9 | 0.7 | 0.9 | 2.4 | 29.7 | 2.0 | 2.6 | 7.1 |
| | 8 p.m. | 83 | 7.6 | 1.1 | 1.4 | 3.9 | 27.9 | 3.0 | 3.9 | 10.5 |

3.4.2. Reduction of the PM Exhaust Emissions through the Use of the ExpoLIS App

The ExpoLIS App provides the exhaust PM emissions of the motorized modes, considering not only the type of fuel, but also the Euro standard. The pollutant emission per passenger, ($E$; g of pollutant emitted per passenger) is calculated by the product between the emission factor ($EF$; g km$^{-1}$) for each motorized transport mode and the total distance of the route ($d$; km) for a two-way trip divided by the number of passengers (p), considering one passenger for cars and motorcycles and fifty passengers for buses (Equation (7)):

$$E = \frac{EF \times d}{p}, \quad (7)$$

The emission factors of the European Monitoring and Evaluation Programme (EMEP) air pollutant emission inventory guidebook published by the European Environment Agency [37] were used in the App calculations.

In this work, using the emission factors presented in Table 5, we compared PM emissions for different vehicles, fuel type, and Euro standard as indicated in Table 6. The highest number of particles was emitted by buses, followed by motorcycles and cars. Euro 5 diesel buses produced 126 times more particles per kilometer than Euro 5 diesel cars, but they typically carry around 50 times more people. Thus, although traveling by car showed the lowest PM2.5 and PM10 inhaled doses among the motorized transport modes, it presented similar emissions per passenger when commuting by bus (depending on the type of fuel considered), considering that 50 people are at each bus. Results also show that in cities, the selection of the bus fleet regarding the type of fuel is crucial for the air quality. The Diesel buses have emissions three times higher than the CNG buses. As motorcycles do not have emission control devices like catalytic converters to neutralize the released pollutants, in general they emit more PM than cars.

**Table 5.** Emission factors of sample motorized transport modes, fuel types, and Euro standards.

| Vehicle | Fuel | Euro Standard | Emission Factor (g PM km$^{-1}$) |
|---|---|---|---|
| passenger car | petrol | Euro 5 | 0.00061 |
| | diesel | Euro 5 | 0.00025 |
| | electric | – | 0 |
| bus | biodiesel | Euro 5 | 0.03138 |
| | diesel | Euro 5 | 0.03138 |
| | electric | - | 0 |
| | CNG | Euro 5 | 0.01 |
| motorcycle | petrol | Euro 5 | 0.00297 |

**Table 6.** PM emissions (g per passenger) of sample motorized transport modes, fuel types, and Euro standards.

| Route | Distance (km) | Emissions (g PM Per Passenger) | | | | | |
|---|---|---|---|---|---|---|---|
| | | Passenger Car | | Bus | | | Motorcycle |
| | | Petrol | Diesel | Biodiesel | Diesel | CNG | Petrol |
| Restauradores—Oriente | 24 | 0.015 | 0.006 | 0.01506 | 0.01506 | 0.0048 | 0.071 |
| Restauradores—Moscavide | 38 | 0.023 | 0.009 | 0.02384 | 0.02384 | 0.0076 | 0.113 |
| Odivelas—Cais do Sodré | 28 | 0.017 | 0.007 | 0.01758 | 0.01758 | 0.0056 | 0.083 |
| Portas de Benfica—Cais do Sodré | 18 | 0.011 | 0.004 | 0.0113 | 0.0.0113 | 0.0036 | 0.054 |

## 4. Conclusions

This study presented the ExpoLIS system which consists of a network of LCS that monitor AQ, and a Health Optimal Routing Service App. In this paper, the authors intended to perform a preliminary evaluation of the ExpoLIS sensor node and application of the Health Optimal Routing Service App.

Under controlled conditions in a lab, the PM LCS had an excellent performance, requiring only a simple correction (through SLR) to overcome the LCS' underestimation of concentrations. However, when exposed to a higher variation of T and RH, the performance of the LCS demonstrated the need to correct the measured concentrations considering these meteorological factors.

In the field test, an exponential increase of the PM10 concentrations with the increase of the RH was observed, and therefore a correction was performed based on the k-Köhler principle. As the exponential impact of the RH was not so pronounced for the PM2.5, it was verified that the best correction is the MLR, considering RH and T.

This work is currently progressing towards extending the field measurements and considering different approaches to calibrate the LCS, such as different algorithms that aim to correct the effect of RH, and machine learning approaches capable of considering several factors that influence the LCS performance.

The ExpoLIS Sensing System was installed in public buses to monitor AQ in the urban area of Lisbon and test their ability to generate data with high spatial and temporal resolution. The results obtained in the tests performed while in motion indicate that PM concentrations measured inside and outside buses depend on the route and period of the day, and that exposure while commuting can be very high. When expanded to the entire city, this system will inform the citizens about the exposure to air pollutants that results from their choices of daily paths, and even the average estimated dose of air pollutants that they inhale. Furthermore, the Health Optimal Routing System App also informs the citizens about the air pollutant emissions that result from their commuting choices, based on the average emission factors of each transport mode and distance. With this information, we intend to inform and make citizens aware of how their daily commuting choices can improve or deteriorate AQ in urban areas.

In conclusion, the results that are presented in this paper demonstrate the importance of the ExpoLIS App to help citizens to make more informed decisions in order to: (i) reduce their exposure to air pollutants by taking the healthiest path between user-specified departure and arrival locations (not necessarily the shortest one) and (ii) reduce the pollutant emissions by selecting environmental-friendly or public transport modes.

**Supplementary Materials:** The following supporting information can be downloaded at: https://www.mdpi.com/article/10.3390/ijerph20054073/s1, Figure S1: Correlation between PM10 concentrations measured by the OPC-N3 sensor and RH; Figure S2: Correlation between PM2.5 concentrations measured by the OPC-N3 sensor and RH.

**Author Contributions:** C.C.: Conceptualization, Formal analysis, Investigation, Resources, Writing—original draft, Writing—review & editing. V.M.: Conceptualization, Formal analysis, Funding acquisition, Investigation, Resources, Writing—original draft, Writing—review & editing. B.M.: Conceptualization, Formal analysis, Investigation, Resources, Writing—original draft, Writing—review & editing. P.S.: Conceptualization, Funding acquisition, Investigation, Methodology, Software, Supervision, Writing—Review & editing. P.M.: Conceptualization, Investigation, Software, Writing—Review & editing. A.A.: Conceptualization, Funding acquisition, Investigation, Resources, Writing—Review & editing. S.M.A.: Conceptualization, Funding acquisition, Formal analysis, Investigation, Methodology, Resources, Supervision, Writing—original draft, Writing—review & editing. All authors have read and agreed to the published version of the manuscript.

**Funding:** This study was developed in the scope of the project ExpoLIS funded by FEDER, through Programa Operacional Regional de Lisboa (LISBOA-01-0145-FEDER-032088), and by national funds (OE), through FCT—Fundação para a Ciência e Tecnologia, I.P. (PTDC/EAM-AMB/32088/2017). Authors also acknowledge the support of FCT through the strategic project UIDB/04349/2020+UIDP/04349/2020, the contract CEECIND/04228/2018 and the PhD grant UI/BD/150996/2021.

**Data Availability Statement:** Data available on request.

**Conflicts of Interest:** The authors declare no conflict of interest.

## References


1. Almeida, S.M.; Silva, A.V.; Sarmento, S. Effects of Exposure to Particles and Ozone on Hospital Admissions for Cardiorespiratory Diseases in SetúBal, Portugal. *J. Toxicol. Environ. Heal. Part A Curr. Issues* **2014**, *77*, 837–848. [CrossRef] [PubMed]
2. Pinault, L.L.; Weichenthal, S.; Crouse, D.L.; Brauer, M.; Erickson, A.; van Donkelaar, A.; Martin, R.V.; Hystad, P.; Chen, H.; Finès, P.; et al. Associations between Fine Particulate Matter and Mortality in the 2001 Canadian Census Health and Environment Cohort. *Environ. Res.* **2017**, *159*, 406–415. [CrossRef] [PubMed]
3. WHO. *Health Risks of Air Pollution in Europe—HRAPIE Project*; WHO: Geneva, Switzerlnd, 2013.
4. Calvo, E.; Ferrer, M. Evaluating the Quality of the Service Offered by a Bus Rapid Transit System: The Case of Transmetro BRT System in Barranquilla, Colombia. *Int. J. Urban Sci.* **2018**, *22*, 392–413. [CrossRef]
5. Gomes de Oliveira, G.; Iano, Y.; Caumo Vaz, G.; Lourenço Chuma, E. Analysis of the Ergonomic Concept of Public Transportation in the City of Campinas (Brazil). In *Advances in Human Aspects of Transportation*; AHFE: New York, NY, USA, 2021; pp. 453–459.
6. Joewono, T.B.; Kubota, H. Safety and Security Improvement in Public Transportation Based on Public Perception in Developing Countries. *IATSS Res.* **2006**, *30*, 86–100. [CrossRef]
7. Snyder, M.; Arunachalam, S.; Isakov, V.; Talgo, K.; Naess, B.; Valencia, A.; Omary, M.; Davis, N.; Cook, R.; Hanna, A. Creating Locally-Resolved Mobile-Source Emissions Inputs for Air Quality Modeling in Support of an Exposure Study in Detroit, Michigan, USA. *Int. J. Environ. Res. Public Health* **2014**, *11*, 12739–12766. [CrossRef] [PubMed]
8. Correia, C.; Martins, V.; Cunha-Lopes, I.; Faria, T.; Diapouli, E.; Eleftheriadis, K.; Almeida, S.M. Particle Exposure and Inhaled Dose While Commuting in Lisbon. *Environ. Pollut.* **2020**, *257*, 113547. [CrossRef] [PubMed]
9. Almeida, S.M.; Manousakas, M.; Diapouli, E.; Kertesz, Z.; Samek, L.; Hristova, E.; Šega, K.; Alvarez, R.P.; Belis, C.A.; Eleftheriadis, K.; et al. Ambient Particulate Matter Source Apportionment Using Receptor Modelling in European and Central Asia Urban Areas. *Environ. Pollut.* **2020**, *266*, 199. [CrossRef] [PubMed]
10. European Environmental Agency. *Air Quality in Europe—2020 Report*; European Environmental Agency: Copenhagen, Denmark, 2020; ISBN 978-92-9480-292-7.
11. Berghmans, P.; Bleux, N.; Panis, L.I.; Mishra, V.K.; Torfs, R.; Van Poppel, M. Exposure Assessment of a Cyclist to PM10 and Ultrafine Particles. *Sci. Total Environ.* **2009**, *407*, 1286–1298. [CrossRef]
12. de Nazelle, A.; Fruin, S.; Westerdahl, D.; Martinez, D.; Ripoll, A.; Kubesch, N.; Nieuwenhuijsen, M. A Travel Mode Comparison of Commuters' Exposures to Air Pollutants in Barcelona. *Atmos. Environ.* **2012**, *59*, 151–159. [CrossRef]
13. Snyder, E.G.; Watkins, T.H.; Solomon, P.A.; Thoma, E.D.; Williams, R.W.; Hagler, G.S.W.; Shelow, D.; Hindin, D.A.; Kilaru, V.J.; Preuss, P.W. The Changing Paradigm of Air Pollution Monitoring. *Environ. Sci. Technol.* **2013**, *47*, 11369–11377. [CrossRef]
14. Jovašević-Stojanović, M.; Bartonova, A.; Topalović, D.; Lazović, I.; Pokrić, B.; Ristovski, Z. On the Use of Small and Cheaper Sensors and Devices for Indicative Citizen-Based Monitoring of Respirable Particulate Matter. *Environ. Pollut.* **2015**, *206*, 696–704. [CrossRef] [PubMed]
15. He, J.; Huang, C.H.; Yuan, N.; Austin, E.; Seto, E.; Novosselov, I. Network of Low-Cost Air Quality Sensors for Monitoring Indoor, Outdoor, and Personal PM2.5 Exposure in Seattle during the 2020 Wildfire Season. *Atmos. Environ.* **2022**, *285*, 119244. [CrossRef]
16. Kortoçi, P.; Motlagh, N.H.; Zaidan, M.A.; Fung, P.L.; Varjonen, S.; Rebeiro-Hargrave, A.; Niemi, J.V.; Nurmi, P.; Hussein, T.; Petäjä, T.; et al. Air Pollution Exposure Monitoring Using Portable Low-Cost Air Quality Sensors. *Smart Health* **2022**, *23*. [CrossRef]
17. Mead, M.I.; Popoola, O.A.M.; Stewart, G.B.; Landshoff, P.; Calleja, M.; Hayes, M.; Baldovi, J.J.; McLeod, M.W.; Hodgson, T.F.; Dicks, J.; et al. The Use of Electrochemical Sensors for Monitoring Urban Air Quality in Low-Cost, High-Density Networks. *Atmos. Environ.* **2013**, *70*, 186–203. [CrossRef]
18. Crilley, L.R.; Shaw, M.; Pound, R.; Kramer, L.J.; Price, R.; Young, S.; Lewis, A.C.; Pope, F.D. Evaluation of a Low-Cost Optical Particle Counter (Alphasense OPC-N2) for Ambient Air Monitoring. *Atmos. Meas. Tech.* **2018**, *11*, 709–720. [CrossRef]
19. Jayaratne, R.; Liu, X.; Thai, P.; Dunbabin, M.; Morawska, L. The Influence of Humidity on the Performance of a Low-Cost Air Particle Mass Sensor and the Effect of Atmospheric Fog. *Atmos. Meas. Tech.* **2018**, *11*, 4883–4890. [CrossRef]
20. Samad, A.; Nuñez, D.R.O.; Castillo, G.C.S.; Laquai, B.; Vogt, U. Effect of Relative Humidity and Air Temperature on the Results Obtained from Low-Cost Gas Sensors for Ambient Air Quality Measurements. *Sensors* **2020**, *20*, 1–29. [CrossRef]
21. Borrego, C.; Ginja, J.; Coutinho, M.; Ribeiro, C.; Karatzas, K.; Sioumis, T.; Katsifarakis, N.; Konstantinidis, K.; De Vito, S.; Esposito, E.; et al. Assessment of Air Quality Microsensors versus Reference Methods: The EuNetAir Joint Exercise—Part II. *Atmos. Environ.* **2018**, *193*, 127–142. [CrossRef]
22. Concas, F.; Mineraud, J.; Lagerspetz, E.; Varjonen, S.; Liu, X.; Puolamäki, K.; Nurmi, P.; Tarkoma, S. Low-Cost Outdoor Air Quality Monitoring and Sensor Calibration: A Survey and Critical Analysis. *ACM Trans. Sens. Netw.* **2021**, *17*, 1–44. [CrossRef]
23. Li, J.; Li, H.; Ma, Y.; Wang, Y.; Abokifa, A.A.; Lu, C.; Biswas, P. Spatiotemporal Distribution of Indoor Particulate Matter Concentration with a Low-Cost Sensor Network. *Build. Environ.* **2018**, *127*, 138–147. [CrossRef]
24. Johnson, N.E.; Bonczak, B.; Kontokosta, C.E. Using a Gradient Boosting Model to Improve the Performance of Low-Cost Aerosol Monitors in a Dense, Heterogeneous Urban Environment. *Atmos. Environ.* **2018**, *184*, 9–16. [CrossRef]
25. Liang, L. Calibrating Low-Cost Sensors for Ambient Air Monitoring: Techniques, Trends, and Challenges. *Environ. Res.* **2021**, *197*, 111163. [CrossRef] [PubMed]
26. Mariano, P.; Almeida, S.M.; Santana, P. On the Automated Learning of Air Pollution Prediction Models from Data Collected by Mobile Sensor Networks. *Energy Sources Part A Recover. Util. Environ. Eff.* **2021**, 1–17. [CrossRef]

27. Santana, P.; Almeida, A.; Mariano, P.; Correia, C.; Martins, V.; Almeida, S.M. Air Quality Mapping and Visualisation: An Affordable Solution Based on a Vehicle-Mounted Sensor Network. *J. Clean. Prod.* **2021**, *315*. [CrossRef]
28. Faria, T.; Martins, V.; Correia, C.; Canha, N.; Diapouli, E.; Manousakas, M.; Eleftheriadis, K.; Almeida, S.M. Children's Exposure and Dose Assessment to Particulate Matter in Lisbon. *Build. Environ.* **2020**, *171*, 106666. [CrossRef]
29. Tagle, M.; Rojas, F.; Reyes, F.; Vásquez, Y.; Hallgren, F.; Lindén, J.; Kolev, D.; Watne, Å.K.; Oyola, P. Field Performance of a Low-Cost Sensor in the Monitoring of Particulate Matter in Santiago, Chile. *Environ. Monit. Assess.* **2020**, *192*. [CrossRef]
30. Pope, F.D. Pollen Grains Are Efficient Cloud Condensation Nuclei. *Environ. Res. Lett.* **2010**, *5*, 15. [CrossRef]
31. Mariano, P.; Almeida, S.; Almeida, A.; Correia, C.; Martins, V.; Moura, J.; Brandão, T.; Santana, P. An Information System for Air Quality Monitoring Using Mobile Sensor Networks. In Proceedings of the 19th International Conference on Informatics in Control, Automation and Robotics, Lisbon, Portugal, 14–16 July 2022; Volume 2022, pp. 238–246. [CrossRef]
32. CML. *Lisboa: O Desafio Da Mobilidade;* Câmara Municipal de Lisboa: Lisbon, Portugal, 2005; ISBN 972-8877-05-6.
33. Bauerová, P.; Šindelářová, A.; Rychlík, Š.; Novák, Z.; Keder, J. Low-Cost Air Quality Sensors: One-Year Field Comparative Measurement of Different Gas Sensors and Particle Counters with Reference Monitors at Tusimice Observatory. *Atmosphere* **2020**, *11*, 1–15. [CrossRef]
34. Buonanno, G.; Giovinco, G.; Morawska, L.; Stabile, L. Tracheobronchial and Alveolar Dose of Submicrometer Particles for Different Population Age Groups in Italy. *Atmos. Environ.* **2011**, *45*, 6216–6224. [CrossRef]
35. Zuurbier, M.; Hoek, G.; Van Den Hazel, P.; Brunekreef, B. Minute Ventilation of Cyclists, Car and Bus Passengers: An Experimental Study. *Environ. Health A Glob. Access Sci. Sour.* **2009**, *8*, 1–10. [CrossRef]
36. Martins, V.; Correia, C.; Cunha-Lopes, I.; Faria, T.; Diapouli, E.; Manousakas, M.I.; Eleftheriadis, K.; Almeida, S.M. Chemical Characterisation of Particulate Matter in Urban Transport Modes. *J. Environ. Sci.* **2021**, *100*, 51–61. [CrossRef] [PubMed]
37. EMEP/EEA EMEP/EEA. *Air Pollutant Emission Inventory Guidebook 2019: Technical Guidance to Prepare National Emission Inventories;* EMEP: Houston, TX, USA, 2019.

International Journal of
*Environmental Research and Public Health*

MDPI

*Article*

# An Educational Game to Teach Children about Air Quality Using Augmented Reality and Tangible Interaction with Sensors

**João Fernandes [1], Tomás Brandão [1,2], Susana Marta Almeida [3] and Pedro Santana [1,2,*]**

1 ISCTE, Instituto Universitário de Lisboa (ISCTE-IUL), Av. das Forças Armadas, 1649-026 Lisboa, Portugal
2 ISTAR—Information Sciences and Technologies and Architecture Research Center, Av. das Forças Armadas, 1649-026 Lisboa, Portugal
3 Centro de Ciências e Tecnologias Nucleares, Instituto Superior Técnico, Universidade de Lisboa, Estrada Nacional 10, 2695-066 Bobadela, Portugal
* Correspondence: pedro.santana@iscte-iul.pt

**Abstract:** Air pollution is known to be one of the main causes of injuries to the respiratory system and even premature death. Gases, particles, and biological compounds affect not only the air we breathe outdoors, but also indoors. Children are highly affected by the poor quality of the air they breathe because their organs and immune systems are still in the developmental stages. To contribute to raising children's awareness to these concerns, this article presents the design, implementation, and experimental validation of an serious augmented reality game for children to playfully learn about air quality by interacting with physical sensor nodes. The game presents visual representations of the pollutants measured by the sensor node, rendering tangible the invisible. Causal knowledge is elicited by stimulating the children to expose real-life objects (e.g., candles) to the sensor node. The playful experience is amplified by letting children play in pairs. The game was evaluated using the Wizard of Oz method in a sample of 27 children aged between 7 and 11 years. The results show that the proposed game, in addition to improving children's knowledge about indoor air pollution, is also perceived by them as easy to use and a useful learning tool that they would like to continue using, even in other educational contexts.

**Keywords:** air quality; augmented reality; child–computer interaction; educational games; serious games; tangible interaction



**Citation:** Fernandes, J.; Brandão, T.; Almeida, S.M.; Santana, P. An Educational Game to Teach Children about Air Quality Using Augmented Reality and Tangible Interaction with Sensors. *Int. J. Environ. Res. Public Health* **2023**, *20*, 3814. https://doi.org/10.3390/ijerph20053814

Academic Editors: Ashok Kumar and Paul B. Tchounwou

Received: 13 January 2023
Revised: 16 February 2023
Accepted: 17 February 2023
Published: 21 February 2023



## 1. Introduction

The World Health Organization (WHO) defines air pollution as the "contamination of the indoor or outdoor environment by any chemical, physical or biological agent that modifies the natural characteristics of the atmosphere" [1]. These compounds affect the air we breathe and are associated with millions of premature deaths and the cases of diseases such as cancer and obstructive pulmonary disease every year [1], as well as increasing infant mortality rates [2]. To achieve both economic development and emission reduction, prioritizing ecological conservation and boosting green development is key [3].

Indoor air represents a considerable part of daily exposure to pollution, since daily life is nowadays mostly spent indoors. In indoor environments, the concentrations of polluting compounds can actually be higher than those outside due to the closer proximity to emitting sources and lower pollutants' dilution. Children are especially vulnerable to air pollution, as their respiratory rate is higher than in adulthood and their immune system is still developing. When exposed to air pollution for long periods of time, the risk of developing or aggravating respiratory pathologies increases considerably [4]. Hence, it is pivotal to protect children from this often invisible and difficult to identify peril.

If we can make children more aware of air pollution and its health impacts, we may be able to promote healthier and environmentally protective behaviors for these children.

In fact, the influence of air pollution on public health diminishes as the education level increases [2]. Raising children's awareness is particularly advantageous in the first years of school activity, as investing in learning at a younger age is known to bring positive effects in later life [5]. Awareness can be raised with traditional learning methods, such as formal classes, books, and documentaries. On the other hand, new interactive digital technologies offer new opportunities for active and customized learning activities towards raising the motivation for learning. It is important to continually innovate in the way teaching methods and tools are created and that they include differentiating factors that induce a surprise factor and help make the experience memorable, especially for primary school children [6]. Investing in quality education during childhood makes children less likely to fail grades and reduces the chances of inclusive and special education needs. Moreover, children are more likely to successfully complete secondary education and be part of the professional market with above-average wages [5].

Educational games are an example of how interactive digital technologies can be used to create immersive experiments that exploit children's intrinsic motivations (e.g., attraction for novelty, play, and stories) to raise their willingness to learn concepts that could be otherwise unpleasant to learn. Using virtual reality (VR), educational games can be designed to immerse the player in a simulated world, amplifying one's sense of presence in that world. However, virtually sending our body to another place can be uncomfortable for some [7]. Moreover, full immersion in a virtual world hampers the ability of students to socially interact in the physical world with their colleagues throughout the experiment. Augmented reality (AR) is a promising add-on to educational games, as it allows students to experience virtual content in the real world that would otherwise be invisible (e.g., air constituents) and to interact with virtual content using physical objects (i.e., tangible interfaces), while physically and socially interacting with their colleagues.

In this article, we present an AR-based educational game that aims to provide an interactive learning experience about indoor air quality for elementary school children, to be played in pairs, in the classroom. The goal is to teach children how everyday objects contribute to households' air quality. To meet this goal, children are asked to present physical objects to a physical sensor node capable of monitoring the air quality. In return, children are able to see the physical sensor node augmented with virtual representations of the object's emanated air pollutants measured by the sensor node, rendering visible the invisible. The game was implemented in Unity/Vuforia and interacts with a sensor node developed in the ExpoLIS project [8].

The game was evaluated with a Wizard of Oz experiment with a sample of 27 children aged between 7 and 11 years. Herein, the air quality measurements were covertly generated in real-time by a human being instead of measured by the sensor node. This option reduces the variance in user testing resulting from the natural stochastic nature of both air diffusion and sensing. The experiments were designed to test four research hypotheses: H1—AR games are able to teach children about the causes of indoor air pollution; H2—AR games are able to teach children about the mechanisms available to clear indoor air pollution; H3—AR games provide a satisfying and emotionally stimulating air pollution learning experience, with high replay value; and H4—AR games benefit from allowing the child to interact with real pollution sources instead of card-based representations. The obtained results confirm the four research hypotheses, showing that the proposed AR-based learning game, in addition to improving children's knowledge about indoor air pollution, is also perceived by them as easy to use and a useful learning tool that they would like to continue using, even in other educational contexts.

This article is organized as follows. Section 2 presents a literature survey of related work. Then, in Section 3, the developed AR-based game is described alongside its key implementation details. Section 4 presents the experimental setup and results. Finally, Section 5 draws conclusions and provides future work directions.

## 2. Related Work

### 2.1. Educational Games in Learning

The use of games in the educational context is an active research topic with especially positive results in younger age groups [9]. Educational games, a sub-genre of serious games, show promise in knowledge transmission as well as students' commitment, motivation, and capacity to retain the acquired knowledge [10]. Combining teaching strategies with game design may guide learners through complex tasks and new concepts in a relaxed and pleasant way. Adequate game-based learning should enable autonomous learning at the user's own pace and boost one's self-motivation [11].

When designing a game, one must need to choose between single-player and multi-player and, for the latter, between cooperative and competitive strategies. Previous studies shed some light about which options are the most likely to deliver the best results when teaching is the ultimate goal of the game. In the study presented in [12], students showed higher rates of participation in classes after playing cooperative games, when compared to after playing competitive games [12]. These data are in line with the study presented in [11], which revealed that students better enjoy games when these are played in groups, cooperatively. Furthermore, [13] investigated the effect of using collaborative methods for learning mathematics in children with and without learning difficulties. With a collaborative group working in pairs and a competitive group working individually, it was observed that the collaborative group achieved the best outcomes. In addition, the peers in the collaborative group established positive relationships with each other, helping and encouraging each other.

Nevertheless, competition encourages personal development and improvement. However, the inseparable need for a winner and a loser leads users to focus not only on their individual success, but also on their opponent's failure. As such, the cooperative vision assumes a more enriching role, since it does not promote tense environments or aggression between the participants. It also reinforces the ability of users to relate positively to each other, creating a suitable empathy and trust environment, and encourages the development of communication skills, which are extremely important for success in today's society. In addition, it helps develop interpersonal relationships and is even associated with more successful professional careers [14].

### 2.2. Augmented Reality in Educational Games

Several studies have demonstrated the ability of computing technologies to persuade and influence their users' behaviors [11]. AR games contribute to this capacity, being especially useful for teaching science, to represent abstract and difficult to visualize subjects [15], and to develop computational thinking skills [16]. Developing computational thinking is important for children to be able to reason about the global consequences of their local actions. AR games have the potential to integrate abstract and difficult-to-interpret information from the real world. This eases the creation of theory–practice links, allowing interaction in real contexts and learning through execution. Problems of traditional education, such as the student's lack of focus and distractions, are important reasons that lead teachers and educators to find new ways for knowledge transmission that adapt to children's needs [17]. There have been many demonstrations that AR is able to compensate for some gaps of traditional teaching [17,18]. AR potentiates a more informal learning environment, allowing learners to interact with the technology as if they were playing, which benefits knowledge acquisition [11]. Learners also feel that AR-based learning is a more efficient and motivating way of acquiring knowledge than traditional teaching methods [19]. Furthermore, AR can assist the learning process for children with disabilities, such as autism, helping them stay focused [20], and to attract them for behavioral therapies [21].

In the educational context, AR is currently most often developed for mobile devices [22], largely due to ease of use and accessibility, as they have a tactile interface. Additionally, mobile devices are affordable as well as easy to acquire and to replace. However, the small size of the screen constrains the amount of information that can be presented. Another disadvantage of mobile devices is the frequency of distractions, especially when used for didactic purposes on child audiences [19,23]. Head-mounted displays (HMD) are an alternative to mobile devices, but their considerable weight and cost are limiting factors for the purpose of educating children. Moreover, HMD may hamper the feeling of co-presence in collaborative scenarios. Therefore, the use of desktop computers equipped with webcams emerges as an interesting alternative to mobile- and HMD-based AR in educational contexts.

Cognitive overload resulting from an excessive presentation of information to the learner, is a well-known key challenge when developing AR-based educational tools [24], contributing to the reduction in student's learning efficiency, focus and motivation. However, when properly designed and validated, AR can be an extremely useful tool to overcome both the information overload and the lack of motivation associated to traditional teaching methods [25,26], serving as creative unlockers [27], improving visualization skills [26], and helping understand difficult to relate concepts [19]. When developing AR-based tools for students, it is also important to bear in mind that the cognitive ability, individual learning style, spatial visualization ability, and previous experiences with these types of learning environments are factors that influence the effectiveness of the teaching techniques [25].

Ideally, the designers of educational interactive tools should be provided with a set of design guidelines, leading the devised tool to meet the adequate teaching effectiveness. In [15], a set of design principles for AR-based learning tools is presented, based on an analysis of existing literature. The first principle, "captivate and then challenge", aims to prevent users from suffering from information overload or feeling that they are not able to deal with the challenge. It is essential to start by outlining strategies to guide users through the most elementary concepts and mechanics of the experience and, only then, to challenge them with more complex problems, ideally adapted to the game's progress phase and the user's performance. The second principle, "guide the experience through the game's story", aims to guide the user's learning process and one's interactions with the system through immersive narratives strategically designed for this purpose, building a bridge between entertainment and the ability to transmit educational content. The narrative should include characters that appear at key moments to provide context, tasks, and guide the user's attention. Scoring systems that reward or penalize their actions, directing you to the idealized outcome, are also important elements to include. The third principle, "see the invisible", is directly related to the basic functionality that AR provides, that is, augmenting the real world with visual representations of content that would otherwise be invisible to the observer. In a metaphorical sense, the AR tool operates as a lenses.

### 2.3. Augmented Reality Meets Sensors

Sensor networks and Internet of Things (IoT) are widely applied in the monitoring of real phenomena, such as air quality [8]. The data generated by these devices contain information relevant to raise children's and young people's awareness of major societal issues, such as environmental monitoring. Providing young people with access to air quality data through immersive 3D environments exhibiting a video game appearance is advantageous when compared to traditional visualization techniques [28]. The use of game engines as explorative, low-entry tools for visualizing complex air pollution datasets is further discussed in [29].

Augmented reality is also recognized as an advantageous means of visualizing, controlling, and interacting with IoT devices [30]. For instance, the awareness of energy consumption among students can be improved using mobile AR to present sensor data gathered from buildings [6]. AR and IoT can also be orchestrated for helping people with reduced mobility to become more independent in performing daily activities, such as choosing products from a supermarket shelf [31]. To enable the observation of electromagnetic radiation emitted by ordinary electronic equipment, present in a room where users can move around and freely explore the environment, an AR experience using an HMD was developed [32]. The results collected by the authors show the validation and appreciation of the participants regarding the use of these techniques for the visualization and learning of invisible content. AR can also be used to teach children about colors by enabling them to point a color sensor at any object in the real world and obtain its color in return [33].

The potential of mobile AR to create attractive and interesting gaming experiences for the presentation of air quality data was demonstrated in [34]. The proposed game consisted of rewarding the user whenever the temperature and $CO_2$ levels from a distant location (where an IoT device is installed) are properly guessed. Clues are provided to the user by the dressing of a virtual character (t-shirt vs. scarf, with or without a gas mask). Users found the experience interesting and fun, promoting awareness of environmental problems. More recently, a study where the residents of a building monitor the air quality of their homes based on the data acquired by an IoT network using AR was carried out [35]. When compared with non-AR users, AR users showed a higher degree of satisfaction regarding quality of experience and effectiveness of the presented information.

### 2.4. User Input in Augmented Reality

In most AR applications, users need to provide some form of input. For instance, the MagicHand project [36] allows users to control IoT equipment via hand gestures. Alternatively, the user may interact with the system by manipulating real objects, i.e., using tangible interfaces. Tangible interfaces take advantage of the user's natural skills for manipulating physical objects, providing a greater degree of immersion by including sensory stimulation mechanisms, such as haptic, weight, texture, and temperature sensations [37,38]. In fact, a comparative study between touchscreens, tangible interfaces, and classic mouse–keyboard interaction concluded that the user preference and interaction speed showed the best results for tangible interfaces [39]. Moreover, in an AR-based educational experiment where children were challenged to relate real plants to their fruits and leaves, the use of real objects was considered to be a factor that provides a strong contribution to the learning experience [40].

Although object detection, segmentation, and tracking is becoming ubiquitous, these remains challenging tasks when objects are being manipulated by users in the wild. To mitigate some of these problems, particularly those related to occlusions, multi-camera settings are often employed [41,42]. Some of the manipulated physical objects can be used as pointers for the user to direct the system's attention. A pointer can be as simple as a stick with a colored sphere for simple vision-based tracking [42]. For a more robust tracking and pose estimation, the tip of the stick can be attached to a visual marker [43]. Visual markers can be used to track the pose of other types of objects, such as books, and even to operate as buttons (pressed by the occlusion of the marker) or sliders (sliding by moving the marker) [44].

Although the appearance of visual markers is often inconsistent with the game's overall aesthetics, which may impact the user's sense of presence if not properly hidden by the virtual augmentations, their detection is well understood, affordable, and simple to implement. Therefore, the marker-based tracking of physical objects still represents the most accessible method for prototyping tangible interfaces and, thus, particularly suited for low-budget classroom contexts.

### 2.5. Discussion

The presented literature survey highlights the value of AR in the design of educational games and of user interfaces for sensor networks and IoT devices. The survey also reveals the advantages of considering physical objects as tangible interfaces. Our educational game, presented herein, combines and extends these findings. In particular, the game provides children with the possibility of interacting with air quality-sensing devices via the manipulation of everyday physical objects. This novel interaction possibility, not addressed by previous work, is intended to render tangible the invisible and, thus, to increase the learning gain. By allying it with a playful gaming experience, the emotional involvement in the learning experience is expected to grow and, hence increasing the chances that the child is willing to fully complete it and even to repeat it. To facilitate accessibility and cooperative gameplay in the classroom context, the game follows the desktop-based AR paradigm and relies on visual markers. The game was designed to be a cooperative multi-player, building upon the identified advantages of collaboration over competition in educational contexts.

## 3. Gamified Experience Description

### *3.1. Overview*

The gamified experience described in this section combines the use of augmented reality with air quality measurements, creating an instrument that allows a child to visualize and interact with elements taken from both the virtual and real worlds. Similarly to microscopes or X-ray machines that allow observations at different scales/spaces, the idea is to create a gaming experience around a customized instrument, which provides users with an augmented real environment representing information that would be otherwise invisible. In this specific case, the objects of observation are air pollutants emitted by everyday objects (e.g., sprays, candles, glue tubes), whose concentrations are measured in real time by a mobile sensor node.

The game is based on a simplified version of the sensor node developed in the ExpoLIS project [8], which is able to analyze the concentrations of carbon monoxide (CO) and nitrogen dioxide ($NO_2$), as well as of particulate matter with an aerodynamic diameter $\leq 1\,\mu m$ (PM1), $\leq 2.5\,\mu m$ (PM2.5), and $\leq 10\,\mu m$ (PM10). Airflow through the sensors is established by a fan attached to the air outlet inside the sensor node's box. Figure 1 depicts the sensor node's box used by the game. Its dimensions are as follows: 42 cm wide, 26 cm high and 9 cm deep. The frontal face of the sensor node's box includes an AR marker that allows the game to track it over time and augment it with virtual graphical representations.

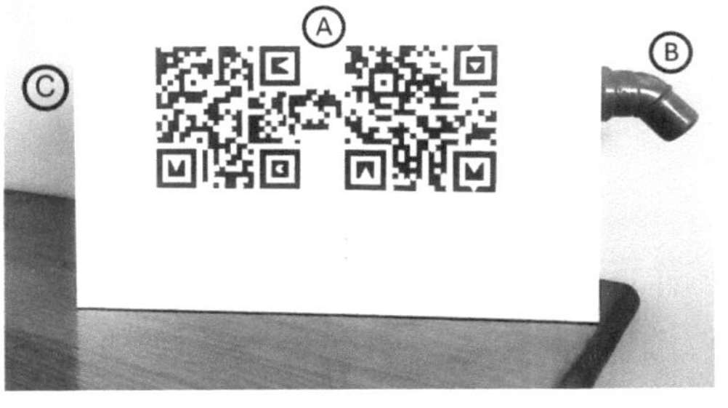


**Figure 1.** Air quality monitoring device. A—AR marker; B—air inlet tube; C—air outlet.

Figure 2 illustrates the gaming experience. The frontal face of the sensor node is rendered transparent using AR, allowing the user to see inside the box, which can represent, for instance, a room. Then, the user is asked to place an everyday object near the air inlet, e.g., a candle. The sensor node's box analyzes the air and the detected pollutants are presented to the user as augmented graphical representations inside the box. Then, the user is asked to clean the air inside the box by using a set of plausible tools, representing actual air quality improvement techniques. These tools are selected and maneuvered with a virtual "wand", which is controlled by the user pointing their hand.

### *3.2. Design Methodology*

Due to the cross-disciplinary nature of serious games, their design and validation should involve people from the application domain (in our case, researchers on air quality), from the educational sciences (in our case, teachers), from computer science and engineering (for the actual development of the game), and, most importantly, representatives of the target user (in our case, children). Therefore, the development of this experience followed a participatory design with short prototyping and testing cycles. It resorts to testing small game components in small groups of users belonging to the target audience. This methodology allows one to collect users' opinions and reactions to the tool, validating small development iterations in a practical and objective way. It avoids major setbacks in later stages of development since problems are identified earlier. It also stimulates new ideas and allows one to understand which features are most valued by the users. Therefore, user tests were carried out according to a task-oriented script during both the *formative* and *summative* evaluation phases [45], hereafter called *formative tests* and *summative tests*, respectively. During the tests, users were encouraged to *think aloud* [45], externalizing their thoughts during the experience.

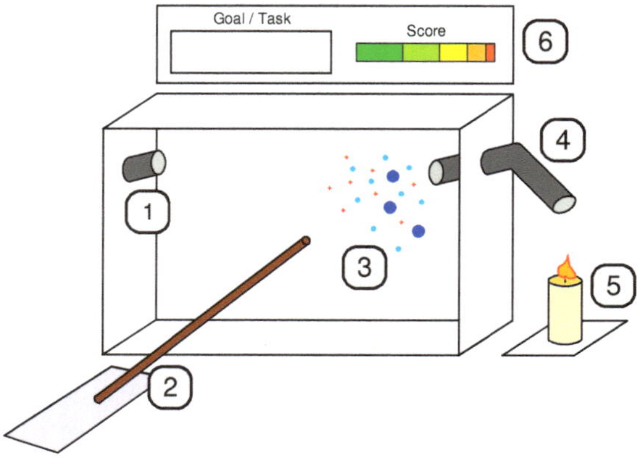


**Figure 2.** Gaming experience illustration. 1—air extraction tube (virtual); 2—"wand" for interaction with the experience (mixed); 3—representation of different air pollution compounds (virtual); 4—air inlet tube (real); 5—area for placing sources of pollution (real); 6—goals/score chart UI (virtual).

The software components were developed using the widely known game engine Unity 3D [46]. Versatility and ease of use are two of its key features that contribute to its popularity. In addition, it is available for free, it has a vast community of active users on web forums, and it is well documented. Vuforia [47] was used for including the augmented reality components of the game. It is a C++ SDK dedicated to the creation of virtual environments interacting with the real world, and can be easily integrated with Unity 3D. Although Vuforia offers a wide variety of tools, only image recognition tools were used in the scope of the game. The goal is to recognize pre-defined markers (patterns) that, among other things, facilitate the sensor node's box detection and its pose estimation. All 3D modeling was performed in Blender [48].

### *3.3. Pollution Compound Representation*

As a result of the participatory design sessions, individualized representations of CO, $NO_2$, PM1, PM2.5, and PM10 were considered excessive for the education level of the target audience (first cycle of basic education). Therefore, the design of the gamified experiment addresses the distinction between gases and particles, without emphasizing their individual classifications, easing the transmission and assimilation of knowledge by the target audience.

The unity particle system was used for creating representations of gases and particles. The particle system filled all the necessary requirements for this experiment since it incorporates physical components with several easily configurable properties: the number of emissions per second, emission velocity, reactions to the application of forces and collisions. The representation of gas and particles can be observed in Figure 3. Both are represented in gray, contrasting with the colored background, associating a negative connotation as if they were an enemy to be eliminated.

(**a**) Gas. (**b**) Particles. (**c**) Gas and particles.

**Figure 3.** Graphical representation of gases and particles.

The emission value in units per second of each pollution element was assigned to one of four possible levels: 1 (low), 2 (moderate), 3 (intense), and 4 (extreme). The emission level assignment is directly related to the measurements retrieved by the sensors and is inspired by the *air quality index* (AQI), defined in [49]. AQI provides information about air quality, using a six-level scale that is easy to interpret. The representation of the pollutants only used the four highest levels from the AQI scale since sensor measurements always fell on those levels in the context of our experiments.

An initial displacement vector is assigned for both the representation of gases and particles. This vector is originated from the air inlet tube and points towards the opposite room wall. To simulate a realistic behavior of the compounds in the air, random displacement vectors are assigned during their lifetime to mimic the suspension of gas and particles in the air, moving slowly according to randomized flows.

### *3.4. User Interaction*

#### 3.4.1. User Roles

As the experience was designed to be experienced by two children simultaneously, it is essential that both participants feel that they take an active part and play a well-defined role. Additionally, it was expected that the collaborative effort between the participants could enhance the outcome of the educational component of the game. Based on these premises, two user roles were defined: `User1`—selection and manipulation of real-world objects close to the sensor box air inlet; and `User2`—selection and manipulation of tools for interacting with the polluting compounds represented on the screen.

The interaction associated with the role of `User1` is to use a set of real-world objects in order to cause a reaction from the sensor box. The available set of objects include: deodorant spray, a dusting cloth, a candle, and a tube of liquid glue. Each object has its own way of being operated in order to spread polluting compounds into the air: the cloth is shaken, the candle is lit, and so on. With the exception of the cloth shaking, which is used as an example in the starting tutorial, the discovery of the actions that cause reactions from the sensor box is left to the users, helping them only if necessary. When correctly operating an object close to the box's air inlet, the representation of the pollution elements released by the object (gas, particles, or both) show up on the screen. The role of `User1` remains the same throughout the experience: the user is responsible for selecting the objects to use at each moment, producing the polluting compounds necessary to achieve the objectives presented during the game run. Since the use of cloth only guarantees the introduction of particles into the environment, the in-game objectives and scoring system will induce the need to further explore the remaining objects in order to discover how to use them and which ones produce gases.

The actions of `User2` begin as soon as the first polluting elements are presented in the screen. This user's role is to clear the air, removing the pollution compounds using a set of tools at one's disposal. These tools are used for directing the gases and particles to the sensor node's air outlet, represented by a virtual window in the augmented sensor node. To learn how to relate and operate each tool with each type of pollutant, the user needs to carry out some experiments. The iterative design of `User2`'s interactions, informed by interleaved formative evaluations, is described in the following paragraphs.

#### 3.4.2. Early Design

During the initial development phase, two kinds of pointing mechanisms were considered: mouse vs. the user's hand. The objective was to test the response of the target audience to this interaction type whilst consuming the least amount of resources and time as possible. The interaction with the virtual environment was achieved through collisions of the element controlled by the pointing device with the game elements, enabling the possibility of selections (e.g., pressing buttons) and object manipulation (e.g., dragging).

The use of a mouse as an interaction device was considered due to its wide spread use and, thus, potential smooth learning curve. The 2D mouse movement was mapped to a 2D movement in a vertical plane aligned with the box, meaning that it did not allow the user to control the depth. Mouse rotation was not considered.

Pointing with one's hand is natural and, thus, potentially more intuitive than using the mouse. As in previous work [50], we track the user's hand using AR makers. Concretely, to track the user's hand when pointing to the sensor node, a rigid AR marker was glued to a ring that could be used in the index finger. The marker was used to track the pose of the finger (position and rotation), interact with other virtual objects according to the laws of physics, and overlay a virtual object representing a *virtual wand* (see Figure 4a).

(**a**) Virtual wand. Left: input image; Right: input image augmented with virtual wand.

(**b**) Virtual hand and laser. A: marker; B: hand; C: laser.

**Figure 4.** Pointing mechanism: early design in (**a**) and final design in (**b**).

Formative evaluation with nine children highlighted a set of flaws in the early design of the marker-based virtual wand (see Figure 5). All participants presented difficulties in perceiving the depth of the virtual objects in relation to the virtual wand. In some cases, participants were not even sure that they were pointing to the sensor node altogether. On the other hand, occasional failures in tracking the markers (box and ring) was well accepted by all participants, which interpreted these events as part of the game challenges. After some trial-and-error, participants devised strategies to avoid failures and facilitate the re-detection of the markers. Behaviors such as bringing the marker closer to the camera and keeping it parallel to the image plane were increasingly frequent.

Six of the nine users reported having a preference for interaction using the mouse over the method with the marker. The children successfully completed the assigned tasks much faster when using the mouse. The learning curve of using the ring with the marker showed to be longer than using the mouse. On the other hand, the experience became less interactive and challenging when using the mouse.

The users' astonishment reaction was evident when the ring method was presented to them; however, this sensation was soon lost due to the difficulty in handling it. By analyzing the results of the formative tests, the poor results for using the marker's ring interaction were not justified by a limitation of the method itself, but by the characteristics of its implementation. The following paragraphs describe the set of improvements that were implemented in order to cope with the found limitations.

**Figure 5.** Two children playing during the formative evaluation.

3.4.3. Final Design

To provide better control over the pointing direction, the context was improved by superposing a virtual hand with a pointing finger on the marker and the virtual wand was substituted by a virtual laser beam. The laser beam provides a continuous visual direction cue from the hand to the screen, facilitating the user perception by requiring less saccades. Figure 4b depicts the final design.

The children's difficulty for dealing with depth during the interaction was tackled by reviewing the mechanics of the experience. In short, interactions along the depth axis were simplified and restricted to a thin slice of the box in the virtual 3D world. Thus, depth variations associated with the movement of gases and particles were set to a much smaller range, making the interactions with those elements closer to a two-dimensional case. Gases and particles bounce back when the boundaries of the box slice are reached.

The manipulation of pollution removal tools along the depth axis was also constrained in order to ease the intersection with the virtual pollutants. For this, an invisible ray is cast along the user's pointing direction and its intersection with the longitudinal plane that splits the slice (defined in the previous paragraph) in half is considered. The tool is then positioned at the intersection point, ensuring that its movement is always performed along the slice splitting plane. Overall, users showed less confusion interacting with the system after these changes took place, and thus, these were included in the final prototype.

### *3.5. Air Pollution Removal Tools*

To guarantee the pertinence and integrity of the information transmitted by the gamified experience, the visual representations of the air pollution removal tools and their interactions with polluting compounds were analyzed in the participatory design sessions. During gameplay, these tools are available to the user on the left side of the sensor node, as illustrated in Figure 6. The importance of aeration in good air quality is taken into account by including a virtual window next to the sensor node's air outlet. Users are expected to remove pollutants through this window.

To select a tool, the user only needs to point to its direction. A tool is replaced if another is selected. Explanatory text associated with the tools was not included, leaving it up to the user to try them out and autonomously discover their features. When a tool is selected, it is coupled to the end point of the virtual laser beam representing the user's pointing direction. The position and orientation of the selected tool is then controlled by the user according to the pose of the finger-mounted marker. Only slight adjustments are applied to the rotation angles relative to the marker in order to make the interaction more natural. As such, the user is able to manipulate the tool to direct the polluting compounds to the extraction point (i.e., the window).

**Figure 6.** Virtual toolbox and sensor box. A—toolbox; B—particle and gas extraction window; and C—sensor air inlet.

Table 1 summarizes the implemented set of tools (filter, fan, and electrostatic) as well as the list of pollution compounds with which each of them is able to interact. The diversity of tools and air pollutants was judiciously selected in order to ensure that, while playing the game, children learned about air pollutants and their countermeasures.

**Table 1.** Tools for pollution compound removal.

| Tools | Pollution Compound | | |
|---|---|---|---|
| | Gases | <PM2.5 | ≥PM2.5 |
| Filter | × | × | ✓ |
| Fan | ✓ | ✓ | ✓ |
| Electrostatic | × | ✓ | ✓ |

With the *filter tool*, the user is able to collect coarser particles, having no effect on gases and finer particles. This binary behavior is a simplification of real-filtering devices, which use meshes to retain particles as a function of their diameter. Figure 7a shows this tool in action.

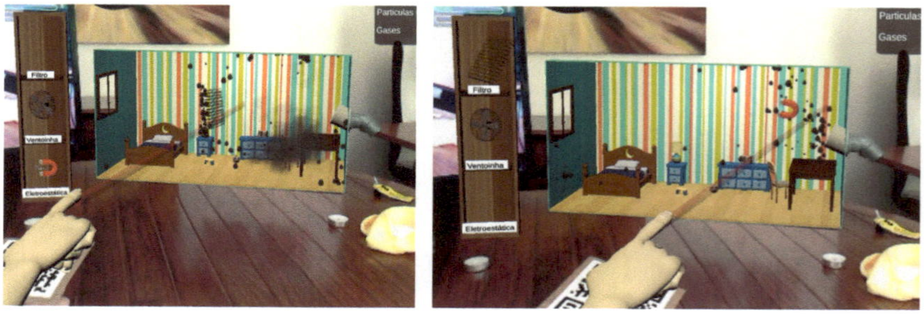

(a) (b)

**Figure 7.** Interaction with the filter tool (**a**) and with the electrostatic tool (**b**).

The *fan tool* creates an air flow along the direction it is pointed in, influencing the movement of all polluting compounds. The tool is implemented as a repulsive force field, of limited range, that interacts with the particle systems controlling the virtual air pollutants so as to push them away along the fan direction. This tool, represented in Figure 8, has a greater effect on gases and smaller particles, as these have a lower mass value.

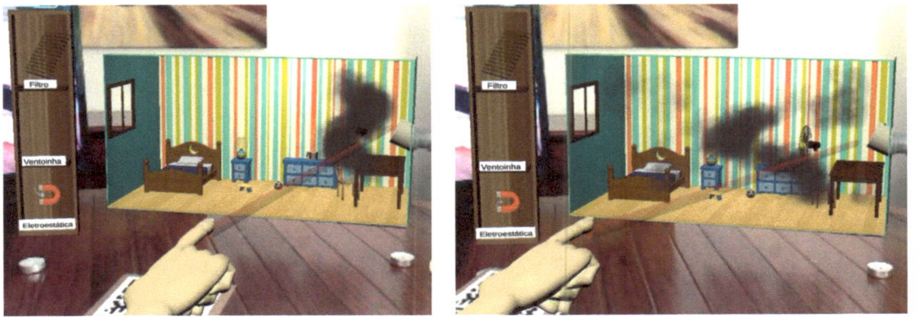

(**a**) Before interacting with gas. (**b**) After interacting with gas.

**Figure 8.** Interaction using the fan tool.

The *electrostatic tool* represents a device, known as electrostatic precipitator, that induces electric charge into the particles, capturing them by electromagnetic attraction. This tool was implemented as an attractive force field, of limited range, influencing the virtual pollutant particles so as to pull them towards the tool. The representation chosen for this technique is the least faithful among the other tools. Instead of considering a graphical representation of an electrostatic precipitator, which is a device not familiar to children, its representation ended up in the form of a magnet. When used, it emits small "sparks" (Figure 7b) to highlight the presence of electrical charges, in order to induce the users to establish a relation between its representation and its operating principles.

### 3.6. Spatial Setup

The setup is composed of a sensor node, a desktop computer, a computer screen, and a USB camera, spatially arranged as depicted in Figure 9. In this layout, the screen and the box are placed side by side, with the camera at the opposite end, facing the box. Among several alternatives tested during the formative evaluation, this spatial arrangement shows to be the most adequate. This decision took into account: (1) position and orientation of the camera so as to capture real-world content and to define the viewing perspective over the virtual elements; (2) position and orientation of the screen where real and virtual world components are combined; and (3) user placement and their interactions with the system.

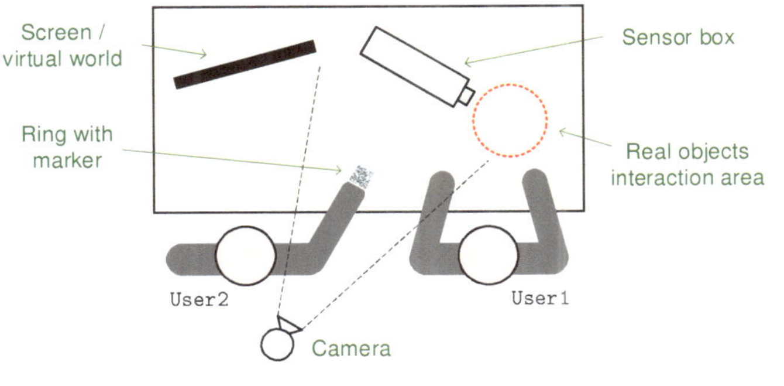


**Figure 9.** Spatial layout.

Both camera and screen were placed at fixed locations. This arrangement frees the children's hands for interaction with real objects and markers. In addition, by keeping the camera still, we avoid a set of technical challenges that could impact upon the user experience, such as distractions emerging from dynamic backgrounds, marker tracking issues resulting from variations in the illumination conditions, and jitter resulting from the varying frame rate due to the increasing complexity of handling dynamic settings.

Given that the screen and the sensor node were placed side-by-side, the former was presented as a "magic mirror", through which it was possible to see things inside the box that are not visible to our eyes. However, mirroring the captured image, for the screen to indeed be a mirror, proved to be an added difficulty for interacting with the marker. For this reason, the mirror metaphor was tested without actually mirroring the image, to which users responded positively. The solution was well accepted, resulting on easier and more intuitive interactions when compared with other tested setup possibilities.

### *3.7. Game Mechanics*

#### 3.7.1. Story Line

To guide the experience and reinforce its didactic content, a non-playable character (NPC) was created. The NPC was a scientist, graphically represented with sprites, who appears only in key moments of the experience. The NPC communicates (in Portuguese) with the user via subtitles, to provide hints on how to interact with the experience and to provide additional data that may help the user understand what is being observed. Figure 10a shows one of these situations, in which the NPC appears with pedagogic information regarding the particles that have just appeared.

(a) (b)

**Figure 10.** Interaction with the NPC. (**a**) A screenshot of the game with the scientist NPC and its subtitles (in Portuguese) at the bottom. Subtitles translation: *Particles can have very different sizes. The larger ones can be expelled by coughing or breathing, but the smaller ones can be harmful to our health.* (**b**) Assistance screen shown when the ring marker is lost. Overlaid text translation: *Ring lost. Place the ring here to continue playing.*

The NPC also intervenes in a small tutorial presented at the beginning of the gaming experience and when problems are detected, namely when the ring marker is not detected for over five seconds. In the latter case, the screen shown in Figure 10b is displayed, telling the user what action must be done to resume marker tracking and, thus, playing. This screen displays an area where the user must place the ring within a strategically chosen position that avoids occluding the sensor node's marker. This display is accompanied by an audio feedback with negative intonation.

#### 3.7.2. Scores and Feedback

The gamified experience comprises a scoring system, which rewards the user whenever gases or particles are directed to the window. This intends to be a simple way of providing feedback and assigning tasks to users, encouraging them to use different objects and tools in order to collect points. The score information is displayed in the upper right corner of the screen (see Figure 11a), and the increment in each score bar is accompanied by audio feedback with positive intonation.

Dynamic difficulty adjustment (DDA) is used to ensure the inclusion of users with less experience or showing greater difficulties adapting to the system, reducing the possibility of frustration. Additionally, DDA helps keep the duration of the experience between 7 and 10 min, which is the empirically found duration for both users to explore the game elements and feel comfortable with the game mechanics to successfully complete the objectives.

Each time the user manages to expel a particle through the virtual window, they earn 2% of the maximum possible score. Conversely, by expelling a gas unit, which is easier than expelling particles, the user earns 0.5% of the maximum possible score. If the user presents significant difficulties in scoring or achieving the objectives assigned during the experience, DDA takes place by: (1) doubling the future particle emission levels; and (2) doubling or even tripling the future earned score increments. Boosting particle emission levels raises the chances of successful interaction with particles, whereas boosting score increments reinforces positive feedback.

### 3.7.3. Gameplay

Once they arrive at the game location, users choose the role they will play in the experience without realizing it through the chair they sit on. In the first contact with the experience, a brief contextualization is made. It is explained that the screen shows the image captured by the camera (which is on their back) and the way in which the markers work, highlighting the connection between the real world and the virtual world by moving the sensor node's box (the term 'box' was used for the sake of simplicity), making it noticeable that it moves similarly in the virtual world. It is then explained that a sensor was placed inside the box that identifies small compounds in the air that we cannot see. If the users have no questions, the experiment starts.

At the beginning of the experiment, the screen is presented as a magic mirror, where it is possible to observe the otherwise invisible polluting compounds. This is followed by the mini tutorial given by the NPC, which suggests to the user that they shake the cloth close to the box's air intake tube (indicated in the screen with an arrow), as can be observed in Figure 11a. As soon as the user does so, the first particles appear in the virtual environment, about which the NPC makes a small theoretical introduction. The interventions of the NPC when the first particles and gases are introduced can be observed in Figure 11b. At this point, the child assuming the role of `User2` has already performed some experiments and realized how one can interact with the virtual elements by pointing at them. The selection of pollution removal tools was typically tried out, as motivated by children's intrinsic curiosity. Even if `User2` does not demonstrate the desired autonomy for selecting a tool and using it to interact with the polluting compounds, some of the compounds eventually end up reaching the window and triggering a positive feedback sound effect. From this moment on, it is expected that both users will explore the tools/objects at their disposal, in order to fill in the score bars completely. Although exploratory freedom is given to the users, these are not allowed to place more than one object simultaneously in the proximity of the sensor's air intake, a situation that was frequently observed during the formative evaluation.

(a) (b)

**Figure 11.** Interaction with NPC. (**a**) NPC giving the initial tutorial. Subtitles translation: *Try shaking the yellow cloth near the tube. Inside the house you can see the pollution that is in the air.* (**b**) NPC providing information after the first release of gas/particles. Subtitles translation: *Gases may be colorless and odorless, but they are toxic and can cause allergies, headaches and more serious illnesses.*

## 4. Evaluation

### 4.1. Method

In the validation phase of the tool, based on summative tests, 27 children aged between 7 and 11 years old, belonging to two primary schools in Sintra, Portugal, participated with parental consent. We sought to obtain a group of participants with a uniform distribution of ages ($\mu = 7.75$, $\sigma = 0.93$) and genders. Although the experiment was carried out in pairs and the questionnaires were answered simultaneously, the participants were sufficiently distanced not to hear each other's answers. An adult read the questions and provided explanations when needed.

To assess satisfaction and what knowledge had been gained about air pollution as a result of playing the game, three questionnaires were answered by the participants: a pre- and post-game questionnaire to assess knowledge about air pollution; a post-game satisfaction and usability questionnaire; and a post-game open-ended questionnaire about participants' opinions and preferences.

To assess the added value of including physical objects in the game, a variant of the game was tested. In this variant, instead of manipulating four physical objects, the user manipulated four markers, i.e., small images printed on paper, each representing one of the physical objects. These markers are automatically detected by the game whenever they fall inside the camera's field of view. All participants experienced both versions of the game in a randomized order, as well as both forms of interaction with each version.

After answering the pre-game questionnaire to assess knowledge about air pollution, a brief explanation about the experience was given. This included using the physical objects and markers, as well as the notion that the box contains air pollution sensors inside, whose observations are graphically represented in the game's screen.

Real-time communication between the game and the sensor node being assured, the validation of the pedagogical component of the game takes priority. To attain this goal, the mechanics of representation and interaction were isolated from the physical box elements. The uncertainty of the readings and the need for calibration are examples of factors that need to be accounted for in the experiments. To ensure the predictability of the system and, consequently, reduce the variance in the experimental data, the Wizard of Oz method [45] was used. This method consists of simulating the reactions of the system under study using covert human actions, leading the user to believe that the reactions are being produced by the system. In our case, the presence of air pollutants emitted by the objects placed by the user near the air inlet tube is covertly reported to the game by a human and not by the actual sensors of the physical box. To apply this method, a wireless keyboard was used and a set of virtual keys were mapped in order to simulate different levels of pollution emissions from each of the elements. The keyboard is used by the person conducting the test, imperceptibly to the user. This allowed to avoid configuration and calibration processes, saving time for the game development, without jeopardizing the gamified experience validation.

#### 4.1.1. Knowledge Validation Questionnaire

The Knowledge Validation Questionnaire, validated by a teacher of basic education, consists of four questions that are answered by the participant before and after playing the game. The goal of this questionnaire is to assess whether playing the game results in knowledge gain regarding air pollution. To reduce stress among the participants, the questions were orally presented to the children, instead of using the written format. The children's oral responses were audio-recorded for offline analysis.

The questionnaire's four questions (translated from Portuguese to English) are as follows:

- Q1: *Do you think there could be air pollution inside our homes?* Response: Yes/No.
- Q2: *What do you think can pollute the air inside our homes?* Response: An enumeration of objects.
- Q3: *What do you think is in the air when it's polluted?* Response: An enumeration of air pollutants.

- Q4: *How do you think we can clean up the air pollution we have inside our homes?* Response: An enumeration of cleaning methods.

  Questions Q2–Q4 were only asked if the participant responded "yes" to Q1.

4.1.2. Satisfaction and Usability Questionnaire

The Satisfaction and Usability Questionnaire (SUQ) was answered by the participants after playing the game with the goal of assessing: (1) their intention to repeat the gaming experience (repetition is important to help consolidate knowledge); (2) their perception regarding how easy/intuitive it is to interact with the game (a non-intuitive experience can hamper learning); and (3) their perception of the game's utility as a learning tool.

SUQ consists of eight statements adapted from the standard questionnaires SUS [51] and TAM [52], which participants were asked to classify using a five-point Likert scale, ranging from "1 (I totally disagree)" to "5 (I totally agree)". A visual representation of the Likert scale was used, as suggested in an adaptation of the SUS for children [53]. The statements composing this questionnaire (translated from Portuguese to English), as well as their goals for the analysis, are:

- S1: *If we had more time, I'd like to keep playing the game.* Goal: assess users' immersion in the game, levels of acceptance, enjoyment, and whether it captivates exploration.
- S2: *I would like my teacher to use these types of games in the classroom.* Goal: assess the participant's perception of the game' utility as a teaching tool compared to traditional teaching methods.
- S3: *If I had this game at home I would like to play it a lot more.* Goal: assess the participant's interest in the game outside the classroom context.
- S4: *I felt confused several times while playing.* Goal: assess the participant's perception of ease of use and interaction, as well as of harmony between the real and virtual world.
- S5: *To play this game I feel like I need an adult's help.* Goal: assess the participant's sense of autonomy during the gaming experience, the learning curve, and the interaction complexity.
- S6: *If I played this game more often, I would learn a lot more about pollution.* Goal: assess the participant's perception of knowledge acquired as a result of playing the game.
- S7: *My friends will really like this game.* Goal: call for a deeper analysis of the levels of satisfaction and enjoyment, asking the user how the game will be perceived by a more general group.
- S8: *My friends will learn a lot about pollution with this game.* Goal: call for a deeper analysis of the game's ability to transmit knowledge, asking the user how the game will be perceived by a more general group.

To avoid biases, statements in the original SUS are posed with both positive and negative formulations. However, the successive change in the connotation of the questions can create some confusion in the user. In fact, during a set of formative tests, we found that this factor is even more evident in children, as they often provided random or nonsensical answers when they did not fully understand the questions. To avoid these issues, our statements are all positively formulated. This is supported by existing evidence that questionnaires can be equally valid when only statements with positive formulations are used [54].

4.1.3. Questionnaire of Opinion and Preference

After answering the SUQ, participants were asked to answer to a final Questionnaire of Opinion and Preference (QOP). This questionnaire is composed of four open questions whose purpose is to validate the data collected with the previous questionnaires, elicit additional information, and assess which components of the gaming experience are most appreciated. The questions composing this questionnaire (translated from Portuguese to English), as well as their goals for the analysis, are:

- Q1: *Do you prefer to play the game with real objects or with cards? Why?* Goal: assess the participant's perception of the importance of including physical objects and sensors in the gaming experience.
- Q2: *Do you think the game is useful for learning? Why?* Goal: assess the participant's perception of knowledge acquired as a result of playing the game.
- Q3: *Would you like to learn other subjects with this type of game? Which ones?* Goal: assess the participant's perception of the utility of AR-based games in learning contexts.
- Q4: *What did you like most and least about the game?* Goal: pinpoint the elements that deserve further attention when developing AR-based learning games.

## 4.2. Results

### 4.2.1. Knowledge Validation Questionnaire

As a result of questions Q2–Q4 of the Knowledge Validation Questionnaire only being asked to participants who answered yes to Q1, those three questions were answered by only 10 and 21 participants in the pre- and post-game questionnaires, respectively.

The answers to *Do you think there could be air pollution inside our homes?* clearly show that the game allowed participants to learn about air pollution. Concretely, Table 2 shows that the proportion of participants acknowledging that air pollution may exist inside their houses is significantly higher ($p < 0.001$, $n = 27$) after playing the game (77.8%) than before playing the game (37.0%). This knowledge gain highlights the importance of learning tools as the one presented. Importantly, none of the participants who answered affirmatively in the pre-game questionnaire changed their opinion in the post-game questionnaire, which seems to indicate that the game did not induce confusion.

**Table 2.** Response distribution (yes/no) to question "Do you think there may be air pollution inside our homes?"

| | Yes | No |
|---|---|---|
| Pre-game | 37.0% | 63.0% |
| Post-game | 77.8% | 22.2% |

The answers to the question *What do you think can pollute the air inside our homes?* were mostly categorized as related to the emission of gases or to particulate matter. For instance, terms like *smoke* and *fireplace* were (mostly) categorized as (related to) *gases*, whereas terms such as*cloth* and *sweep* were categorized as *particles*. The goal was to determine the extent to which participants were aware that the air quality is influenced by objects and people's behaviors that do not necessarily result in the production of gases. Table 3 shows that all participants were unaware of this relationship before playing the game. Remarkably, after playing the game, 22.2% of the participants provided answers using expressions categorized as particles. In the same line, the percentage of participants mentioning expressions related to gases roughly doubled after playing the game. The table also shows that the percentage of participants that did not know how to respond, provided invalid responses, or failed to recognize that there is pollution inside our homes, was roughly halved after playing the game.

The answers to the question *What do you think is in the air when it's polluted?*, summarized in Table 4, are aligned with the answers provided to the previous question and revealed a considerable knowledge gain after playing the game. The percentage of participants that did not know how to respond, provided invalid responses, or failed to recognize that there is pollution inside our homes, reduced by approximately two thirds after playing the game. Additionally, the percentage of participants that provided answers using expressions categorized as particles and gases was roughly six and five times higher after playing the game, respectively. Hence, Tables 3 and 4 show that the game contributed to the enhancement of participants' knowledge and self-confidence.

**Table 3.** Percentage of the 27 participants whose answers to the question "What do you think can pollute the air inside our homes?" mostly included expressions related to either particles (first column) or gases (second column). Some participants included expressions related to both types of air pollutants in their responses. The third column presents the percentage of participants which reported that they did not know how to respond (D), provided invalid responses (I), or failed to recognize that there is pollution inside our homes (U).

| | Particles | Gases | D + I + U |
|---|---|---|---|
| Pre-game | 0.0% | 29.6% | 70.4% |
| Post-game | 22.2% | 63.0% | 37.0% |

**Table 4.** Percentage of the 27 participants whose answers to the question "What do you think is in the air when it's polluted?" mostly included expressions related to either particles (first column) or gases (second column). Some participants included expressions related to both types of air pollutants in their responses. The third column presented the percentage of participants that reported they did not know how to respond, provided invalid responses, or that failed to recognize that there is pollution inside our homes.

| | Particles | Gases | D + I + U |
|---|---|---|---|
| Pre-game | 3.7% | 11.1% | 85.2% |
| Post-game | 22.2% | 55.6% | 29.6% |

The percentage of participants providing valid answers to question *How do you think we can clean up the air pollution we have inside our homes?* increased from 22.2% before playing the game to 55.6%, after playing the game. The results show that the children were able to improve their knowledge on this topic. Concretely, the references to *ventilation* and related words used by the children raised from 7.4% (pre-game) to 22.2% (post-game). This is aligned with references to the explicit use of a *fan*, which increased from 0% (pre-game) to 18.5% (post-game). Unfortunately, the other tools were explicitly mentioned by only one of the participants. This means that the fan is notoriously the most memorable tool in the game.

#### 4.2.2. Satisfaction and Usability Questionnaire

Table 5 presents the 95% confidence intervals associated with SUQ responses. The closer to five the score in each statement of the questionnaire is, the better the game has been experienced by the participants, except for two statements (marked with * in the table), for which a score closer to one is better. All obtained scores are between the ideal scale extreme and the scale's mid-point, meaning that the game is globally perceived as satisfying, usable, and educationally enriching. This conclusion is reinforced by the positive results obtained around the three aggregating factors presented in Table 6: (1) intention to use again; (2) perceived learning utility; and (3) perceived easiness of use. Interestingly, it was observed that all pairs of participants explored all objects and all tools, even when only some of them were required to complete the given game goals.

#### 4.2.3. Questionnaire of Opinion and Preference

To the question *Do you prefer to play the game with real objects or with cards?*, the majority of participants (85.2%) stated to prefer the version of the experience with real objects over the version in which these objects were replaced by paper card representatives (see Table 7). This result confirms the value of the tangible interface of the game based on the interaction with physical objects. When asked why they prefer physical objects over paper cards, participants revealed four main reasons summarized in Table 8: more realistic (33.3%), better control (29.6%), easier perception, (14.8%), more fun (11.1%), while 11.1% were unable to provide a reason. Conversely, the participants who stated to prefer to interact with paper cards over physical objects (14.8%) justified this preference due to the easiness of interaction.

**Table 5.** 95% Confidence intervals (CI) for the results obtained with the Satisfaction and Usability Questionnaire, answered with a five-point Likert scale (from 1 to 5). Note that, in opposition to all other statements, the lower the values obtained for S4 and S5, the better. These are annotated with *.

| Statement (Summarized) | | 95% CI |
|---|---|---|
| S1 | *I'd like to keep playing this game* | [4.61, 4.95] |
| S2 | *I'd like my teacher to use these types of games* | [4.17, 4.94] |
| S3 | *I'd like to play this game again at home* | [4.08, 4.66] |
| S4 | *I felt confused while playing this game* | * [1.70, 2.60] |
| S5 | *I feel I need adult's help to play this game* | * [1.36, 1.90] |
| S6 | *I'd learn more if I play this game more* | [4.19, 4.70] |
| S7 | *My friends will like to play this game* | [4.42, 4.91] |
| S8 | *My friends will learn with this game* | [4.10, 4.64] |

**Table 6.** Results of the Satisfaction and Usability Questionnaire presented in Table 5 averaged (AVG) by factors. Note that, in opposition to all other factors, the lower the value obtained for *perceived easiness of use*, the better. This inverted value is annotated with *.

| Factor | Statements | AVG |
|---|---|---|
| Intention to use again | 1, 3, 7 | 4.61 |
| Perceived learning utility | 2, 6, 8 | 4.46 |
| Perceived easiness of use | 4, 5 | * 1.88 |

**Table 7.** Distribution of the responses to the question "Do you prefer to play the game with real objects or with cards?".

| Paper Cards | Real Objects |
|---|---|
| 14.8% | 85.2% |

**Table 8.** Distribution of the reasons presented by the participants to the question "Why did you prefer to use real objects over paper cards?"

| Realism | Control | Perception | Fun | Do Not Know |
|---|---|---|---|---|
| 33.3% | 29.6% | 14.8% | 11.1% | 11.1% |

To the question *Do you think the game is useful for learning?*, all participants answered affirmatively. These results are in agreement with the results obtained in the SUQ. The same trend has been observed in the answers to the question *Would you like to learn other subjects with this type of game?*, to which all participants but one answered affirmatively. Hence, we can conclude that participants perceive playing AR-based learning games to be an enriching learning activity whose application spectrum is not limited to air quality. The topics that participants reported to be ones they would like to see addressed by AR-based learning games were mostly related to pollution and waste processing, but also animals and forest. Other less frequently reported preferences include the human body, the universe, mathematics, dinosaurs, and cooking.

Participants were unable to provide a significant amount of answers to the question *What did you like most and least about the game?*. This may mean that the question is too open for the participants' age group or that they were already tired in the end of the testing session. Nevertheless, we believed it was worth analyzing the provided responses.

The most frequent responses regarding the most liked aspects of the game included the terms *fan*, *candle*, *ring*, *smoke*, and *winning*. These highlight the interest in the augmented reality (*ring*), the gaming component (*winning*), the toy-like characteristics of the game (*fan* used for clearing *smoke*), and the importance of using interesting real-world objects (*candle*).

The least liked component of the game was the *filter*, possibly because it was the less familiar item. *Interaction with cards* was also mentioned by participants, since they strongly preferred playing using real-world objects. This is a positive result that supports the importance of considering real objects as tangible interface. *Little time to play* was also among least liked aspects. This was also an interesting response, as it suggests that participants were willing to play for longer periods.

### *4.3. Discussion*

The results show that playing the AR game results in knowledge gain regarding the causes of indoor air pollution (objects and people behaviors) and about its mitigation mechanisms, thus confirming research hypotheses H1 and H2. This knowledge gain is particularly evident regarding the influence of particulate matter on air quality (in comparison to gases) and on the importance of house ventilation to improve air quality.

The results also confirm research hypothesis H3 by showing that the AR game is perceived by children as useful for learning, easy to use, and providing a satisfying experience that they are willing to repeat. These results are important because learning to be effective requires repetition and a game with low replay value is not able to deliver this.

All participants explored all objects and tools, even though they were not required to complete the game. We believe that this desirable exploratory behavior is at least in part due to the physics-based mechanics of the game, which was included to leverage on the children's intrinsic motivation towards toys. If provided with toy-like properties, a game should become more interesting to play with, even before the player knows exactly what game's goals are [7].

Finally, the results show that participants prefer using real objects as a tangible interface rather than card-based representatives, thus confirming the research hypothesis H4. The realism and easiness of interaction were the most indicated reasons. This shows the value of augmented reality, enhancing the physical reality with digital content, rather than substituting it altogether.

## 5. Conclusions

A serious game for teaching children about air pollution was presented. More than simply conveying the dangers of air pollution, the game aims to inform children about the several types of air pollutants, what causes them, and which mechanisms are available to clear polluted air. To accommodate these goals, the children are invited to interact with a physical sensor node in order to playfully correlate everyday objects and the air pollution emitted by them. This tangible form of interaction aims to render the learning of messages more memorable and plausible. The game exploits augmented reality to provide children with the possibility to effortlessly view the invisible, that is, the air pollutants emitted by the everyday objects manipulated by them, further reinforcing the learned messages. To boost learning and enjoyment via socialization, children play the game in pairs.

A participatory design process was followed with intermediate formative evaluation moments to ensure that the design meets the desirable goals. The final design was subject summative evaluation with a sample of 27 children aged between 7 and 11 years using the Wizard of Oz method. The results show that playing the game is an effective learning experience. Concretely, comparing pre- and post-game data, the results show that playing the game allowed most children to better recognize that air pollution may exist inside our homes, air pollution is not only composed of gases but also of particulate matter, garbage and air pollution are not the same, and that ventilating our homes is important. The results also show that most children felt that the game was useful, usable, and enjoyable, resulting in an intention play again. In fact, all participants stated that they would like to use this type of game to learn other subjects. Finally, the results also show that most children preferred playing with real everyday life objects than with card surrogates, mostly due to realism and control.

The game was tested with a sample size of 27, which is close to the usually considered threshold for the central limit theorem to hold ($n \geq 30$) [55]. Nevertheless, given that it is better to have a larger and more diverse sample size, we plan to expand our user study. As future work, we also intend to assess the potential long-term learning benefits provided by the game. For this, it will be necessary to design an experimental protocol in order to test the children's knowledge after a prolonged period from the moment they played the game. We would also like to repeat the experiments without using the Wizard of Oz method, that is, using real data automatically collected by the physical sensor node. We also intend to expand the game in order to teach children about the impact of other human activities and behaviors on indoor air quality (e.g., cooking, house cleaning, incense burning, and fireplace use) and outdoor air quality (e.g., people and goods transportation, industrial activity). Raising children's awareness of this impact is important because these activities emit air pollutants and are common in households and cities. From a technical standpoint, we plan to study the use of hand tracking techniques that do not require the use of a ring marker. We also consider it important to explore the use of visualization devices that would allow the game to be played on the move (e.g., AR glasses), beyond the tabletop, enabling the inclusion of location-based mechanics. Finally, we would like to expand the game with additional narratives, capable of attracting children of different age groups. Ideally, these narratives are inclusive and personalized to the children.

**Author Contributions:** Conceptualization, P.S., J.F., T.B. and S.M.A.; methodology, P.S., T.B. and S.M.A.; software, J.F.; validation, P.S., T.B. and J.F.; formal analysis, J.F., P.S. and T.B; investigation, J.F., P.S. and T.B.; resources, J.F.; data curation, J.F.; writing—original draft preparation, P.S. and T.B.; writing—review and editing, P.S., T.B., S.M.A. and J.F.; supervision, P.S. and T.B.; project administration, P.S. and S.M.A.; funding acquisition, P.S. and S.M.A. All authors have read and agreed to the published version of the manuscript.

**Funding:** This work was developed within the ExpoLIS project, funded by FEDER, through Programa Operacional Regional de Lisboa (LISBOA-01-0145-FEDER-032088), and by national funds (OE), through Fundação para a Ciência e a Tecnologia (FCT), I.P., Portugal (PTDC/EAM-AMB/32088/2017). Part of this work was also supported by FCT under ISTAR projects UIDB/04466/2020 and UIDP/0446 6/2020 and C2TN projects UIDB/04349/2020 and UIDP/04349/2020.

**Informed Consent Statement:** Informed consent was obtained from all subjects involved in this study.

**Data Availability Statement:** The datasets used and/or analyzed during the current study are available from the authors on reasonable request.

**Conflicts of Interest:** The authors declare no conflict of interest.

## Abbreviations

The following abbreviations are used in this manuscript:

| | |
|---|---|
| AQI | Air Quality Index |
| AR | Augmented Reality |
| CI | Confidence Interval |
| CO | Carbon Monoxide |
| DDA | Dynamic Difficulty Adjustment |
| HMD | Head Mounted Display |
| IoT | Internet of Things |
| QOP | Questionnaire of Opinion and Preference |
| $NO_2$ | Nitrogen Dioxide |
| NPC | Non-Playable Character |
| PM1 | Particulate Matter with an Aerodynamic Diameter $\leq 1$ µm |
| PM2.5 | Particulate Matter with an Aerodynamic Diameter $\leq 2.5$ µm |
| PM10 | Particulate Matter with an Aerodynamic Diameter $\leq 10$ µm |

| | |
|---|---|
| SUQ | Satisfaction and Usability Questionnaire |
| SUS | System Usability Scale |
| TAM | Technology Acceptance Model |
| VR | Virtual Reality |
| WHO | World Health Organization |

## References


1. WHO. Air Pollution. Available online: https://www.who.int/health-topics/air-pollution (accessed on 25 November 2021).
2. Zhang, Z.; Zhang, G.; Su, B. The spatial impacts of air pollution and socio-economic status on public health: Empirical evidence from China. *Socio-Econ. Plan. Sci.* **2022**, *83*, 101167. [CrossRef]
3. Zhang, Z.; Zhang, J.; Feng, Y. Assessment of the carbon emission reduction effect of the air pollution prevention and control action plan in China. *Int. J. Environ. Res. Public Health* **2021**, *18*, 13307. [CrossRef] [PubMed]
4. WHO. *Effects of Air Pollution on Children's Health and Development: A Review of the Evidence Special Programme on Health and Environment*; Technical Report; World Health Organization: Geneva, Switzerland, 2005.
5. McCoy, D.C.; Yoshikawa, H.; Ziol-Guest, K.M.; Duncan, G.J.; Schindler, H.S.; Magnuson, K.; Yang, R.; Koepp, A.; Shonkoff, J.P. Impacts of early childhood education on medium-and long-term educational outcomes. *Educ. Res.* **2017**, *46*, 474–487. [CrossRef] [PubMed]
6. Mylonas, G.; Triantafyllis, C.; Amaxilatis, D. An augmented reality prototype for supporting IoT-based educational activities for energy-efficient school buildings. *Electron. Notes Theor. Comput. Sci.* **2019**, *343*, 89–101. [CrossRef]
7. Schell, J. *The Art of Game Design: A Book of Lenses*; CRC Press: Boca Raton, FL, USA, 2008.
8. Santana, P.; Almeida, A.; Mariano, P.; Correia, C.; Martins, V.; Almeida, S.M. Air quality mapping and visualisation: An affordable solution based on a vehicle-mounted sensor network. *J. Clean. Prod.* **2021**, *315*, 128194. [CrossRef]
9. Blunt, R. Does game-based learning work? Results from three recent studies. In Proceedings of the Interservice/Industry Training, Simulation, & Education Conference, Orlando, FL, USA, 26–29 November 2007; pp. 945–955.
10. Partovi, T.; Razavi, M.R. The effect of game-based learning on academic achievement motivation of elementary school students. *Learn. Motiv.* **2019**, *68*, 101592. [CrossRef]
11. Zhu, Z.T.; Yu, M.H.; Riezebos, P. A research framework of smart education. *Smart Learn. Environ.* **2016**, *3*, 4. [CrossRef]
12. Creighton, S.; Szymkowiak, A. The effects of cooperative and competitive games on classroom interaction frequencies. *Procedia-Soc. Behav. Sci.* **2014**, *140*, 155–163. [CrossRef]
13. Xin, J.F. Computer-assisted cooperative learning in integrated classrooms for students with and without disabilities. *Inf. Technol. Child. Educ. Annu.* **1999**, *1999*, 61–78.
14. Green, V.A.; Rechis, R. Children's cooperative and competitive interactions in limited resource situations: A literature review. *J. Appl. Dev. Psychol.* **2006**, *27*, 42–59. [CrossRef]
15. Dunleavy, M. Design principles for augmented reality learning. *TechTrends* **2014**, *58*, 28–34. [CrossRef]
16. Theodoropoulos, A.; Lepouras, G. Augmented Reality and programming education: A systematic review. *Int. J. Child-Comput. Interact.* **2021**, *30*, 100335. [CrossRef]
17. Khan, T.; Johnston, K.; Ophoff, J. The impact of an augmented reality application on learning motivation of students. *Adv. Hum.-Comput. Interact.* **2019**, *2019*, 7208494. [CrossRef]
18. Kurilovas, E. Evaluation of quality and personalisation of VR/AR/MR learning systems. *Behav. Inf. Technol.* **2016**, *35*, 998–1007. [CrossRef]
19. Cai, S.; Liu, E.; Shen, Y.; Liu, C.; Li, S.; Shen, Y. Probability learning in mathematics using augmented reality: Impact on student's learning gains and attitudes. *Interact. Learn. Environ.* **2020**, *28*, 560–573. [CrossRef]
20. Escobedo, L.; Tentori, M.; Quintana, E.; Favela, J.; Garcia-Rosas, D. Using augmented reality to help children with autism stay focused. *IEEE Pervasive Comput.* **2014**, *13*, 38–46. [CrossRef]
21. Almurashi, H.; Bouaziz, R.; Alharthi, W.; Al-Sarem, M.; Hadwan, M.; Kammoun, S. Augmented reality, serious games and picture exchange communication system for people with ASD: Systematic literature review and future directions. *Sensors* **2022**, *22*, 1250. [CrossRef]
22. Carmigniani, J.; Furht, B.; Anisetti, M.; Ceravolo, P.; Damiani, E.; Ivkovic, M. Augmented reality technologies, systems and applications. *Multimed. Tools Appl.* **2011**, *51*, 341–377. [CrossRef]
23. Tuli, N.; Mantri, A. Evaluating usability of mobile-based augmented reality learning environments for early childhood. *Int. J. Hum.-Comput. Interact.* **2021**, *37*, 815–827. [CrossRef]
24. Zhang, H.; Cui, Y.; Shan, H.; Qu, Z.; Zhang, W.; Tu, L.; Wang, Y. Hotspots and trends of virtual reality, augmented reality and mixed reality in education field. In Proceedings of the 6th International Conference of the Immersive Learning Research Network (iLRN), San Luis Obispo, CA, USA, 21–25 June 2020; IEEE: Piscataway, NJ, USA, 2020; pp. 215–219.
25. Sun, R.; Wu, Y.J.; Cai, Q. The effect of a virtual reality learning environment on learners' spatial ability. *Virtual Real.* **2019**, *23*, 385–398. [CrossRef]

26. Habig, S. Who can benefit from augmented reality in chemistry? Sex differences in solving stereochemistry problems using augmented reality. *Br. J. Educ. Technol.* **2020**, *51*, 629–644. [CrossRef]
27. Huang, H.L.; Hwang, G.J.; Chang, C.Y. Learning to be a writer: A spherical video-based virtual reality approach to supporting descriptive article writing in high school Chinese courses. *Br. J. Educ. Technol.* **2020**, *51*, 1386–1405. [CrossRef]
28. Teles, B.; Mariano, P.; Santana, P. Game-like 3d visualisation of air quality data. *Multimodal Technol. Interact.* **2020**, *4*, 54. [CrossRef]
29. Meyer, U.; Becker, J.; Broscheit, J. Visualising Air Pollution Datasets with Real-Time Game Engines. In *Proceedings of the World Conference on Information Systems and Technologies*; Springer: Berlin/Heidelberg, Germany, 2019; pp. 304–312.
30. Kim, J.C.; Laine, T.H.; Åhlund, C. Multimodal interaction systems based on internet of things and augmented reality: A systematic literature review. *Appl. Sci.* **2021**, *11*, 1738. [CrossRef]
31. Rashid, Z.; Melià-Seguí, J.; Pous, R.; Peig, E. Using Augmented Reality and Internet of Things to improve accessibility of people with motor disabilities in the context of Smart Cities. *Future Gener. Comput. Syst.* **2017**, *76*, 248–261. [CrossRef]
32. Vrellis, I.; Delimitros, M.; Chalki, P.; Gaintatzis, P.; Bellou, I.; Mikropoulos, T.A. Seeing the unseen: User experience and technology acceptance in Augmented Reality science literacy. In Proceedings of the IEEE 20th International Conference on Advanced Learning Technologies (ICALT), Tartu, Estonia, 6–9 July 2020; IEEE: Piscataway, NJ, USA, 2020; pp. 333–337.
33. Mahmoudi, M.T.; Zeraati, F.Z.; Yassini, P. A color sensing AR-based interactive learning system for kids. In Proceedings of the 12th Iranian and 6th International Conference on e-Learning and e-Teaching (ICeLeT), Tehran, Iran, 4 March 2018; IEEE: Piscataway, NJ, USA, 2018; pp. 13–20.
34. Pokric, B.; Krco, S.; Drajic, D.; Pokric, M.; Rajs, V.; Mihajlovic, Z.; Knezevic, P.; Jovanovic, D. Augmented Reality Enabled IoT Services for Environmental Monitoring Utilising Serious Gaming Concept. *J. Wirel. Mob. Netw. Ubiquitous Comput. Dependable Appl.* **2015**, *6*, 37–55.
35. Hadj Sassi, M.S.; Chaari Fourati, L. Architecture for visualizing indoor air quality data with augmented reality based cognitive internet of things. In *Proceedings of the International Conference on Advanced Information Networking and Applications*; Springer: Berlin/Heidelberg, Germany, 2020; pp. 405–418.
36. Sun, Y.; Armengol-Urpi, A.; Kantareddy, S.N.R.; Siegel, J.; Sarma, S. Magichand: Interact with iot devices in augmented reality environment. In Proceedings of the IEEE Conference on Virtual Reality and 3D User Interfaces (VR), Osaka, Japan, 23–27 March 2019; IEEE: Piscataway, NJ, USA, 2019; pp. 1738–1743.
37. Ishii, H. The tangible user interface and its evolution. *Commun. ACM* **2008**, *51*, 32–36. [CrossRef]
38. Joyce, R.D.; Robinson, S. Passive haptics to enhance virtual reality simulations. In Proceedings of the AIAA Modeling and Simulation Technologies Conference, Grapevine, TX, USA, 9–13 January 2017; p. 1313.
39. Besançon, L.; Issartel, P.; Ammi, M.; Isenberg, T. Mouse, tactile, and tangible input for 3D manipulation. In Proceedings of the 2017 CHI Conference on Human Factors in Computing Systems, Denver, CO, USA, 6–11 May 2017; pp. 4727–4740.
40. Alakärppä, I.; Jaakkola, E.; Väyrynen, J.; Häkkilä, J. Using nature elements in mobile AR for education with children. In Proceedings of the 19th International Conference on Human–Computer Interaction with Mobile Devices and Services, Vienna, Austria, 4–7 September 2017; pp. 1–13.
41. Martens, J.B.; Qi, W.; Aliakseyeu, D.; Kok, A.J.; van Liere, R. Experiencing 3D interactions in virtual reality and augmented reality. In Proceedings of the 2nd European Union symposium on Ambient Intelligence, Eindhoven, The Netherlands, 8–11 November 2004; pp. 25–28.
42. Chamzas, D.; Moustakas, K. 3D Augmented Reality Tangible User Interface using Commodity Hardware. *arXiv* **2020**, arXiv:2003.01092.
43. Teng, C.H.; Peng, S.S. Augmented-Reality-Based 3D Modeling System Using Tangible Interface. *Sens. Mater.* **2017**, *29*, 1545–1554.
44. Cardoso, J.C.; Ribeiro, J.M. Tangible VR book: Exploring the design space of marker-based tangible interfaces for virtual reality. *Appl. Sci.* **2021**, *11*, 1367. [CrossRef]
45. Dix, A.; Finlay, J.; Abowd, G.D.; Beale, R. *Human–Computer Interaction*; Pearson Education: Upper Saddle River, NJ, USA, 2004.
46. Unity. Unity Real-Time Development Platform. Available online: https://unity.com/ (accessed on 23 October 2021).
47. Vuforia. Vuforia Developer Portal. Available online: https://developer.vuforia.com/ (accessed on 23 October 2021).
48. Blender. Blender 2.93 Reference Manual. Available online: https://docs.blender.org/manual/en/latest/ (accessed on 1 November 2021).
49. EPA. *Technical Assistance Document for the Reporting of Daily Air Quality—The Air Quality Index (AQI)*; Technical Report; U.S. Environmental Protection Agency: Washington, DC, USA, 2018.
50. Dias, M.; Jorge, J.; Carvalho, J.; Santos, P.; Luzio, J. Usability evaluation of tangible user interfaces for augmented reality. In Proceedings of the IEEE International Augmented Reality Toolkit Workshop, Tokyo, Japan, 7 October 2003; IEEE: Piscataway, NJ, USA, 2003; pp. 54–61.
51. Brooke, J. SUS-A quick and dirty usability scale. *Usability Eval. Ind.* **1996**, *189*, 4–7.
52. Davis, F.D. Perceived usefulness, perceived ease of use, and user acceptance of information technology. *MIS Q.* **1989**, *13*, 319–340. [CrossRef]
53. Putnam, C.; Puthenmadom, M.; Cuerdo, M.A.; Wang, W.; Paul, N. Adaptation of the system usability scale for user testing with children. In Proceedings of the Extended Abstracts of the 2020 CHI Conference on Human Factors in Computing Systems, Honolulu, HI, USA, 25–30 April 2020; pp. 1–7.

54. Sauro, J.; Lewis, J.R. When designing usability questionnaires, does it hurt to be positive? In Proceedings of the SIGCHI Conference on Human Factors in Computing Systems, Vancouver, BC, Canada, 7–12 May 2011; pp. 2215–2224.
55. Sauro, J.; Lewis, J.R. *Quantifying the User Experience: Practical Statistics for User Research*, 2nd ed.; Morgan Kaufmann: San Francisco, CA, USA, 2016.

International Journal of
*Environmental Research and Public Health*

*Article*

# Citizens' Perception on Air Quality in Portugal—How Concern Motivates Awareness

**Nuno Canha [1,*], Ana Rita Justino [1], Carla A. Gamelas [1,2] and Susana Marta Almeida [1]**

[1] Centro de Ciências e Tecnologias Nucleares, Instituto Superior Técnico, Universidade de Lisboa, Estrada Nacional 10, Km 139.7, 2695-066 Bobadela, Portugal
[2] Instituto Politécnico de Setúbal, Escola Superior de Tecnologia de Setúbal, Centro de Investigação em Energia e Ambiente, IPS Campus, 2914-508 Setúbal, Portugal
* Correspondence: nunocanha@ctn.tecnico.ulisboa.pt

**Abstract:** This study aimed to understand the knowledge of Portuguese citizens about air quality and the extent to which the concerns about specific environmental problems can motivate their acquaintance of information. Moreover, this study also allowed to understand which information about air quality needs further dissemination to provide the citizens with all the available tools and the correct knowledge. For this, a national online survey about air quality perception was conducted, where 1131 answers were obtained and two different populations were compared: the general population and a sub-population from an urban-industrial area of Lisbon metropolitan area that had experienced frequent air pollution events in the past. Air pollution was considered the environmental topic of higher concern among this sub-population (61.4%), while in the general population it ranked thirdly (27.4%). Generally, the sub-population showed higher knowledge about air quality than the general population, with 61% being able to identify at least one air pollutant. The perception of the local air quality was also very different between populations, with 61% of the sub-population considering it poor or very poor, while only 14% of the general population had the same perception, which highlights the different levels of concern between populations. A weak knowledge about air pollutants (50% of the general population could not identify any air pollutant) and an erroneous perception of the contribution of the different pollution sources to air quality levels were found. More than 50% of the respondents of both populations were considered to not have enough information regarding the air quality in their area of residence, with the national air quality database being unknown to almost everyone. Overall, strong efforts should be made to increase the awareness about the importance of air quality, which may promote a higher acceptance of the implementation of future actions to improve air quality.

**Keywords:** air pollution; air pollutants; sources; perception; air quality; citizens' awareness; Portugal



**Citation:** Canha, N.; Justino, A.R.; Gamelas, C.A.; Almeida, S.M. Citizens' Perception on Air Quality in Portugal—How Concern Motivates Awareness. *Int. J. Environ. Res. Public Health* **2022**, *19*, 12760. https://doi.org/10.3390/ijerph191912760

Academic Editor: Paul B. Tchounwou

Received: 18 July 2022
Accepted: 30 September 2022
Published: 5 October 2022



## 1. Introduction

Ambient air pollution has become a growing concern, mostly due to the rapid urbanization, industrialization, and traffic. In Europe, it is perceived as the second biggest concern (after climate change) and it is the most relevant environmental risk to human health [1]. Exposure to ambient air pollution is associated with a variety of health impacts such as cardiovascular diseases, cancer, respiratory diseases and mortality [2].

People's understanding and response to ambient air pollution are vital to recognize the best mitigation measures concerning the protection of public health [3]. Thus, it is important to consider people's perception and which factors may promote their behavioral changes, which may vary among groups and individuals [4]. The studies of air quality perception have not shown an association between the perceived air quality and the concentration of measured pollutants [5]. Instead, air quality perception seems to be influenced by sensory experience, awareness, and knowledge, the emotions it provides (such as nuisance), communication, and risk perception (which takes into account the psychological, social and cultural factors) [3,5]. Moreover, empowered populations with knowledge about air quality are typically more politically active to request actions from authorities to promote the control of air pollution.

For instance, a study developed in seven European countries (Austria, Belgium, Germany, Italy, Poland, Sweden and the United Kingdom) comparing the public perception of air pollution sources and the real-world situation through a survey involving 16,101 participants found that it existed a very high underestimation of the contribution of the agriculture sector to air pollution [6]. Moreover, this trend was common in all seven countries but demonstrated a small influence of gender, age, and socio-economic status of the respondents.

The present study aims to understand the awareness levels of Portuguese citizens regarding air quality along with their knowledge about this topic. Moreover, considering that some populations may be aware of the impact of local pollution sources in their daily life, this study also aimed to understand to which extension their concerns influence their awareness and knowledge about air quality. To achieve these goals, a national online questionnaire was made available during two months and the overall population was compared with a sub-population from an urban-industrial area of Portugal that is known to be aware of air pollution due to occasional settled dust events in the area [7,8].

## 2. Materials and Methods

### *2.1. Study Area*

This study was conducted in Portugal and it compared the general Portuguese population and the population of a parish from a specific urban-industrial area where the inhabitants have presented several public complaints about the air quality (due to the operation of the industries located within its limits) [9]. This site is the parish of União das Freguesias do Seixal, Arrentela e Aldeia de Paio Pires (UFSAAPP), located in the municipality of Seixal (Portugal), which is integrated in the Metropolitan Area of Lisbon (Portugal), next to the Nature Reserve of the Tagus Estuary (Figure 1). With 167,294 inhabitants in 95.5 km$^2$, the municipality of Seixal is one of the most densely populated municipalities in Portugal [10].

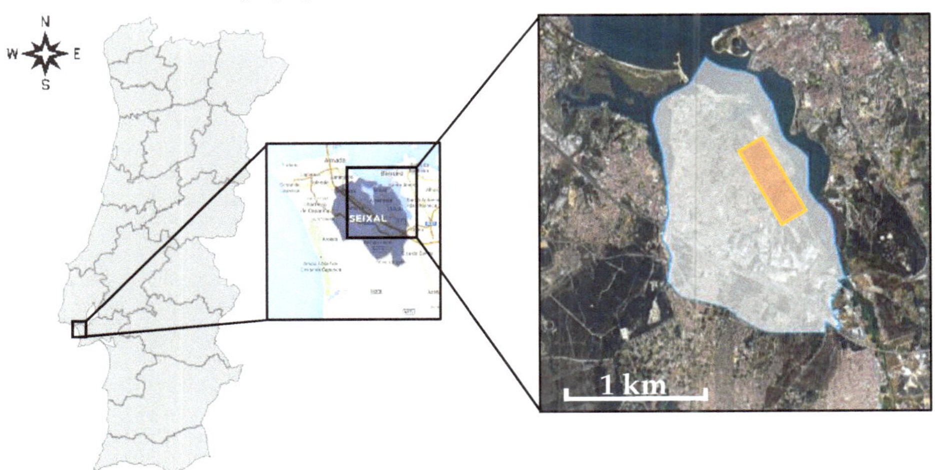


**Figure 1.** Location of the UFSAAPP area: (**left**) Framework of the study area (black rectangle) in Portugal mainland; (**center**) location of Seixal municipality highlighted; (**right**) location of UFSAAPP (whitish area with blue border represents the limits of the parish and orange represents the industrial area within it).

UFSAAPP is considered a typical urban-industrial area that comprises a densely populated residential area and a significant network of highways and national roads (EN10, A2-IP7 and A33), along with an industrial park where the facilities of different small and medium-sized industries are located (e.g., steelworks, lime factory and metal waste management and treatment [11–13]).

### 2.2. Questionnaires

A questionnaire was created to assess the perception of the Portuguese population regarding air quality. The questionnaire had five different sections: (i) participants' demographic characteristics; (ii) perception towards air quality, including knowledge about air pollutants and their sources; (iii) perception of the impacts of air quality, (iv) sources of information, and (v) knowledge of applicable regulations for controlling air quality. Overall, the questionnaire had a total of 36 questions and it was created based in the questionnaires developed by the project RISKAR LX [14]. The questionnaire is available in the Supplementary Materials (English version).

The questionnaire was disseminated by social media and an invitation was sent to all Portuguese municipalities to share it with their citizens. The questionnaire was available online from 1 February to 26 June 2020 for anonymous answer and a total of 1131 answers were gathered. Only participants with 18 or more years old were considered.

The survey implementation and data handling were conducted according to the guidelines of the Declaration of Helsinki. The online questionnaires were presented to the participants alongside a short introductory summary, which defined the objectives of the study and ensured the anonymity and confidentiality of the provided information. For data analysis, all answers were codified and treated anonymously.

### 2.3. Statistical Analysis

Statistical tests were carried out using the supplement XLSTAT of Microsoft Excel (Addinsoft, Paris, France). Chi-square tests of independence were used to examine if participants' gender, age, educational level, monthly income, among others, were associated with the participant's level of concern about air quality and the evaluation of the air quality (with a significance level of 0.050).

## 3. Results and Discussion

### 3.1. Characteristics of Respondents

Table 1 presents the answer rate obtained by Portuguese district, where answers from all the Portuguese districts were obtained. Excluding the answers from the parish of UFSAAPP, the distribution among districts is more even, despite a great contribution from Lisbon district (37.3% instead of only 22.0%) and a lower contribution from Porto district (3.8% instead of 17.3%).

Table 2 provides the general demographic information of the respondents of the survey for both populations. For the general population, which included 1004 respondents, 61.9% were female and 37.7% were male, with the age groups of 21–25 and 41–45 years old as the ones with a higher percentage of respondents, namely, 17.6% and 17.2%, respectively. The most common school level was degree (47.9%), followed by high school (29.2%), while the most prevalent working status was active (78.3%) and the most common monthly income was between 1001–2000€ (35.8%). The demographic information of UFSAAPP population has a similar structure (despite not being statistical equal) and gathered a total of 127 respondents. Regarding the gender balance of this population, 63.0 % were female and 34.6% were male, with the age groups of 41–45 and above 60 years old being more represented (with 26.0 and 14.2%, respectively). Similar to the general population, the UFSAAPP population was mainly characterized by respondents with degree (44.9%) or high school (42.5%), being active regarding their working status (77.2%), with a monthly income of between 1001 and 2000€ (40.2%).

**Table 1.** Portuguese population per district and answer rate to the survey per Portuguese district, considering all the answers received and excluding the ones from UFSAAPP parish ($n$ = 129).

| | Population | | All Answers to Survey | | Answers to Survey, Excluding from Seixal Municipality | |
|---|---|---|---|---|---|---|
| **District** | **Inhabitants** | % | $n$ | % | $n$ | % |
| Aveiro | 700,964 | 6.8 | 16 | 1.4 | 16 | 1.6 |
| Beja | 144,410 | 1.4 | 4 | 0.4 | 4 | 0.4 |
| Braga | 846,515 | 8.2 | 61 | 5.4 | 61 | 6.1 |
| Bragança | 122,833 | 1.2 | 11 | 1.0 | 11 | 1.1 |
| Castelo Branco | 177,912 | 1.7 | 32 | 2.8 | 32 | 3.2 |
| Coimbra | 408,631 | 3.9 | 10 | 0.9 | 10 | 1.0 |
| Évora | 152,436 | 1.5 | 5 | 0.4 | 5 | 0.5 |
| Faro | 467,495 | 4.5 | 97 | 8.6 | 97 | 9.7 |
| Guarda | 143,019 | 1.4 | 5 | 0.4 | 5 | 0.5 |
| Leiria | 458,679 | 4.4 | 109 | 9.6 | 109 | 10.9 |
| Lisboa | 2,275,591 | 22.0 | 377 | 33.0 | 374 | 37.3 |
| Portalegre | 104,989 | 1.0 | 11 | 1.0 | 11 | 1.1 |
| Porto | 1,786,656 | 17.3 | 38 | 3.4 | 38 | 3.8 |
| Região Autónoma da Madeira | 251,060 | 2.4 | 2 | 0.2 | 2 | 0.2 |
| Região Autónoma dos Açores | 236,657 | 2.3 | 5 | 0.4 | 5 | 0.5 |
| Santarém | 425,431 | 4.1 | 32 | 2.9 | 33 | 3.3 |
| Setúbal | 875,656 | 8.5 | 275 | 24.7 | 148 | 14.8 |
| Viana do Castelo | 231,488 | 2.2 | 3 | 0.3 | 3 | 0.3 |
| Vila Real | 185,878 | 1.8 | 5 | 0.4 | 5 | 0.5 |
| Viseu | 351,592 | 3.4 | 33 | 2.9 | 33 | 3.3 |
| Total | 10,347,892 | 100.0 | 1131 | 100.0 | 1002 | 100.0 |

**Table 2.** Sociodemographic characterisation of the respondents, where $n$ is the number of individuals in each category.

| | | Population | | | |
|---|---|---|---|---|---|
| | | General | | UFSAAPP | |
| **Characteristic** | **Category** | $n$ | % | $n$ | % |
| Gender | Female | 621 | 61.9 | 80 | 63.0 |
| | Male | 379 | 37.7 | 44 | 34.6 |
| | Prefer not to answer | 4 | 0.4 | 3 | 2.4 |
| Age | <20 | 40 | 4.0 | 2 | 1.6 |
| | 21–25 | 177 | 17.6 | 7 | 5.5 |
| | 26–30 | 60 | 6.0 | 2 | 1.6 |
| | 31–35 | 62 | 6.2 | 13 | 10.2 |
| | 36–40 | 100 | 10.0 | 9 | 7.1 |
| | 41–45 | 173 | 17.2 | 33 | 26.0 |

**Table 2.** *Cont.*

| | | Population | | | |
|---|---|---|---|---|---|
| | | General | | UFSAAPP | |
| Characteristic | Category | *n* | % | *n* | % |
| Age | 46–50 | 142 | 14.1 | 17 | 13.4 |
| | 51–55 | 98 | 9.8 | 12 | 9.4 |
| | 56–60 | 92 | 9.2 | 14 | 11.0 |
| | >60 | 60 | 6.0 | 18 | 14.2 |
| School Level | Primary School (4 years) | 5 | 0.5 | 1 | 0.8 |
| | Basic school (6 years) | 11 | 1.1 | 0 | 0.0 |
| | Middle school (9 years) | 31 | 3.1 | 3 | 2.4 |
| | High school (12 years) | 293 | 29.2 | 54 | 42.5 |
| | Degree | 481 | 47.9 | 57 | 44.9 |
| | Master | 159 | 15.8 | 12 | 9.4 |
| | PhD | 24 | 2.4 | 0 | 0.0 |
| Working Status | Student | 152 | 15.1 | 6 | 4.7 |
| | Active | 786 | 78.3 | 98 | 77.2 |
| | Retired | 32 | 3.2 | 17 | 13.4 |
| | Unemployed | 25 | 2.5 | 4 | 3.1 |
| | Others | 9 | 0.9 | 2 | 1.6 |
| Monthly income | <300 € | 17 | 1.7 | 0 | 0.0 |
| | 301–635 € | 66 | 6.6 | 8 | 6.3 |
| | 636–900 € | 186 | 18.5 | 18 | 14.2 |
| | 901–1000 € | 132 | 13.1 | 11 | 8.7 |
| | 1001–2000 € | 359 | 35.8 | 51 | 40.2 |
| | 2001–3000 € | 54 | 5.4 | 11 | 8.7 |
| | >3000 € | 14 | 1.4 | 0 | 0.0 |
| | Not applicable | 120 | 12.0 | 8 | 6.3 |
| | Prefer not to answer | 56 | 5.6 | 20 | 15.7 |
| | Total | 1004 | 100.0 | 127 | 100.0 |

As described in the methodology, this survey gathered information regarding five different topics and the results were evaluated comparing the general population ($n = 1004$, all individuals that answered the survey, except the individuals from the parish of UFSAAPP) and the population of UFSAAPP ($n = 127$), the study area that is urban-industrial and whose population is known to be aware about air pollution due to occasional settled dust events in the area [9].

### 3.2. Issues of Concern

The main issues of concern of the two populations were assessed to understand how the air quality is perceived as an important issue within different topics. For that, the participants were asked to rate from 1 (no concern) to 4 (major concern) different environmental topics that could potentially affect their life. Figures 2 and 3 present the percentages of the level of concern for the different environmental topics for the general and UFSAAPP populations, respectively.

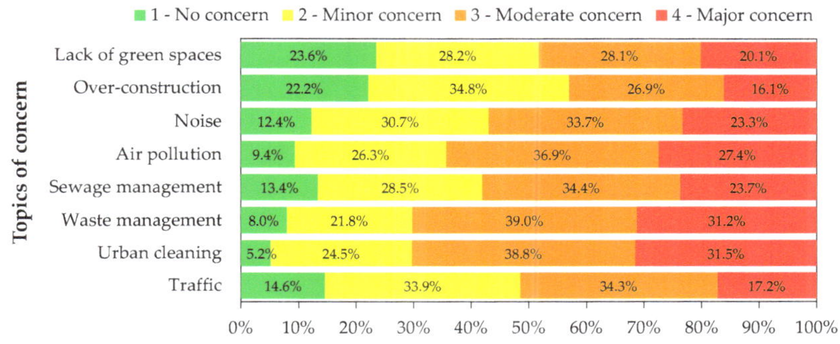


**Figure 2.** Level of concern of the general population regarding different environmental topics.

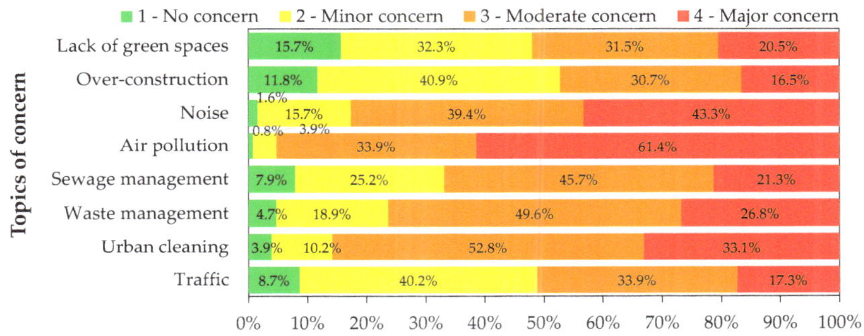


**Figure 3.** Level of concern of the UFSAAPP population regarding different environmental topics.

For the general population, the environmental topics that gather the highest levels of concern are urban cleaning (31.5%) and waste management (31.2%), followed in third position by air pollution (27.4%). This rank changes greatly when focusing on the UFSAAPP population, where air pollution gathers the highest level of concern (61.4%), followed by noise (43.3%) and then, urban cleaning (33.1%).

The association of the top-two main concern issues revealed by UFSAAPP population (air pollution and noise) was already identified previously [15] and usually it is difficult to distinguish them since they have common sources [16], such as transports, industry, agriculture and, with minimal contribution, household and neighborhood. Similar trend was already verified elsewhere [17], with the noise annoyance felt by citizens being directly related with their perception of a worst air quality.

The potential association between the level of concern regarding air quality and sociodemographic characteristics was assessed by $\chi^2$—test. In the general population, the level of concern was found dependent on the education level ($\chi^2 = 37.6$, $p$-value = 0.004) and the district of residence ($\chi^2 = 138.7$, $p$-value < 0.0001). Individuals with the basic school (six years) presented a higher level of concern regarding air quality (considering the categories "moderate concern" and "major concern" together, totalizing 91% of the individuals with the basic school) than the remaining individuals with different scholar levels (with an average of 66% for the same categories). However, it is important to highlight that the population with basic school only represents 1% of the total participants. In fact, the educational level is an important factor and, for instance, improving the level of education per capita has been shown to promote the gradual decrease of the impact of increased air pollution on public health damage [18].

The district of residence was also a factor of influence on the concern level, since the individuals that lived in districts in the country inland or islands (which are districts of low population density) presented a lower level of concern than the ones that lived in districts with higher population density (such as Braga, Guarda, Lisbon, and Setúbal), whereas the individuals of Setúbal presented the highest level of concern among all (with 83% of the respondents from Setúbal reporting "moderate concern" and "major concern").

### 3.3. Perception of Air Quality

Figure 4 provides the perception of state of the air quality by the citizens in different levels (country, municipality and neighborhood), considering the two types of study populations: general and UFSAAPP. As described above, the general population considers the citizens that reside in all districts of the country, except in the UFSAAPP area, while the UFSAAPP population refers exclusively to the inhabitants that reside UFSAAPP parish.

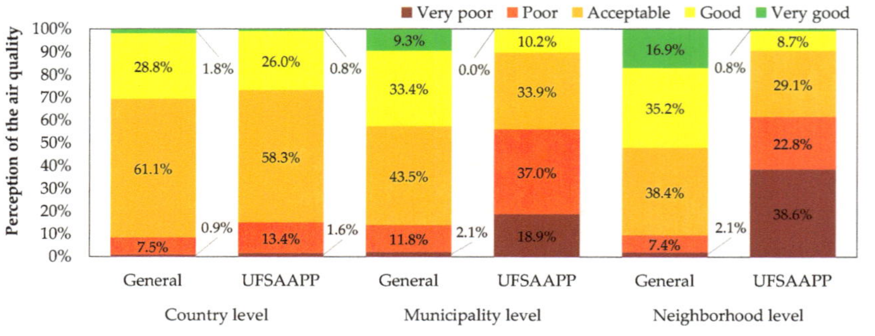


**Figure 4.** Assessment of the perception on air quality in Portugal, at country, municipality and neighborhood levels for the general and UFSAAPP populations.

At a country level, both populations have a similar perception of the air quality of the country, with "acceptable" air quality being the most common perception (61.1 % for the general population and 58.3% for the UFSAAPP population), followed by "good" air quality (28.8% for the general population and 26.0% for the UFSAAPP population) and then, by "poor" air quality (7.5% for the general population and 13.4% for the UFSAAPP population). The perception of good or very good air quality is similar between populations: 30.6% and 26.8% for the general and UFSAAPP populations, respectively. However, when assessing the air quality of the municipality or the neighborhood, the perceptions differ greatly between populations, with the UFSAAPP population clearly having a worst perception. Both at municipality and neighborhood levels, 55.9% and 61.4% of the UFSAAPP respondents have a "very poor" or "poor" perception of the air quality, respectively. At the municipality level, 13.9% of the general population have the perception of "poor" or "very poor" air quality, while this perception decreases to 9.5% when focusing on the neighborhood level.

This great difference on the perception of the air quality between the general and UFSAAPP populations at the local level highlights the concerns of the UFSAAPP population, reflected by the high level of awareness and sensibility of the UFSAAPP population toward this topic [9].

### 3.4. Identification of Pollution Sources

The survey also aimed to identify the knowledge of the participants regarding pollution sources and air pollutants. Figure 5 provides the main air pollution sources identified by the general and UFSAAPP populations in their area of residence.

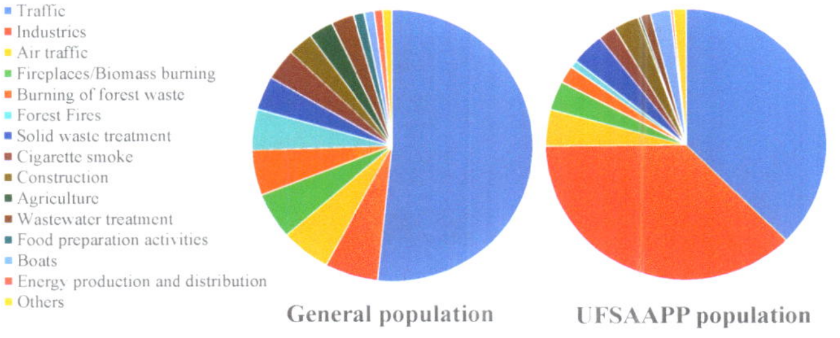


**Figure 5.** Air pollution sources identified by the general and the UFSAAPP populations.

For the general population, traffic was the most frequently identified pollution source, being referred by 51.7% of the respondents in the general population, which agrees with other studies where traffic was highlighted as one of the main sources of air pollution: China (with 78.5%) [19], Malaysia (where "motor vehicle emissions" were ranked as the most significant contributor to air pollution in their residential areas) [20], Mexico (with 50% highlighting cars and trucks) [21], and in seven European countries (ranging from 29.1% in Germany to 42.4% in Sweden) [6]. Industry (6.2%) and air traffic (5.7%) were considered as the second and third main pollution sources in our study. However, it is relevant to highlight that the study developed in seven European countries [6] (Austria, Belgium, Germany, Italy, Poland, Sweden, and United Kingdom) showed industry as the main contributor to the air pollution, while traffic was identified only as the second main pollution source. In the present study, it was possible to identify a pattern where the participants from districts of the country inland indicated pollution sources more associated with the rural areas, such as the burning of green wastes, forest fires, and fireplaces/biomass burning (for home heating).

The main air pollution sources perceived by the UFSAAPP population were traffic and industries (both with 37.4%), followed in third by air traffic (4.4%). In the UFSAAPP population, industries are considered one of the two main sources of air pollution (six times more than in the general population), probably due to the fact that UFSAAPP has suffered from several punctual pollution events (such as events of deposition of coarse particles in the area [7,9]), which have been attributed to the local industries [8] by the population. The proximity to an industrial area may affect the individual perception [22], which can be confirmed by the higher levels of concern of the UFSAAPP population since the heavy industry park is at walking distance from the residential area of UFSAAPP, along with the existence of many small and medium sized industries dispersed by the area. Another relevant issue is that, typically, the individual control of pollution sources from industries is minimum, which results in a higher degree of attention to those sources, including an increase of exposure reports from the population [23]. Therefore, taking into account that the UFSAAPP population has previous concerns and complaints regarding the local industries, it would be expected that industries would be considered by them one of the main pollution sources.

However, considering $PM_{2.5}$ as a proxy for air pollution, since it is the air pollutant responsible for more premature deaths in EU countries than any other pollutants (for instance, a total of 4900 premature deaths were attributable to $PM_{2.5}$ exposure in Portugal in 2019 [24]), the real direct emission sources for $PM_{2.5}$ may differ from the citizens' perception. Figure 6 presents the emission sources of $PM_{2.5}$ in 2019 for EU-27 and Portugal, along with the main air pollution sources identified by the general and the UFSAAPP populations (using a categorization of only seven pollution sources, where, for instance, traffic, air traffic and boat were framed in a single category called "transport").

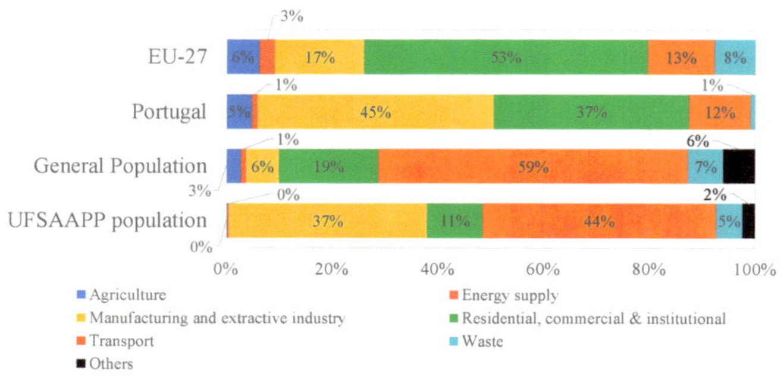


**Figure 6.** Emission sources of $PM_{2.5}$ in 2019 for EU-27 and Portugal [24], and the main air pollution sources perceived by the general and the UFSAAPP populations in the present study.

The major differences between the official and real-world data regarding the $PM_{2.5}$ emissions sources (both from EU-27 and Portugal) and the sources identified by the two studied populations are:

(i) The overestimation of the contribution of transport sector to air pollution by the public perception (59% and 44% for the general and UFSAAPP populations, respectively, against 12% of real contribution of transports to $PM_{2.5}$ levels in Portugal);

(ii) The underestimation of the "residential, commercial and institutional" source by the public perception (19% and 11% for the general and UFSAAPP populations, respectively, against 37% of the contribution of this source to the $PM_{2.5}$ levels in Portugal)

(iii) The great underestimation of the "manufacturing and extractive industry" contribution by the general population (6%) regarding the real contribution of 45% in Portugal (where the average contribution in the EU-27 is 17%). The UFSAAPP population indicated a value of 37% (probably, as described previously, due to their own concerns regarding the local industries), which is close to the real contribution verified in Portugal. However, it is relevant to highlight that the real contribution is higher than the perception of both populations, which indicates that the common citizen is not aware of the impact of industry in the air quality.

(iv) The negligible contribution of agriculture to air pollution, perceived by the general (3%) and the UFSAAPP (0%) populations, while the real world data indicate a higher contribution of 5% in Portugal and of 6% in the EU-27. However, if considering the secondary PM sources, such as $SO_2$ and $NO_X$ from the industry contribution and $NO_X$ from traffic emissions, combined with ammonia emission (from the agricultural sector), the solo contribution of agriculture may be very significant to air pollution levels and, typically, it is neglected by the public perception [6].

### *3.5. Identification of Air Pollutants*

The citizens' knowledge about air pollutants was also assessed in this survey, where the participants indicated which air pollutants they knew about (open question). For the identification of air pollutants and summarizing the results, the main air pollutants were considered individually (namely, $CH_4$, CO, $CO_2$, $NO_X$, $O_3$, PM, $SO_X$), while other identified air pollutants were considered as "Others". When a respondent did not supply any answer or answered "I do not know", it was quantified as "DK/NO". Answers that considered pollution sources (e.g., traffic) instead of air pollutants were not considered. For the category "PM", all answers related with particulate matter were considered (such as "particles", "dust", "aerosols", "black carbon", "$PM_{2.5}$", and "$PM_{10}$", among others).

Figure 7 provides the air pollutants identified by both populations. Half of the general population (50%) indicated that they could not identify any air pollutant, while only 39% of the UFSAAPP population provided the same answer. The higher knowledge about air pollutants in the UFSAAPP population is probably due to their higher sensibility and awareness regarding the topic of air pollution, as described previously, since that population has experienced several air pollution events in the past, which promoted their need to acquire knowledge about air pollutants.

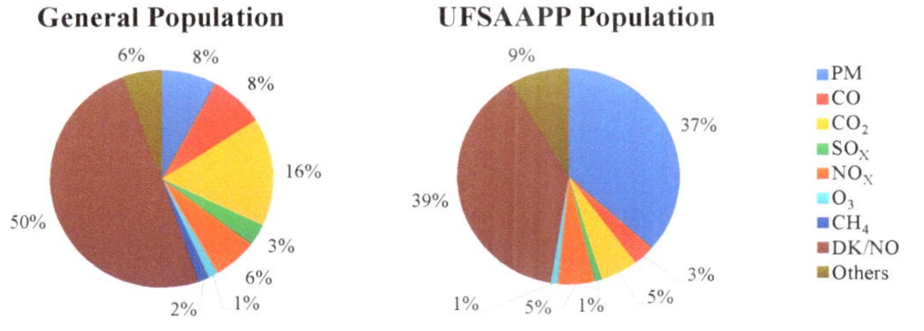


**Figure 7.** Air pollutants identified by the two studied populations. DK/NO stands for "Do not know/No answer".

For the general population, the air pollutant most frequently identified was $CO_2$ (16%), followed by PM and CO (both with 8%). Carbon dioxide is typically known by the citizens since it is one of the most discussed air pollutants in the media (mainly television), especially due to its association with climate change and all the global efforts that are being conducted to carbon neutrality [25]. The air pollutant more identified by the UFSAAPP population was particulate matter (37%), followed by "Others" (9%) and, in third, $CO_2$ and $NO_X$ (both with 5%). In the UFSAAPP population, settled dust events are one of the main problems affecting the population (which they assume to be related to the local industries) and, therefore, it seems natural that the air pollutant that most respondents are familiar to is PM. Moreover, it is relevant to highlight that the concerns of the UFSAAPP population represent probably one of the main triggers for their environmental awareness regarding air pollutants.

### 3.6. Information about Air Quality

Information about air quality and, consequently, about air quality indexes, has to be reliable and understandable to all the population, since it is through trustworthy information that it is possible to increase the awareness of the general public regarding this issue. In order to understand if the citizens feel that they are informed about the local air quality, all participants were inquired about "Do you feel informed about the air quality in your area of residence?" and asked to answer, ranging from "Not at all" to "Very much". Figure 8 provides the results for both populations.

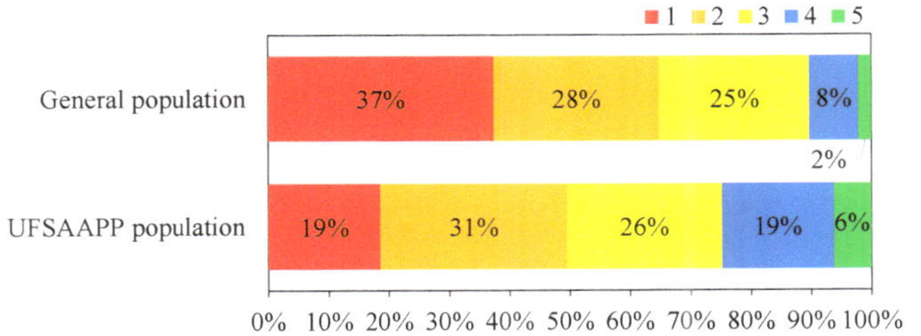


**Figure 8.** Level of how much citizens feel informed about air quality in their area of residence, ranging from 1 ("not at all") to 5 ("very much").

The majority of the general population (65%) feels that they do not have enough information about air quality in their area of residence (considering the grades "1" and "2"), with only 10% answering that they felt well informed about it (considering the grades "4" and "5"). Half of the UFSAAPP population (50%) also considers not to have sufficient information about air quality (considering the grades "1" and "2"), but an increase of the informed population is observed with 25% stating that they felt well informed about air quality in their local area (considering the grades "4" and "5"). The feeling of insufficient information about air quality (considering grades "1" and "2") in the present study is slightly lower than the expressed by Portuguese citizens in a European survey carried out on September 2019 [26], where a mean of 54% of the respondents from 28 member states of European Union considered that they did not feel well-informed about air quality problems in their country, ranging from 75% in Portugal to 18% in Finland. This survey concluded that only 25% of the Portuguese population considered that was well-informed about air quality problems in their country, a similar result to the one obtained for the UFSAAPP population.

Figure 9 provides the main sources of information of both studied populations. In the general population, 13.7% of the participants stated that they have difficulties accessing information about the local air quality, while only 4.0% of the UFSAAPP population reported the same problem. For both populations, the main source of information is the internet with around 26%. For the general population, the second main source of information is TV (18.6%), followed by "Newspapers and magazines" (7.6%). For the UFSAAPP population, the second main source of information is "Environmental groups" (14.9%), followed by "Town hall" (13.2%).

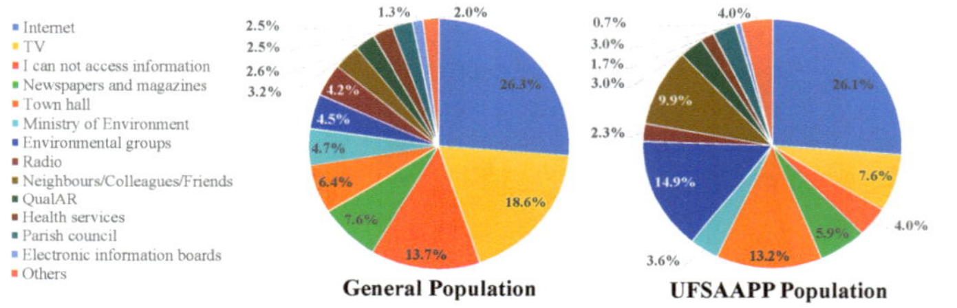


**Figure 9.** Sources of information about air quality for general and UFSAAPP populations.

A study conducted in China concluded that the main sources of information regarding air pollution were TV/radio (20.8%), Internet (18.9%) and Newspapers/magazines (18.7%) [27], a similar trend to the one found in the present study for the general population. In a study developed in the United Kingdom [25], the main sources of information about air quality that were identified by the participants were internet (44.7%), local council (29.3%), and the media services, namely, radio/TV/newspapers (13.2%). The main difference registered in our study is the contribution of "Environmental groups" as information source for the UFSAAPP population. However, this may be due to the very specific characteristics of this population since they have experienced several air quality problems in the past and, therefore, they seek information on the environmental groups that target their issue (both locally and nationally), along with the local Town hall. Some of the environmental groups were created by citizens specifically due to the air pollution events that occurred in the UFSAAPP area, and others are environmental non-governmental organizations of national range [28]. Once again, this highlights that the concern regarding local air quality problems potentiates the citizens to seek information and, typically, using a more local approach.

However, these results also show that the Portuguese governmental online database of air quality (created by the Portuguese Environment Agency), which is of free access and regularly updated, called "QualAR" (https://qualar.apambiente.pt/ accessed on 17 July 2022) is not a common source of information for the citizens. Only 2.6% and 3.0% of the participants from the general and the UFSAAPP populations, respectively, identified QualAR as a source of information. This fact highlights the need for the governmental stakeholders to promote the awareness of this tool, to empower the general public with knowledge about their local air quality.

### 3.7. Impacts of Air Quality in the Daily Life

Both populations were asked if they felt that, at some moment, they were already affected by air quality. In the general population, 52.9% revealed that they had already felt affected by air quality, while in the UFSAAPP population this percentage reached 86.8%, which highlights the strong feeling of the UFSAAPP population that their residence area is affected by local air quality problems. Figure 10 presents the incidence of the main impacts of air quality problems perceived by the respondents.

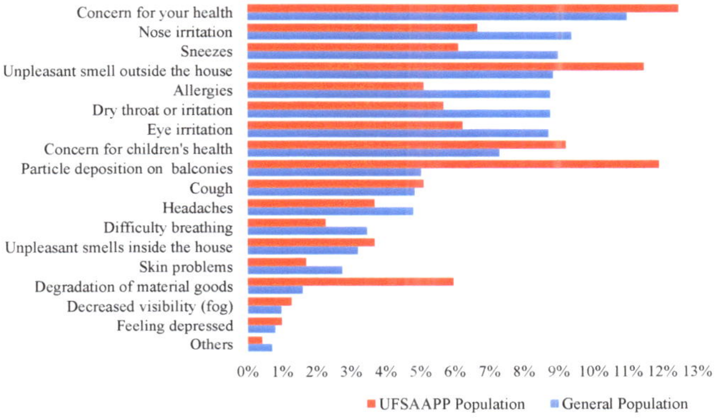


**Figure 10.** Perception of the impacts of air quality problems on the general and UFSAAPP populations.

Both populations considered that air quality problems had an impact on their own health, with 11.0% of the general population and 12.5% of the UFSAAPP population reporting this. For the general population, the main issues of impact of air quality problems were health-related issues, such as nose irritation (9.4%) and sneezes (9.0%). The UFSAAPP population focused on the remaining main impacts of air quality problems, such as on events of particle deposition on balconies (11.9%) and unpleasant smell outside the house (11.5%), which are directly related to the previous complaints by this population, as already reported above. Additionally, in this population, the concern regarding the deterioration of material goods due to air pollution was also very high (6.0%), when comparing with the general population (1.6%).

When asked if the participants have made changes in their daily life when they felt that were being affected by air pollution, only 31% of the respondents of the general population answered affirmatively, while 69% of the UFSAAPP population reported the same behavior. Figure 11 describes the main changes made by citizens in their daily life when they felt being affected by air pollution.

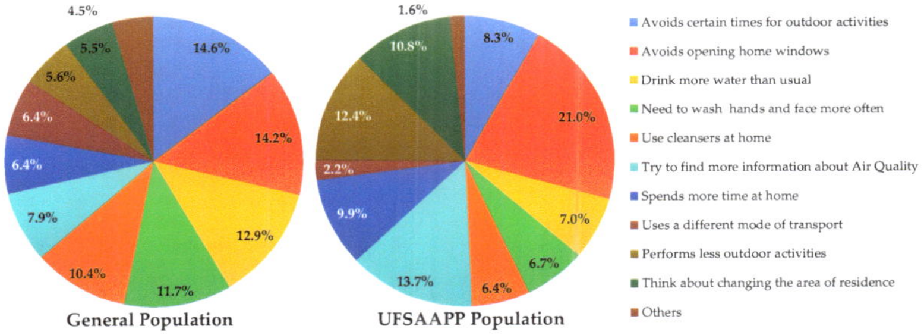


**Figure 11.** Changes made in the daily life when citizens felt affected by air pollution, for the general population (**left**) and the UFSAAPP population (**right**).

For the general population, the main changes were to avoid certain time periods for outdoor activities (14.6%), to avoid opening home windows (14.2%), and to drink more water than usual (12.9%). For the UFSAAPP population, the main changes were to avoid opening the home windows (21.0%), to try to find more information about air quality (13.7%) and to perform less outdoor activities (12.4%). It is relevant to highlight that 10.8% of the respondents from the UFSAAPP population revealed considering changing their area of residence due to air pollution issues, which is the almost the double verified in the general population (5.5%). This issue reveals that air quality may be a relevant issue to consider in the real estate market.

### *3.8. To Which Mitigation Measures to Improve Air Quality Are the Citizens More Favorable to?*

The study conducted by the EU identified that 67% of the Portuguese population considered that the public authorities were not doing enough to promote good air quality to their citizens [26]. To understand which mitigation measures the citizens are more receptive to adopt or to support, all participants were asked the degree of priority (ranging from "1" as minimum priority to "5" as maximum priority) that they attributed to different measures to improve air quality. Figures 12 and 13 present the results obtained for the general and UFSAAPP populations, respectively.

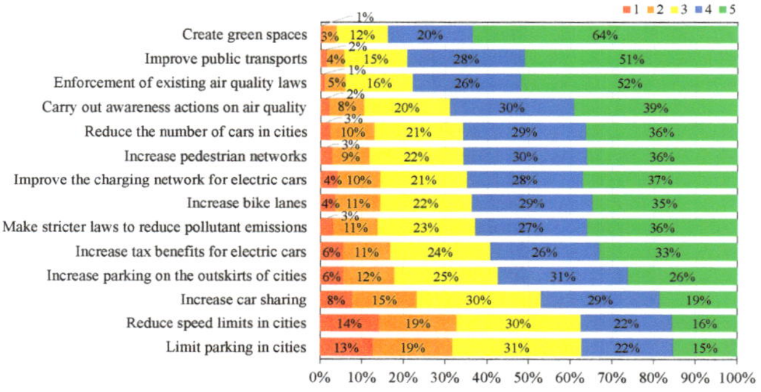


**Figure 12.** Degree of priority given by the citizens of the general population to different measures to improve air quality (ranging from 1 for "minimum priority" to 5 for "maximum priority").

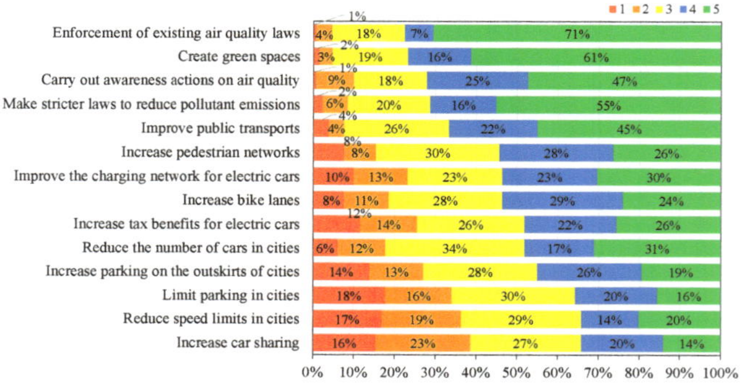


**Figure 13.** Degree of priority given by the citizens of the UFSAAPP population to different measures to improve air quality (ranging from 1 for "minimum priority" to 5 for "maximum priority").

The general population considered that the measures with higher priority (classifications of "4" and "5") were to create green spaces (84%), to improve public transports (79%), and to enforce the existing air quality laws (78%). The UFSAAPP population considered that the measures with higher priority were the enforcement of existing air quality laws

(78%), the creation of green spaces (77%) and to carry out awareness actions on air quality (72%). The difference on the priority rank of the mitigation measures considered by the UFSAAPP may be influenced again by their own experience of previous episodes of air pollution events and also by living in an industrial-urban area, which makes them more favorable to the enforcement of existing air quality legislation, in terms of compliance of the industry emissions with their environmental licenses and legislated limit values. Comparing with the European study, the EU-28 citizens (including the Portuguese participants) considered that the most effective way to tackle the air quality problems would be to apply stricter pollution controls on industrial and energy-production activities (44% of the EU-28 participants and 43% of the Portuguese participants) [26].

### 3.9. Considerations

Although the survey conducted in this study gathered answers from all the Portuguese districts, it is important to highlight that the adequate percentage of representativeness among regions was not achieved for the general population of Portugal. Moreover, the stratification of the characteristics between the studied populations are not equal, which may influence the analysis. The gender variability of participants is also not representative of the Portuguese population (52% Female and 48% Male) [10], with the present study gathering around 62–63% of answers from females. The influence of gender on air quality perception is known with females generally perceiving more air pollution [29]. However, the present results still provide new and valuable insights regarding the perception of the citizens on air quality issues, which may be determinant on the approaches chosen by public authorities to maximize their dissemination strategies targeting the citizens. The comparison with a specific population that is more aware of air pollution issues (which is reflected by their higher level of knowledge regarding the air pollution issues) highlights that real life concerns potentiate the search for information and empowerment of the citizens regarding this topic (which is the case of the specific population since they have been subject to occasional air pollution events, which motivated their search for knowledge to understand the potential implications of those events). Some main ideas are important to retain and to highlight, such as:

- Great differences were found when comparing both studied populations. It was found that UFSAAPP population showed a higher concern regarding air pollution in comparison with the general population (with 61.4% of the UFSAAPP population considering it the main environmental concern). This higher concern was also demonstrated by their significant knowledge of possible pollutants and higher need to search for information about the topic. Furthermore, the UFSAAPP population considered industry as the main source of air pollution (with 37.4% of the answers) along with traffic (37.4%). This trend was not found in the general population, where traffic was appointed as the main pollution source (51.7%), followed by the industry (only with 6.2%). This fact highlights the concern and awareness that the UFSAAPP population has regarding industry as a pollution source.
- A great part of the Portuguese population feels that are not suitably informed regarding the air quality levels in their area (65%), with only 10% stating that they feel well informed about it. It would be important that reliable and easily understandable information about air quality could be of easy access and widespread throughout the country; this would empower the citizens regarding air quality and promote their future engagement in mitigation actions to improve air quality. This is a crucial area that policy makers and governmental bodies should focus on in order to decrease the national environmental illiteracy regarding this topic and promote behavioral actions in the population that can lead to an improvement in the air quality.
- Unfortunately, this study revealed that the governmental online and free database of information about local air quality (QualAR) is almost unknown by the Portuguese population (being acknowledged to be a source of information by only 3% of the participants). This highlights that the current dissemination strategies are not working

or are not enough to reach the general public, which should be targeted to maximize all the potentialities of this available tool.

- Due to the citizens' awareness about air pollution and its health and daily life impacts, air quality is a relevant issue in the real estate market (10.8% of citizens consider to change their area of residence when under air pollution events). This fact potentiates the engagement of local governmental authorities to implement measures to improve local air quality, to improve the quality of life of their citizens, to attract new inhabitants to the area and to improve the touristic potential of their municipalities.

A follow-up study of this survey should be conducted in the future targeting to achieve a higher participation (to obtain a higher sample number and representativeness of the Portuguese population), in order to evaluate the changes on the perception of citizens, which will provide insights regarding whether the current strategies of awareness regarding air pollution are effective or not. Moreover, it would also be important to allow the participants to provide their opinions on some specific issues, such as their best suggestions/strategies to improve air quality.

It is important also to highlight that strategies to promote a higher public acceptance of mitigation policies rely on actions with the population to promote their empowerment and engagement, such as health literacy programs, awareness raising campaigns, and public participation activities [30].

## 4. Conclusions

This study allowed to assess the perception of the air quality by the population of an area affected by air pollution events and to compare it with the general population. This sub-population showed to have a higher level of knowledge and awareness regarding the topic of air pollution, considering it as the main environmental concern (while the general population ranked it only in third). Moreover, the sub-population also showed more knowledge about air pollutants than the general population, with 61% being able to identify at least one air pollutant, while half of the general population did not manage to identify any air pollutant. The perception of the local air quality (at the neighborhood level) was also very different between populations, with 61.4% of the sub-population considering it poor or very poor, while only 9.5% of the general population had the same perception, which highlights the different levels of concern between populations.

Some issues related to air quality still demonstrate a weak knowledge among the studied populations, namely the identification of air pollutants (50% of the general population could not identify any air pollutant) and an erroneous perception of the contribution of the different pollution sources to air quality levels (traffic was identified as the main source by both populations, with an overestimation of its impact between four and five times higher than the real emissions impact on air quality levels).

Despite both populations having some knowledge about air quality, still more than 50% of both populations feel that they do not have enough information regarding the air quality in their area of residence. The current tools of air quality information made available by governmental bodies are unknown to most of the population. Therefore, it is important and crucial to invest in the empowerment of the populations regarding environmental knowledge (especially air quality), since it is a strategy to increase and obtain their receptivity to the implementation of future actions to improve air quality.

The main actions considered by both populations to improve air quality to which the citizens are more receptive to include enforcement of existing air quality laws, creation of green spaces, implementation of awareness actions on air quality, and improvement of public transportations.

Overall, our study highlights several weaknesses regarding the citizens' knowledge about air quality and the need for the governmental stakeholders to promote the awareness of the importance of air quality, including which tools are available regarding information and monitoring.

**Supplementary Materials:** The following supporting information can be downloaded at: https://www.mdpi.com/article/10.3390/ijerph191912760/s1, File S1: Questionnaire on the perception of air quality in Portugal.

**Author Contributions:** N.C., A.R.J., C.A.G. and S.M.A. were responsible for the design of the online survey, and N.C. and A.R.J. for its implementation; A.R.J. performed the data and results' analysis, including statistical analysis and creation of raw figures; N.C. wrote the original draft; N.C., C.A.G. and S.M.A. contributed to the review and editing of the original draft. All authors have read and agreed to the published version of the manuscript.

**Funding:** N. Canha acknowledges the funding by national funds through FCT—Fundação para a Ciência e Tecnologia, I.P. (Portugal) for his IST-ID contract (IST-ID/098/2018) and the contract 2021.00088.CEECIND. The FCT support is also acknowledged by C2TN/IST authors (UIDB/04349/2020 + UIDP/04349/2020).

**Informed Consent Statement:** Informed consent was obtained from all subjects involved in the study. Participation in the study was completely voluntary and anonymous. Implicit consent for the survey was assumed when study participants completed the survey. Participants received no compensation.

**Conflicts of Interest:** The authors declare no conflict of interest.

## References


1. European Environment Agency. *Air Quality in Europe: 2019 Report*; EEA Report No 10/2019; Publications Office of the European Union: Luxembourg, 2019.
2. Manisalidis, I.; Stavropoulou, E.; Stavropoulos, A.; Bezirtzoglou, E. Environmental and Health Impacts of Air Pollution: A Review. *Front. Public Health* **2020**, *8*, 14. [CrossRef] [PubMed]
3. Oltra, C.; Sala, R. Perception of Risk from Air Pollution and Reported Behaviors: A Cross-Sectional Survey Study in Four Cities. *J. Risk Res.* **2018**, *21*, 869–884. [CrossRef]
4. Pantavou, K.; Lykoudis, S.; Psiloglou, B. Air Quality Perception of Pedestrians in an Urban Outdoor Mediterranean Environment: A Field Survey Approach. *Sci. Total Environ.* **2017**, *574*, 663–670. [CrossRef]
5. Brody, S.D.; Peck, B.M.; Highfield, W.E. Examining Localized Patterns of Air Quality Perception in Texas: A Spatial and Statistical Analysis. *Risk Anal.* **2004**, *24*, 1561–1574. [CrossRef]
6. Maione, M.; Mocca, E.; Eisfeld, K.; Kazepov, Y.; Fuzzi, S. Public Perception of Air Pollution Sources across Europe. *Ambio* **2021**, *50*, 1150–1158. [CrossRef] [PubMed]
7. Justino, A.R.; Canha, N.; Gamelas, C.; Coutinho, J.T.; Kertesz, Z.; Almeida, S.M. Contribution of Micro-PIXE to the Characterization of Settled Dust Events in an Urban Area Affected by Industrial Activities. *J. Radioanal. Nucl. Chem.* **2019**, *322*, 1953–1964. [CrossRef]
8. Abecasis, L.; Gamelas, C.A.; Justino, A.R.; Dionísio, I.; Canha, N.; Kertesz, Z.; Almeida, S.M. Spatial Distribution of Air Pollution, Hotspots and Sources in an Urban-Industrial Area in the Lisbon Metropolitan Area, Portugal—A Biomonitoring Approach. *Int. J. Environ. Res. Public Health* **2022**, *19*, 1364. [CrossRef]
9. Chaíça, I. Poeira Negra Que Cobre Paio Pires Não é Nociva Para a Saúde. Origem Permanece Desconhecida. *Público*, 11 May 2019. Available online: https://www.publico.pt/2019/05/11/local/noticia/poeira-negra-cobre-paio-pires-nao-inalavel-sao-precisosestudos-saber-onde-vem-1872369 (accessed on 1 July 2022).
10. PORDATA. Base de Dados Portugal Contemporâneo. Available online: http://www.pordata.pt/Municipios/Ambiente+de+Consulta/Tabela (accessed on 22 September 2021).
11. Agência Portuguesa do Ambiente. *Licença Ambiental*; Lusosider Aços Planos, S.A.: Lisbon, Portugal, 2008.
12. Agência Portuguesa do Ambiente. *Licença Ambiental LA N° 658_1.1_2017*; SN Seixal_Siderurgia Nacional, SA: Lisbon, Portugal, 2017.
13. Agência Portuguesa do Ambiente. *Licença Ambiental Microlime*; Produtos de Cal e Derivados, S.A.: Lisbon, Portugal, 2011.
14. Schmidt, L.; Guerra, J. *RISKAR LX—Qualidade Do Ar: Poluição Atmosférica, Perceções e Vulnerabilidades Em Lisboa*; OBSERVA: Lisbon, Portugal, 2013.
15. European Commission. Science for environment policy. In *Links between Noise and Air Pollution and Socioeconomic Status*; European Commission: Brussels, Belgium, 2016; Volume 29, ISBN 9789279457340.
16. Shepherd, D.; Dirks, K.; Welch, D.; McBride, D.; Landon, J. The Covariance between Air Pollution Annoyance and Noise Annoyance, and Its Relationship with Health-Related Quality of Life. *Int. J. Environ. Res. Public Health* **2016**, *13*, 792. [CrossRef] [PubMed]
17. Lercher, P.; Schmitzberger, R.; Kofler, W. Perceived Traffic Air Pollution, Associated Behavior and Health in an Alpine Area. *Sci. Total Environ.* **1995**, *169*, 71–74. [CrossRef]
18. Zhang, Z.; Zhang, G.; Su, B. The Spatial Impacts of Air Pollution and Socio-Economic Status on Public Health: Empirical Evidence from China. *Socioecon. Plann. Sci.* **2022**, *83*, 101167. [CrossRef]
19. Liao, X.; Tu, H.; Maddock, J.E.E.; Fan, S.; Lan, G.; Wu, Y.; Yuan, Z.K.K.; Lu, Y. Residents' Perception of Air Quality, Pollution Sources, and Air Pollution Control in Nanchang, China. *Atmos. Pollut. Res.* **2015**, *6*, 835–841. [CrossRef]

20. Chin, Y.S.J.; De Pretto, L.; Thuppil, V.; Ashfold, M.J. Public Awareness and Support for Environmental Protection—A Focus on Air Pollution in Peninsular Malaysia. *PLoS ONE* **2019**, *14*, e0212206. [CrossRef]
21. Zárate Valencia, A.R.; Reyes Umaña, M.; Arellano Wences, H.J.; Rodríguez Rosales, A.A.; Rodríguez Alviso, C.; González González, J. The Air Quality Perception of Residents in the Metropolitan Zone of Acapulco Who Live Around Intersections with Intense Traffic. *Environments* **2020**, *7*, 21. [CrossRef]
22. Howel, D.; Moffatt, S.; Bush, J.; Dunn, C.E.; Prince, H. Public Views on the Links between Air Pollution and Health in Northeast England. *Environ. Res.* **2003**, *91*, 163–171. [CrossRef]
23. Saksena, S. Public Perceptions of Urban Air Pollution Risks. *Risk Hazards Cris. Public Policy* **2011**, *2*, 19–37. [CrossRef]
24. European Environment Agency. *Air Quality in Europe 2021—Report No. 15/2021*; European Environment Agency: Brussels, Belgium, 2021.
25. Smallbone, K. *Individuals Interpretation of Air Quality Information: Customer Insight and Awareness Study*; University of Brighton: Brighton, UK, 2012.
26. European Commission. *Special Eurobarometer 497—Attitudes of Europeans towards Air Quality*; European Commission: Brussels, Belgium, 2019; p. 100.
27. Liu, X.; Zhu, H.; Hu, Y.; Feng, S.; Chu, Y.; Wu, Y.; Wang, C.; Zhang, Y.; Yuan, Z.; Lu, Y. Public's Health Risk Awareness on Urban Air Pollution in Chinese Megacities: The Cases of Shanghai, Wuhan and Nanchang. *Int. J. Environ. Res. Public Health* **2016**, *13*, 845. [CrossRef]
28. Gromicho, I. Grupo de Moradores Recolhe Provas de Poluição em Paio Pires No Seixal. *Ambiente Magazine*, 13 February 2019. Available online: https://www.ambientemagazine.com/grupo-de-moradores-recolhe-provas-de-poluicao-em-paio-pires-no-seixal (accessed on 1 July 2022).
29. Lou, B.; Barbieri, D.M.; Passavanti, M.; Hui, C.; Gupta, A.; Hoff, I.; Lessa, D.A.; Sikka, G.; Chang, K.; Fang, K.; et al. Air Pollution Perception in Ten Countries during the COVID-19 Pandemic. *Ambio* **2022**, *51*, 531–545. [CrossRef]
30. Cori, L.; Donzelli, G.; Gorini, F.; Bianchi, F.; Curzio, O. Risk Perception of Air Pollution: A Systematic Review Focused on Particulate Matter Exposure. *Int. J. Environ. Res. Public Health* **2020**, *17*, 6424. [CrossRef]

International Journal of *Environmental Research and Public Health*

MDPI

*Article*

# Are House Prices Affected by $PM_{2.5}$ Pollution? Evidence from Beijing, China

**Wenhao Xue [1], Xinyao Li [2,*], Zhe Yang [1] and Jing Wei [3]**

[1] School of Economics, Qingdao University, Qingdao 266071, China; xuewh@mail.bnu.edu.cn (W.X.); yz69env@163.com (Z.Y.)
[2] Business School, Beijing Normal University, Beijing 100875, China
[3] Department of Atmospheric and Oceanic Science, Earth System Science Interdisciplinary Center, University of Maryland, College Park, MD 20742, USA; weijing_rs@163.com
* Correspondence: xinyao_0121@163.com

**Abstract:** With the progress of high-quality development in China, residents have begun to focus on the air quality of their residential areas in an effort to reduce the health threats of air pollution. Gradually, the risk associated with air pollution has become an important factor affecting housing prices. To quantitatively analyze the impact of air pollution on house prices, panel data, including data for fine particulate matter ($PM_{2.5}$) concentrations, house prices and other auxiliary variables from 2009 to 2018, were collected from 16 districts in Beijing, China. Based on this dataset, ordinary least squares (OLS), moderating effect and threshold effect models were constructed for empirical investigation. Within the studied decade, $PM_{2.5}$ pollution shows a significant decreasing trend of $-3.79$ μg $m^{-3}$ $yr^{-1}$ ($p < 0.01$). For house prices, the opposite trend was found. The empirical results indicate that $PM_{2.5}$ pollution has a negative effect on house prices and that every 1% increase in $PM_{2.5}$ causes an approximately 0.541% decrease in house prices. However, the inhibition of $PM_{2.5}$ on housing prices is moderated by regional educational resources, especially in areas with high education levels. In addition, per capita disposable income can also cause heterogeneities in the impact of $PM_{2.5}$ on house prices, whereby the threshold is approximately CNY 101,185. Notably, the endogeneity problems of this study are solved by the instrumental variable method, and the results are robust. This outcome suggests that the coordinated control of air pollution and balanced educational resources among regions are required for the future sustainable development of the real estate market.

**Keywords:** $PM_{2.5}$ pollution; house price; China; educational resources; ordinary least squares

check for updates

**Citation:** Xue, W.; Li, X.; Yang, Z.; Wei, J. Are House Prices Affected by $PM_{2.5}$ Pollution? Evidence from Beijing, China. *Int. J. Environ. Res. Public Health* **2022**, *19*, 8461. https://doi.org/10.3390/ijerph19148461

Academic Editor: Ling Tim Wong

Received: 20 May 2022
Accepted: 6 July 2022
Published: 11 July 2022



## 1. Introduction

At present, urbanization and economic growth are accelerating in China, especially in metropolises and city clusters. However, China's rapid modernization has been accompanied by serious air pollution, a problem closely related to human health [1–3]. Among all air pollutants, fine particulate matter (with an aerodynamic diameter of less than 2.5 μm, $PM_{2.5}$) is particularly harmful. Long-term exposure to high $PM_{2.5}$ loading significantly increases the risk of developing cancer as well as cardiovascular and respiratory diseases [4,5]. In addition, $PM_{2.5}$ pollution poses a threat to ecological security [6–8]. At the same time, in addition to providing feedback regarding the quality of the atmospheric environment, such pollution can affect the living circumstances of urban residents [9,10]. Especially for areas experiencing high economic development, the economic losses caused by air pollution are substantial [11,12]. Meanwhile, with the increase in media publicity regarding the problem and rising income levels, low air pollution risk has become a goal for urban Chinese people. An increasing number of people tend to live in areas with low $PM_{2.5}$ loading, resulting in counter urbanization, which has become a new challenge for environmental management and urban planning at the regional and national scales [13].

For urban residents, housing conditions are the foundation of high-quality development. Since the financial crisis of 2008, a long-term housing boom has appeared in China. According to the statistical data from the National Bureau of Statistics, the total sales of commercial housing in China were only approximately CNY 4399.5 billion in 2009. In contrast, they surpassed CNY 14,997.3 billion in 2018, accounting for ~16.3% of the gross domestic product (GDP). During this period, house prices increased substantially in China's megacities, especially in Beijing, and real-averaged house prices increased by ~41.8% from 2012 to 2015 [14]. There are many factors that cause fluctuations in house prices in China [15–19]. Liu et al. used a demand-supply framework and annual data from 31 provinces across China from 2000 to 2018 to argue that five variables (i.e., land prices, real estate developer loans, per capita savings and the proportion of individuals with college or higher educational degrees) accounted for ~72% of the increase in house prices [20]. In addition, other phenomena, such as education level, entertainment facilities, wedding time, monetary factors and policy orientation, have been found to be potential influencing factors in China [21–25].

Since 2015, concomitant with the proposal by the United Nations of sustainable development goals (SDGs), the Chinese government has attached substantial importance to environmental protection [26]. Meanwhile, the awareness of environmental protection has gradually improved at the government, enterprise and individual scales. At present, environmental risk factors have gradually been incorporated into the considerations of house buyers. Subsequently, several researchers have examined the negative externality of the environmental quality of residential areas with respect to house prices [27–30]. Among all potential influence factors, air pollution, especially for $PM_{2.5}$ pollution, plays an important role in house prices. Dai et al. found that higher $PM_{2.5}$ risks were accompanied by lower house prices in Nanjing, China [9]. Similarly, the negative effects of $PM_{2.5}$ concentration on housing prices were also found across cities in China, and the heterogeneity is also captured [31]. Sun and Yang, using the quantile regression and geographically weighted quantile regression, found the presence of asymmetric and spatial non-stationary effects among $PM_{2.5}$ and housing prices in China [32]. Furthermore, the neighboring pollution also can lead to changes in local home prices [33]. Despite this wealth of literature, the mechanism of the effect of air pollution on the real estate market is less analyzed. More importantly, with the proposal of the 'three-child policy' in 2021 [34], the relationship between housing prices and the environment has become a matter of concern for every family, one that bears importance on the healthy growth of children in China.

The goal of this study was to investigate the impact of $PM_{2.5}$ pollution on house prices. For this purpose, we collected panel data in all of Beijing's administrative divisions from 2009 to 2018, including data on house prices, $PM_{2.5}$ concentrations and auxiliary variables. Based on these datasets, the spatiotemporal variations in $PM_{2.5}$ pollution and house prices were analyzed in detail. Then, an ordinary least squares (OLS) model was used as an econometric model to quantify the impact of $PM_{2.5}$ concentrations on house prices. In addition, considering actual local/regional circumstances, the moderating effects of educational resources are explained. Finally, the per capita disposable incomes of local residents were used as a threshold effect to accurately identify the heterogeneities of the impact of $PM_{2.5}$ concentrations on house prices. This study will provide a strong scientific basis and literature reference for regional urban planning, environmental protection and the benign development of the real estate market.

## 2. Study Area

China's capital, Beijing, is the country's center of political, cultural, scientific and technological innovation. Figure 1 shows the geographical location of Beijing in northern China at longitudes and latitudes ranging from 115.7°–117.4° E and 39.4°–41.6° N, respectively. There are currently 16 administrative divisions in Beijing: Changping (CP), Chaoyang (CY), Daxing (DX), Dongcheng (DC), Fangshan (FS), Fengtai (FT), Haidian (HD), Huairou (HR), Mentougou (MTG), Miyun (MY), Pinggu (PG), Shijingshan (SJS), Shunyi (SY), Tongzhou

(TZ), Xicheng (XC) and Yanqing (YQ). The CY, DC, FT, HD, SJS and XC districts are usually identified as the central urban areas. Figure 1 also shows the distribution of population density in Beijing. Generally, high population densities were captured in the central urban areas. The highest population density was captured in the XC district, with a value of 26,603 persons km$^{-2}$, followed by the DC, HD, CY, SJS and FT districts, with mean population densities ranging from 7385 to 19,946 persons km$^{-2}$. In contrast, the population density in the YQ district is the lowest (~173 persons km$^{-2}$).

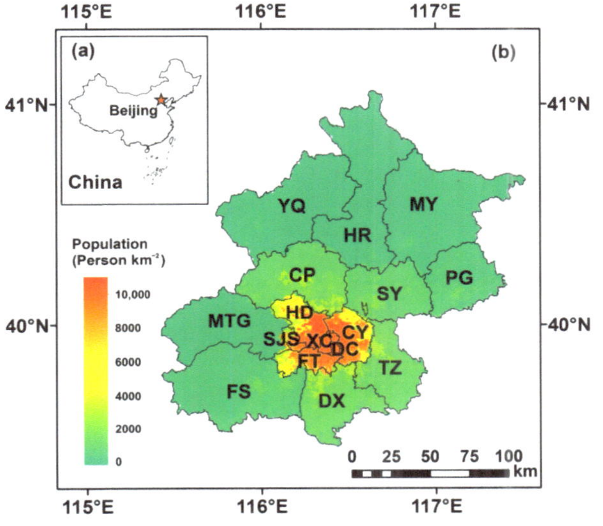


**Figure 1.** Beijing's 16 administrative divisions. (**a**) is the geographical location of Beijing in China; (**b**) is the distribution of 16 districts in Beijing. The background colors indicate the distribution of population density (persons km$^{-2}$); these data are available from the resource and environmental science and data center (https://www.resdc.cn/, accessed on 1 January 2022).

## 3. Dataset and Methodology

### *3.1. Dataset*

#### 3.1.1. House Prices

The annual average house prices (unit: yuan) were collected in the 16 districts of Beijing from 2009 to 2018. These data were sourced from Anjuke, Inc., a real estate information service enterprise in China. The data are available at https://www.anjuke.com/fangjia/ (accessed on 1 January 2022). However, because of inflation effects, authentic house prices will be overvalued. Therefore, to calculate the authentic house prices of each district in Beijing, we used the provided house prices divided by a GDP deflator (i.e., the ratio of nominal GDP to real GDP).

#### 3.1.2. $PM_{2.5}$ Concentrations

The ground-level $PM_{2.5}$ concentrations were collected from the ChinaHigh$PM_{2.5}$ datasets (http://doi.org/10.5281/zenodo.3987359, accessed on 1 January 2022), with a high horizontal resolution of 1 km. In this dataset, the daily $PM_{2.5}$ concentrations were generated using a linear mixed effect (LME) model combined with the Moderate Resolution Imaging Spectroradiometer (MODIS) Multi-Angle Implementation of Atmospheric Correction (MAIAC) aerosol optical depth products and meteorological factors, including boundary layer height (BLH), evaporation (ET), total precipitation (PRE), relative humidity (RH), surface pressure (SP), 2-m temperature (TEM), wind direction (WD) and wind speed (WS), which are potentially relevant variables in PM2.5 pollution. The build processes were explained in our previous study in detail [35]. Here, we extracted the $PM_{2.5}$ concentrations of the corresponding pixels in each district of Beijing from 2009 to 2018. To avoid errors, we omitted pixels with values missing for more than ten days. Then, the monthly and annual concentrations were collected to analyze $PM_{2.5}$ temporal trends and the impacts on house prices, respectively.

#### 3.1.3. Control Variables

According to previous research, greening facilities and socioeconomic factors can affect house prices. The descriptive statistics of all the control variables used in our study are listed in Table 1. In addition to annual $PM_{2.5}$ concentrations, we selected seven other indices as control variables to reflect the greening facilities and socioeconomic status of all districts in Beijing during 2009–2018. These variables include GDP, the gross output value of residential services and other services (Service), per capita disposable income (Income), the gross output value of the construction industry (Industry), the normalized difference vegetation index (NDVI), the registered population (Population) and the number of private cars (Traffic). Similar to house prices, the sample capacities of all the control variables are 160 samples. In addition, the mean values and the standard deviation (Std) of all the control variables are provided. All other control variables were drawn from the regional statistical yearbook of Beijing with the temporal and spatial resolution of the district and annual level. In particular, in order to eliminate the influence of the heteroscedasticity of the model, we logarithmize all the continuous variables. Meanwhile, we made a collinearity diagnosis. The Variance Inflation Factor (VIF) values are all less than 10, which indicates that there is no collinearity problem [36].

**Table 1.** The descriptive statistic of all control variables ($n$ = 160).

| Abbreviation | Control Variable | Unit | Mean | Std |
|---|---|---|---|---|
| $PM_{2.5}$ | Annual averaged concentration of $PM_{2.5}$ | $\mu g\ m^{-3}$ | 67.33 | 15.95 |
| GDP | Gross domestic product | Billion yuan | 98.07 | 118.00 |
| Service | Gross output value of residential services and other services | Million yuan | 71,202.03 | 78,697.96 |
| Income | Per capita disposable income | Yuan | 32,551.19 | 9659.59 |
| Industry | Gross output value of the construction industry | Million yuan | 379.56 | 368.15 |
| NDVI | Normalized Difference Vegetation Index | - | 0.39 | 0.09 |
| Population | Registered population | Thousand person | 920.60 | 572.60 |
| Traffic | Number of private cars | Set | 254,223.80 | 226,015.90 |

#### 3.1.4. Other Variables

Due to the uneven distribution of urban educational resources, a derivative, i.e., school district housing, occurs in the real estate market across Beijing, which could also impact house prices on a local scale. Therefore, to investigate the moderating effect of educational resources on the relationship between $PM_{2.5}$ and house prices, the gross regional product of education from 2009 to 2018 in all districts was selected as the index to reflect the education level. These data were collected from the regional statistical yearbook of Beijing. Furthermore, to avoid the endogeneity problems caused by missing variables, the temperature was selected as the instrumental variable. Temperature can affect the generation and diffusion of $PM_{2.5}$. In addition, it has no direct relationship with house prices. Here, the annual averaged temperature during 2009–2018 was adopted from the fifth-generation European Center for Medium-Range Weather Forecasts (ECMWF) atmospheric reanalysis dataset of the global climate (ERA5), with a horizontal resolution of $0.25^{\circ} \times 0.25^{\circ}$ [37].

### 3.2. *Methodology*

#### 3.2.1. Benchmark Regression Model

To authenticate the test of the impact of $PM_{2.5}$ concentrations on house prices, we selected the OLS model as the benchmark regression model. This model is a mature model in econometrics, environmental economics and other economic-related research, and it was proven to be one of the econometric models that can identify the causal relationship of variables [38–46]. Meanwhile, considering the possible endogeneity problem of the model and in order to compare the results of similar models, we selected temperature as an instrumental variable and used the two-stage least squares (2SLS) method to test and compare the results, which can also prove the robustness of our model [47]. The OLS model was set as follows.

$$\begin{aligned} HP_{it} = \alpha_0 + \alpha_1 \text{PM}_{2.5_{it}} + \alpha_2 \text{GDP}_{it} + \alpha_3 \text{Service}_{it} + \alpha_4 \text{Income}_{it} + \alpha_5 \text{Industry}_{it} \\ + \alpha_6 \text{NDVI}_{it} + \alpha_7 \text{Population}_{it} + \alpha_8 \text{Traffic}_{it} + \varepsilon 1_{it} \end{aligned} \tag{1}$$

where $i$ indicates the district, and $t$ indicates the year. $HP$ represents house prices. $PM_{2.5}$ indicates the regional average $PM_{2.5}$ concentration. The coefficient $\alpha_1$ captures the effect of $PM_{2.5}$ on house prices. $\alpha_0$ is the constant term, and $\varepsilon 1_{it}$ is the error term. $\alpha_2, \alpha_3, \ldots, \alpha_8$ represent the effects of other control variables on house prices, including GDP, Service, Income, Industry, NDVI, Population and Traffic.

#### 3.2.2. Moderating Effect Models

We further explain the regulating role of education in the relationship between $PM_{2.5}$ concentration and house prices. A moderating variable, i.e., education, was selected and added to the basic econometric model. In addition, the interaction term of education and $PM_{2.5}$ is calculated as a new explanatory variable in Equation (2).

$$\begin{aligned} HP_{it} = \beta_0 + \beta_1 \text{PM}_{2.5_{it}} + \beta_2 \text{GDP}_{it} + \beta_3 \text{Service}_{it} + \beta_4 \text{Income}_{it} + \beta_5 \text{Industry}_{it} \\ + \beta_6 \text{NDVI}_{it} + \beta_7 \text{Population}_{it} + \beta_8 \text{Traffic}_{it} + \beta_9 \text{Education}_{it} \\ + \beta_{10} \text{Education}_{it} \times \text{PM}_{2.5_{it}} + \varepsilon 2_{it} \end{aligned} \tag{2}$$

where $\text{Education}_{it}$ refers to the education level in district $i$ in year $t$. $\beta_0$ is the constant term. $\beta_1, \beta_2, \ldots, \beta_9$ represent the effects of the control variables on house prices. In addition to the control variables in Equation (1), education was also considered here. $\text{Education}_{it} \times \text{PM}_{2.5_{it}}$ is the interaction term of education and $PM_{2.5}$. $\beta_{10}$ indicates the coefficient of the regulatory effect of education. $\varepsilon 2_{it}$ is the error term.

$$\frac{\partial (HP)_{it}}{\partial (\text{PM}_{2.5})_{it}} = \beta_1 + \beta_{10} \text{Education}_{it} \tag{3}$$

Then, by deriving $PM_{2.5}$ as Equation (3), the boundary effects of education on the relationship between $PM_{2.5}$ and house prices are quantified, whereby $\beta_1$ represents the direct effect of $PM_{2.5}$ on house prices, and $\beta_{10}$ indicates the interactions.

#### 3.2.3. Threshold Effect

However, the relationship between $PM_{2.5}$ and house prices may be nonlinear, and the regional per capita disposable income level may be an important determinant. Therefore, we use a threshold effect model and adopt per capita disposable income as a threshold variable to portray this relationship. The threshold effect model was established as follows:

$$\begin{aligned} HP_{it} = {} & \gamma_0 + \gamma_1 \text{PM}_{2.5}(thr < \theta)_{it} + \gamma_2 \text{PM}_{2.5}(thr \geqslant \theta)_{it} + \gamma_3 \text{GDP}_{it} + \gamma_4 \text{Service}_{it} \\ & + \gamma_5 \text{Industry}_{it} + \gamma_6 \text{NDVI}_{it} + \gamma_7 \text{Population}_{it} + \gamma_8 \text{Traffic}_{it} \\ & + \varepsilon 3_{it} \end{aligned} \tag{4}$$

where $thr$ represents the threshold variable, and $\theta$ is the estimated threshold value. $\gamma_0$ indicates the constant term. $\gamma_1$ and $\gamma_2$ represent the influence coefficient of $PM_{2.5}$ on the dependent variable house prices under the level of per capita disposable income $< \theta$ and $\geqslant \theta$ in district $i$ in year $t$, respectively. $\gamma_3, \gamma_4 \ldots \gamma_8$ represent the effects of the control variables on house prices. The control variables include GDP, Service, Industry, NDVI, Population and Traffic. $\varepsilon 3_{it}$ is the error term. The overall framework for analyzing the impact of $PM_{2.5}$ on house prices is shown in Figure 2.

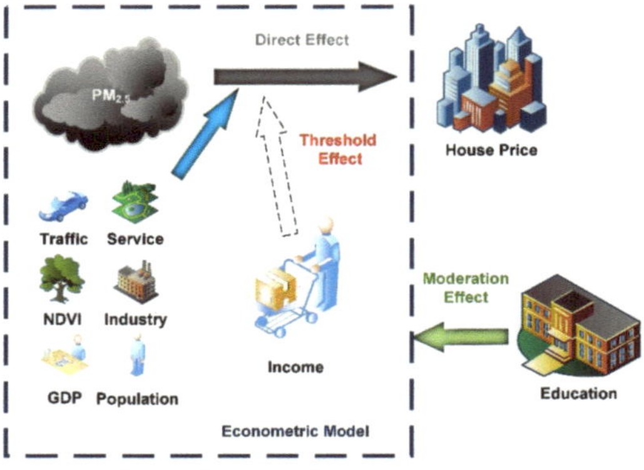


**Figure 2.** The analytical framework for the impact of $PM_{2.5}$ on house prices.

#### 3.2.4. Temporal Trend and Correlation Analysis

To examine $PM_{2.5}$ pollution characteristics in the BTH region, we used the monthly deseasonalized temporal trend analysis method, first calculating the monthly anomalous $PM_{2.5}$ concentrations at a horizontal resolution of 1 km. Then, the linear $PM_{2.5}$ trend was calculated in each pixel based on the OLS fitting method [48]. To evaluate the accuracy of the temporal trend analysis, a paired-samples T-test was used. In addition, to confirm the relationship between the control variables and house prices, the Pearson correlation coefficient (r) was calculated before the construction of the econometric model. The T-test was selected as the significance level test in this study.

## 4. Results and Discussion

### *4.1. Spatiotemporal Characteristics of $PM_{2.5}$ and House Prices*

Figure 3a shows the spatial distribution of annual average house prices across Beijing from 2009 to 2018. Generally, the mean house price in Beijing is approximately CNY 25,654.34 $m^{-2}$. However, significant spatial heterogeneities in house prices occur in the city. The high-level house price areas are mainly concentrated in the central urban districts, with a mean value of CNY 40,088.39 $m^{-2}$. Among all the districts, high values were captured in the XC, DC and HD districts, with house prices of CNY 55,899.58 $m^{-2}$, CNY 49,026.04 $m^{-2}$ and CNY 42,665.04 $m^{-2}$, respectively. In contrast, low-level house price areas are found in the YQ, MY and PG districts, with values of CNY 11,077.95 $m^{-2}$, CNY 12,603.18 $m^{-2}$ and CNY 12,681.34 $m^{-2}$, respectively. Figure 3 also shows the temporal variation characteristics of house prices in Beijing during 2009–2018. With the development of the local economic level and changes in the supply-demand relationship, significant increasing trends are found in every district. Generally, the average house prices in Beijing increased by approximately 2.58 times in one decade, and the increasing trend was CNY 3350.0 $m^{-2}$ per $year^{-1}$. Similarly, the increasing rates of each district are quite different. The largest increase was captured in the XC district, with a trend of approximately CNY 7754.7 $m^{-2}$ per $year^{-1}$ (an increase of 2.63 times). The YQ district exhibited the lowest enhancement—the increasing trend was only CNY 1143.9 $m^{-2}$ per $year^{-1}$. Nevertheless, in the YQ district, the house prices in 2018 were still 2.84 times those in 2009. Throughout the past decade, there have been two periods of accelerated increase: 2009–2010 and 2012–2013, in which the mean house price growth in Beijing was 47.0% and 35.6%, respectively. Notably, reflecting the market and policy regulation mechanism, two periods of decrease are captured, with decreases of 3.0% and 5.7% during 2011–2012 and 2017–2018, respectively.

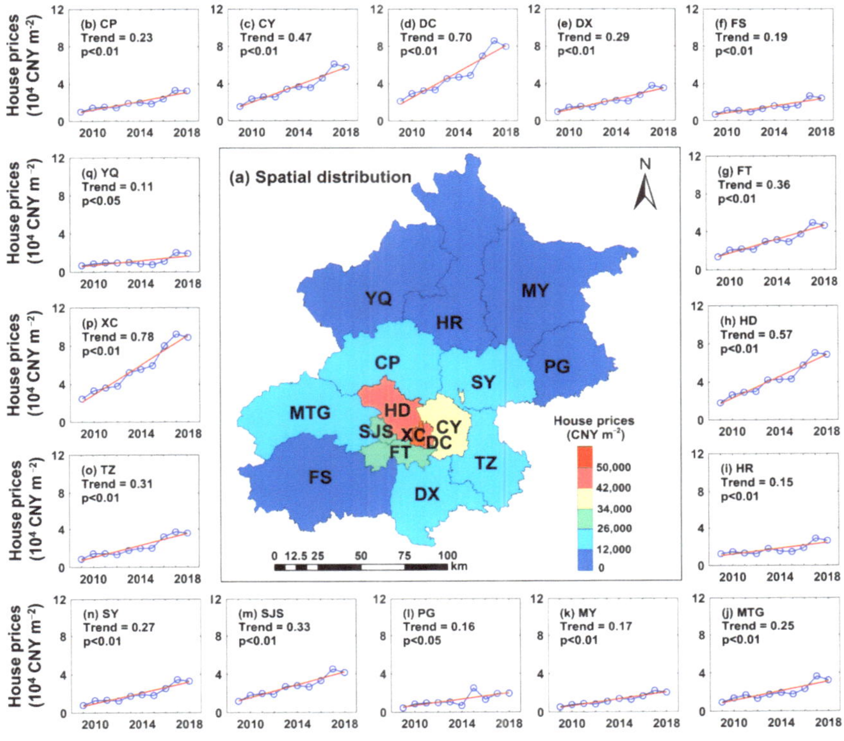


**Figure 3.** Spatiotemporal characteristics of house prices across Beijing from 2009 to 2018. (**a**) is the spatial distribution; (**b–q**) represent the temporal changes in the 16 districts.

Figure 4 shows the mean $PM_{2.5}$ spatial distributions during the study period in Beijing. Generally, the annual average $PM_{2.5}$ concentration is 67.33 ± 15.95 μg m$^{-3}$, and in nearly all of the districts, it was higher than 35 μg m$^{-3}$ (the second level of the ambient air quality standards in China). The highest average concentration can reach approximately 83.18 μg m$^{-3}$. However, the concentration changed obviously on a spatial scale and was extremely high in the southern regions, especially in the DX district (82.38 ± 14.81 μg m$^{-3}$), followed by the TZ district (79.80 ± 15.12 μg m$^{-3}$). Conversely, low $PM_{2.5}$ loading was captured in the northern regions, especially in the YQ, HR and MY districts, with annual averaged $PM_{2.5}$ concentrations of 49.95 ± 11.64 μg m$^{-3}$, 51.05 ± 12.12 μg m$^{-3}$ and 54.16 ± 12.97 μg m$^{-3}$, respectively. Significantly, the spatial pattern of $PM_{2.5}$ concentrations in Beijing is consistent with the topography and distribution of the secondary industry [35]. Figure 4 also presents the temporal trends of $PM_{2.5}$ concentrations. Overall, the $PM_{2.5}$ concentrations decreased significantly across Beijing during this period, with a trend of −3.79 μg m$^{-3}$ yr$^{-1}$ ($p < 0.01$). An accelerated decreasing trend was captured after 2014 (−5.58 μg m$^{-3}$ yr$^{-1}$, $p < 0.01$), which reflects the implementation of the Air Pollution Prevention and Control Action Plan in the Beijing-Tianjin-Hebei region [49]. Regarding the spatial distributions of temporal trends, most Beijing districts show significant decreasing $PM_{2.5}$ pollution ($p < 0.05$), especially in the southeastern region, i.e., the DX (−4.46 μg m$^{-3}$ yr$^{-1}$, $p < 0.01$), TZ (−4.64 μg m$^{-3}$ yr$^{-1}$, $p < 0.01$) and CY (−4.41 μg m$^{-3}$ yr$^{-1}$, $p < 0.01$) districts, respectively. In contrast, slight weakening trends were observed in several western areas in Beijing (e.g., the MTG and CP districts).

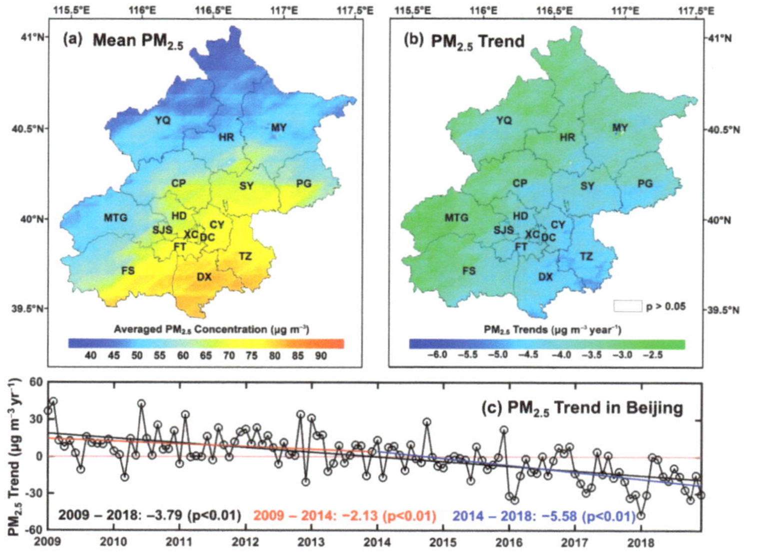


**Figure 4.** Annual averaged concentrations and trends in $PM_{2.5}$ distributions across Beijing from 2009 to 2018. (**a**) is the spatial distribution of annual averaged $PM_{2.5}$; (**b**) is the spatial distribution of the temporal trends; (**c**) is the overall trend of $PM_{2.5}$ concentration in Beijing.

### 4.2. Spatial Correlation

In addition, we also calculated the spatial relationship between $PM_{2.5}$ and HP. Firstly, we vectorized the geographical map of the Beijing region to obtain latitude and longitude information. Then, we constructed the regional inverse distance weight matrix as Equation (5). It assumes that the strength of the spatial effect depends on the distance, and the closer the spatial effect between spatial units is, the stronger the spatial effect. $W_{ij}$ indicates the inverse distance weight matrix, and $d_{ij}$ is the distance for each district.

$$W_{ij} = \begin{cases} \frac{1}{{d_{ij}}^2} \\ 0 \end{cases} \tag{5}$$

Then, the spatial autocorrelation analysis was used in this study to verify the spatial correlation between $PM_{2.5}$ and house prices in Beijing. The $PM_{2.5}$ concentration of different districts may have some spatial correlation for two main reasons. First, $PM_{2.5}$ in one region will diffuse to other districts through atmospheric transport. Second, $PM_{2.5}$ may receive shocks from other non-environmental factors such as economic and political factors and thus exhibit spatial correlation. Therefore, the global Moran's I is used to study the overall spatial correlation as follows:

$$I = \frac{\sum_{i=1}^{n} \sum_{j=1}^{n} W_{ij} \left(X_i - \overline{X}\right)\left(X_j - \overline{X}\right)}{S^2 \sum_{i=1}^{n} \sum_{j=1}^{n} W_{ij}} \tag{6}$$

where $n$ is the number of districts, and $X_i$ and $X_j$ are the $PM_{2.5}$ of district $i$ and district $j$, respectively. $\overline{X}$ is the average distance among all districts. $W_{ij}$ is the spatial weight matrix, and $s^2$ is the variance value of $PM_{2.5}$. At a certain level of significance, the larger the absolute value of Moran's I is, the higher the spatial correlation is. The significance of Moran's I was tested as follows:

$$Z_I = \frac{I - E[I]}{\sqrt{V[I]}} \tag{7}$$

where $E[I]$ is the expectation of Moran's I and $V[I]$ denotes the standard deviation of the variable. Meanwhile, the spatial autocorrelation of house prices was also explored (Table 2). The Moran's I values are all positive ($p < 0.01$), which indicates that the $PM_{2.5}$ distribution in different areas has a high spatial correlation. The spatial autocorrelation of house prices also shows the same characteristics. Therefore, it can be seen that $PM_{2.5}$ and house prices have a spatial correlation.

**Table 2.** Results of the Moran test.

| Year | 2009 | 2010 | 2011 | 2012 | 2013 |
|---|---|---|---|---|---|
| $PM_{2.5}$ | 0.108 *** | 0.081 *** | 0.102 *** | 0.105 *** | 0.130 *** |
| HP | 0.175 *** | 0.206 *** | 0.216*** | 0.208 *** | 0.205 *** |
| **Year** | **2014** | **2015** | **2016** | **2017** | **2018** |
| $PM_{2.5}$ | 0.101 *** | 0.068 *** | 0.108 *** | 0.137 *** | 0.138 *** |
| HP | 0.211 *** | 0.187 *** | 0.203 *** | 0.214 *** | 0.211 *** |

*** indicate $p < 0.01$.

## 4.3. Correlation Analysis

Figure 5 shows the r values between house prices and the other variables used. Generally, the r values between the control variables and house prices are all significant ($p < 0.01$). Here, $PM_{2.5}$ and NDVI show a negative correlation with house prices, with coefficients of −0.21 ($p < 0.01$) and −0.52 ($p < 0.01$), respectively, which proves that there are associations between house prices and air pollution. Notably, the correlation coefficients among the control variables are significant, but the values are heterogeneous, indicating that the other variables exist independently from one another to varying degrees and can be used for modeling here.

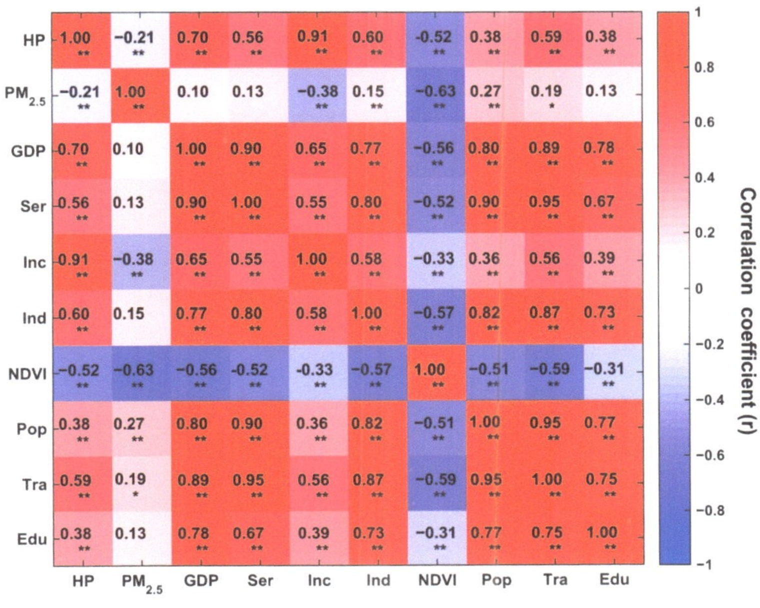


**Figure 5.** Correlations among house prices, control variables and moderating variables. * and ** indicate significance levels of $p$ less than 0.05 and 0.01, respectively.

### 4.4. Impact of $PM_{2.5}$ on House Prices

Table 3 lists the regression results of the OLS estimation. The results indicate that the $PM_{2.5}$ concentration could reduce house prices when other control variables (e.g., GDP, Service, Income) are controlled, and the significance $p$ is at the 1% level. Specifically, according to the estimation coefficient of our econometric model, when the annual mean $PM_{2.5}$ concentration increases by 1%, the average house prices decrease by 0.541% across Beijing. As the economy continues to develop rapidly, environmental pollution is becoming increasingly serious, resulting in an increasing marginal cost caused by environmental pollution. Currently, although rapid economic development at the expense of the environment can improve the material consumption level of residents, it can also offset the satisfaction of local residents brought by the increase in economic income. Due to the negative externality of the environment, house prices will decline. Currently, residents favor residential districts with better air quality to avoid the health risk of long-term exposure to air pollution. This phenomenon can be explained by the strong susceptibility of residents to the air pollution level of their residential area. During our study period, although the $PM_{2.5}$ concentration showed a significant decreasing trend, it remained at a high level. This phenomenon enhances the demand for housing with higher environmental quality by local residents, which will eventually promote the rise in house prices.

**Table 3.** The regression results of the OLS estimation, moderating effect and threshold effect.

| Variables | (1)<br>OLS Model<br>House Prices | (2)<br>Moderating Effect<br>House Prices | (3)<br>Threshold Regression<br>House Prices |
|---|---|---|---|
| $PM_{2.5}$ | −0.541 ***<br>(−3.20) | −0.349 **<br>(−2.14) | |
| GDP | 0.128 ***<br>(2.66) | 0.157 ***<br>(3.04) | 1.140 ***<br>(4.73) |
| Service | 0.093<br>(1.60) | 0.060<br>(1.02) | 0.235 **<br>(1.99) |
| Income | 1.020 ***<br>(8.10) | 1.085 ***<br>(8.63) | |
| Industry | 0.101 **<br>(2.53) | 0.110 ***<br>(2.70) | 0.180 **<br>(2.51) |
| NDVI | −0.879 ***<br>(−6.51) | −0.784 ***<br>(−5.61) | −1.041 ***<br>(−2.62) |
| Population | −0.212 ***<br>(−4.14) | −0.283 ***<br>(−4.97) | −0.090<br>(−1.42) |
| Traffic | −0.036<br>(−0.62) | −0.009<br>(−0.13) | 0.228<br>(1.46) |
| Education | | 0.005<br>(0.12) | |
| Education × $PM_{2.5}$ | | 0.243 ***<br>(3.00) | |
| PM2.5 (Income < $\theta$) | | | −0.425 *<br>(−1.93) |
| PM2.5 (Income ≥ $\theta$) | | | −0.461 **<br>(−2.08) |
| Constant | −1.423<br>(−0.83) | −3.086 *<br>(−1.83) | −12.622 ***<br>(−3.11) |
| Observations | 160 | 160 | 160 |
| R-squared | 0.901 | 0.906 | 0.897 |

***, ** and * indicate $p < 0.01$, $p < 0.05$ and $p < 0.1$.

The regression results for the other control variables also have practical significance. Here, significantly positive coefficients are captured for GDP, Income and Industry. Every 1% increase in GDP will lead to an increase in house prices of 0.128% ($p < 0.01$). This result illustrates the direct impact of regional economic strength on house prices, which is consistent with previous studies [50]. Regarding Income, the estimated coefficient is approximately 1.020 ($p < 0.01$), indicating that the improvement of per capita disposable income will significantly increase the pursuit of living quality for local residents. Notably, a 1% increase in the gross output value of the construction industry will also raise house prices by approximately 0.101% ($p < 0.05$). As a megacity in northern China, Beijing, with its high-quality social and public resources and good employment opportunities, exerts a strong attraction to the floating population. However, increasing the floating population could also cause a housing shortage. Moreover, the space available for housing in the central urban area is limited, resulting in numerous real estate developments in surrounding areas. Meanwhile, the construction cost of housing is also enhanced. These two factors will eventually lead to higher home prices. In contrast, significant negative impacts were captured for NDVI and Population with respect to house prices, with coefficients of $-0.879$ ($p < 0.01$) and $-0.212$ ($p < 0.01$), respectively. The areas with high NDVIs are mainly concentrated in Beijing's suburban regions, while the house prices in those areas are generally low, resulting in negative effects. The registered population is closely related to local house price control policies, e.g., preferential policies for the first house, price limits and purchase restrictions, which can encourage highly skilled individuals to settle in the locality.

We also found that the estimated coefficients of Service and Traffic are not significant. Nevertheless, they still possess practical significance. Following the implementation of a policy to strengthen livelihoods, the investment in public facilities in all districts has increased, and local residents enjoy good infrastructure and services across Beijing. Therefore, Service is no longer a factor that must be considered when purchasing a house. Regarding Traffic, on the one hand, because of the increase in per capita GDP, private car ownership generally increased across all districts during the study period, resulting in the improvement of the convenience of local resident travel. On the other hand, with the construction of basic transportation services, public transportation is available in all of Beijing's districts. These two items indicated that the traffic situation is no longer the main factor affecting house prices in the city.

### 4.5. Moderating Effect of Education

Table 3 also shows the results of the moderating effect of education on the relationship between $PM_{2.5}$ and house prices. Here, the estimation coefficient of $PM_{2.5}$ concentrations on house prices is negative, with a significance level of $p < 0.05$. In contrast, the estimation coefficient interactive item of Education and $PM_{2.5}$ was significantly positive ($p < 0.01$), indicating that educational resources may positively adjust the relationship between house prices and $PM_{2.5}$ concentrations. Specifically, local residents will be more inclined to allocate their purchasing power to housing that includes access to superior educational resources, resulting in the enhancement of house prices near preferred school districts. Figure 6 intuitively shows the moderating mechanism of education level on the relationship between $PM_{2.5}$ pollution and house prices. Generally, $PM_{2.5}$ concentrations can cause decreases in house prices, and the reduction will increase with the aggravation of the $PM_{2.5}$ pollution level. However, with the moderating influence of educational resources, this negative effect will be relieved effectively. In addition, the mitigation will be enhanced with an increase in the level of educational resources. During the study period, the policy for the next generation of the compulsory education stage (before high school) in Beijing mainly consisted of students attending schools nearby their homes. This policy not only strengthens the equity of educational resource allocation but also makes the location of residential areas the decisive factor in the allocation of basic educational resources. Currently, education level is a primary development factor for the younger generation.

Therefore, houses with access to high-level educational resources have become popular among Beijing homebuyers. Such houses, which can be termed "school district houses", eventually result in increased house prices for the residents of the preferred districts.

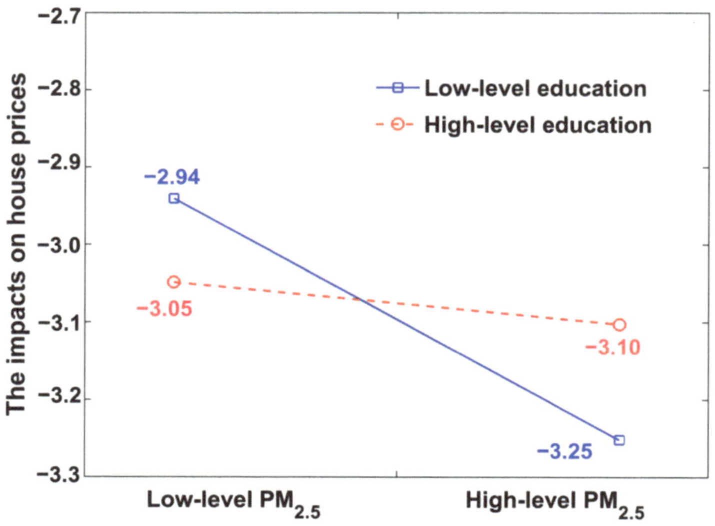


**Figure 6.** Moderating effect of education on the impact of $PM_{2.5}$ on house prices.

## 4.6. Threshold Regression Result

Significant heterogeneities may occur in the relationship between $PM_{2.5}$ pollution and house prices. Therefore, per capita disposable income was selected as a threshold variable to explain those heterogeneities. Initially, to determine the threshold effect, we first estimated the threshold condition according to Formula 4 under single, double and triple thresholds. Then, the joint hypotheses test (F statistic) was used to determine whether the model parameters were suitable for estimation. Additionally, the significance level $p$ was calculated according to the bootstrap method [51]. Overall, only a single threshold effect was statistically significant and reached a significant level. Table 4 shows the test results of the single threshold regression. Here, the F statistic and relevant critical values in this table are the results of repeated sampling (1000 times) with the bootstrap method. The F statistic was significant in the single threshold model, with the $p$ value at the 5% significance level ($p$ = 0.017). Furthermore, the threshold value of per capita disposable income $\theta$ was calculated to be approximately CNY 101,185.

**Table 4.** The test results of the threshold regression.

| | F | $p$ | RSS | MSE | Ctrit10 | Ctrit5 | Ctrit1 |
|---|---|---|---|---|---|---|---|
| Single ($\theta$ = 101,185) | 14.430 | 0.017 | 2.673 | 0.018 | 9.368 | 11.123 | 14.909 |

Table 3 shows the results of the threshold regression. Obvious differences in the impact of $PM_{2.5}$ pollution on house prices were captured under high- and low-level per capita disposable income. Generally, the estimation coefficients are both significantly negative, indicating that $PM_{2.5}$ pollution will reduce house prices. However, significant differences occurred in the quantitative estimation. When per capita disposable income is less than $\theta$ (Income < CNY 101,185), for every 1% increase in $PM_{2.5}$ concentration, house prices will decrease by 0.425%. Then, with the growth of per capita disposable income (Income $\geq$ CNY 101,185), the impact of $PM_{2.5}$ pollution on house prices will intensify.

Under such circumstances, 1% $PM_{2.5}$ pollution aggravation can cause house prices to decrease by approximately 0.461%. Higher per capita disposable income often means that residents will spend more on improving their quality of life. High $PM_{2.5}$ pollution loading can seriously threaten human health, which can increase the health risks for surrounding residents. At the same time, the awareness of environmental risks among local residents is also gradually growing. Therefore, they may spend more money to live in an environment with lower air pollution loading, resulting in changes in the relationship between $PM_{2.5}$ concentrations and house prices in Beijing.

### 4.7. Robustness Test

#### 4.7.1. Endogeneity Problems

Although we attempted to reduce the endogeneity problems by using control variables, the bias of omitted variables could not be completely avoided, which may lead to deviations in the estimation results. Therefore, the instrumental variable method was used to overcome the influence of endogeneity in our econometric model. The annual mean temperature was selected as the instrumental variable, and a 2SLS model was constructed. The first and second columns of Table 5 show the results of the 2SLS model. Generally, the instrumental variable is valid, and its estimation coefficient of the regression is significantly negative in the first stage ($p < 0.05$). A temperature increase can lead to an increase in the mixing layer height, which creates good air-diffusion conditions. With an intensification of atmospheric flow, air pollution is diluted, eventually resulting in a decrease in $PM_{2.5}$ concentration. In addition, the F value is greater than 10 and at the significance level of the 1% confidence interval, indicating that no weak instrumental variable problem occurred in the model. Among the results of the two-stage regression, the estimation coefficient of $PM_{2.5}$ is also significantly negative ($p < 0.05$). This outcome means that $PM_{2.5}$ pollution can still decrease housing prices even when fully considering the influence of the housing price background and control variables. Specifically, for every 1% increase in $PM_{2.5}$ concentration, house prices will decrease by approximately 0.897% across Beijing. Compared with the results of the basic econometric model, the estimation coefficient of the 2SLS model is 1.66 times that of the OLS model, indicating that the OLS model may slightly underestimate the impacts of $PM_{2.5}$ concentrations on house prices. In addition, the positive and negative conditions for the estimated coefficients of the other control variables in the 2SLS model have not changed, which can also prove the robustness of the results obtained in our study.

**Table 5.** The results of the robust test.

| Variables | (1) Stage1 House Prices | (2) Stage2 House Prices | (3) Robust House Prices |
|---|---|---|---|
| PM2.5 | | −0.897 ** (−2.20) | −0.513 *** (−3.06) |
| GDP | 0.183 *** (4.06) | 0.093 * (1.71) | 0.133 *** (2.82) |
| Service | 0.027 (0.45) | 0.119 * (1.89) | 0.085 (1.51) |
| Income | 1.314 *** (13.13) | 0.828 *** (3.38) | 1.040 *** (8.31) |
| Industry | 0.091 * (1.95) | 0.131 ** (2.34) | 0.099 ** (2.49) |
| NDVI | −0.674 *** (−6.46) | −1.079 *** (−4.38) | −0.866 *** (−6.21) |
| Population | −0.331 *** (−6.32) | −0.152 ** (−2.10) | −0.200 *** (−3.84) |

**Table 5.** *Cont.*

| Variables | (1) Stage1 House Prices | (2) Stage2 House Prices | (3) Robust House Prices |
|---|---|---|---|
| Traffic | 0.053 | −0.066 | −0.047 |
| | (0.86) | (−0.92) | (−0.82) |
| Temperature | −14.255 ** | | |
| | (−2.10) | | |
| Constant | 73.330 * | 2.058 | −1.632 |
| | (1.92) | (0.50) | (−0.95) |
| Observations | 160 | 160 | 160 |
| Cragg-Donald Wald F statistics | 43.320 | - | - |
| R-squared | 0.898 | 0.897 | 0.904 |

***, ** and * indicate $p < 0.01$, $p < 0.05$ and $p < 0.1$.

#### 4.7.2. Winsorized Robust Measures

Winsorization was also used to further determine the robustness of our results. Because of the limited sample size, we winsorized 3% of the samples. The third column of Table 5 shows the results of the winsorized robustness test. Overall, our econometric model is robust, and the positive and negative characteristics of the estimated coefficient of all controls are consistent with the results of the basic regression. Additionally, the level of significance did not change. Only the quantification of the estimated coefficients changed, extremely subtly in our case. In summary, combining the results of the IV method and the winsorized robustness test, our econometric model is robust. The impact of $PM_{2.5}$ pollution on house prices in Beijing was real during the study period.

## 5. Conclusions

At present, health risks, especially $PM_{2.5}$ pollution risks, in residential areas have gradually become the focus of local residents in China. In this study, the panel data for 16 districts in Beijing from 2009 to 2018 were used to investigate the impact of $PM_{2.5}$ pollution on house prices from theoretical and empirical perspectives. Through an econometric model analysis, we found that $PM_{2.5}$ pollution can curb the increase in house prices in Beijing. For every 1% increase in annual mean $PM_{2.5}$ concentration, house prices decrease by 0.541%. However, this impact can be suppressed by the moderating effects of the level of education. Moreover, as educational resources increase, this moderating effect will gradually increase. Meanwhile, the effects of $PM_{2.5}$ concentration on house prices are also nonlinear and influenced by disposable income per capita. When per capita disposable income is less than CNY 101,185, house prices will decrease by 0.425% for every 1% increase in $PM_{2.5}$; otherwise, house prices will decrease by 0.460% for every 1% increase in $PM_{2.5}$. Significantly, endogeneity problems are solved here by the instrumental variable method, and the conclusions of the paper are demonstrated to be robust.

## 6. Policy Suggestions

In this research, we provide a new perspective for understanding the economic consequences of air pollution. Our findings have important policy implications for promoting a win-win relationship between air pollution control and a stable real estate market across China. Based on our results, we offer several policy suggestions. First, the government should further optimize the environmental governance system and promote interregional collaborative governance. $PM_{2.5}$ pollutants usually exert spillover effects. Therefore, it is necessary to establish an interregional joint prevention and control system to decrease air pollution. Second, local governments should establish a house price control system to strictly control the inefficient expansion of urbanization. Additionally, housing planning should be strengthened to promote the sustainable development of a livable environment.

Through the rationalized control of house prices and the optimization of the spatial layout, the impact of air pollution on house prices can be reduced, and the win-win goal of green development and house price regulation can be achieved. Finally, the local government should balance educational resources among all districts of the city. The matching degree of the supply and demand of such resources should be improved, which can effectively reduce the spatial difference of house prices. These suggestions will contribute to achieving the high-quality development goals of balancing the housing price market, reducing resident health risks, reasonably improving the education level of the next generation and ameliorating urban planning.

**Author Contributions:** Conceptualization, W.X. and Z.Y.; methodology, W.X. and X.L.; software, W.X. and X.L.; validation, W.X., X.L. and Z.Y.; formal analysis, W.X.; data curation, W.X. and J.W.; writing—original draft preparation, W.X. and X.L.; writing—review and editing, W.X., X.L., Z.Y. and J.W. All authors have read and agreed to the published version of the manuscript.

**Funding:** This study was supported by the National Natural Science Foundation of China (41575144), the National Key R&D Program of China (2017YFA0603603) and the Qingdao Social Science Planning Project (QDSKL2101073).

**Institutional Review Board Statement:** Not applicable.

**Informed Consent Statement:** Not applicable.

**Conflicts of Interest:** The authors declare no conflict of interest.

## References


1. Geng, G.; Zheng, Y.; Zhang, Q.; Xue, T.; Zhao, H.; Tong, D.; Zheng, B.; Li, M.; Liu, F.; Hong, C.; et al. Drivers of PM2.5 air pollution deaths in China 2002–2017. *Nat. Geosci.* **2021**, *14*, 645–650. [CrossRef]
2. Wei, J.; Li, Z.; Lyapustin, A.; Sun, L.; Peng, Y.; Xue, W.; Su, T.; Cribb, M. Reconstructing 1-km-resolution high-quality PM2.5 data records from 2000 to 2018 in China: Spatiotemporal variations and policy implications. *Remote Sens. Environ.* **2020**, *252*, 112136. [CrossRef]
3. Wei, J.; Li, Z.; Li, K.; Dickerson, R.; Pinker, R.; Wang, J.; Liu, X.; Sun, L.; Xue, W.; Cribb, M. Full-coverage mapping and spatiotemporal variations of ground-level ozone ($O_3$) pollution from 2013 to 2020 across China. *Remote Sens. Environ.* **2022**, *270*, 112775. [CrossRef]
4. Cohen, A.J.; Brauer, M.; Burnett, R.; Anderson, H.R.; Frostad, J.; Estep, K.; Balakrishnan, K.; Brunekreef, B.; Dandona, L.; Dandona, R.; et al. Estimates and 25-year trends of the global burden of disease attributable to ambient air pollution: An analysis of data from the Global Burden of Diseases Study 2015. *Lancet* **2017**, *389*, 1907–1918. [CrossRef]
5. Wei, J.; Liu, S.; Li, Z.; Liu, C.; Qin, K.; Liu, X.; Pinker, R.; Dickerson, R.; Lin, J.; Boersma, K.; et al. Ground-level $NO_2$ surveillance from space across China for high resolution using interpretable spatiotemporally weighted artificial intelligence. *Environ. Sci. Technol.* **2022**. [CrossRef]
6. Miao, W.; Huang, X.; Song, Y. An economic assessment of the health effects and crop yield losses caused by air pollution in mainland China. *J. Environ. Sci.* **2017**, *56*, 102–113. [CrossRef] [PubMed]
7. Peng, C.; Li, B.; Nan, B. An analysis framework for the ecological security of urban agglomeration: A case study of the Beijing-Tianjin-Hebei urban agglomeration. *J. Clean. Prod.* **2021**, *315*, 128111. [CrossRef]
8. Xue, W.; Zhang, J.; Ji, D.; Che, Y.; Lu, T.; Deng, X.; Li, X.; Tian, Y.; Wei, J. Aerosol-induced direct radiative forcing effects on terrestrial ecosystem carbon fluxes over China. *Environ. Res.* **2021**, *200*, 111464. [CrossRef]
9. Dai, J.; Lv, P.; Ma, Z.; Bi, J.; Wen, T. Environmental risk and housing price: An empirical study of Nanjing, China. *J. Clean. Prod.* **2019**, *252*, 119828. [CrossRef]
10. Jiao, L.; Liu, Y. Geographic Field Model based hedonic valuation of urban open spaces in Wuhan, China. *Landsc. Urban Plan.* **2010**, *98*, 47–55. [CrossRef]
11. Gao, M.; Guttikunda, S.K.; Carmichael, G.R.; Wang, Y.; Liu, Z.; Stanier, C.O.; Saide, P.E.; Yu, M. Health impacts and economic losses assessment of the 2013 severe haze event in Beijing area. *Sci. Total Environ.* **2015**, *511*, 553–561. [CrossRef] [PubMed]
12. Yang, Y.; Luo, L.; Song, C.; Yin, H.; Yang, J. Spatiotemporal Assessment of PM2.5-Related Economic Losses from Health Impacts during 2014–2016 in China. *Int. J. Environ. Res. Public Health* **2018**, *15*, 1278. [CrossRef] [PubMed]
13. Chen, S.; Jin, H. Pricing for the clean air: Evidence from Chinese housing market. *J. Clean. Prod.* **2018**, *206*, 297–306. [CrossRef]
14. Zhang, L.; Yi, Y. What contributes to the rising house prices in Beijing? A decomposition approach. *J. Hous. Econ.* **2018**, *41*, 72–84. [CrossRef]
15. Huang, M.; Lu, B. Measuring the Housing Market Demand Elasticity in China—Based on the Rational Price Expectation and the Provincial Panel Data. *Open J. Soc. Sci.* **2016**, *4*, 21–25. [CrossRef]

16. Hui, E.C.-M.; Wang, X.-R.; Jia, S.-H. Fertility rate, inter-generation wealth transfer and housing price in China: A theoretical and empirical study based on the overlapping generation model. *Habitat Int.* **2016**, *53*, 369–378. [CrossRef]
17. Shen, Y.; Liu, H. Housing prices and economic fundamentals: A cross city analysis of china for 1995–2002. *Econ. Res. J.* **2004**, *6*, 78–86.
18. Du, H.; Ma, Y.; An, Y. The impact of land policy on the relation between housing and land prices: Evidence from China. *Q. Rev. Econ. Financ.* **2011**, *51*, 19–27. [CrossRef]
19. Wang, Y.; Wang, S.; Li, G.; Zhang, H.; Jin, L.; Su, Y.; Wu, K. Identifying the determinants of housing prices in China using spatial regression and the ge-ographical detector technique. *Appl. Geogr.* **2017**, *79*, 26–36. [CrossRef]
20. Liu, M.; Ma, Q.-P. Determinants of house prices in China: A panel-corrected regression approach. *Ann. Reg. Sci.* **2021**, *67*, 47–72. [CrossRef]
21. Zhang, Y.; Hua, X.; Zhao, L. Exploring determinants of housing prices: A case study of Chinese experience in 1999–2010. *Econ. Model.* **2012**, *29*, 2349–2361. [CrossRef]
22. Feng, H.; Lu, M. School quality and housing prices: Empirical evidence based on a natural experiment in shanghai, china. *J. Hous. Econ.* **2013**, *22*, 291–307. [CrossRef]
23. Li, H.; Wei, Y.D.; Wu, Y.; Tian, G. Analyzing housing prices in shanghai with open data: Amenity, accessibility and urban structure. *Cities* **2019**, *91*, 165–179. [CrossRef]
24. Ouyang, Y.; Cai, H.; Yu, X.; Li, Z. Capitalization of social infrastructure into China's urban and rural housing values: Empirical evidence from Bayesian Model Averaging. *Econ. Model.* **2022**, *107*, 105706. [CrossRef]
25. Yu, H. China's house price: Affected by economic fundamentals or real estate policy? *Front. Econ. China* **2010**, *5*, 25–51. [CrossRef]
26. Xue, W.; Zhang, J.; Zhong, C.; Ji, D.; Huang, W. Satellite-derived spatiotemporal pm2.5 concentrations and variations from 2006 to 2017 in china. *Sci. Total Environ.* **2020**, *712*, 134577. [CrossRef]
27. Ossokina, I.V.; Verweij, G. Urban traffic externalities: Quasi-experimental evidence from housing prices. *Reg. Sci. Urban Econ.* **2015**, *55*, 1–13. [CrossRef]
28. Tsui, W.H.K.; Tan, D.T.W.; Shi, S. Impacts of airport traffic volumes on house prices of New Zealand's major regions: A panel data approach. *Urban Stud.* **2017**, *54*, 2800–2817. [CrossRef]
29. Le Boennec, R.; Salladarre, F. The impact of air pollution and noise on the real estate market. The case of the 2013 European green capital: Nantes, France. *Ecol. Econ.* **2017**, *138*, 82–89. [CrossRef]
30. Hao, Y.; Zheng, S. Would environmental pollution affect home prices? An empirical study based on china's key cities. *Environ. Sci. Pollut. Res.* **2014**, *24*, 24545–24561. [CrossRef]
31. Chen, D.; Chen, S. Particulate air pollution and real estate valuation: Evidence from 286 Chinese prefecture-level cities over 2004–2013. *Energy Policy* **2017**, *109*, 884–897. [CrossRef]
32. Sun, B.; Yang, S. Asymmetric and Spatial Non-Stationary Effects of Particulate Air Pollution on Urban Housing Prices in Chinese Cities. *Int. J. Environ. Res. Public Health* **2020**, *17*, 7443. [CrossRef] [PubMed]
33. Zheng, S.; Cao, J.; Kahn, M.E.; Sun, C. Real Estate Valuation and Cross-Boundary Air Pollution Externalities: Evidence from Chinese Cities. *J. Real. Estate Financ. Econ.* **2013**, *48*, 398–414. [CrossRef]
34. Megan, T. China's three-child policy. *Lancet* **2021**, *397*, 2238.
35. Xue, W.; Zhang, J.; Zhong, C.; Li, X.; Wei, J. Spatiotemporal PM2.5 variations and its response to the industrial structure from 2000 to 2018 in the Beijing-Tianjin-Hebei region. *J. Clean. Prod.* **2020**, *279*, 123742. [CrossRef]
36. Nachtsheim, C.; Neter, J.; Kutner, M.; Wasserman, W. *Applied Linear Statistical Models*, 5th ed.; McGraw-Hill: New York, NY, USA, 2005.
37. Hersbach, H.; Bell, B.; Berrisford, P.; Hirahara, S.; Horányi, A.; Muñoz-Sabater, J.; Nicolas, J.; Peubey, C.; Abdalla, S.; Abellan, X.; et al. The era5 global reanalysis. *Q. J. R. Meteorol. Society.* **2020**, *146*, 1999–2049. [CrossRef]
38. David, C. Estimating the return to schooling: Progress on some persistent econometric problems. *Econometrica* **2001**, *69*, 1127–1160.
39. Frankel, J.A.; Romer, D. Does trade cause growth? *Am. Econ. Rev.* **1999**, *89*, 379–399. [CrossRef]
40. Jakob, M.; Haller, M.; Marschinski, R. Will history repeat itself? Economic convergence and convergence in energy use patterns. *Energy Econ.* **2012**, *34*, 95–104. [CrossRef]
41. Jeffrey, M.; Wooldridge. Cluster-sample methods in applied econometrics. *Am. Econ. Rev.* **2003**, *93*, 133–138.
42. Newey, W.; West, K. Autocovariance lag selection in covariance matrix estimation. *Rev. Econ. Stud.* **1994**, *61*, 613–653. [CrossRef]
43. Schultz, T.P. Wage Gains Associated with Height as a Form of Health Human Capital. *Am. Econ. Rev.* **2002**, *92*, 349–353. [CrossRef]
44. Lechene, V.; Pendakur, K.; Wolf, A. Ordinary Least Squares Estimation of the Intrahousehold Distribution of Expenditure. *J. Political Econ.* **2022**, *130*, 681–731. [CrossRef]
45. York, R.; Gossard, M.H. Cross-national meat and fish consumption: Exploring the effects of modernization and ecological context. *Ecol. Econ.* **2004**, *48*, 293–302. [CrossRef]
46. Horrace, W.C.; Oaxaca, R.L. Results on the bias and inconsistency of ordinary least squares for the linear probability model. *Econ. Lett.* **2006**, *90*, 321–327. [CrossRef]
47. Kelejian, H.H.; Prucha, I.R. A Generalized Spatial Two-Stage Least Squares Procedure for Estimating a Spatial Autoregressive Model with Autoregressive Disturbances. *J. Real Estate Financ. Econ.* **1998**, *17*, 99–121. [CrossRef]
48. Wei, J.; Peng, Y.; Mahmood, R.; Sun, L.; Guo, J. Intercomparison in spatial distributions and temporal trends derived from multi-source satellite aerosol products. *Atmos. Chem. Phys.* **2019**, *19*, 7183–7207. [CrossRef]

49. Cai, S.; Wang, Y.; Zhao, B.; Wang, S.; Xing, C.; Hao, J. The impact of the "air pollution prevention and control action plan" on PM2.5 concentrations in jing-jin-ji region during 2012–2020. *Sci. Total Environ.* **2017**, *580*, 197–209. [CrossRef]
50. Xu, T. The Relationship between Interest Rates, Income, GDP Growth and House Prices. *Res. Econ. Manag.* **2016**, *2*, 30. [CrossRef]
51. Efron, B.; Tibshirani, R. Bootstrap methods for standard errors, confidence intervals, and other measures of statistical accuracy. *Stat. Sci.* **1986**, *1*, 54–75. [CrossRef]

International Journal of
*Environmental Research and Public Health*

MDPI

*Article*

# Impact of Different Air Pollutants ($PM_{10}$, $PM_{2.5}$, $NO_2$, and Bacterial Aerosols) on COVID-19 Cases in Gliwice, Southern Poland

**Ewa Brągoszewska [1] and Anna Mainka [2,*]**

[1] Department of Technologies and Installations for Waste Management, Faculty of Energy and Environmental Engineering, Silesian University of Technology, 18 Konarskiego St., 44-100 Gliwice, Poland

[2] Department of Air Protection, Silesian University of Technology, 22B Konarskiego St., 44-100 Gliwice, Poland

* Correspondence: anna.mainka@polsl.pl

**Abstract:** Many studies have shown that air pollution may be closely associated with increased morbidity and mortality due to COVID-19. It has been observed that exposure to air pollution leads to reduced immune response, thereby facilitating viral penetration and replication. In our study, we combined information on confirmed COVID-19 daily new cases (DNCs) in one of the most polluted regions in the European Union (EU) with air-quality monitoring data, including meteorological parameters (temperature, relative humidity, atmospheric pressure, wind speed, and direction) and concentrations of particulate matter ($PM_{10}$ and $PM_{2.5}$), sulfur dioxide ($SO_2$), nitrogen oxides (NO and $NO_2$), ozone ($O_3$), and carbon monoxide (CO). Additionally, the relationship between bacterial aerosol (BA) concentration and COVID-19 spread was analyzed. We confirmed a significant positive correlation ($p < 0.05$) between $NO_2$ concentrations and numbers of confirmed DNCs and observed positive correlations ($p < 0.05$) between BA concentrations and DNCs, which may point to coronavirus air transmission by surface deposits on bioaerosol particles. In addition, wind direction information was used to show that the highest numbers of DNCs were associated with the dominant wind directions in the region (southern and southwestern parts).

**Keywords:** air pollution; $PM_{2.5}$; $PM_{10}$; $NO_2$; bioaerosols; COVID-19; meteorological parameters; atmospheric air; human health

check for updates

**Citation:** Brągoszewska, E.; Mainka, A. Impact of Different Air Pollutants ($PM_{10}$, $PM_{2.5}$, $NO_2$, and Bacterial Aerosols) on COVID-19 Cases in Gliwice, Southern Poland. *Int. J. Environ. Res. Public Health* **2022**, *19*, 14181. https://doi.org/10.3390/ijerph192114181

Academic Editor: Paul B. Tchounwou

Received: 27 September 2022
Accepted: 27 October 2022
Published: 30 October 2022



## 1. Introduction

Respiratory infections are the leading cause of epidemics, causing about 5 million deaths per year around the world [1]. In 2020, we became participants in an unprecedented international public health challenge. As a result of the coronavirus-associated acute respiratory syndrome (SARS-CoV-2), with COVID-19 disease as a symptom, both educational and commercial systems, as well as the general well-being of societies, have suffered [2]. In Poland, the first case of SARS-CoV-2 infection was diagnosed on March 4, while on March 12 the WHO regional director for Europe identified the region as the center of the pandemic. On March 17, every country in Europe had at least one confirmed case of COVID-19 [3].

Poland is a country with one of the largest air pollution problems in the European Union (EU). The Silesia voivodeship is the most polluted region in Poland, a country with 36 cities in a ranking of the 50 most polluted cities in the EU [4]. It has been estimated that, due to exposure to air pollution, the life expectancy of the average Polish citizen is shortened by around nine months, and 48,000 people die prematurely every year due to air pollution [5]. Epidemiological data and pathophysiological mechanisms suggest that ambient air pollution affects both the spread of COVID-19 disease and its severity [6]. Therefore, it is crucial to define the role that air pollutants play in the increase in morbidity and mortality due to COVID-19 [7].

Air pollution is known to damage many organs and systems of the human body. Of particular importance is the reduction in immunity to bacterial or viral infections of the respiratory system and the effect on the functioning of the cardiovascular system [8]. Moreover, after the outbreak of COVID-19, people are infected more often by SARS-CoV-2 in areas with high levels of air pollution than in less polluted areas. Air contamination disables airway mucosal functioning, including the production of fluid that lines the airway surface and contains respiratory host defense peptides, as well as mucus production and the tight junctions between epithelial cells. That is why air pollution can provoke cilia dysfunction, with changed surfactant composition and higher permeability of the airway epithelium [9–11]. Consequently, impairment of the mucosal barrier impairs lung defense against inhaled pathogens, such as SARS-CoV-2 [12].

Studies over the past two years on regions with high levels of air pollution have shown correlations with COVID-19 mortality. Regions in Northern Italy, including Lombardy, Veneto, and Emilia-Romagna, can be used as examples [13]. Similar trends have been observed in other regions with high air pollution, such as the Wuhan region of China and the United States, where poor air quality is correlated with a high incidence of COVID-19 and positive results of COVID-19 tests [10]. Currently, research has shown a relationship between routinely measured air pollutants, for example, particulate matter ($PM_{10}$ and $PM_{2.5}$) and nitrogen oxides ($NO_x$), and increased numbers of COVID-19 cases [13–19]. The impact of meteorological conditions has also been analyzed [20–24]. However, there is still a lack of reports on the relationship between bacterial aerosols (BAs) in ambient air and incidence of COVID-19. Worldwide studies have revealed that BA concentrations vary among different types of outdoor environments, with considerable seasonal variations as well [25–27]. BA concentration and composition in outdoor air can be influenced by specific micro- and macroscale determinants, such as land use, emission sources, air humidity, temperature, and UV radiation [26,28,29].

Bacterial aerosol particles are significant health risk factors, and exposure to these particles is associated with a varied range of health effects, including three major groups: infections, toxic reactions, and allergic reactions [30,31]. Therefore, the main aim of our study was to determine the impact of bacterial aerosols present in ambient air on the increase in COVID-19 cases in Gliwice in the Upper Silesia region of Poland, which is one of the most polluted areas in the EU [31].

This research is a contribution to the public debate on whether ambient particles can transport viruses that cause COVID-19 [16]. We believe that increasing knowledge of the relationship between air pollution and the incidence of COVID-19 symptoms can be beneficial in informing public health measures all around the world.

## 2. Materials and Methods

### 2.1. Sampling Sites

The study was carried out in Gliwice (50°17′37.1″ N 18°40′54.9″ E). Gliwice is a typical representative of a city located in the industrial area of Upper Silesia, Poland, with 178.186 thousand occupants (Figure 1). It is a densely populated and highly industrialized region of Poland and is responsible for the highest level of coal production. There are numerous coal-fired power plants, coking plants, and steel mills. Due to high levels of air pollution, the Silesia region has the shortest life expectancy and the highest incidence of premature births as well as genetic birth defects in Poland [32].

### 2.2. Measurements of Ambient Air Pollutants

The ambient air pollutants measured included bacterial aerosols (BAs), $PM_{2.5}$, $PM_{10}$, $SO_2$, and $NO_x$, including NO and $NO_2$, as well as $O_3$ and CO; various meteorological parameters, such as relative humidity (RH), air temperature (t), atmospheric pressure (P), and wind speed and direction, were also measured. All measurements were carried out during March 2021 from Monday to Friday. Additionally, an analysis of the impact of $PM_{2.5}$

and $PM_{10}$ concentrations on COVID-19 daily new cases (DNCs) was conducted during the winter season (from November 2020 to February 2021).

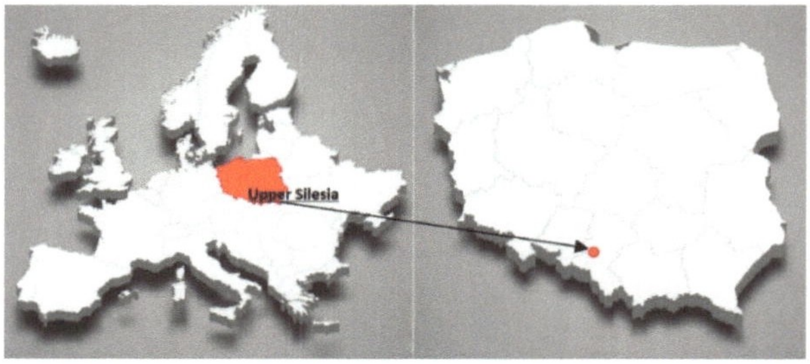


**Figure 1.** Localization of the measurement point in Gliwice, Upper Silesia, Southern Poland.

The data on $PM_{10}$ concentrations, as well as all gaseous and meteorological parameters, were gathered by the mobile air quality station for air pollutant emission measurements located at the Silesian University of Technology in Gliwice. The measuring equipment includes continuous automatic certificated monitors for $PM_{10}/PM_{2.5}$ particulate matter (Beta Attenuation Monitor BAM1020 Met One Instruments, Inc., Grants Pass, OR, USA), $SO_2$ (fluorescence analyzer—T100/API-Teledyne, San Diego, CA, USA), $NO_x$ (chemiluminescence analyser—T200/API Teledyne, San Diego, CA, USA), $O_3$ (UV absorption analyzer—T400/API-Teledyne, San Diego, CA, USA), and CO (infrared energy absorption analyzer—T300/API-Teledyne, San Diego, CA, USA), as well as the meteorological station (Meteo set WS 500 Lufft, G. Lufft Mess- und Regeltechnik GmbH, Fellbach, Germany). Additionally, data on $PM_{10}$ and $PM_{2.5}$ ambient levels were taken from the air monitoring station nearest to the Silesian University of Technology (at a distance of about 2500 m) at Mewy Street. The monitoring station belongs to the National Inspectorate of Environmental Protection in the Upper Silesia voivodeship [33]. The BA concentrations were measured using an Air Ideal (bioMérieux, France) one-stage impactor with an air flow rate of 100 $dm^3/min$, at a height of about 1.5 m to simulate aspiration from the human breathing zone, with the same operational details as in our previous studies [34,35]. Air pollutant levels and meteorological parameters are 24 h averages.

In addition, after a 24 h incubation, single colonies of BAs were passaged on a Biolog Universal Growth Agar (24 h incubation at 37 °C). Characterization of the isolates was performed using Gram staining and cell morphological analysis. In the next step, selected strains were then identified using the Biolog OmniLog system (Biolog, Haward, CA, USA) and a GEN III MicroPlate™, as in our previous research [31,35].

### 2.3. Measurements of SARS-CoV-2 Cases

The official data for SARS-CoV-2 infections in Poland are published daily by the Polish Ministry of Health [36]. All cases are diagnosed as positive based on polymerase chain reaction tests for SARS-CoV-2. We collected the cumulative number of cases for the district of Gliwice in Upper Silesia, Poland, from 23 November 2020 (the first day of available data) up to 31 March 2021. The data on the daily new cases (DNCs) due to COVID-19 were obtained from publicly available databases; hence, ethical approval was not required.

### 2.4. Statistical Analyses

The data were analyzed using Statistica software (TIBCO Software Inc. Palo Alto, CA, USA), version 13.3 for Windows, and a $p$-value $< 0.05$ was considered statistically significant. To determine whether a small data set ($n < 50$) was normally distributed, two tests were used: the Lilliefors test and the Shapiro–Wilk test. Table 1 presents the results of the normality tests of the random distributions of the measured parameters. Normality was revealed for total bacteria levels, daily new SARS-CoV-2 cases, concentrations of $NO_2$

and $O_3$, as well as all meteorological parameters, except ambient temperature. In the case of these parameters, linear regression could be used. For the other parameters, Spearman's rank correlation was used to test whether there was concordance (strength and direction) between the total bacteria levels, ambient air pollutants, meteorological parameters, and SARS-CoV-2 cases. Spearman correlations, not Pearson correlations, were used due to the non-normal distribution of the obtained variables (PM fractions, $SO_2$, NO, $NO_x$, CO, and temperature) generated from the daily time series data.

**Table 1.** Mean, median, minimum, and maximum values of parameters for daily new cases (DNCs) and all variables included in the analysis for March 2021.

| Parameters | Mean | Median | SD | Min | Max | Lilliefors Test | W | Shapiro–Wilk Test |
|---|---|---|---|---|---|---|---|---|
| SARS-CoV-2 cases | 112 | 100 | 74 | 14 | 279 | $p < 0.1$ | 0.92 | 0.07 |
| BAs, CFU/$m^3$ | 703 | 690 | 161 | 410 | 980 | $p > 0.2$ | 0.97 | 0.73 |
| $PM_{2.5}$ [1], μg/$m^3$ | 33.1 | 24.2 | 20.9 | 10.9 | 76.1 | $p < 0.01$ | 0.87 | 0.01 |
| $PM_{10}$ [1], μg/$m^3$ | 45.3 | 36.6 | 25.8 | 17.2 | 102.9 | $p < 0.15$ | 0.89 | 0.02 |
| $PM_{10}$ [2], μg/$m^3$ | 38.9 | 28.1 | 26.5 | 11.4 | 106.0 | $p < 0.01$ | 0.85 | 0.01 |
| $SO_2$, ppb | 3.0 | 2.2 | 2.4 | 0.5 | 9.1 | $p < 0.01$ | 0.79 | <0.01 |
| NO, ppb | 5.0 | 4.3 | 3.6 | 1.5 | 15.5 | $p < 0.05$ | 0.85 | <0.01 |
| $NO_2$, ppb | 10.8 | 9.5 | 5.6 | 2.9 | 23.2 | $p < 0.15$ | 0.93 | 0.09 |
| $NO_x$, ppb | 14.8 | 14.3 | 7.9 | 5.0 | 35.1 | $p < 0.2$ | 0.91 | 0.05 |
| $O_3$, ppb | 22.5 | 23.6 | 5.9 | 5.5 | 31.4 | $p > 0.2$ | 0.94 | 0.15 |
| CO, ppm | 0.4 | 0.3 | 0.2 | 0.2 | 0.9 | $p < 0.05$ | 0.86 | <0.01 |
| t, °C | 4.4 | 3.4 | 3.9 | −0.3 | 13.3 | $p < 0.15$ | 0.91 | 0.03 |
| RH, % | 73.3 | 71.5 | 6.9 | 61.5 | 88.8 | $p > 0.2$ | 0.97 | 0.58 |
| P, hPa | 996.9 | 995.6 | 7.4 | 985.9 | 1013.4 | $p < 0.1$ | 0.94 | 0.17 |
| Wind speed, m/s | 1.5 | 1.3 | 0.7 | 0.5 | 2.8 | $p > 0.2$ | 0.95 | 0.34 |
| Wind direction, ° | 185.8 | 195.4 | 69.4 | 17.8 | 324.5 | $p < 0.05$ | 0.94 | 0.19 |

[1] Monitoring station. [2] Mobile air quality station at the sampling site.

## 3. Results and Discussion

### *3.1. Particulate Matter (PM) Concentrations and SARS-CoV-2 Daily New Cases (DNCs)*

Long-term chronic exposure to air pollutants might play a significant role in the spread of COVID-19 [37]. In addition, short-term exposure to high levels of ground PM concentrations found in ambient air is associated with reduction in lung function and induction of respiratory symptoms, including cough, shortness of breath, and pain on deep inspiration [38,39]. New systematic reports have emphasized a possible association between the transmission of the virus in exposed populations and the level of PM in the atmosphere. However, confounding effects may be present, such as gender, age, smoking status, and high population density, as potential risk factors for higher morbidity and mortality due to COVID-19 [40,41]. Therefore, caution has to be taken in translating values of conventional indicators, such as $PM_{2.5}$ and $PM_{10}$ levels, into measures of vulnerability to COVID-19.

In our study, we observed a relationship between PM concentrations and daily new cases (DNCs). Figures 2 and 3 present the similarity in the daily course of $PM_{10}$ and $PM_{2.5}$ levels and DNCs during the winter season (from November 2020 to February 2021). The plots are consistent with other results showing a relationship between higher air pollution levels and COVID-19 cases [4,7,16,42].

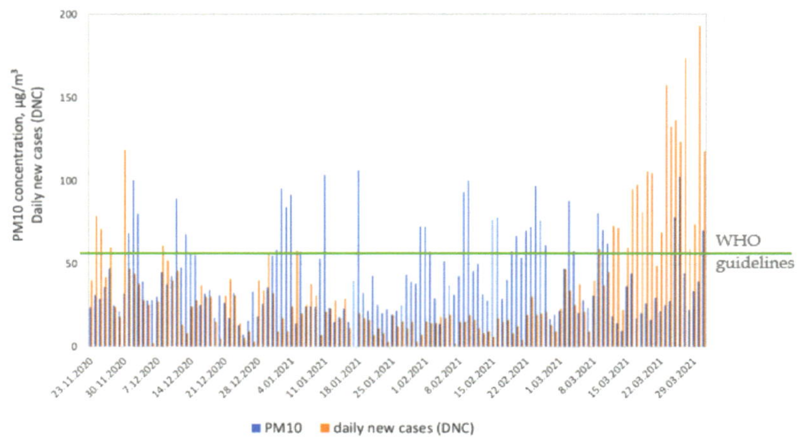


**Figure 2.** $PM_{10}$ concentrations and SARS-CoV-2 daily new cases (DNCs).

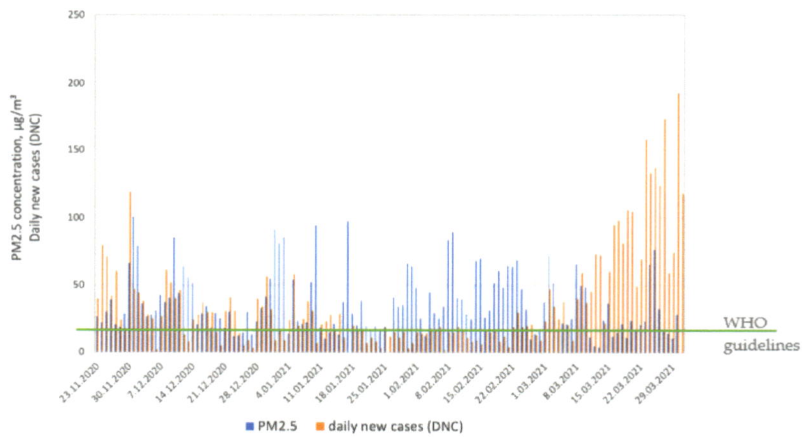


**Figure 3.** $PM_{2.5}$ concentrations and SARS-CoV-2 daily new cases (DNCs).

In Poland, just as in many other countries in Central and Eastern Europe, high levels of two PM fractions ($PM_{10}$ and $PM_{2.5}$) are observed every winter season. This is due to the high share of solid fuels in the primary energy source structure and the large share of low communal emissions [43,44]. However, what interested us and became an inspiration for further research was that we observed that the relationship between PM and DNCs had been weakening since March (spring season, end of the heating season), despite the continuous increase in the number of cases of SARS-CoV-2. Table 1 presents the means, medians, and ranges (min–max) of parameters monitored in March 2021.

Following the newest WHO global air quality guidelines [45], the recommended 24 h concentration of $PM_{2.5}$ is 15 $\mu g/m^3$, that of $PM_{10}$ is 45 $\mu g/m^3$, that of sulfur dioxide ($SO_2$) is 40 $\mu g/m^3$, that of nitrogen dioxide ($NO_2$) is 25 $\mu g/m^3$, and that of carbon monoxide (CO) is 4 $\mu g/m^3$, while, for ozone ($O_3$), the recommended 8 h average concentration is 100 $\mu g/m^3$. All gaseous pollutants were found to be below the level recommended by the WHO, while PM fractions exceeded recommended levels. The highest concentrations of major air pollutants monitored during the selected month were observed for both $PM_{2.5}$ and $PM_{10}$. These two fractions determined overall air quality in March 2021.

Figure 4 shows that the contribution of the air quality index (AQI) during March 2021 was mainly moderate. The correlation matrix (Table 2) for SARS-CoV-2 daily new cases (DNCs), ambient air pollutant concentrations, and bacterial aerosol concentrations during March 2021 suggests that, in moderate ambient air conditions, DNCs are significantly correlated with bacterial aerosols (BAs) and $NO_2$. The correlation coefficients ($r$) were 0.903 and 0.724, respectively.

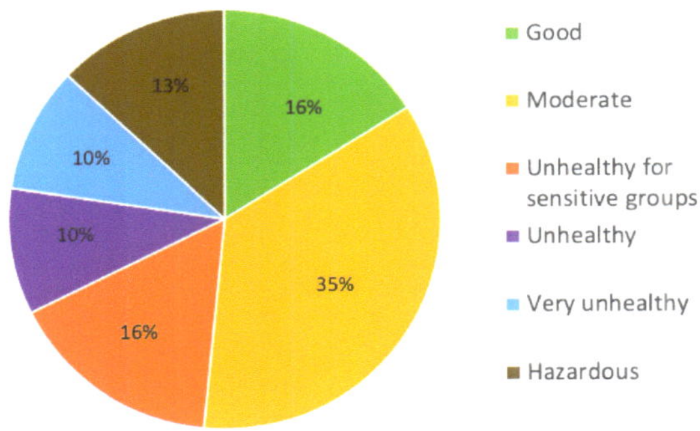


**Figure 4.** Contribution of Air Quality Index (AQI) during March 2021 in Gliwice.

**Table 2.** Correlation matrix for SARS-CoV-2 daily new cases (DNCs), bacterial aerosol concentrations, and ambient air pollution levels during March 2021.

| Parameters | DNCs | BAs | $PM_{2.5}$ [1] | $PM_{10}$ [1] | $PM_{10}$ [2] | $SO_2$ | NO | $NO_2$ | $NO_x$ | $O_3$ | CO |
|---|---|---|---|---|---|---|---|---|---|---|---|
| DNCs | 1 | **0.903** | −0.060 | 0.062 | −0.205 | 0.377 | **0.477** | **0.724** | **0.595** | −0.325 | 0.266 |
| BA CFU/$m^3$ | | 1 | −0.014 | <0.01 | −0.237 | 0.384 | 0.319 | **0.632** | **0.479** | −0.247 | 0.207 |
| $PM_{2.5}$ [1] µg/$m^3$ | | | 1 | **0.869** | **0.891** | **0.513** | 0.086 | 0.072 | −0.007 | −0.344 | **0.614** |
| $PM_{10}$ [1] µg/$m^3$ | | | | 1 | **0.868** | **0.708** | 0.329 | 0.320 | 0.245 | **−0.491** | **0.828** |
| $PM_{10}$ [2] µg/$m^3$ | | | | | 1 | **0.527** | −0.244 | −0.005 | −0.324 | −0.226 | **0.602** |
| $SO_2$ | | | | | | 1 | **0.460** | **0.622** | **0.556** | **−0.517** | **0.776** |
| NO | | | | | | | 1 | **0.853** | **0.925** | **−0.809** | **0.506** |
| $NO_2$ | | | | | | | | 1 | **0.970** | **−0.607** | **0.521** |
| $NO_x$ | | | | | | | | | 1 | **−0.681** | **0.465** |
| $O_3$ | | | | | | | | | | 1 | **−0.637** |

Correlation coefficients with $p < 0.05$ are in bold. [1] Monitoring station. [2] Mobile air monitoring lab at the sampling site.

### 3.2. Meteorological Conditions and SARS-CoV-2 Daily New Cases (DNCs)

Table 3 shows the results of a correlation analysis of meteorological parameters and DNCs as well as BAs. Interestingly, the analysis revealed a significant negative correlation between wind direction and DNCs ($r = -0.477$), as well as between BAs and atmospheric pressure. Figure 5 presents a wind rose diagram for March 2021 derived from the monitoring by a mobile air quality station located in Gliwice. The results of the study indicated that the wind in the studied area dominantly blew towards the south and southwest, and the wind speed values were low, in a range from 0.5 to 2.8 m/s (Table 1). As can be seen, the highest numbers of COVID-19 cases correspond to wind directions.

**Table 3.** Correlation matrix for SARS-CoV-2 daily new cases (DNCs), bacterial aerosol (BA) concentrations, and meteorological conditions during March 2021.

| Parameters | DNCs | BAs | Temperature | RH | Atmospheric Pressure | Wind Speed | Wind Direction |
|---|---|---|---|---|---|---|---|
| Temperature | 0.387 | 0.251 | 1 | −0.309 | 0.278 | −0.089 | −0.024 |
| RH | −0.377 | −0.386 | −0.309 | 1 | 0.007 | 0.194 | 0.409 |
| Atmospheric pressure | −0.148 | **−0.426** | 0.278 | 0.007 | 1 | −0.367 | −0.010 |
| Wind speed | 0.016 | 0.041 | −0.089 | 0.194 | −0.367 | 1 | 0.005 |
| Wind direction | **−0.477** | −0.288 | 0.024 | 0.409 | −0.010 | 0.005 | 1 |

Correlation coefficients with $p < 0.05$ are in bold.

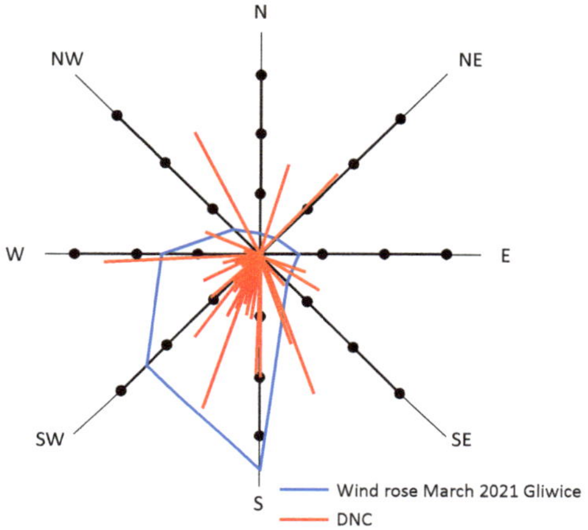


**Figure 5.** Wind rose vs. daily new cases (DNCs).

### *3.3. Bacterial Aerosol (BA) Concentrations and SARS-CoV-2 Daily New Cases (DNCs)*

To our knowledge, no research has previously been carried out to evaluate the link between concentrations of bacterial aerosols (BAs) in the outdoor air and numbers of cases of SARS-CoV-2. These results seem even more interesting given that, for a 10-year period in Poland, we recorded the maximum average concentration of BAs in the spring season (the time these analyses were conducted) and the lowest in the winter. During winter, extreme conditions, such as decreases in temperature and the heaviest rainfall and snowfall of the year, might contribute to the decrease in BA levels. On the other hand, in the summer, it would seem that the most favorable conditions for the growth of bacteria that we observed decreased BA concentrations. The reason for this decline may be the extremely high temperatures and strong UV radiation from the sun noted at this time.

The median BA concentration was 690 CFU/m$^3$ and varied in a range from 410 to 980 CFU/m$^3$ (Table 1). Figure 6 shows that the BA concentrations were linked to increased numbers of new SARS-CoV-2 cases. Table 2 presents a matrix of correlation coefficients ($r$) for daily new cases (DNCs) and all variables included in the analysis, which suggests that BA concentrations during March 2021 were highly correlated with DNCs ($r = 0.903$) and that the relationship was linear ($R^2 = 0.758$).

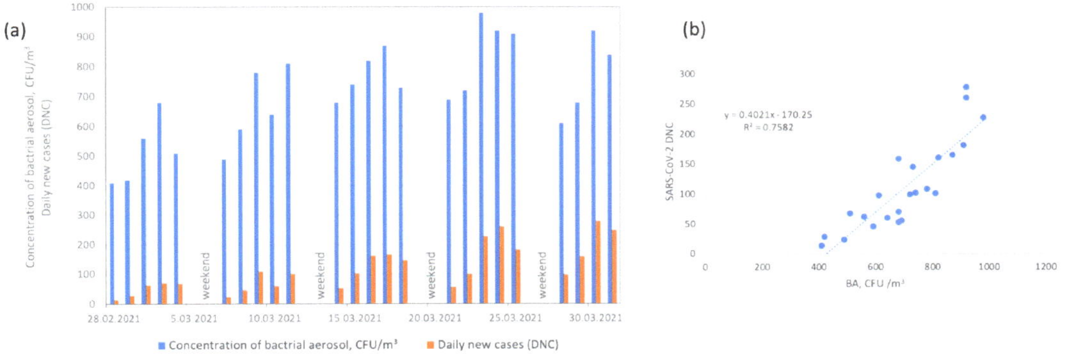


**Figure 6.** (**a**) Bacterial aerosol (BA) concentrations and SARS-CoV-2 daily new cases (DNCs). (**b**) Linear regression coefficients between bacterial aerosol (BA) concentrations and SARS-CoV-2 daily new cases (DNCs).

We suggest that bacterial infections may cause increases in the numbers of COVID-19 patients. However, the collection of respiratory samples from this type of patient is complicated because of the elevated risks associated with aerosol generation procedures. Consequently, bacterial respiratory tract infections are likely to be under-detected in patients hospitalized with COVID-19. There are only a limited number of papers that have reported species identities or sampling times, making it impossible to determine whether patients had bacterial infections at the time of hospital admission [46].

A significant correlation ($p < 0.05$) was found between BA concentrations and SARS-CoV-2 in a hospital in Iran, where the obtained results implied that contact with bioaerosols generated through COVID-19 patients', healthcare workers', and visitors' exhalations in hospital wards may pose a serious health threat, especially to susceptible individuals [47]. Zhou et al. found that bacterial infections (bacteraemia and pneumonia) were more common in fatal COVID-19 cases compared with recovered cases in Wuhan, China [48].

There is a suspicion that pollen bioaerosols can also affect coronavirus survival [49,50]. Considering the summer incidence of coronavirus during June 2022, under suitable environmental conditions, simultaneous or co-exposure to SARS-CoV-2 (via other infected human carriers) and airborne aerosols might promote viral infection. Therefore, we must detect the seasonal patterns of bioaerosols and airborne viruses, including COVID-19, based on environmental factors.

The most commonly isolated bacterial group in our research was that of the Gram-positive rods that form endospores, among which *Bacillus* was the most frequently isolated genus (Table 4).

**Table 4.** Bacterial species identifications.

| Species of Isolated Bacteria |
| --- |
| *Bacillus cereus* |
| *Bacillus subtilis* |
| *Bacillus flexus* |
| *Bacillus licheniformis* |
| *Paenibacillus barengoltzii* |
| *Micrococcus luteus* |
| *Macrococcus equipercicus* |
| *Macrococcus brunensis* |
| *Nocardia alba* |
| *Lactobacillus crispatus* |

The spores of *Bacillus* have remarkable resistance to chemical and physical factors. This genus of bacteria is commonly found in soil and water and is a component of the normal flora of the skin and mucous membranes of humans and animals [51]. This result corresponds with our previous findings [26,35] and is common to other studies [52–54].

### 3.4. Nitrogen Dioxide ($NO_2$) Concentrations and SARS-CoV-2 Daily New Cases (DNCs)

Nitrogen dioxide ($NO_2$) is another important air pollutant toxic to human respiratory systems when present at higher concentrations in the atmosphere [55]. In our study, the median $NO_2$ concentration was 9.5 ppb (Table 1). Figure 7 shows that $NO_2$ concentrations were linked to increased numbers of SARS-CoV-2 cases. The relation was linear ($R^2$ = 0.597), as shown by regression analysis, and the correlation was high ($r$ = 0.724); on the other hand, NO and $NO_x$ concentrations (0.477 and 0.595, respectively) were linked to increased numbers of SARS-CoV-2 cases to a lesser extent.

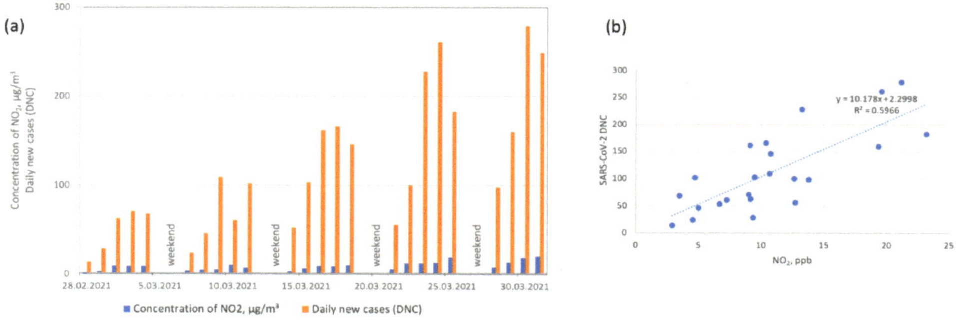


**Figure 7.** (**a**) $NO_2$ concentrations and SARS-CoV-2 daily new cases (DNCs). (**b**) Linear regression coefficients between $NO_2$ concentrations and SARS-CoV-2 daily new cases (DNCs).

Similar results were found in Wuhan, China, where Li et al. (2020) found a significant linear correlation between SARS-CoV-2 DNCs and $NO_2$ concentrations ($R^2$ = 0.329, $p$ < 0.001) [56]. In Spain, Italy, France, and Germany, it was observed that out of 4443 fatalities observed at the beginning of the pandemic of COVID-19, 3487 deaths, accounting for 78% of the total deaths, were confined to areas where $NO_2$ pollution was predominant [57].

## 4. Conclusions

Understanding the airborne route of SARS-CoV-2 transmission is essential for infection prevention and control, and improvements in terms of air pollution, lifestyle, and the environment will help to prevent future viral pandemics.

Our study found that the role of ambient air pollution given moderate air quality is largely unknown, necessitating further epidemiological studies. Although the current study was conducted only in Gliwice, Poland, it points to the as yet unrepresented implication that bacterial aerosol (BA) concentrations in the period characterized by moderate air quality were significantly associated with SARS-CoV-2 daily new cases (DNCs).

In conclusion, we think that our analyses of the correlations between bacterial aerosol (BA) concentrations and new COVID-19 cases are foundations for further, wider research. However, the drastic mutational nature of the virus makes it difficult to predict which mechanisms and ecological parameters will affect its growth and prevalence.

**Author Contributions:** Conceptualization, E.B.; methodology, E.B.; formal analysis, E.B.; investigation, E.B.; resources, E.B. and A.M.; data curation, A.M.; writing—original draft preparation, E.B. and A.M.; writing—review and editing, E.B. and A.M.; visualization, A.M.; supervision, E.B.; project administration, E.B.; funding acquisition, A.M. and E.B. All authors have read and agreed to the published version of the manuscript.

**Funding:** Research funded by the statutory research of the Faculty of Energy and Environmental Engineering, Silesian University of Technology.

**Institutional Review Board Statement:** Not applicable.

**Informed Consent Statement:** Not applicable.

**Data Availability Statement:** All data used in the paper are publicly available. We may supply the data that we gathered from public sources upon the making of a reasonable request to the corresponding author.

**Conflicts of Interest:** The authors declare no conflict of interest.

## References

1. Zumla, A.; Niederman, M.S. Editorial: The Explosive Epidemic Outbreak of Novel Coronavirus Disease 2019 (COVID-19) and the Persistent Threat of Respiratory Tract Infectious Diseases to Global Health Security. *Curr. Opin. Pulm. Med.* **2020**, *26*, 193–196. [CrossRef] [PubMed]
2. WHO. Organization Mundial de la Salud (OMS). WHO | Coronavirus Disease (COVID-19) Pandemic-Emergency Use Listing Procedure (EUL) Open for in Vitro Diagnostics. 2020. Available online: https://extranet.who.int/pqweb/vitro-diagnostics/coronavirus-disease-covid-19-pandemic-%E2%80%94-emergency-use-listing-procedure-eul-open (accessed on 20 September 2022).
3. Gujski, M.; Raciborski, F.; Jankowski, M.; Nowicka, P.M.; Rakocy, K.; Pinkas, J. Epidemiological Analysis of the First 1389 Cases of COVID-19 in Poland: A Preliminary Report. *Med. Sci. Monit.* **2020**, *26*, e924702. [CrossRef] [PubMed]
4. Semczuk-Kaczmarek, K.; Rys-Czaporowska, A.; Sierdzinski, J.; Kaczmarek, L.D.; Szymanski, F.M.; Platek, A.E. Association between Air Pollution and COVID-19 Mortality and Morbidity. *Intern. Emerg. Med.* **2021**, *17*, 467–473. [CrossRef] [PubMed]
5. Pikala, M.; Maniecka-Bryła, I. Fifteen-Year Mortality Trends in Poland Analysed with the Use of Standard Expected Years of Life Lost, 2000–2014. *Sci. Rep.* **2017**, *7*, 8730. [CrossRef] [PubMed]
6. Tacconelli, E.; Carrara, E.; Savoldi, A.; Harbarth, S.; Mendelson, M.; Monnet, D.L.; Pulcini, C.; Kahlmeter, G.; Kluytmans, J.; Carmeli, Y.; et al. Discovery, Research, and Development of New Antibiotics: The WHO Priority List of Antibiotic-Resistant Bacteria and Tuberculosis. *Lancet Infect. Dis.* **2018**, *18*, 318–327. [CrossRef]
7. Comunian, S.; Dongo, D.; Milani, C.; Palestini, P. Air Pollution and Covid-19: The Role of Particulate Matter in the Spread and Increase of Covid-19's Morbidity and Mortality. *Int. J. Environ. Res. Public Health* **2020**, *17*, 4487. [CrossRef]
8. Urrutia-Pereira, M.; Mello-da-Silva, C.A.; Solé, D. COVID-19 and Air Pollution: A Dangerous Association? *Allergol. Immunopathol.* **2020**, *48*, 496–499. [CrossRef]
9. Albano, G.D.; Montalbano, A.M.; Gagliardo, R.; Anzalone, G.; Profita, M. Impact of Air Pollution in Airway Diseases: Role of the Epithelial Cells (Cell Models and Biomarkers). *Int. J. Mol. Sci.* **2022**, *23*, 2799. [CrossRef]
10. van der Valk, J.P.M.; In't Veen, J.C.C.M. The Interplay Between Air Pollution and Coronavirus Disease (COVID-19). *J. Occup. Environ. Med.* **2021**, *63*, e163–e167. [CrossRef]
11. Ciencewicki, J.; Jaspers, I. Air Pollution and Respiratory Viral Infection. *Inhal. Toxicol.* **2007**, *19*, 1135–1146. [CrossRef]
12. Huff, R.D.; Carlsten, C.; Hirota, J.A. An Update on Immunologic Mechanisms in the Respiratory Mucosa in Response to Air Pollutants. *J. Allergy Clin. Immunol.* **2019**, *143*, 1989–2001. [CrossRef] [PubMed]
13. Conticini, E.; Frediani, B.; Caro, D. Can Atmospheric Pollution Be Considered a Co-Factor in Extremely High Level of SARS-CoV-2 Lethality in Northern Italy? *Environ. Pollut.* **2020**, *261*, 114465. [CrossRef] [PubMed]
14. Wu, X.; Nethery, R.C.; Sabath, M.B.; Braun, D.; Dominici, F. Exposure to Air Pollution and COVID-19 Mortality in the United States: A Nationwide Cross-Sectional Study. *Sci. Adv.* **2020**, *6*, 1–6. [CrossRef] [PubMed]
15. Setti, L.; Passarini, F.; De Gennaro, G.; Barbieri, P.; Perrone, M.G.; Borelli, M.; Palmisani, J.; Di Gilio, A.; Torboli, V.; Fontana, F.; et al. SARS-CoV-2RNA Found on Particulate Matter of Bergamo in Northern Italy: First Evidence. *Environ. Res.* **2020**, *188*, 109754. [CrossRef] [PubMed]
16. Jiang, Y.; Xu, J. The Association between COVID-19 Deaths and Short-Term Ambient Air Pollution/Meteorological Condition Exposure: A Retrospective Study from Wuhan, China. *Air Qual. Atmos. Health* **2021**, *14*, 1–5. [CrossRef] [PubMed]
17. Yao, Y.; Pan, J.; Liu, Z.; Meng, X.; Wang, W.; Kan, H.; Wang, W. Ambient Nitrogen Dioxide Pollution and Spreadability of COVID-19 in Chinese Cities. *Ecotoxicol. Environ. Saf.* **2021**, *208*, 111421. [CrossRef]
18. Travaglio, M.; Yu, Y.; Popovic, R.; Selley, L.; Leal, N.S.; Martins, L.M. Links between Air Pollution and COVID-19 in England. *Environ. Pollut.* **2021**, *268*, 115859. [CrossRef]
19. Accarino, G.; Lorenzetti, S.; Aloisio, G. Assessing Correlations between Short-Term Exposure to Atmospheric Pollutants and COVID-19 Spread in All Italian Territorial Areas. *Environ. Pollut.* **2021**, *268*, 115714. [CrossRef]
20. Bochenek, B.; Jankowski, M.; Gruszczynska, M.; Nykiel, G.; Gruszczynski, M.; Jaczewski, A.; Ziemianski, M.; Pyrc, R.; Figurski, M.; Pinkas, J. Impact of Meteorological Conditions on the Dynamics of the COVID-19 Pandemic in Poland. *Int. J. Environ. Res. Public Health* **2021**, *18*, 3951. [CrossRef]
21. Rendana, M.; Idris, W.M.R. New COVID-19 Variant (B.1.1.7): Forecasting the Occasion of Virus and the Related Meteorological Factors. *J. Infect. Public Health* **2021**, *14*, 1320–1327. [CrossRef]

22. Sarmadi, M.; Rahimi, S.; Evensen, D.; Kazemi Moghaddam, V. Interaction between Meteorological Parameters and COVID-19: An Ecological Study on 406 Authorities of the UK. *Environ. Sci. Pollut. Res.* **2021**, *28*, 67082–67097. [CrossRef] [PubMed]
23. Rendana, M. Impact of the Wind Conditions on COVID-19 Pandemic: A New Insight for Direction of the Spread of the Virus. *Urban. Clim* **2020**, *34*, 100680. [CrossRef] [PubMed]
24. Mao, N.; Zhang, D.; Li, Y.; Li, Y.; Li, J.; Zhao, L.; Wang, Q.; Cheng, Z.; Zhang, Y.; Long, E. How Do Temperature, Humidity, and Air Saturation State Affect the COVID-19 Transmission Risk? *Environ. Sci. Pollut. Res. Int.* **2022**. *online ahead of print*. [CrossRef] [PubMed]
25. Haas, D.; Unteregger, M.; Habib, J.; Galler, H.; Marth, E.; Reinthaler, F.F. Exposure to Bioaerosol from Sewage Systems. *Water Air Soil Pollut.* **2010**, *207*, 49–56. [CrossRef]
26. Brągoszewska, E.; Pastuszka, J.S. Influence of Meteorological Factors on the Level and Characteristics of Culturable Bacteria in the Air in Gliwice, Upper Silesia (Poland). *Aerobiologia* **2018**, *34*, 241–255. [CrossRef] [PubMed]
27. Woo, A.C.; Brar, M.S.; Chan, Y.; Lau, M.C.Y.; Leung, F.C.C.; Scott, J.A.; Vrijmoed, L.L.P.; Zawar-Reza, P.; Pointing, S.B. Temporal Variation in Airborne Microbial Populations and Microbially-Derived Allergens in a Tropical Urban Landscape. *Atmos. Environ.* **2013**, *74*, 291–300. [CrossRef]
28. Ruiz-Gil, T.; Acuña, J.J.; Fujiyoshi, S.; Tanaka, D.; Noda, J.; Maruyama, F.; Jorquera, M.A. Airborne Bacterial Communities of Outdoor Environments and Their Associated Influencing Factors. *Environ. Int.* **2020**, *145*, 106156. [CrossRef]
29. Šantl-Temkiv, T.; Gosewinkel, U.; Starnawski, P.; Lever, M.; Finster, K. Aeolian Dispersal of Bacteria in Southwest Greenland: Their Sources, Abundance, Diversity and Physiological States. *FEMS Microbiol. Ecol.* **2018**, *94*, fiy031. [CrossRef]
30. Chegini, F.M.; Baghani, A.N.; Hassanvand, M.S.; Sorooshian, A.; Golbaz, S.; Bakhtiari, R.; Ashouri, A.; Joubani, M.N.; Alimohammadi, M. Indoor and Outdoor Airborne Bacterial and Fungal Air Quality in Kindergartens: Seasonal Distribution, Genera, Levels, and Factors Influencing Their Concentration. *Build. Environ.* **2020**, *175*, 106690. [CrossRef]
31. Brągoszewska, E.; Biedroń, I. Efficiency of Air Purifiers at Removing Air Pollutants in Educational Facilities: A Preliminary Study. *Front. Environ. Sci.* **2021**, *9*, 370. [CrossRef]
32. Dabrowiecki, P.; Adamkiewicz, Ł.; Mucha, D.; Czechowski, P.O.; Soliński, M.; Chciałowski, A.; Badyda, A. Impact of Air Pollution on Lung Function among Preadolescent Children in Two Cities in Poland. *J. Clin. Med.* **2021**, *10*, 2375. [CrossRef] [PubMed]
33. GIOS. Chief Inspectorate for Environmental Protection. Available online: https://www.cleanenergywire.org/experts/chief-inspectorate-environmental-protection-gios-poland (accessed on 15 September 2022).
34. Brągoszewska, E.; Mainka, A.; Pastuszka, J. Bacterial and Fungal Aerosols in Rural Nursery Schools in Southern Poland. *Atmosphere* **2016**, *7*, 142. [CrossRef]
35. Brągoszewska, E.; Biedroń, I. Indoor Air Quality and Potential Health Risk Impacts of Exposure to Antibiotic Resistant Bacteria in an Office Rooms in Southern Poland. *Int. J. Environ. Res. Public Health* **2018**, *15*, 2604. [CrossRef]
36. Government of Poland Report of Coronavirus Infection (SARS-CoV-2)-Coronavirus: Information and Recommendations—Portal Gov.Pl. Available online: https://www.gov.pl/web/koronawirus/wykaz-zarazen-koronawirusem-sars-cov-2 (accessed on 28 October 2021).
37. Khan, Z.; Ualiyeva, D.; Khan, A.; Zaman, N.; Sapkota, S.; Khan, A.; Ali, B.; Ghafoor, D. A Correlation among the COVID-19 Spread, Particulate Matters, and Angiotensin-Converting Enzyme 2: A Review. *J. Environ. Public Health* **2021**, *2021*, 1–8. [CrossRef]
38. Seposo, X.; Ueda, K.; Sugata, S.; Yoshino, A.; Takami, A. Short-Term Effects of Air Pollution on Daily Single- and Co-Morbidity Cardiorespiratory Outpatient Visits. *Sci. Total Environ.* **2020**, *729*, 138934. [CrossRef] [PubMed]
39. Kelly, F.J.; Fussell, J.C. Air Pollution and Airway Disease. *Clin. Exp. Allergy* **2011**, *41*, 1059–1071. [CrossRef] [PubMed]
40. Contini, D.; Costabile, F. Does Air Pollution Influence COVID-19 Outbreaks? *Atmosphere* **2020**, *11*, 377. [CrossRef]
41. Pansini, R.; Fornacca, D. Early Spread of COVID-19 in the Air-Polluted Regions of Eight Severely Affected Countries. *Atmosphere* **2021**, *12*, 795. [CrossRef]
42. Meo, S.A.; Almutairi, F.J.; Abukhalaf, A.A.; Alessa, O.M.; Al-Khlaiwi, T.; Meo, A.S. Sandstorm and Its Effect on Particulate Matter $PM_{2.5}$, Carbon Monoxide, Nitrogen Dioxide, Ozone Pollutants and SARS-CoV-2 Cases and Deaths. *Sci. Total Environ.* **2021**, *795*, 148764. [CrossRef]
43. Nidzgorska-Lencewicz, J.; Czarnecka, M. Thermal Inversion and Particulate Matter Concentration in Wrocław in Winter Season. *Atmosphere* **2020**, *11*, 1351. [CrossRef]
44. Zajusz-Zubek, E.; Filipek, J.; Mainka, A. The Impact of Sources Burning Solid Fuels on Environmental Pollution with $PM_{10}$ on the Example of the City Tarnowskie Góry. *Eng. Prot. Environ.* **2019**, *22*, 15–26. [CrossRef]
45. World Health Organization. *WHO Global Air Quality Guidelines. Particulate Matter ($PM_{2.5}$ and $PM_{10}$), Ozone, Nitrogen Dioxide, Sulfur Dioxide and Carbon Monoxide*; WHO: Geneva, Switzerland, 2021.
46. Farrell, J.M.; Zhao, C.Y.; Tarquinio, K.M.; Brown, S.P. Causes and Consequences of COVID-19-Associated Bacterial Infections. *Front. Microbiol.* **2021**, *12*, 1–6. [CrossRef] [PubMed]
47. Hemati, S.; Mobini, G.R.; Heidari, M.; Rahmani, F.; Soleymani Babadi, A.; Farhadkhani, M.; Nourmoradi, H.; Raeisi, A.; Ahmadi, A.; Khodabakhshi, A.; et al. Simultaneous Monitoring of SARS-CoV-2, Bacteria, and Fungi in Indoor Air of Hospital: A Study on Hajar Hospital in Shahrekord, Iran. *Environ. Sci. Pollut. Res.* **2021**, *28*, 43792–43802. [CrossRef] [PubMed]
48. Zhou, F.; Yu, T.; Du, R.; Fan, G.; Liu, Y.; Liu, Z.; Xiang, J.; Wang, Y.; Song, B.; Gu, X.; et al. Clinical Course and Risk Factors for Mortality of Adult Inpatients with COVID-19 in Wuhan, China: A Retrospective Cohort Study. *Lancet* **2020**, *395*, 1054–1062. [CrossRef]

49. Ravindra, K.; Goyal, A.; Mor, S. Does Airborne Pollen Influence COVID-19 Outbreak? *Sustain. Cities Soc.* **2021**, *70*, 02887. [CrossRef]
50. Shammi, M.; Rahman, M.M.; Tareq, S.M. Distribution of Bioaerosols in Association With Particulate Matter: A Review on Emerging Public Health Threat in Asian Megacities. *Front. Environ. Sci.* **2021**, *9*, 328. [CrossRef]
51. Mentese, S.; Arisoy, M.; Rad, A.Y.; Güllü, G. Bacteria and Fungi Levels in Various Indoor and Outdoor Environments in Ankara, Turkey. *Clean* **2009**, *37*, 487–493. [CrossRef]
52. Aydogdu, H.; Asan, A.; Tatman Otkun, M. Indoor and outdoor airborne bacteria in child day-care centers in Edirne City (Turkey), seasonal distribution and influence of meteorological factors. *Environ. Monit. Assess.* **2009**, *164*, 53–66. [CrossRef]
53. Bartlett, K.H.; Kennedy, S.M.; Brauer, M.; van Netten, C.; Dill, B. Evaluation and Determinants of Airborne Bacterial Concentrations in School Classrooms. *J. Occup. Environ. Hyg.* **2004**, *1*, 639–647. [CrossRef]
54. Fang, Z.; Ouyang, Z.; Zheng, H.; Wang, X.; Hu, L. Culturable Airborne Bacteria in Outdoor Environments in Beijing, China. *Microb. Ecol.* **2007**, *54*, 487–496. [CrossRef]
55. Ali, N.; Islam, F. The Effects of Air Pollution on COVID-19 Infection and Mortality—A Review on Recent Evidence. *Front. Public Health* **2020**, *8*, 580057. [CrossRef] [PubMed]
56. Li, H.; Xu, X.L.; Dai, D.W.; Huang, Z.Y.; Ma, Z.; Guan, Y.J. Air Pollution and Temperature Are Associated with Increased COVID-19 Incidence: A Time Series Study. *Int. J. Infect. Dis.* **2020**, *97*, 278–282. [CrossRef] [PubMed]
57. Ogen, Y. Response to the Commentary by Alexandra A. Chudnovsky on 'Assessing Nitrogen Dioxide ($NO_2$) Levels as a Contributing Factor to Coronavirus (COVID-19) Fatality'. *Sci. Total Environ.* **2020**, *740*, 138605. [CrossRef] [PubMed]

International Journal of
*Environmental Research and Public Health*

MDPI

Article

# Connection between Weather Types and Air Pollution Levels: A 19-Year Study in Nine EMEP Stations in Spain

**Nuria Pardo [1], Samuel Sainz-Villegas [2,3], Ana I. Calvo [2,*], Carlos Blanco-Alegre [2] and Roberto Fraile [2]**

1 Department of Applied Physics, Faculty of Sciences, University of Valladolid, Paseo de Belén, 7, 47011 Valladolid, Spain
2 Department of Physics, University of León, 24071 León, Spain
3 IHCantabria-Instituto de Hidráulica Ambiental de la Universidad de Cantabria, 39011 Santander, Spain
* Correspondence: aicalg@unileon.es

**Abstract:** This study focuses on the analysis of the distribution, both spatial and temporal, of the $PM_{10}$ (particulate matter with a diameter of 10 µm or less) concentrations recorded in nine EMEP (European Monitoring and Evaluation Programme) background stations distributed throughout mainland Spain between 2001 and 2019. A study of hierarchical clusters was used to classify the stations into three main groups with similarities in yearly concentrations: GC (coastal location), GNC (north–central location), and GSE (southeastern location). The highest $PM_{10}$ concentrations were registered in summer. Annual evolution showed statistically significant decreasing trends in $PM_{10}$ concentration in all the stations covering a range from −0.21 to −0.50 µg m$^{-3}$/year for Barcarrota and Víznar, respectively. Through the Lamb classification, the weather types were defined during the study period, and those associated with high levels of pollution were identified. Finally, the values exceeding the limits established by the legislation were analyzed for every station assessed in the study.

**Keywords:** atmospheric pollution; EMEP; $PM_{10}$; trends; weather types



**Citation:** Pardo, N.; Sainz-Villegas, S.; Calvo, A.I.; Blanco-Alegre, C.; Fraile, R. Connection between Weather Types and Air Pollution Levels: A 19-Year Study in Nine EMEP Stations in Spain. *Int. J. Environ. Res. Public Health* **2023**, *20*, 2977. https://doi.org/10.3390/ijerph20042977

Academic Editor: Paul B. Tchounwou

Received: 13 January 2023
Revised: 3 February 2023
Accepted: 6 February 2023
Published: 8 February 2023



## 1. Introduction

The World Health Organization (WHO) states that suspended particulate matter (PM) is one of the atmospheric pollutants that produce the greatest effect on human health [1]. The concentration of aerosols depends on different variables, including weather conditions and atmospheric stability [2]. There are air quality studies based on meteorological conditions that show that the levels of the different pollutants vary depending on the weather types [3,4].

The transport of atmospheric pollutants acquires great relevance in air pollution studies, since the important role of meteorological phenomena comes into play. This transport is particularly evident in remote stations, located far from sources of anthropogenic pollution. The importance of these sampling points is highlighted through the creation of different programs focused on solving transboundary air pollution problems, among other tasks. Thus, in 1979, European authorities created the Convention on Long-Range Transboundary Air Pollution (LRTAP) [5]. This agreement addresses problems related to eutrophication and acidification, tropospheric ozone, heavy metals, persistent organic compounds, and PM. In Spain, the network with this mission is known as EMEP/VAG/CAMP related to AEMET (National Agency for Meteorology, in its Spanish acronym). The network is integrated by various programs: EMEP (European Monitoring and Evaluation Programme), VAG program (Vigilancia Atmosférica Global), and CAMP (Comprehensive Atmospheric Monitoring Programme).

Because fluctuations in atmospheric pollutant concentrations are closely related to meteorological parameters [6], the movement of air masses is a crucial aspect for the study of air pollution. Weather type classifications constitute an important approach for dealing with atmospheric circulation [7]. For this reason, including these classifications in the studies of atmospheric pollution transport is crucial. In recent years, several authors have published different automatic classification models for the Iberian Peninsula [8–12].

The circulation of the atmosphere can be studied from several approaches. One of them is the analysis of the weather types that affect a specific place at a specific time. The classification into weather types was initially based on the daily synoptic situation, using the shape of the isobars (which entails a certain level of subjectivity). However, currently, circulation is classified according to sea level pressure at several points close to the study area [10,13,14]. This type of classification has also been applied to the Iberian Peninsula [7–9,12,15–17], mostly in studies on precipitation.

In addition to purely meteorological studies, the classification into weather types has more recently been used to search for the contribution of atmospheric circulation to some pollution episodes [18], such as the Saharan dust intrusions in the Iberian Peninsula [4]. More recently, this classification has also been used, along with other techniques, for the control of aerosol pollution [19], for the search for sources and sinks of atmospheric aerosol [2,20,21] and for analyzing the influence of weather on health [22,23].

Long-term studies are of particular interest when continuous pollution data are available for long periods of time. With these studies, the effectiveness of national and international policies and regulations carried out to reduce levels of air pollution can be verified. So far, there are several studies that show a progressive decreasing trend in atmospheric pollutants, such as CO, $NO_x$, $SO_2$, $O_3$, and PM. Most of those studies are focused on urban areas: Guerreiro et al. [24] focus on 38 European cities during the period 2002–2011, and Gualtieri et al. [25] study the trends of different pollutants in the city of Florence, Italy, between 1993 and 2012. Querol et al. [26] focus on different Spanish cities between the years 2001 and 2012. However, studies related to background regions are still scarce [27,28]. Two different PMs are usually studied, $PM_{10}$ (PM with a diameter of 10 μm or less) and $PM_{2.5}$ (PM with a diameter of 2.5 μm or less). Both of them are pollutants that can lead to major environmental and public health problems. The European directives and regulations indicate that a daily $PM_{10}$ concentration of 50 $\mu g\ m^{-3}$ (European Union Official Diary, Directive 2008/50/EC of the European Parliament and of the Council of 21 May 2008) should not be surpassed more than 35 days a year. The threshold applied in the $PM_{2.5}$ case makes reference to a mean annual value of these concentrations instead of daily value. Although both pollutants have great relevance in air quality and atmospheric contamination studies, the analysis of $PM_{10}$ concentrations is more suitable when the daily exceedances or the relationship between $PM_{10}$ values and (daily) weather types needs to be studied since the threshold set by regulations includes also a daily value. In the context of the actual climatic situation, the mentioned long-term studies, as the one described here, are of great importance in the evaluation of pollution levels in areas far from emission sources in order to verify the real decrease in the concentrations of the main pollutants, particularly $PM_{10}$, as a result of the application of European emission reduction policies and plans of regional and local proposals by public environmental authorities.

Therefore, the present study focuses on the analysis of the daily $PM_{10}$ values as a means of finding the relationship between weather types and $PM_{10}$ concentration in several background stations located in mainland Spain from 2001 to 2019. The temporal trend of $PM_{10}$ has been analyzed during this period, and a cluster analysis has been carried out to group air quality stations with similar $PM_{10}$ concentrations. Finally, the relationship between $PM_{10}$ exceedances and weather types has also been evaluated.

## 2. Study Site

The study area covers the whole of mainland Spain through nine air quality stations [29] belonging to the Spanish network EMEP/VAG/CAMP between 2001 and 2019. All the stations analyzed in the study are background stations located far away from any possible anthropogenic source of pollution (Table 1, Figure 1). The geographic distribution of the stations is rather different. Four of them are not far from the Mediterranean coast

(Cabo de Creus, Els Torms, Zarra, and Víznar), two of them in the north (Cabo de Creus and Els Torms) and the other ones in the south. Niembro is located in the Cantabrian coast, while the remaining four stations (O Saviñao, Peñausende, Campisábalos, and Barcarrota) could be considered inland stations. In mainland Spain, most of the time, the effects of the North Atlantic anticyclone cause local thermal circulations to appear, such as coastal or mountain breezes [30]. On the other hand, the proximity with the north of Africa and Central and Eastern Europe makes particle-laden air masses come into the peninsula through these areas, which could be considered a potential source of anthropogenic pollution, particularly in the cases of Central and Eastern Europe [31], or natural pollution as in the case of Saharan dust intrusions from Northern Africa [32]. In addition, in the Mediterranean coast, where four stations are located (Figure 1), breeze episodes could favor the dispersion of pollutants during most of the year [33].

**Table 1.** Features of the background stations belonging to the network EMEP/VAG/CAMP and used in the study.

| No. | National Cod. | International Cod. | Name | Region | Latitude (deg.) | Longitude (deg.) | Altitude (m a.s.l) |
|---|---|---|---|---|---|---|---|
| 1 | 33036999 | ES0008R | Niembro | Asturias | 43.439 | −4.850 | 134 |
| 2 | 27058999 | ES0016R | O Saviñao | Lugo | 42.635 | −7.705 | 506 |
| 3 | 17032999 | ES0010R | Cabo de Creus | Girona | 42.319 | 3.316 | 23 |
| 4 | 25224999 | ES0014R | Els Torms | Lleida | 41.394 | 0.735 | 470 |
| 5 | 19061999 | ES0009R | Campisábalos | Guadalajara | 41.274 | −3.143 | 1360 |
| 6 | 49149999 | ES0013R | Peñausende | Zamora | 41.239 | −5.898 | 985 |
| 7 | 46263999 | ES0012R | Zarra | Valencia | 39.083 | −1.101 | 885 |
| 8 | 06016999 | ES0011R | Barcarrota | Badajoz | 38.473 | −6.924 | 393 |
| 9 | 18189999 | ES0007R | Víznar | Granada | 37.237 | −3.534 | 1230 |

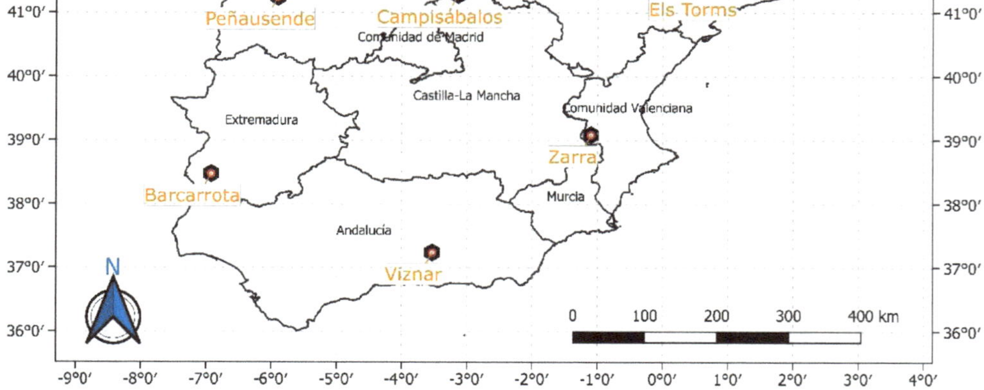


**Figure 1.** Location of the background stations used in the study. Administrative boundaries' layer retrieved from IGN (https://www.ign.es/web/rcc-info-delimitaciones (accessed on 1 September 2022)) and modified through QGIS.

## 3. Materials and Methods

### 3.1. Databases

Daily $PM_{10}$ data were obtained from the EBAS website (http://ebas.nilu.no/ (accessed on 2 September 2022)), developed and operated by the Norwegian Institute for Air Research (NILU). Information related to European stations used in the study (location, instrumentation, etc.) is retrieved from the EMEP website. All the stations are equipped with the same instrumentation, and the data are collected and analyzed following the same protocol established by the EMEP [34]. Daily data were collected at 0700 UTC and sent once a week to the Carlos III Health Institute, to the Atmospheric Pollution Department. A high-volume collector was used for $PM_{10}$ samplings, and its concentration determined by the gravimetric method.

The daily data of sea level pressure used to determine the weather types have been obtained from the website of the National Center for Atmospheric Research, available at https://ncar.ucar.edu/ (accessed on 2 September 2022).

### 3.2. Data Analysis

Two different statistical analyses were applied. First, a univariate analysis for describing the data and their distribution by calculating kurtosis, standard deviation, skewness, minimum, maximum, median, and variance (based on daily data). Second, a multivariate analysis to find similarities between stations by grouping them according to similar $PM_{10}$ values. This analysis was carried out using MATLAB to obtain a dendrogram, following the Ward aggregation method jointly with the Euclidean distance [35]. The smaller distances indicate a greater relationship between stations grouped in the same cluster [36,37].

### 3.3. Lamb Classification in Weather Types

The Lamb classification evaluates the daily circulation patterns in a way to find a method able to classify them. The methodology followed by Jenkinson and Collinson [38] and Jones et al. [13] to objectively define different weather types in the British Isles, and based on the Lamb classification, has been applied in mainland Spain. In this area, the application of this classification was developed by Spellman [10] and Trigo and DaCamara [11], who established the location of the network of points in the Iberian Peninsula. The application relies on the determination of several indices associated with the direction and vorticity of the geostrophic flux [39], resulting in a classification into 26 weather types, as shown in Table 2. For the whole study period, every single day was characterized by a specific weather type. This classification of the synoptic conditions was then used to find a possible relationship between $PM_{10}$ concentrations and weather types.

**Table 2.** Lamb classification of weather types.

| Weather Types | | | | | |
|---|---|---|---|---|---|
| Anticyclonic | | Directional | | Cyclonic | |
| A | Anticyclonic | | | C | Cyclonic |
| ANE | Anticyclonic northeasterly | NE | Northeasterly | CNE | Cyclonic northeasterly |
| AE | Anticyclonic easterly | E | Easterly | CE | Cyclonic easterly |
| ASE | Anticyclonic southeasterly | S | Southerly | CSE | Cyclonic southeasterly |
| AS | Anticyclonic southerly | SE | Southeasterly | CS | Cyclonic southerly |
| ASW | Anticyclonic southwesterly | SW | Southwesterly | CSW | Cyclonic southwesterly |
| AW | Anticyclonic westerly | W | Westerly | CW | Cyclonic westerly |
| ANW | Anticyclonic northwesterly | NW | Northwesterly | CNW | Cyclonic northwesterly |
| AN | Anticyclonic northerly | N | Northerly | CN | Cyclonic northerly |

### 3.4. Regulations, Limit Values, and Intrusions

For each station, daily values of $PM_{10}$ concentration exceeding the daily limit value of 50 $\mu g\ m^{-3}$ established by European directives were identified. Thanks to the information included in the annual reports provided by the Ministry for Ecological Transition and Demographic Challenge regarding to the identification of natural episodes of transboundary contributions of particles (African episodes), and other types of natural episodes, the previously identified days exceeding the daily $PM_{10}$ limit value have been linked or not to Saharan intrusion episodes. The methodology followed in those reports to identify days related to intrusion episodes coming from the Sahara desert, applying the HYSPLIT model together with other tools, is widely described at https://www.miteco.gob.es/es/calidad-y-evaluacion-ambiental/temas/atmosfera-y-calidad-del-aire/metodologiaparaepisodiosnaturales-revabril2013_tcm30-186522.pdf (accessed on 18 October 2022).

## 4. Results and Discussion

### 4.1. Analysis of the Representative Values for Each Station

Results for the univariate statistical analysis of $PM_{10}$ data are shown in Table 3. Víznar was the station with the lowest missing data for the whole period in contrast to Campisábalos, which presented almost 20% of missing data. Among all the stations included in the study, Cabo de Creus, Niembro, and Víznar stood out for presenting high $PM_{10}$ mean concentrations, with around 17 $\mu g\ m^{-3}$ or higher. Minimum averaged values were obtained for most of the stations located inland, such as Campisábalos, O Saviñao, and Peñausende. The highest daily $PM_{10}$ concentration was reached in Zarra, with 320 $\mu g\ m^{-3}$. However, this station was not the one presenting the highest mean concentration, which was recorded in Víznar. This fact proves that maximum values can greatly differ from mean values due to isolated episodes of high concentrations. Mean values are also affected by those isolated episodes of high concentrations, something that is corroborated by the high values of the standard deviation shown in Table 3. Therefore, median values better represent characteristic concentration values for each station. Stations presenting high maximum values were all located in the southern area of the peninsula (Zarra, Víznar, and Barcarrota). All stations present similar positive skewness values, which means all distributions are right-skewed. Similarly, all stations present positive kurtosis values (Table 3). This means that the concentrations registered are closely grouped together around a narrow interval close to the median.

**Table 3.** Averaged concentration (±standard deviation), maximum, minimum, median, kurtosis, skewness, and variance values of the data series covering the period from 2001 to 2019 for each weather station. Percentages of missing data per station are also shown.

| *STATION* / *POLLUTANT* | *BARCARROTA* | *CABO DE CREUS* | *CAMPISÁBALOS* | *ELS TORMS* | *NIEMBRO* | *O SAVIÑAO* | *PEÑAUSENDE* | *VÍZNAR* | *ZARRA* |
|---|---|---|---|---|---|---|---|---|---|
| **$PM_{10}$ ($\mu g\ m^{-3}$)** | **15.47 ± 11.48** | **17.54 ± 8.20** | **10.11 ± 10.09** | **14.89 ± 9.75** | **16.90 ± 9.27** | **11.35 ± 8.16** | **10.13 ± 8.75** | **18.31 ± 15.32** | **13.39 ± 10.75** |
| *Maximum* | 246 | 133 | 200 | 169 | 104 | 157 | 197 | 309 | 320 |
| *Minimum* | 0 | 0 | 0.5 | 2.0 | 1.0 | 1.0 | 1.0 | 1.0 | 1.0 |
| *Median* | 12 | 16 | 8 | 13 | 15 | 9 | 8 | 15 | 11 |
| *Kurtosis* | 52.38 | 18.46 | 51.86 | 23.35 | 6.76 | 33.48 | 61.85 | 44.29 | 126.45 |
| *Skewness* | 4.64 | 2.56 | 4.90 | 3.16 | 1.84 | 3.78 | 5.11 | 4.33 | 6.62 |
| *Variance* | 131.80 | 67.28 | 101.82 | 95.15 | 85.94 | 66.61 | 76.48 | 234.96 | 115.64 |
| *% Missing Values* | 13.3% | 9.6% | 18.5% | 12.6% | 10.5% | 14.8% | 12.7% | 7.2% | 9.5% |

Results of the multivariate statistical analysis led to the clustering shown in the dendrogram in Figure 2. Average values for each month and station were taken into consideration in the dendrogram. Finally, three main groups of stations located in the coastal (Cabo de Creus and Niembro), north–central (Campisábalos, O Saviñao, and Peñausende), and southeastern (Barcarrota, Els Torms, Víznar, and Zarra) areas were obtained. For later references, these groups were named as GC (coastal), GNC (north–central), and GSE (southeastern). The results of the dendrogram vary considerably, depending on the input values that are used for the calculation. Results for dendrogram tests with different input values are not shown in this paper. However, there are three stations that always remain in the same cluster regardless of whether the input values were monthly or yearly or included other types of pollutants. These stations are Campisábalos, O Saviñao, and Peñausende. As seen in Table 1 and Figure 1, Niembro and Cabo de Creus are the stations nearest to the coast and characterized by the minimum altitude, their location being almost at sea level.

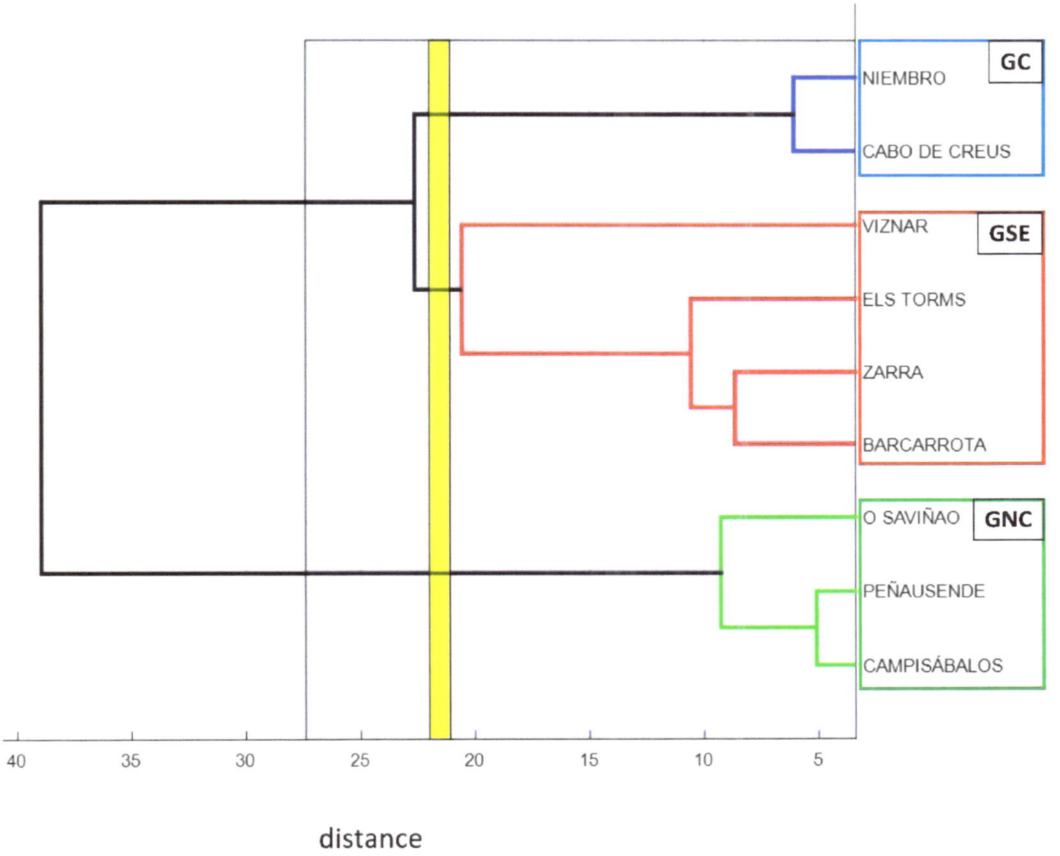


**Figure 2.** Dendrogram showing the clusters regarding similarities in $PM_{10}$ monthly concentration values.

### 4.2. Monthly and Annual Evolution

Figure 3 shows the monthly and yearly evolution of $PM_{10}$ for the different groups established. This evolution is markedly different between the different groups of stations. Figure 3b, where GC is represented, shows a practically constant value during all the months with a slight decrease in winter. In contrast, GNC (Figure 3a) and GSE (Figure 3c) present a monthly evolution with a slight increase in the concentrations from January to March, followed by a slight decrease in April, which gives way to a continuous increase from May to July/August when the maximum value is reached. The slight decrease in April can be explained by the increasing frequencies of the advections of Atlantic air masses associated with high rates of precipitation [30] during this month.

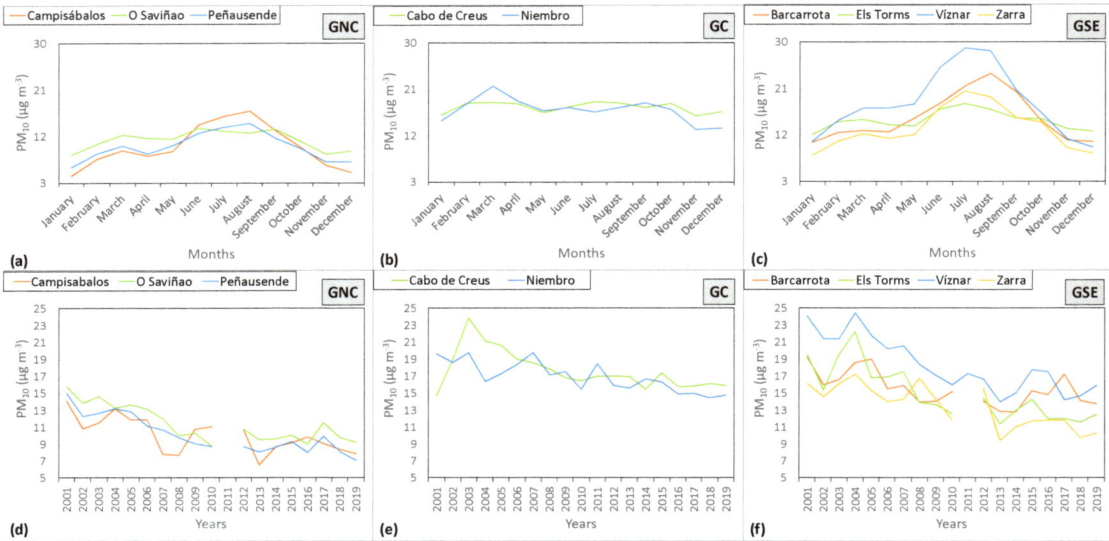


**Figure 3.** Monthly (**a–c**) and yearly (**d–f**) evolution of $PM_{10}$ for the clustering group and station for the study period 2001–2019.

Considering a common range for all the groups (Figure 3) covering from 3 to 30 μg $m^{-3}$, for the monthly evolution, GNC (Figure 3a) maintains its values practically within the first third range, GC (Figure 3b) presents almost a constant monthly evolution with values corresponding to the second third, and GSE (Figure 3c) covers the entire range of values. GNC and GSE present minimum values in winter and maximum ones in summer. There is a clear difference between them, the maximum value (with the exceptions of O Saviñao and Barcarrota) is reached in August for GNC and in July for GSE, reaching maximum values greater than those stations belonging to GSE, particularly Víznar with a value of 28.97 μg $m^{-3}$ in July.

$PM_{10}$ concentrations are closely related to meteorological conditions. During summer, when the maximum concentration is reached, anticyclonic situations are frequent with a low capacity of air mass renewal, and consequently, a low dispersion of pollutants is registered. These conditions also favor the resuspension of soil particles. During winter, these conditions are much less frequent, and therefore, the mean concentrations are much lower [30]. Besides, the intrusions of Saharan dust notably affect the stations located in the south and east of the Iberian Peninsula. There is a special incidence of intrusions during summer [40,41], and this must be taken into account when analyzing the results obtained and presented in Figure 3.

Annual averaged $PM_{10}$ values were calculated for each station and the whole study period, and the results are shown in Figure 3d–f. The differences between groups rely on the concentration levels, GNC being the one with a lower level, as mentioned above (Figure 3a–c). Although the trend has been downward, some stations have increased their concentrations in the past few years, particularly the stations in GSE. The quantification of these downward trends has been computed using a linear regression (Figure 4). Decreasing rates seem not to be related to the established groups, and differences could be found between stations even in the same group. All stations have shown a statistically significant decreasing trend in their $PM_{10}$ concentrations with values for the determination coefficient ($R^2$) from 0.35 to 0.78 for Cabo de Creus and Peñausende, respectively.

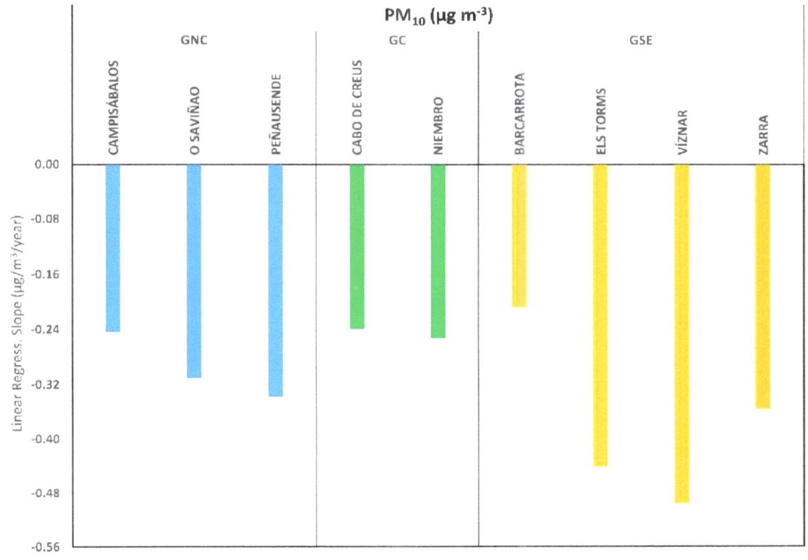


**Figure 4.** Annual trends by clustering group and station for the study period 2001–2019.

Figure 4 presents the stations alphabetically ordered within their group. Even within the same group, not all stations show the same trend in magnitude, and it should be noted that a common behavior cannot be found when analyzing decreasing trends. The values show that the station with the strongest trend is Víznar with a decrease of approximately 0.5 µg $m^{-3}$ per year, followed by Els Torms and Zarra, all of them in the same group (GSE). Similar results were obtained by Querol et al. [26] in the same EMEP stations for a study period between 2001 and 2012.

In the Iberian Peninsula, $PM_{10}$ is generally more affected by African intrusions, while $PM_{2.5}$ is more affected by anthropogenic emissions into the atmosphere [26]. Therefore, the application of European strategies to reduce pollution, as well as the impact of the financial crisis on southern Europe that has originated a sharp decrease in pollutant emissions [42,43], could have more impact on $PM_{2.5}$ concentrations than on $PM_{10}$ concentrations, even when the decreasing rates for $PM_{10}$ are quite high (Figure 4).

### 4.3. *$PM_{10}$ and Regulations*

Taking into consideration the threshold set by the European directives and regulations, for the study period of 19 years, none of the nine stations surpassed the value of 35 days a year in the $PM_{10}$ concentration case. The maximum number of days surpassing that value was around 10 for almost half the stations, with the exception of Peñausende and O Saviñao (both in GSE) with lower values. Only two stations, Els Torms and Víznar, registered more days with values over 50 µg $m^{-3}$, 15 and 25 days, respectively, in 2003. When considering the whole study period, Víznar stood out as the station with the highest number of days surpassing the threshold value (196 days), far from the ones registered in the other stations. However, its mean concentration during these days remains similar to the ones registered in the other stations with a value of 73.9 µg $m^{-3}$. On the other hand, the station with the smallest number of exceedances was O Saviñao with 25 days in the total study period and a mean concentration of 71.9 µg $m^{-3}$, followed by Cabo de Creus, with 38 days surpassing the daily threshold value, and with a mean concentration of 66.9 µg $m^{-3}$.

In the background stations used for this study, the anthropogenic influence is minimal, so the isolated cases when the daily $PM_{10}$ concentration threshold was surpassed could be related to Saharan dust intrusions [32] or to the presence of PM coming from Central

Europe [44]. Consequently, the episodes of Saharan dust intrusions were studied, as they constitute one of the main causes of the exceedances in the daily threshold value. Figure 5 presents the percentage of exceedances related to intrusions in each station for the period 2005–2019. Only this period has been analyzed due to the lack of data from 2001 to 2004. For the 2005–2019 period, both exceedances and intrusions data were available in the URL of the Spanish Ministry for Ecological Transition and Demographic Challenge [45].

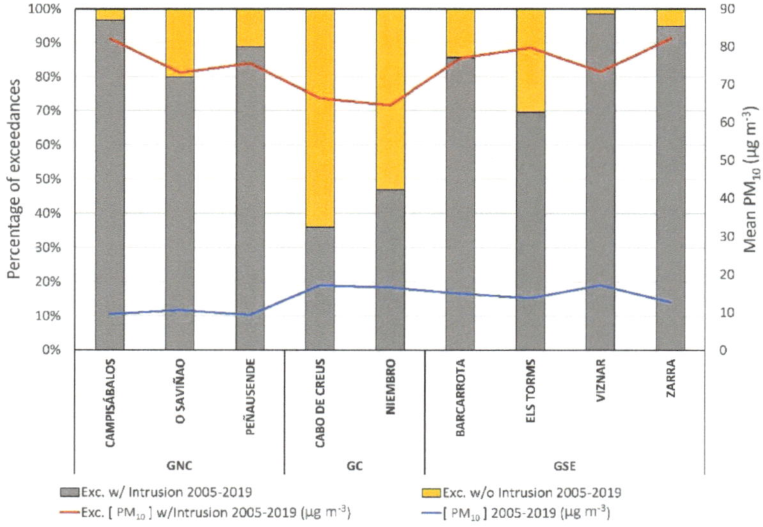


**Figure 5.** Relationship between the percentage of days exceeding the daily threshold value related (gray column) or not (orange column) with Saharan dust intrusions for the period 2005–2019. Mean $PM_{10}$ concentrations for the days exceeding the daily threshold value for the whole measuring period (blue line) and for the values related to intrusions (red line) are also shown.

The mean $PM_{10}$ concentrations computed for the total sampling period or only for the days exceeding the daily threshold value greatly differ, and results are also shown in Figure 5. For the total period, mean $PM_{10}$ ranged between 9.3 and 17.2 µg m$^{-3}$, with small differences among stations. However, in the case of exceedances, $PM_{10}$ concentration soared, reaching values of 66.2 to 82.1 µg m$^{-3}$. Complementarily, the analysis of the exceedances of the threshold value combined with the episodes of intrusions has been carried out. Except in the case of the stations belonging to GC, in the rest of the cases, it has been observed that the exceedances are highly linked to episodes of Saharan dust intrusions (Figure 5). Particularly, stations located in the south of mainland Spain, such as Víznar and Zarra (Figure 5), were greatly affected by those Saharan intrusions [46]. Two clear exceptions can be seen, Cabo de Creus and Niembro, both in GC, for which over 50% of the exceedances were not related to episodes of Saharan dust intrusions (Figure 5). It is in these two stations where the average values (with daily $PM_{10}$ over 50 µg m$^{-3}$) were also lower than in the rest of the stations (red line in Figure 5).

### *4.4. Relationship between $PM_{10}$ and Weather Types*

Figure 6 shows the total frequency of the Lamb weather types for the period 2001–2019 in the study area. The different types are classified into three main groups: anticyclonic, directional, and cyclonic. The prevalent type is A, accounting 22% of the total for the whole study period, followed by NE and N types, with 12% and 8%, respectively. The most dominant weather type in the area is, thus, the A type, followed by almost all directional types. Cyclonic types are less frequent, with frequencies below 2%, with the exception of the C type.

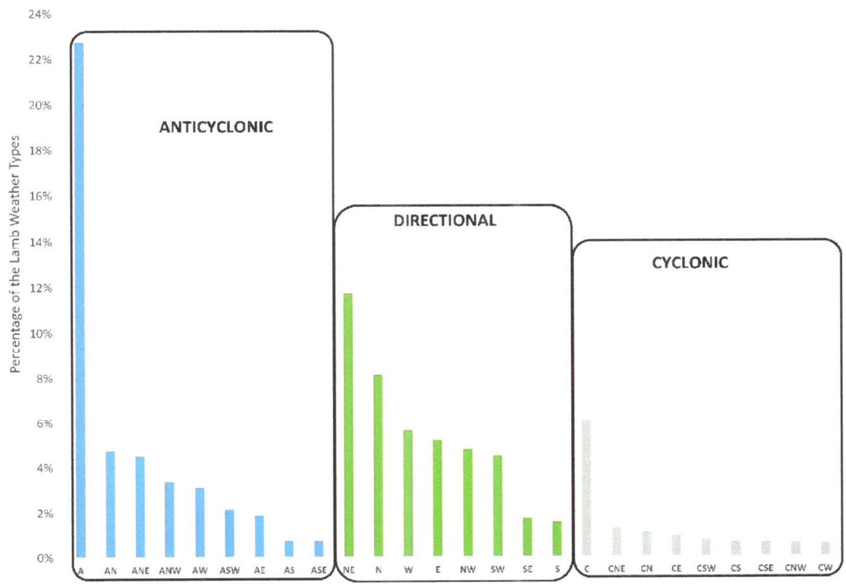


**Figure 6.** Frequency of each weather type for the period 2001–2019.

In this study, the three most frequent weather types (A, NE, and N) are characterized by atmospheric stability. A similar result was obtained by Grimalt et al. [47], who studied the distribution of the weather types in the Mediterranean basin, where several of the stations studied in this paper are located, finding out as a result that the most frequent type was A, followed by C. The weather types depend on various meteorological parameters and help us study their influence on the concentration of pollutants. Among all those types, for example, the anticyclonic is the one that originates the most relevant pollution scenarios [48], being the weather type prevailing in the Iberian Peninsula during the study period (Figure 6).

The results of the analysis of the relationship between $PM_{10}$ concentrations and weather types are shown in Figure 7. This figure shows the maximum and minimum averaged concentration values for each station and weather type. Mean $PM_{10}$ values were calculated for every weather type and station for the period 2005–2019, and maximum and minimum mean $PM_{10}$ values were identified for every station. Only weather types corresponding to some maximum or minimum value are presented in the graphs (Figure 7). Almost half of the highest concentrations were measured under CE conditions, with the remaining ones a bit more dispersed between different weather types, depending on the station. The weather type that presents the minimum $PM_{10}$ mean concentration is the CW type for more than half of the stations except for those in GC: Niembro (NW type) and Cabo de Creus (ANW). The other two exceptions are Campisábalos (W) and Víznar (CSW). This result shows that minimum and maximum mean concentrations tend to concentrate around the same weather type.

Spellman [10] states that the C-type during summer is related to low pressure systems coming from central Sahara. However, the results presented in Figure 7 show that the C type does not stand out in any station as one of the types where maximum or minimum concentrations were measured.

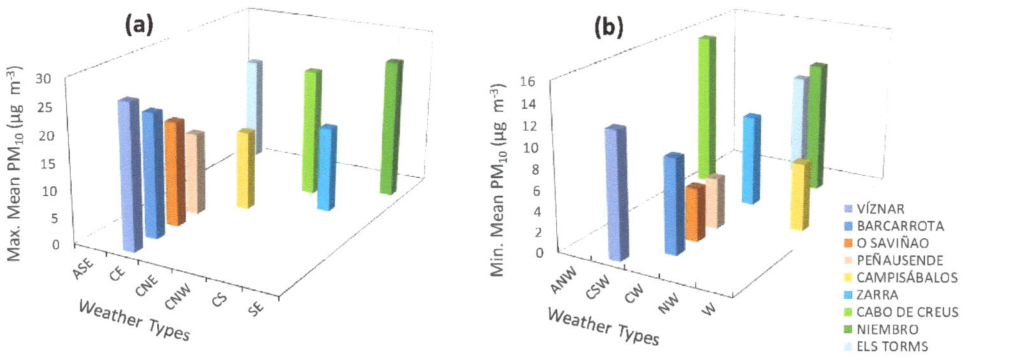


**Figure 7.** Distribution of the weather types related to maximum (**a**) and minimum (**b**) mean $PM_{10}$ for the period 2005–2019. Each color indicates a single station. Only weather types with the maximum or minimum values have been represented.

A detailed analysis of the exceedances in the daily $PM_{10}$ concentration (>50 µg $m^{-3}$) has also been carried out, considering the relationship between those exceedances and the weather types. The cases where the exceedances are linked to Saharan dust intrusions have also been investigated. Figure 8 shows the relationship between the number of episodes for each weather type and the total number of $PM_{10}$ exceedances during the whole study period. As in the results shown in Figure 6, the weather type with the highest number of episodes is A, followed by the NE, C, SW, and N types with values of 53, 41, 41, 32, and 32, respectively. Since the number of episodes related to each weather type is not the same, the fact that the A type is the one that presents the greatest number of exceedances is not indicative of a strong linkage to $PM_{10}$ exceedances. Because A is the most frequent type, it was expected to present the highest number of exceedances. For this reason, an intercomparison of the weight of each weather type against the number of exceedances and the $PM_{10}$ concentration must be performed.

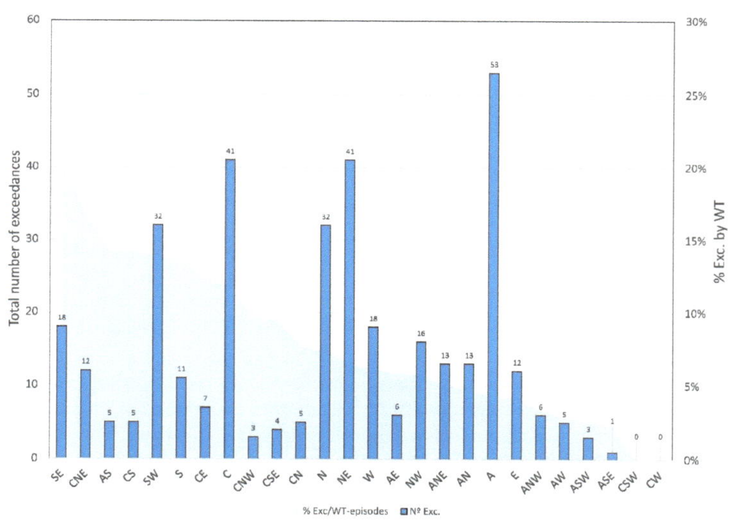


**Figure 8.** Total number of exceedances ($PM_{10}$ > 50 µg $m^{-3}$) for each weather type in the period 2005–2019, and ratio (in percentage) between the number of exceedances and the total number of episodes for each weather type.

Taking into consideration the real weight of each weather type in the exceedance episodes, a more real representativeness of each weather type related to $PM_{10}$ exceedances can be obtained. Figure 8 shows the ratio between the exceedances and the frequency of weather types (blue area). The type with the highest normalized number of exceedances is SE, stating that 20.2% of the episodes of this weather type were associated with exceedances of the daily $PM_{10}$. The A type is very far from this value, with less than 5% of the episodes of this weather type related to $PM_{10}$ exceedances. It should be noted that there are three weather types that have presented one or no exceedance at all during the whole study period: ASE, CSW, and CW. There are other weather types that present similar characteristics, but still present exceedances and a high ratio when the normalized values are analyzed, as shown in Figure 8 (CNW, AS, CS, and CSE). Similar results for the weather types related to C have been obtained by other authors when the relationship between this type and rainfall has been studied [11]. Fernández-González et al. [7] also studied the relationship between weather and precipitation, stating that the C, W, and SW types provide more rainfall than the remaining types.

### 4.5. Relationship between $PM_{10}$ and Other Variables

First, an analysis seeking a possible relationship between $PM_{10}$ and altitude, longitude, and latitude was carried out. For the period 2005–2019, the number of $PM_{10}$ exceedances of the daily threshold value was counted, and the mean $PM_{10}$ concentrations above that limit were computed for each air quality station, only for episodes of Saharan dust intrusions. In the case of longitude, no correlation was found between these variables ($R^2 < 0.03$). However, latitude and altitude presented a correlation with the number of days of exceedances ($R^2 = 0.68$) and with the average $PM_{10}$ (without intrusion) concentrations ($R^2 = 0.44$), respectively, both statistically significant for a significance level of 0.05. The correlation found is positive in the case of altitude (slope = +0.01 μg $m^{-3}$/m.a.s.l.) and negative in the case of latitude (slope = −14.73 number of exceedances with intrusion/deg.), which seems coherent: the greater the distance from the Sahara, the lower the frequency of intrusions.

Second, the relationship between wind direction and $PM_{10}$ was analyzed. The various weather types classified in Table 2 were grouped according to the prevailing wind direction into eight categories, excluding the A and C types. The number of exceedances under the presence of an intrusion episode was related to those categories, and the results are shown in Figure 9. For the stations belonging to GC, there is no common prevailing direction related to exceedances of the $PM_{10}$ daily threshold value. However, for GSE and GNC, the NE–SW direction seems to be the prevailing direction, although for GSE, also the weather types characterized by a north component are related to exceedance episodes, especially in Víznar.

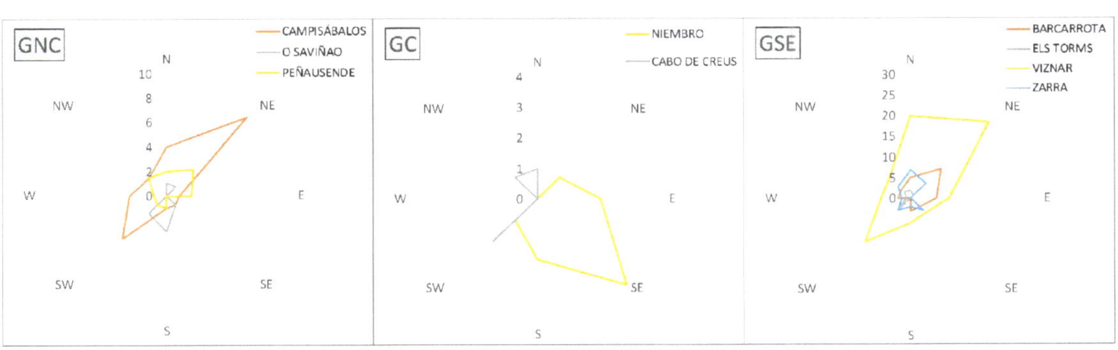


**Figure 9.** Wind roses showing the number of $PM_{10}$ exceedances related to Saharan intrusions for the period 2005–2019.

## 5. Conclusions

A full analysis of $PM_{10}$ has been made for nine background EMEP stations covering the whole of mainland Spain. Three groups have been established, clustering the stations with similar characteristics and pollutant concentrations using a dendrogram: GC (coastal location), GNC (north–central location), and GSE (south–eastern location).

The lowest mean concentrations were registered for stations in GNC, and the highest concentrations were registered in stations belonging to the GSE group. Maximum values for $PM_{10}$ were reached during summer, mainly influenced by meteorological conditions and Saharan dust intrusion episodes. In addition, a constant negative trend was found for each station analyzed from the beginning of the data series in 2001 until 2019.

The two stations that stand out for presenting the highest number of exceedances related to intrusions were Campisábalos and Víznar. In contrast, the two stations with the highest number of exceedances without direct relation to an intrusion were Cabo de Creus and Niembro.

The prevailing weather type in mainland Spain is anticyclonic; as a result, this is the weather type presenting the highest number of exceedances. Since it is the dominant weather type, it is logical to expect that it will be the type that also presents the highest number of exceedances. Despite this, the anticyclonic weather type is not the one characterized by most exceedances. When taking into consideration the weight of every weather type regarding the total of measurements, it is the southeasterly directional type, the one for which every episode is related to exceedances, followed by the cyclonic northeasterly type.

When computing mean $PM_{10}$ concentrations for every weather type and station, results state that minimum and maximum mean concentrations tend to concentrate around the same weather type, cyclonic westerly and cyclonic easterly, respectively.

**Author Contributions:** Conceptualization, R.F. and A.I.C.; methodology, N.P., S.S.-V., R.F. and A.I.C.; formal analysis, N.P. and S.S.-V.; investigation, N.P., S.S.-V., C.B.-A., R.F. and A.I.C.; resources, N.P., S.S.-V., C.B.-A., R.F. and A.I.C.; data curation, N.P. and S.S.-V.; writing—original draft preparation, N.P.; writing—review and editing, N.P., R.F. and A.I.C.; visualization, N.P.; supervision, R.F. and A.I.C.; project administration, R.F. and A.I.C.; funding acquisition, R.F. and A.I.C. All authors have read and agreed to the published version of the manuscript.

**Funding:** This work was partially supported by the Junta de Castilla y Leon (Grant LE025P20) and by the AEROHEALTH and C3HEALTH projects (Ministry of Science and Innovation, Grants PID2019-106164RBI00 and TED2021-132292B-I00). The study was also cofinanced with FEDER funds.

**Institutional Review Board Statement:** Not applicable.

**Informed Consent Statement:** Not applicable.

**Data Availability Statement:** The results presented in this study have been processed from the data available on the websites mentioned in the main text.

**Acknowledgments:** Data about Saharan dust intrusions, owned by the Spanish Ministry for Ecological Transition and Demographic Challenge, were provided under the "Commission of the Ministry for ecological transition to the State Agency Higher Council for Scientific Research for the detection of natural episodes of cross-border particle inputs and other sources of contamination of particulate matter, and tropospheric ozone formation". Data about air pollutant concentrations were provided by EBAS and EMEP. Noelia Ramón patiently checked the English version.

**Conflicts of Interest:** The authors declare no conflict of interest. The funders had no role in the design of the study; in the collection, analyses, or interpretation of data; in the writing of the manuscript; or in the decision to publish the results.

## References


1. Henschel, S.; Chan, G. *Health Risks of Air Pollution in Europe-HRAPIE Project: New Emerging Risks to Health from Air Pollution-Results from the Survey of Experts*; World Health Organization Regional Office for Europe: Copenhagen, Denmark, 2013; p. 65.
2. Oduber, F.; Calvo, A.I.; Blanco-Alegre, C.; Castro, A.; Alves, C.; Cerqueira, M.; Lucarelli, F.; Nava, S.; Calzolai, G.; Martin-Villacorta, J.; et al. Towards a model for aerosol removal by rain scavenging: The role of physical-chemical characteristics of raindrops. *Water Res.* **2021**, *190*, 116758. [CrossRef] [PubMed]
3. Saturtún, A.; González-Hidalgo, J.C.; Sánchez-Lorenzo, A.; Zarrabeitia, M.T. Surface ozone concentration trends and its relationship with weather types in Spain (2001–2010). *Atmos. Environ.* **2015**, *101*, 10–22. [CrossRef]
4. Russo, A.; Sousa, P.M.; Durão, R.M.; Ramos, A.M.; Salvador, P.; Linares, C.; Díaz, J.; Trigo, R.M. Saharan dust intrusions in the Iberian Peninsula: Predominant synoptic conditions. *Sci. Total Environ.* **2020**, *717*, 137041. [CrossRef] [PubMed]
5. EMEP. EMEP: History and Structure. Available online: http://emep.int/emep_overview.html (accessed on 12 September 2022).
6. Padró-Martínez, L.T.; Patton, A.P.; Trull, J.B.; Zamore, W.; Brugge, D.; Durant, J.L. Mobile monitoring of particle number concentration and other traffic-related air pollutants in a near-highway neighbourhood over the course of a year. *Atmos. Environ.* **2012**, *61*, 253–264. [CrossRef]
7. Fernández-González, S.; del Río, S.; Castro, A.; Penas, A.; Fernández-Raga, M.; Calvo, A.I.; Fraile, R. Connection between NAO, weather types and precipitation in León, Spain. *Int. J. Climatol.* **2012**, *32*, 2181–2196. [CrossRef]
8. Goodess, C.M.; Palutikof, J.P. Development of daily rainfall scenarios for southeast Spain using a circulation-type approach to downscalling. *Int. J. Climatol.* **1998**, *18*, 1051–1083. [CrossRef]
9. Corte-Real, J.; Qian, B.; Xu, H. Circulation patterns, daily precipitation in Portugal and implications for climate change simulated by the second Hadley Centre GCM. *Clim. Dyn.* **1999**, *15*, 921–935. [CrossRef]
10. Spellman, G. The application of an objective weather-typing system to the Iberian Peninsula. *Weather* **2000**, *55*, 375–385. [CrossRef]
11. Trigo, R.M.; DaCamara, C.C. Circulation weather types and their influence on the precipitation regime in Portugal. *Int. J. Climatol.* **2000**, *20*, 1559–1581. [CrossRef]
12. Goodess, C.M.; Jones, P.D. Links between circulation and changes in the characteristics of Iberian rainfall. *Int. J. Climatol.* **2002**, *22*, 1593–1615. [CrossRef]
13. Jones, P.D.; Hulme, M.; Briffa, K.R. A comparison of Lamb circulation types with an objective classification scheme. *Int. J. Climatol.* **1993**, *13*, 655–663. [CrossRef]
14. Huth, R.; Beck, C.; Philipp, A.; Demuzere, M.; Ustrnul, Z.; Cahynova, M.; Kysely, J.; Tveito, O.E. Classifications of atmospheric circulation patterns: Recent Advances and Applications. In Trends and directions in climate research. *Ann. N. Y. Acad. Sci.* **2008**, *1146*, 150–152. [CrossRef]
15. Paredes, D.; Trigo, R.M.; García-Herrera, R.; Trigo, I.F. Understanding precipitation changes in Iberia in early spring: Weather typing and storm-tracking approaches. *J. Hydrometerol.* **2006**, *7*, 101–113. [CrossRef]
16. Lorenzo, M.N.; Taboada, J.J.; Gimeno, L. Links between circulation weather types and teleconnection patterns and their influence on precipitation patterns in Galicia (NW Spain). *Int. J. Climatol.* **2008**, *28*, 1493–1505. [CrossRef]
17. Queralt, S.; Hernández, E.; Barriopedro, D.; Gallego, D.; Ribera, P.; Casanova, C. North Atlantic Oscillation influence and weather types associated with winter total and extreme precipitation events in Spain. *Atmos. Res.* **2009**, *94*, 675–683. [CrossRef]
18. Kou, X.; Peng, Z.; Zhang, M.; Zhang, N.; Lei, L.; Zhao, X.; Miao, S.; Li, Z.; Ding, Q. Assessment of the Meteorological Impact on Improved $PM_{2.5}$ Air Quality Over North China During 2016–2019 Based on a Regional Joint Atmospheric Composition Reanalysis Data-Set. *J. Geophys. Res. Atmos.* **2021**, *126*, 1325–1337. [CrossRef]
19. Yan, Y.; Zhou, Y.; Kong, S.; Lin, J.; Wu, J.; Zheng, H.; Zhang, Z.; Song, A.; Bai, Y.; Ling, Z.; et al. Effectiveness of emission control in reducing $PM_{2.5}$ pollution in central China during winter haze episodes under various potential synoptic controls. *Atmos. Chem. Phys.* **2021**, *21*, 3143–3162. [CrossRef]
20. Calvo, A.I.; Pont, V.; Olmo, F.J.; Castro, A.; Alados-Arboledas, L.; Vicente, A.M.; Fernández-Raga, M.; Fraile, R. Air masses and weather types: A useful tool for characterizing precipitation chemistry and wet deposition. *Aerosol Air Qual. Res.* **2012**, *12*, 856–878. [CrossRef]
21. Oduber, F.; Calvo, A.I.; Castro, A.; Blanco-Alegre, C.; Alves, C.; Calzolai, G.; Nava, S.; Lucarelli, F.; Nunes, T.; Barata, J.; et al. Characterization of aerosol sources in León (Spain) using Positive Matrix Factorization and weather types. *Sci. Total Environ.* **2021**, *754*, 142045. [CrossRef]
22. Fernández-Arróyabe, P.; Marti-Ezpeleta, A.; Royé, D.; Santurtún-Zarrabeitia, A. Effects of circulation weather types on influenza hospital admissions in Spain. *Int. J. Biometeorol.* **2021**, *65*, 1325–1337. [CrossRef]
23. Li, J.; Ma, Y.; Cheng, B.; Zhang, Y.; Guo, Y.; Zhao, Y. Circulation weather types and hospital admissions for cardiovascular disease in Changchun, China. *Environ. Geochem. Health* **2022**, *44*, 2799–2813. [CrossRef]
24. Guerreiro, C.; Foltescu, V.; de Leeuw, F. Air quality status and trends in Europe. *Atmos. Environ.* **2014**, *98*, 376–384. [CrossRef]
25. Gualtieri, G.; Crisci, A.; Tartaglia, M.; Toscano, P.; Vagnoli, C.; Andreini, B.P.; Gioli, B. Analysis of 20-year air quality trends and relationship with emission data: The case of Florence (Italy). *Urban Clim.* **2014**, *10 P3*, 530–549. [CrossRef]
26. Querol, X.; Alastuey, A.; Pandol, M.; Reche, C.; Pérez, N.; Minguillón, M.C.; Moreno, T.; Viana, M.; Escudero, M.; Orio, A.; et al. 2001–2012 trends on air quality in Spain. *Sci. Total Environ.* **2014**, *490*, 957–969. [CrossRef] [PubMed]
27. Calvo, A.I.; Olmo, F.J.; Lyamani, H.; Alados-Arboledas, L.; Castro, A.; Fernández-Raga, M.; Fraile, R. Chemical composition of wet precipitation at the background EMEP station in Víznar (Granada, Spain) (2002–2006). *Atmos. Res.* **2010**, *96*, 408–420. [CrossRef]

28. Rua Vieites, A.; Hernández, E.; Parras, J.; Martín, I.; Gimeno, L. Sources of $SO_2$, $SO_4^{2-}$, $NO_x$, and $NO_{3-}$ in the air of four Spanish remote stations. *J. Air Waste Manag. Assoc.* **2011**, *48*, 838–845. [CrossRef]
29. MITECO (2022) Análisis de la Calidad Del Aire en España. Available online: https://www.miteco.gob.es/es/calidad-y-evaluacion-ambiental/temas/atmosfera-y-calidad-del-aire/calidad-del-aire/evaluacion-datos/redes/ (accessed on 12 September 2022).
30. Querol, X.; Alastuey, A.; Rodríguez, S.; Viana, M.; Artíñano, B.; Salvador, P.; Mantilla, E.; Do Santos, S.G.; Patier, R.F.; De La Rosa, J.; et al. Levels of particulate matter in rural, urban and industrial sites in Spain. *Sci. Total Environ.* **2004**, *334–335*, 359–376. [CrossRef]
31. Rodríguez, S.; Querol, X.; Alastuey, A.; Kallos, G.; Kakaliagou, O. Saharan dust contributions to $PM_{10}$ and TSP levels in Southern and Eastern Spain. *Atmos. Environ.* **2001**, *35*, 2433–2447. [CrossRef]
32. Krasnov, H.; Katra, I.; Friger, M. Increase in dust storm related $PM_{10}$ concentrations: A time series analysis of 2001–2015. *Environ. Pollut.* **2016**, *213*, 36–42. [CrossRef]
33. Gangoiti, G.; Millán, M.M.; Salvador, R.; Mantilla, E. Long-range transport and re-circulation of pollutants in the western Mediterranean during the project Regional Cycles of Air Pollution in the West-Central Mediterranean Area. *Atmos. Environ.* **2001**, *35*, 6267–6276. [CrossRef]
34. EMEP. *Manual for Sampling and Chemical Analysis*; Norwegian Meteorological Institute: Oslo, Norway, 2001.
35. Pérez, C. *Técnicas de Análisis de Datos Con SPSS.*; Pearson Educación Pearson Prentice Hall: Madrid, Spain, 2009.
36. Giri, D.; Murthy, V.K.; Adhikary, P.R.; Khanal, S.N. Cluster analysis applied to atmospheric $PM_{10}$ concentration data for determination of sources and spatial patterns in ambient air-quality of Kathmandu Valley. *Curr. Sci. India* **2007**, *93*, 684–688.
37. Jato-Espino, D.; Castillo-Lopez, E.; Rodriguez-Hernandez, J.; Ballester-Muñoz, F. Air quality modelling in Catalonia from a combination of solar radiation, surface reflectance and elevation. *Sci. Total Environ.* **2018**, *624*, 189–200. [CrossRef]
38. Jenkinson, A.F.; Collison, F.P. An Initial Climatology of Gales over the North Sea. In *Synoptic Climatology Branch Memorandum*; Meteorological Office: Bracknell, UK, 1977; Volume 62, p. 18.
39. Fernández-Raga, M.; Fraile, R.; Palencia, C.; Marcos, E.; Castañón, A.M.; Castro, A. The role of weather types in assessing the rainfall key factors for erosion in two different climatic regions. *Atmosphere* **2020**, *11*, 443. [CrossRef]
40. Querol, X.; Pey, J.; Pandolfi, M.; Alastuey, A.; Cusack, M.; Pérez, N.; Moreno, T.; Viana, M.; Mihalopoulos, N.; Kallos, G.; et al. African dust contributions to mean ambient $PM_{10}$ mass-levels across the Mediterranean Basin. *Atmos. Environ.* **2009**, *43*, 4266–4277. [CrossRef]
41. Pay, M.T.; Jiménez-Guerrero, P.; Jorba, O.; Basart, S.; Querol, X.; Pandolfi, M.; Baldasano, J.M. Spatio-temporal variability of concentrations and speciation of particulate matter across Spain in the CALIOPE modeling system. *Atmos. Environ.* **2012**, *46*, 376–396. [CrossRef]
42. Cusack, M.; Alastuey, A.; Pérez, N.; Pey, J.; Querol, X. Trends of particulate matter ($PM_{2.5}$) and chemical composition at a regional background site in the Western Mediterranean over the last nine years (2002–2010). *Atmos. Chem. Phys.* **2012**, *12*, 8341–8357. [CrossRef]
43. Cerro, J.C.; Cerdà, V.; Pey, J. Trends of air pollution in the Western Mediterranean Basin from a 13-year database: A research considering regional, suburban and urban environments in Mallorca (Balearic Islands). *Atmos. Environ.* **2015**, *103*, 138–146. [CrossRef]
44. Escudero, M.; Querol, X.; Ávila, A.; Cuevas, E. Origin of the exceedances of the European daily PM limit value in regional background areas of Spain. *Atmos. Environ.* **2007**, *41*, 730–744. [CrossRef]
45. MITECO. Available online: https://www.miteco.gob.es/es/calidad-y-evaluacion-ambiental/temas/atmosfera-y-calidad-del-aire/calidad-del-aire/documentacion-oficial/Analisis-CA.aspx (accessed on 30 August 2021).
46. Pey, J.; Querol, X.; Alastuey, A.; Forastiere, F.; Stafoggia, M. African dust outbreaks over the Mediterranean Basin during 2001-2011: $PM_{10}$ concentrations, phenomenology and trends, and its relation with synoptic and mesoscale meteorology. *Atmos. Chem. Phys.* **2013**, *13*, 1395–1410. [CrossRef]
47. Grimalt, M.; Tomàs, M.; Alomar, G.; Martin-Vide, J.; Moreno-García, M.C. Determination of the Jenkinson and Collinson's weather types for the western Mediterranean basin over the 1948–2009 period. *Temporal Anal. Atmós.* **2013**, *26*, 75–94.
48. Dharshana, K.G.T.; Kravtsov, S.; Kahl, J.D.W. Relationship between synoptic weather disturbances and particulate matter air pollution over the United States. *J. Geophys. Res.* **2010**, *115*, e0187933. [CrossRef]

International Journal of
*Environmental Research and Public Health*

MDPI

*Article*

# Physico-Chemical Properties and Deposition Potential of $PM_{2.5}$ during Severe Smog Event in Delhi, India

**Sadaf Fatima [1,2], Sumit Kumar Mishra [1,2,*], Ajit Ahlawat [3] and Ashok Priyadarshan Dimri [4,5]**

1 CSIR-National Physical Laboratory, New Delhi 110012, India
2 Academy of Scientific and Innovative Research (AcSIR), Ghaziabad 201002, India
3 Atmospheric Chemistry Department, Leibniz Institute for Tropospheric Research (TROPOS), Permoserstraße, 04318 Leipzig, Germany
4 School of Environmental Sciences, Jawaharlal Nehru University, New Delhi 110067, India
5 Indian Institute of Geomagnetism, Navi Mumbai 410206, India
* Correspondence: sumitkumarm@gmail.com

**Abstract:** The present work studies a severe smog event that occurred in Delhi (India) in 2017, targeting the characterization of $PM_{2.5}$ and its deposition potential in human respiratory tract of different population groups in which the $PM_{2.5}$ levels raised from 124.0 μg/m$^3$ (pre-smog period) to 717.2 μg/m$^3$ (during smog period). Higher concentration of elements such as C, N, O, Na, Mg, Al, Si, S, Fe, Cl, Ca, Ti, Cr, Pb, Fe, K, Cu, Cl, P, and F were observed during the smog along with dominant organic functional groups (aldehyde, ketones, alkyl halides (R-F; R-Br; R-Cl), ether, etc.), which supported potential contribution from transboundary biomass-burning activities along with local pollution sources and favorable meteorological conditions. The morphology of individual particles were found mostly as non-spherical, including carbon fractals, aggregates, sharp-edged, rod-shaped, and flaky structures. A multiple path particle dosimetry (MPPD) model showed significant deposition potential of $PM_{2.5}$ in terms of deposition fraction, mass rate, and mass flux during smog conditions in all age groups. The highest $PM_{2.5}$ deposition fraction and mass rate were found for the head region followed by the alveolar region of the human respiratory tract. The highest mass flux was reported for 21-month-old ($4.7 \times 10^2$ μg/min/m$^2$), followed by 3-month-old (49.2 μg/min/m$^2$) children, whereas it was lowest for 21-year-old adults (6.8 μg/min/m$^2$), indicating babies and children were more vulnerable to $PM_{2.5}$ pollution than adults during smog. Deposition doses of toxic elements such as Cr, Fe, Zn, Pb, Cu, Mn, and Ni were also found to be higher (up to $1 \times 10^{-7}$ μg/kg/day) for children than adults.

**Keywords:** $PM_{2.5}$; chemical composition; morphology; deposition potential; smog; health effects



**Citation:** Fatima, S.; Mishra, S.K.; Ahlawat, A.; Dimri, A.P. Physico-Chemical Properties and Deposition Potential of $PM_{2.5}$ during Severe Smog Event in Delhi, India. *Int. J. Environ. Res. Public Health* **2022**, *19*, 15387. https://doi.org/10.3390/ijerph192215387

Academic Editors: Nuno Canha, Marta Almeida and Evangelia Diapouli

Received: 30 September 2022
Accepted: 13 November 2022
Published: 21 November 2022



## 1. Introduction

The presence of primary and secondary anthropogenic PM together with favorable meteorological conditions can form a thick layer of haze/smog. Major European cities have suffered from severe smog conditions due to high PM concentrations [1]. During autumn and winter seasons, various Hungarian cities have witnessed smog conditions due to the combined effect of higher PM concentrations, favorable weather conditions, and geographical location [2]. In the Czech Republic, smog episodes were characterized by the highest $PM_{2.5}$ concentrations and organic compounds, poor dispersion conditions, lower temperature, and temperature inversion conditions [3]. The northern parts of India have also experienced severe haze/smog conditions due to increased PM emissions from a variety of emission sources, including transportation, industrial, residential energy usage, and biomass burning [4]. The variations in meteorological conditions during post-monsoon and winter seasons in north India are favorable for PM build-up due to lower temperature, higher relative humidity, lower wind speeds, and significant changes in wind directions, leading to lower visibility conditions during smog episodes [5]. Due to frequent crop-residue-burning activities taking place in the northwest Indian states such as Punjab and Haryana, air quality degrades in Delhi and nearby areas, which also extends up to other Indo-Gangetic Plains (IGP) regions of India such as Uttar Pradesh, Bihar, and West Bengal [5,6]. Higher concentrations of $PM_{2.5}$ cause serious respiratory and cardiopulmonary diseases such as asthma, bronchitis, etc., in the residents of Delhi [7]. During episodic cases, these particles consist of a different chemical composition which leads to haze/smog conditions increasing these illnesses and allergic diseases [8]. Each year smog conditions in Delhi lead to emergency shut-down of schools and large open gatherings, interruptions/cancellation of transportation services such as flight/train/road traffic, cancellation of games such as cricket matches, and heavy losses to the economic sector [9]. During 2017, a severe smog event was observed in Delhi during the month of November (post-monsoon) due to major crop-burning activities which took place in agricultural areas of nearby states such as Punjab, Haryana, and Uttar Pradesh [10]. During Smog-2017, air quality degraded to severe condition marked by poor air visibility, which badly affected daily life of Delhiites.

Since chemical constituents of PM play a major role in determining its source identification, it is necessary to study variations in chemical signature of PM during episodic cases such as smog formation. Anthropogenic PM contains different constituents, including sulfates, nitrates, mineral dust, metals, organics, black carbon, fly ash, etc. [11]. Inorganic constituents cover up to ~70% of PM mass, whereas organic compounds constitute almost 30% of the fine particulate mass [12]. Smog is mainly contributed by fine PM such as $PM_{2.5}$, and ~70% mass of $PM_{2.5}$ is composed of carbon (C), nitrogen (N), and sulphur (S) [13]. In Delhi, higher $PM_{2.5}$ concentrations with increased N and S constituents were observed during winter/post-monsoon seasons as a result of fossil fuels combustion and increased biomass-burning activities. Moreover, higher N and S species formation gives rise to increased $PM_{2.5}$ concentrations by the process of gas-to-particle conversions/oxidation/destruction of the primary aerosol particles during transboundary pollutant transfer in IGP regions such as Delhi [14]. During smog conditions, various PM bound metals were found in increased concentrations such as N, S, Cl, K, Si, Al, Zn, Pb, Fe, Mn, and Na [15], which may cause airway injury and inflammation through Fenton reaction [16]. In addition, the presence of transition metals such as Cr, Fe, Zn, Ni, Cu, Cd, etc., increase the production of reactive oxygen species (ROS) which cause oxidative stress as a result of which cells and tissues are damaged, leading to inflammation [17]. Moreover, these transition metals also cause genotoxic effects [18]. Fine particles ($PM_{2.5}$) consisting of Cu, Fe, and elemental carbon (soot) are found strongly associated with death from heart attack and Chronic Obstructive Pulmonary Disease (COPD) [19], whereas the presence of heavy metals such as Pd, Cd, and Hg in aerosol particles may adversely affect the central nervous system [20]. Chemical compositions of inhaled PM may also cause pro-inflammatory response in nervous tissues that lead to neurodegenerative diseases [21].

In addition, morphological parameters of PM also play a major role in aggravating the adverse health effects due to their fine sizes and non-spherical shapes [22]. When sharp-edged fine particles enter into our lungs, they reach the alveoli and get retained in the lung parenchyma [23]. The chemical composition of these fine-sized deposited particles may adversely affect the physiological and biochemical processes within our body once they reach the cell system of our body. After inhalation, PM gets deposited in various parts of the human lungs depending on their size ranges. PM with diameter <2.5 μm may reach the respiratory bronchioles and finally the alveoli, which are common sites for gaseous exchange within our body [24]. These particles can penetrate the alveoli and affect gaseous exchange processes, and due to their fine sizes, they may even reach the bloodstream and adversely affect human health [25]. Particles with diameter <1 μm with transition metals behave similarly to gas molecules, and, therefore, by the process of diffusion, they penetrate deep down to the alveoli and through systematic circulation processes, reach cells/tissues and get translocated there [26]. In addition, along with the bloodstream, they may even become a part of the circulation system [23] and cause damage to other organs, such as the brain [27]. Deposition potential of various sizes of particles including $PM_1$, $PM_{2.5}$, and $PM_{10}$ are found variable in different parts of human respiratory tract such as the head, trachea–bronchial, and pulmonary regions. Studies on deposition potential of PM for different regions of human lungs were carried out at Dehradun city, India [28], Chennai, India [29], and Xi'an City, China [30], whereas studies on deposition doses of different elements were studied at Kanpur [31], Dhanbad [32], and Delhi, India [33], showing the significance of PM deposition in human lungs during inhalation.

Worldwide limited studies are available on variations in PM concentrations, chemical compositions [1–3,15], and adverse health effects of PM [8,34] during smog episodes. In addition, very few studies are available on morphology of particles during smog conditions around the globe [35,36]. Since, limited studies are available for smog episodes in different parts of India [4,6,37], and extremely limited studies are available in Delhi for variations in PM concentrations [38] and inorganic components of PM (C, N, S components only) [5] during smog, this study is very useful for providing detailed variations in PM's elemental and organic compositions along with morphological variations during smog episodes in Delhi. In addition, we found no such study on exposure assessment for deposition doses of PM (using the MPPD model) and associated elements during smog episodes available in Delhi and other parts of India. Therefore, the present study will provide new insights for a holistic approach towards study of physico-chemical parameters of PM and health effects in terms of its deposition potential during smog conditions.

The present study addresses the following objectives:

i. Variations in $PM_{2.5}$ concentration and meteorological parameters pre-, during, and post-Smog Event-2017.
ii. $PM_{2.5}$ analysis for elemental composition, organic functional groups, and morphology during Smog Event-2017.
iii. Variations in $PM_{2.5}$ deposition potential and elemental deposition doses for different age groups pre-, during, and post- Smog Event-2017.

## 2. Study Area and Methodology

Delhi is a part of the IGP belt with an area of approx. 1485 km$^2$ (latitude: 28°24′17″ N to 28°53′00″ N; longitude: 76°50′24″ E to 77°20′37″ E), predominantly surrounded by two Indian states, namely Uttar Pradesh and Haryana, where the former lies in the East, and the latter covers other directions sharing their boundaries with Delhi, together known as National Capital Territory (NCT) regions. As per the 2011 census, Delhi's population was estimated to be more than 11 million [39]. Delhi lies in a semi-arid zone of India where meteorological conditions, including temperature, relative humidity, and rainfall greatly vary. Delhi has a typical humid subtropical IGP climate with hot summers generally affected by a frequent number of dust storms and dry and mild winter seasons, having significant smog/fog conditions.

### 2.1. Sample Collection

The sampling site selected for the study is CSIR-National Physical Laboratory (CSIR-NPL), which is situated in the central part of Delhi (Figure 1). The site characteristics include thick vegetation cover from one side and busy road from another side where heavy traffic can be seen every day. $PM_{2.5}$ samples were collected every 24 h for pre-, during, and post-smog events during the year 2017 starting from 1 November 2017 to 15 November 2017; details are provided in Table 1. $PM_{2.5}$ aerosol particles samples were collected using a high-volume (~16.67 L/min flow rate) air sampler (Model-APM-550; Envirotech®, New Delhi, India) placed on a rooftop at a height of 10 m from the ground, away from any other interferences, which is in accordance with an earlier study [40]. The samples were collected on Polytetrafluoroethylene (PTFE) filter paper (47 mm diameter; Pall®, New York, NY, USA) which were desiccated for 24 h to remove moisture and weighed before and after each sampling to obtain gravimetric data.

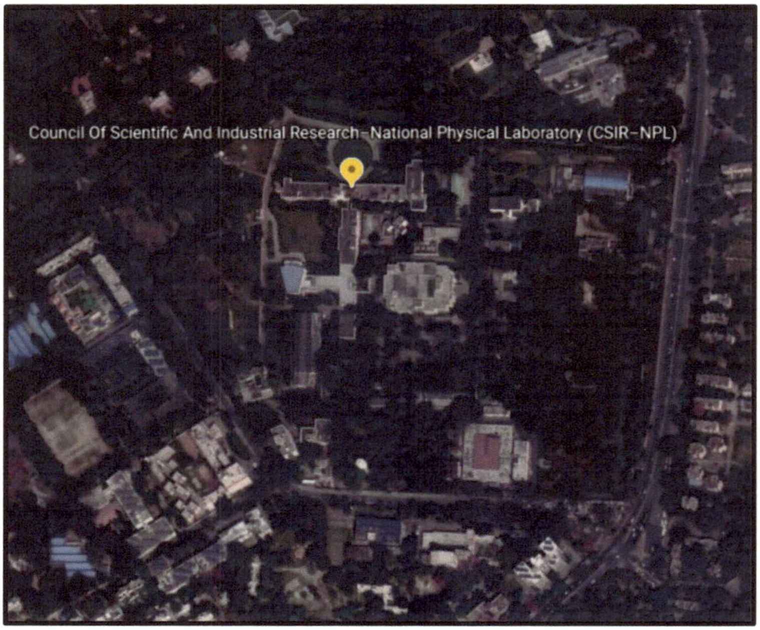


**Figure 1.** Sampling location CSIR-NPL at New Delhi, India [41].

**Table 1.** Details of sampling duration for $PM_{2.5}$ samples collected during Smog-2017.

| Sr. No. | Sampling Dates | Details |
|---|---|---|
| 1 | 1 November 2017 | Pre-smog |
| 2 | 8 November 2017 | Smog |
| 3 | 9 November 2017 | Post-smog |
| 4 | 10 November 2017 | Post-smog |
| 5 | 11 November 2017 | Post-smog |
| 6 | 13 November 2017 | Post-smog |
| 7 | 14 November 2017 | Post-smog |
| 8 | 15 November 2017 | Post-smog |

### 2.2. Individual Particles Collection and Measurement

To collect individual particles for morphological analysis, tin substrates (purity: >99%; size: $1 \times 1$ mm$^2$; thickness: ~0.1 mm) with marked exposure sides were placed on PTFE filters during sampling, as discussed in Section 2.1. The collected individual particles were further characterized for physico-chemical properties. The scanning electron microscopy with energy-dispersive X-ray spectroscopy (SEM-EDS) imaging technique (SEM Model: ZEISS; EDS Model: Oxford Link ISIS 300; facility present at CSIR-NPL, New Delhi, India) was used to obtain SEM micrographs for morphological and chemical signatures of individual particles collected on tin substrates. SEM technique provides high resolution images (~3 nm) of individual particles, whereas EDS gives elemental composition (beryllium (Be) to uranium (U)) present in the sample. Similar methodology for individual particles sample collection were followed in other studies in Delhi [40,42], but to the best of our knowledge, we found very limited studies on morphology of particles in Delhi and other parts of India, especially during episodic cases such as smog.

### 2.3. Quantitative and Qualitative Measurement of Bulk $PM_{2.5}$ Samples

The PTFE filters of $PM_{2.5}$ samples were analyzed for inorganic and organic signatures using wavelength-dispersive X-ray fluorescence (WD-XRF) and open path-Fourier transform infrared spectroscopy (OP-FTIR), respectively. Since, OP-FTIR is a non-destructive technique, filter samples were first analyzed for organic functional groups using this technique, followed by WD-XRF for obtaining inorganic chemical signatures present in the filter samples.

Organic functional groups present in $PM_{2.5}$ aerosol were analyzed using OP-FTIR (Model: Bruker IFS 125M; facility present at CSIR-NPL, New Delhi, India). The mid infrared band (MIR) used as a light source and the transmitted radiation were detected using a wide-band mercury cadmium telluride (MCT) detector continuously cooled with liquid nitrogen. Nitrogen purging was performed in the sample chamber during each analysis to remove moisture and $CO_2$ interference in the samples. Organic functional groups in the wavenumber range of 500–4700 $cm^{-1}$ were detected in the sample.

$PM_{2.5}$ filter papers were analyzed using WD-XRF (Model: Rigaku ZSX primus; facility present at CSIR-NPL, New Delhi, India). During sample analysis, filter samples were exposed to the X-ray tube of XRF, which excites atoms present in the sample and produces photons of characteristic wavelength, which were detected by detectors providing elemental composition of the samples. Data of blank filters were subtracted as a background during inorganic and organic analysis of filter samples. To the best of our knowledge, limited studies [5] are available related to inorganic and organic composition of PM during smog episodes in Delhi.

### 2.4. Secondary Data Collection

The meteorological parameters (temperature, relative humidity, wind speed, and air visibility) data were provided by India Meteorological Department (IMD), New Delhi. The backward wind trajectories were used for studying probable source contributions from transboundary pollution transfer using a NOAA HYSPLIT trajectory model [43]. The sources of fire count data that confirmed biomass-burning activities include NASA's FIRMS data [44] published in the Indian newspaper, The Hindustan Times, published on 09 November 2017 [45].

### 2.5. Deposition Potential Calculation

The deposition potential of $PM_{2.5}$ was calculated using the multiple path particle dosimetry model (MPPD)-Version 3.04 [46]. The MPPD model gives output data for particle deposition in different regions of the human respiratory tract (HRT), e.g., head, trachea–bronchial (TB) region, pulmonary (P) region, and total (Head + TB + P) deposition. This model considers three types of particle deposition processes, including diffusion, impaction, and sedimentation for the calculation of deposition fraction with three principal input sections described as follows:

(i) Airway morphometry: Out of eight different airway morphometry models, we selected "Yeh-Schum age-specific model", which considers the different structure of lungs in relation to respective age groups. As asymmetric branching structure of human lungs greatly causes bias in both airflow and particle deposition in HRT, we selected this model, as it provides multiple path (all airways) particle deposition. Age groups (children and adults) that were selected for the study included 3-month, 21-month, 28-month, 3-year, 8-year, 14-year, 18-year, 21-year, and 30-year. Values for other input parameters in the airway morphometry category were set as default specific to the respective age category (e.g., functional residual capacity (FRC) and upper respiratory tract (URT) volume).

(ii) Inhalation properties: This included input parameters such as aspect ratio, particle diameter, density, etc. The value of aspect ratio for $PM_{2.5}$ were set as 1.3 [42,47], particle diameter for $PM_{2.5}$ was set as 2.5 μm, and particle density for $PM_{2.5}$ was set as 1.5 $g/cm^3$ for the calculations [48].

(iii) Exposure conditions: Two types of exposure conditions can be chosen in the model as constant and variable. For our study we selected 'constant exposure' for estimating 24-h PM depositions of a given concentration ($mg/m^3$) at a constant rate. Other input parameters included acceleration of gravity ($981.0\ cm/s^2$); body orientation (upright); breathing frequency (per min); tidal volume (in mL); inspiratory fraction (0.5); pause fraction (0); and breathing scenario (nasal) for each age group.

Deposition potential parameters studies included deposition fraction; deposited mass rate (μg/min); and deposited mass flux ($\mu g/min/m^2$) of $PM_{2.5}$ for all age groups pre-, during, and post-Smog Event-2017. The MPPD model was used in previous reported studies [28,29]. However, we could not find many studies incorporating the MPPD model for PM deposition potential, especially during smog episodes in Delhi and other parts of India, which provides novelty to the present study.

*2.6. Exposure Assessment Calculation for $PM_{2.5}$ Associated Elements*

Health risk associated with the exposure of elements present in ambient $PM_{2.5}$ were calculated for 18 elements using the USEPA numerical model [49]. Average daily dose (μg/kg/day) of different elements present in $PM_{2.5}$ were calculated for children and adults during different days of sampling for Smog Event-2017 using Equation (1) as given below:

$$(\text{ADDinh}, \mu g/kg/day) = C \times \text{InhR} \times \text{EF} \times \text{ED}/\text{BW} \times \text{AT} \times \text{PEF} \quad (1)$$

Here,

C = Metal concentrations in $PM_{2.5}$ ($\mu g/m^3$);
InhR = Inhalation rate ($m^3/day$) (7.63 for adults and 20 for children);
EF = Exposure frequency (365 days/year);
ED = Exposure duration (24 year for adults and 6 year for children);
BW = Body weight (70 kg for adults and 15 kg for children);
PEF = Particle emission factor ($1.36 \times 109\ m^3/kg$);
AT = Averaging time for non-carcinogens (365 days/year).

Exposure assessment studies using the above USEPA numerical model have been carried out in earlier studies in Delhi [33] and other parts of India [31,32]. However, we could not find many detailed studies on exposure assessment of different elements during smog events in Delhi.

## 3. Results & Discussion

$PM_{2.5}$ samples were collected every 24 h for pre- (1 November 2017), during (8 November 2017), and post- (9 to 15 November 2017) Smog-2017, which were further analyzed for gravimetric, inorganic, organic, and morphological parameters as follows:

*3.1. Variations in $PM_{2.5}$ Concentrations and Meteorological Parameters during Smog Event-2017*

The variations in $PM_{2.5}$ concentration pre-, during, and post- Smog-2017 are shown in Figure 2. During pre-smog sampling (1 November 2017) the $PM_{2.5}$ concentration was reported as 124.0 $\mu g/m^3$, which drastically increased up to 717.2 $\mu g/m^3$ on the first day of Smog Event-2017 (8 November 2017) which was ~6 times higher than that of pre-smog $PM_{2.5}$ concentration (Figure 2). The previous study over Delhi reported the average $PM_{2.5}$ concentration during Smog Event-2016 as 793 (±27.8) $\mu g/m^3$ [5]. The increased $PM_{2.5}$ concentration during Smog-2017 was reported ~29 times and ~12 times higher than the 24 h average $PM_{2.5}$ permissible limit set by World Health Organization (WHO) and National Ambient Air Quality Standards (NAAQS), India, which are 25 $\mu g/m^3$ and 60 $\mu g/m^3$, respectively. Such a higher concentration of $PM_{2.5}$ may cause adverse health effects even in the short-term and aggravate respiratory diseases such as asthma, severely affecting sensitive communities such as children and older persons. Although, $PM_{2.5}$ concentrations decreased on consecutive days after the smog event (9/10 November 2017) but were still higher than the pre-smog $PM_{2.5}$ concentrations due to post-smog effects. The post-smog effects included pollution build-up coupled with favorable meteorological conditions, such as lower temperature, lower wind speed, and higher relative humidity (Table 2). Similar findings were reported for Smog-2016 where post-smog $PM_{2.5}$ concentrations increased with the dominance of secondary formed particulate ammonium sulphate via gas-to-particle conversion along with favorable meteorological conditions [5]. On the post-smog day (11 November 2017), higher $PM_{2.5}$ concentration of 459.0 $\mu g/m^3$ was observed which decreased on further consecutive days of post-smog event period (13/14/15 November 2017) (Figure 2). The increase in $PM_{2.5}$ concentration on 11 November 2017 may be due to secondary particle formation leading to increase in fine PM concentration. It has been reported that secondary $PM_{2.5}$ formation takes places due to photochemical reactions among multiple emitted pollutants, which causes severe haze conditions as reported in many cities around the world [50,51]. In addition, increases in N and S species on 11 November 2017 (Table 3) confirm the formation of secondary particles, such as ammonium sulphate, as reported during Smog-2016 [5].

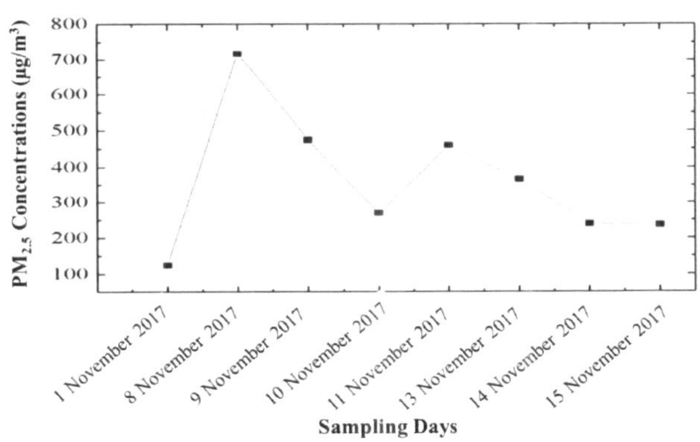


**Figure 2.** Variations in $PM_{2.5}$ concentration ($\mu g/m^3$) during Smog Event-2017.

**Table 2.** Day-wise variations in meteorological parameters during Smog Event-2017.

| Sampling Date | WS (m/s) | Visibility (m) | T (°C) | RH (%) |
|---|---|---|---|---|
| 1 November 2017 | 2.36 | 1942 | 23.3 | 75.9 |
| 8 November 2017 | 0.71 | 385 | 19.8 | 81.9 |
| 9 November 2017 | 0.86 | 709 | 20.0 | 76.3 |
| 10 November 2017 | 0.60 | 940 | 19.8 | 76.1 |
| 11 November 2017 | 0.79 | 795 | 19.7 | 77.3 |
| 13 November 2017 | 2.07 | 942 | 19.1 | 79.9 |
| 14 November 2017 | 3.00 | 1123 | 20.0 | 75.5 |
| 15 November 2017 | 2.48 | 1806 | 19.4 | 75.0 |

**Table 3.** Variations in $PM_{2.5}$ associated elements (µg/m$^3$) during Smog-2017.

| Sr. No. | Elements | 01 November 2017 | 08 November 2017 | 09 November 2017 | 10 November 2017 | 11 November 2017 | 13 November 2017 | 14 November 2017 | 15 November 2017 |
|---|---|---|---|---|---|---|---|---|---|
| 1 | N | 10.0 | 58.0 | 38.4 | 21.8 | 37.2 | 29.6 | 19.3 | 19.1 |
| 3 | Na | 0.2 | 1.0 | 0.7 | 0.4 | 0.7 | 0.5 | 0.3 | 0.3 |
| 4 | Mg | 0.2 | 1.1 | 0.7 | 0.4 | 0.7 | 0.6 | 0.4 | 0.4 |
| 5 | Al | 0.8 | 4.8 | 3.2 | 1.8 | 3.1 | 2.4 | 1.6 | 1.6 |
| 6 | Si | 2.1 | 12.3 | 8.1 | 4.6 | 7.9 | 6.3 | 4.1 | 4.0 |
| 7 | P | 0.2 | 1.1 | 0.7 | 0.4 | 0.7 | 0.6 | 0.4 | 0.4 |
| 8 | S | 16.3 | 94.2 | 62.3 | 35.4 | 60.4 | 48.1 | 31.4 | 31.0 |
| 9 | Cl | 0.7 | 3.9 | 2.6 | 1.5 | 2.5 | 2.0 | 1.3 | 1.3 |
| 10 | K | - | 10.5 | 6.9 | 3.9 | 6.7 | 5.3 | 3.5 | 1.8 |
| 11 | Ca | 1.1 | 6.1 | 4.0 | 2.3 | 3.9 | 3.1 | 2.0 | 2.0 |
| 12 | Cr | 2.9 | 17.0 | 11.3 | 6.4 | 10.9 | 8.7 | 5.7 | 5.6 |
| 13 | Mn | 0.2 | 0.9 | 0.6 | 0.3 | 0.6 | 0.5 | 0.3 | 0.3 |
| 14 | Fe | 2.4 | 13.8 | 9.1 | 5.2 | 8.8 | 7.0 | 4.6 | 4.5 |
| 15 | Ni | 0.1 | 0.7 | 0.5 | 0.3 | 0.4 | 0.4 | 0.2 | 0.2 |
| 16 | Zn | 1.6 | 9.3 | 6.2 | 3.5 | 6.0 | 4.8 | 3.1 | 3.1 |
| 17 | Pb | 0.5 | 2.9 | 1.9 | 1.1 | 1.9 | 1.5 | 1.0 | 1.0 |
| 18 | Cu | 0.2 | 1.2 | 0.8 | 0.4 | 0.7 | 0.6 | 0.4 | 0.4 |
| 19 | Br | - | 0.8 | 0.6 | 0.3 | 0.5 | 0.4 | 0.3 | 0.3 |
| 20 | Ti | 0.2 | 1.1 | 0.7 | 0.4 | 0.7 | 0.5 | 0.4 | 0.4 |

The variations in meteorological parameters, such as temperature, relative humidity (RH), wind speed, and air visibility conditions in pre-, during, and post-Smog-2017 are shown in Table 2. On the pre-smog day, wind speed and temperature were found to be higher, 2.36 m/s and 23.3 °C, respectively. Higher wind speed leads to dispersion of pollutants, whereas higher temperature causes breakdown of pollutants in the presence of sunlight, leading to lower $PM_{2.5}$ concentration [52,53], as also found in the present study. On the contrary, during Smog Event-2017, both wind speed and temperature were comparatively lower and RH was higher (Table 2), favoring the accumulation of particulate matter. In the presence of higher RH, PM acts as a nucleus for the condensation of water vapor present in the air, forming a dense mass known as fog. When this fog is mixed with other pollutants, such as S and N species from burning, it leads to the formation of smog [5]. During the severe haze period, heterogeneous reactions become the major formation pathway of secondary aerosol particles under high RH conditions [51]. The presence of lower $PM_{2.5}$ concentration on pre-smog day also instigated good atmospheric visibility of 1942 m, whereas during smog conditions, visibility greatly reduced up to 385 m, which was lowest among all days of sampling. This signifies that meteorological parameters also played a significant role in PM pollution build up during Smog Event-2017.

Fire count data on 09 November 2017 [45] and backward air mass trajectory on smog day, i.e., 08 November 2017 (Figure 3) revealed that biomass-burning activities took place in the last week of October and the first week of November, 2017, during which air parcels moved from these areas to Delhi. This trans-boundary PM transport along with favorable meteorological conditions contributed to the severe smog event during the year 2017. In addition, the Punjab Pollution Control Board reported 39,686 burning events that took place in Punjab after 15 October 2017 [53]. The estimated total biomass burnt during 2017 until the Smog Event-2017 was reported as 23 million tons altogether from adjoining areas of Delhi, including Punjab, Haryana, and Uttar Pradesh [10]. Every year biomass-burning events take place during October-November months in these areas to make room for winter crops, causing severe smog conditions during post-monsoon/winter season.

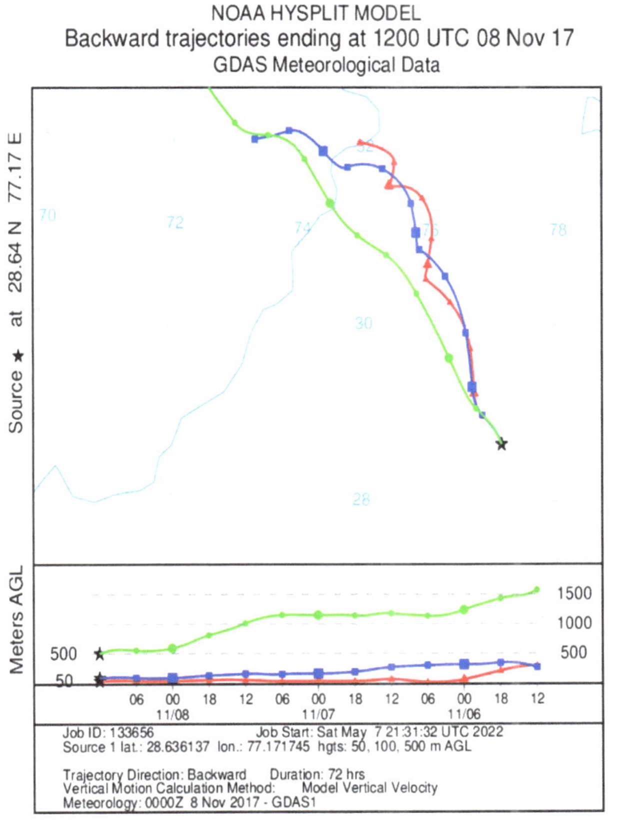


**Figure 3.** Backward air mass trajectory on 8 November 2017.

### *3.2. Variations in $PM_{2.5}$ Associated Elements, Organic Functional Groups, and Morphology during Smog Event-2017*

#### 3.2.1. Elemental Composition

The variations in $PM_{2.5}$ associated elements during Smog Event-2017 are shown in Table 3. On pre-smog day, the higher $PM_{2.5}$ associated elemental concentration (>1 $\mu g/m^3$) were reported for N (10.0 $\mu g/m^3$), Si (2.1 $\mu g/m^3$), S (16.3 $\mu g/m^3$), Ca (1.1 $\mu g/m^3$), Cr (2.9 $\mu g/m^3$), and Zn (1.6 $\mu g/m^3$), whose concentrations drastically increased during the smog event (Table 3). During the smog period, the $PM_{2.5}$-associated elemental concentrations were reported much higher for elements such as N (58.0 $\mu g/m^3$), Na (1.0 $\mu g/m^3$), Mg (1.1 $\mu g/m^3$), Al (4.8 $\mu g/m^3$), Si (12.3 $\mu g/m^3$), P (1.1 $\mu g/m^3$), S (94.2 $\mu g/m^3$), K (10.5 $\mu g/m^3$), Ca (6.1 $\mu g/m^3$), Cr (17.0 $\mu g/m^3$), Fe (13.8 $\mu g/m^3$), Zn (9.3 $\mu g/m^3$), Pb (2.9 $\mu g/m^3$), Cu (1.2 $\mu g/m^3$), Br (0.9 $\mu g/m^3$), and Ti (1.1 $\mu g/m^3$) (Table 3). Higher concentrations of elements such as N, S, Cl, K, Cr, Fe, Zn, Pb, Cu, and Br during the smog event showed the signature of biomass-burning activities [2]. The presence of higher Al and Si during the smog event may be attributed to dust transport from the Punjab area along with air parcel movement to Delhi, as well as local road dust suspension contribution. During the smog episode, higher Cu and Zn concentrations found may be associated with brake/tire abrasion or emitted from lubricating oil [54]. In addition, higher concentrations of Zn and Pb in the droplet mode may be emitted from traffic and industrial sources [55]. Major elements found during the smog episode are S and N, which basically form smog. These elements greatly contribute to secondary gas-to-particle conversion processes, such as conversion of particulate $NO_3^-$ and $SO_4^{2-}$ from gaseous $NO_x$ and $SO_x$ in the presence of high humidity and low photochemical activity, increasing fine PM concentrations [5]. PM-bound S contributes to the formation of secondary inorganic aerosols containing $NH_4NO_3$ and $(NH_4)_2SO_4$ under smog conditions [56]. In addition, sulfate formation via oxidation of $SO_2$ is catalyzed by the presence PM metal ions such as Fe during the smog phase, increasing haziness and thus reducing visibility [5]. Higher concentrations of the elements in the present study may also be attributed to lower wind speed along with higher source contribution during the smog episode (Table 2). Typically, three constituents were found contributing to haze conditions in Shanghai, including secondary inorganic pollution, dust, and biomass-burning constituents [57].

### 3.2.2. Organic Functional Groups

The qualitative variations in $PM_{2.5}$-associated organic functional groups during Smog Event-2017 are shown in Table 4. FTIR analysis (qualitative) of $PM_{2.5}$ provided variations in 17 organic functional groups present in pre-, during, and post-Smog Event-2017 which were identified by using the National Institute of Standards and Technology (NIST) library. On the basis of the source/mechanism of formation, organic functional groups have been categorized as [58]:

- Biogenic functional groups (ether, carbohydrates, hydroxyl groups, amino acids, and amines functional groups);
- Oxygenated functional groups (carboxylic acid, aldehydes, ketones, esters, lactone, and acid anhydride);
- Aliphatic hydrocarbon functional groups (aliphatic CH, alkenes, methyl, and methylene functional groups), and,
- Aromatic hydrocarbon functional groups, etc.

**Table 4.** Variations in $PM_{2.5}$-associated organic functional groups (in absorbance units) during Smog Event-2017.

| Sr. No. | Functional Groups | 01 November 2017 | 08 November 2017 | 09 November 2017 | 10 November 2017 | 11 November 2017 | 13 November 2017 | 14 November 2017 | 15 November 2017 |
|---|---|---|---|---|---|---|---|---|---|
| 1 | Alkyl halides (R-I) | 0.49 | 0.66 | 0.87 | 0.79 | 0.44 | 0.98 | 0.98 | 0.78 |
| 2 | Alkyl halides (R-Br) | 0.48 | 1.01 | 0.87 | 0.72 | 0.525 | 0.85 | 0.67 | 0.73 |
| 3 | Alkyl halides (R-F) | 0.06 | 0.24 | 0.41 | 0.28 | 0.46 | 0.22 | 0.39 | 0.14 |
| 4 | Alkyl halides (R-Cl) | - | 0.3 | - | - | - | - | - | - |
| 5 | Alcohol | 0.06 | 0.24 | 0.41 | 0.52 | 0.46 | 0.41 | 0.26 | 0.49 |
| 6 | Ethers | 0.21 | 0.24 | 0.41 | 0.28 | 0.46 | 0.16 | 0.26 | 0.14 |
| 7 | Esters | 0.41 | 0.42 | 0.44 | 0.45 | 0.54 | 0.41 | 0.52 | 0.37 |
| 8 | Organonitrates | 0.31 | 0.42 | 0.44 | 0.45 | 0.54 | 0.41 | - | 0.37 |
| 9 | Phenol | 0.41 | 0.42 | 0.302 | 0.45 | 0.54 | 0.41 | 0.52 | 0.37 |
| 10 | Amino acids/Amines | 0.41 | 0.42 | 0.44 | 0.45 | 0.54 | 0.41 | 0.52 | 0.37 |
| 11 | Aldehydes | 0.079 | 0.41 | 0.501 | 0.36 | 0.22 | 0.365 | 0.26 | 0.25 |
| 12 | Ketones | 0.079 | 0.41 | 0.501 | 0.36 | 0.22 | 0.365 | 0.26 | 0.25 |
| 13 | Carbonyl carbon | 0.079 | 0.41 | 0.501 | 0.36 | 0.22 | 0.365 | 0.26 | 0.25 |
| 14 | Alkanes and Alkyls | 0.36 | 0.54 | 0.47 | 0.62 | 0.1 | 0.55 | 0.52 | 0.53 |
| 15 | Carboxylic acids | 0.39 | 0.54 | 0.52 | 0.62 | 0.14 | 0.55 | 0.64 | 0.53 |
| 16 | Amides | - | 0.39 | 0.53 | 0.67 | 0.11 | - | 0.67 | 0.62 |
| 17 | Alkenes | - | - | - | 0.64 | 0.34 | - | 0.62 | - |

Table 4 shows lowest absorbance of organic functional groups on pre-smog day, whereas increased concentration during smog conditions. Dominant organic functional groups (absorbance > 0.10) on pre-smog day included alkyl halides (R-I) > alkyl halides (R-Br) > esters ≥ phenol ≥ amino acids/amines > carboxylic acid > alkane and alkyls > organonitrates > ether (Table 4). Dominant organic functional groups (absorbance > 0.10) during smog conditions included alkyl halides (R-Br) > alkyl halides (R-I) > alkane and alkyls ≥ carboxylic acids > esters ≥ phenol ≥ amino acids/amines ≥ organonitrates > aldehydes ≥ ketones ≥ carbonyl groups > amide > alcohol ≥ ethers ≥ alkyl halides (R-F) (Table 4). The presence of dominant species during smog showed their contribution from fossil fuel combustion including gasoline- and diesel-powered vehicles (alkyne, alcohols, nitro-compounds), burning activities (alkyl halides (R-F), alkyl halides (R-Br)), biogenic emission (ether), and oxidation processes (esters) [58]. In addition, aldehyde and ketones found to be originated from biomass burning, and ketones may also be formed from alkane oxidation processes [59]. Amine and amide groups found are mostly biogenic in emissions [60]. The percentage (%) contribution of different organic functional groups are presented in Figure 4. Higher % contribution of species, such as alkyl halides (R-I), alkyl halides (R-Cl), amines, amides, aldehydes, and ketones, confirm biomass-burning and biogenic emissions are dominant sources during Smog Event-2017. The presence of dominant organic functional groups during smog are found to be similar to those found in the ambient PM samples affected by wildfires and wood-burning activities [36]. Even after the smog event, significant absorbance of organic functional groups was found during the post-smog period (08/09/10 November 2017). This may due to the fact that biomass burning produces a significant number of organic compounds (VOCs and carbonaceous matter), which after oxidation with hydroxyl (OH) radical, become further increased in concentration during transboundary pollution transfer [61]. In addition, during haze conditions, organic compounds contribute to particle growth processes which are driven by secondary aerosol formation processes and air mass origin [62]. The organic compounds also contributed from moderate to severe haze conditions reducing visibility conditions in northeast China [15].

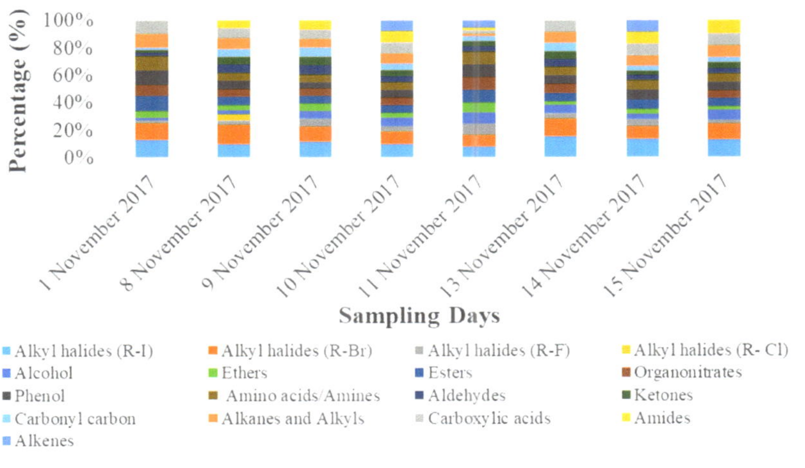


**Figure 4.** Percentage (%) contribution of different organic functional groups in $PM_{2.5}$.

### 3.2.3. Morphology

SEM-EDS micrographs shows the presence of individual particles in both spherical and non-spherical shapes during Smog Event-2017 (Figure 5). During the studied smog episode, non-spherical particles were abundantly dominant, followed by some spherical/nearly spherical particles. Round, flaky structures, solid and irregular-shaped particles, carbon fractals, sticky aggregates, sharp-edged particles, and rod-shaped and rectangular particles were abundantly found during the smog event. A variety of elements with variable mass % were found associated with these particles, such as C, N, O, Na, Mg, Al, Si, S, Fe, Cl, Ca, Ti, Cr, Pb, Fe, K, Cu, Cl, P, and F. The presence of elements such as C, N, S, Mg, Cl, K, F, Cr, Pb, Cu, etc., shows signatures from biomass-burning activities. The chained and branched structures (carbon fractals) rich in elements such as C, N, O, Na, and Fe were found, which may have originated from combustion processes. Aggregated and agglomerated particles rich in elements such as C, N, O, Na, S, Pb, Cl, Fe, K, and Cu were found in dominance showing biomass burning signatures. The irregular and sharp-edged particles with C, O, Na, P, S, and K were also found which may cause injury to inner tissues of lungs after inhalation. In addition, during smog, higher mass % of toxic elements in some individual particles such as Cu (up to 70%), Pb (up to 32%), and Cr (up to 25%) shows greater concern in terms of human health during smog conditions in Delhi. Metals such as Cr and Cu are common toxicants to lungs, and Pb causes neurological disorders while present in PM [63]. Similar studies for morphology and chemical composition of individual particles were conducted at Delhi, India [42], Haryana, India [64], Jaipur, India [65], Kanpur, India [66], Los Angeles, USA [67], and New Mexico [68], but limited studies on morphology of particles are available for episodic cases such as dust storms [40] and smog in Delhi.

### 3.3. Variations in $PM_{2.5}$ Deposition Potential during Smog Event-2017

The variations in $PM_{2.5}$ deposition potential for different age groups including children and adults, pre- during, and post-Smog Event-2017 using MPPD model are shown in Tables 5 and 6. The MPPD model provides output data for deposition of particles in different regions of human respiratory tract (HRT), e.g., the head, TB, P, and total deposition. Deposition fraction is the ratio of the number of aerosol particles of a specific size (e.g., for $PM_{2.5}$ particle diameter = 2.5 μm) deposited in a specific respiratory airway to the number of the same size entering the overall respiratory tract. Deposition rate determines the number of particles deposited in HRT per unit time and is represented in units as μg/min, whereas deposition flux determines the number of particles deposited in HRT per unit time per unit area of human lungs and is represented in units as μg/min/$m^2$.

**Table 5.** Comparative analysis of $PM_{2.5}$ deposition fraction for different age groups.

| | Present Study | Age Groups | | | | | | | | |
|---|---|---|---|---|---|---|---|---|---|---|
| Sr. No. | Deposition Fraction | 3-Month | 21-Month | 28-Month | 3-Year | 8-Year | 14-Year | 18-Year | 21-Year | 30-Year |
| 1 | Head deposition fraction | 0.24 | 0.25 | 0.29 | 0.28 | 0.27 | 0.26 | 0.42 | 0.41 | 0.47 |
| 2 | TB deposition fraction | 0.12 | 0.14 | 0.06 | 0.05 | 0.06 | 0.06 | 0.05 | 0.05 | 0.06 |
| 3 | Pulmonary deposition fraction | 0.30 | 0.29 | 0.21 | 0.28 | 0.40 | 0.33 | 0.28 | 0.31 | 0.20 |
| 4 | Total deposition fraction | 0.66 | 0.69 | 0.57 | 0.62 | 0.73 | 0.66 | 0.75 | 0.77 | 0.73 |
| | Manojkumar et al., 2019 [29] | | | | | | | | | |
| 5 | Head deposition fraction | 0.23 | - | 0.29 | 0.28 | 0.27 | 0.26 | 0.42 | 0.41 | - |
| 6 | TB deposition fraction | 0.12 | - | 0.06 | 0.05 | 0.06 | 0.06 | 0.05 | 0.05 | - |
| 7 | Pulmonary deposition fraction | 0.30 | - | 0.21 | 0.28 | 0.40 | 0.33 | 0.28 | 0.31 | - |
| 8 | Total deposition fraction | 0.65 | - | 0.56 | 0.61 | 0.73 | 0.65 | 0.74 | 0.77 | - |

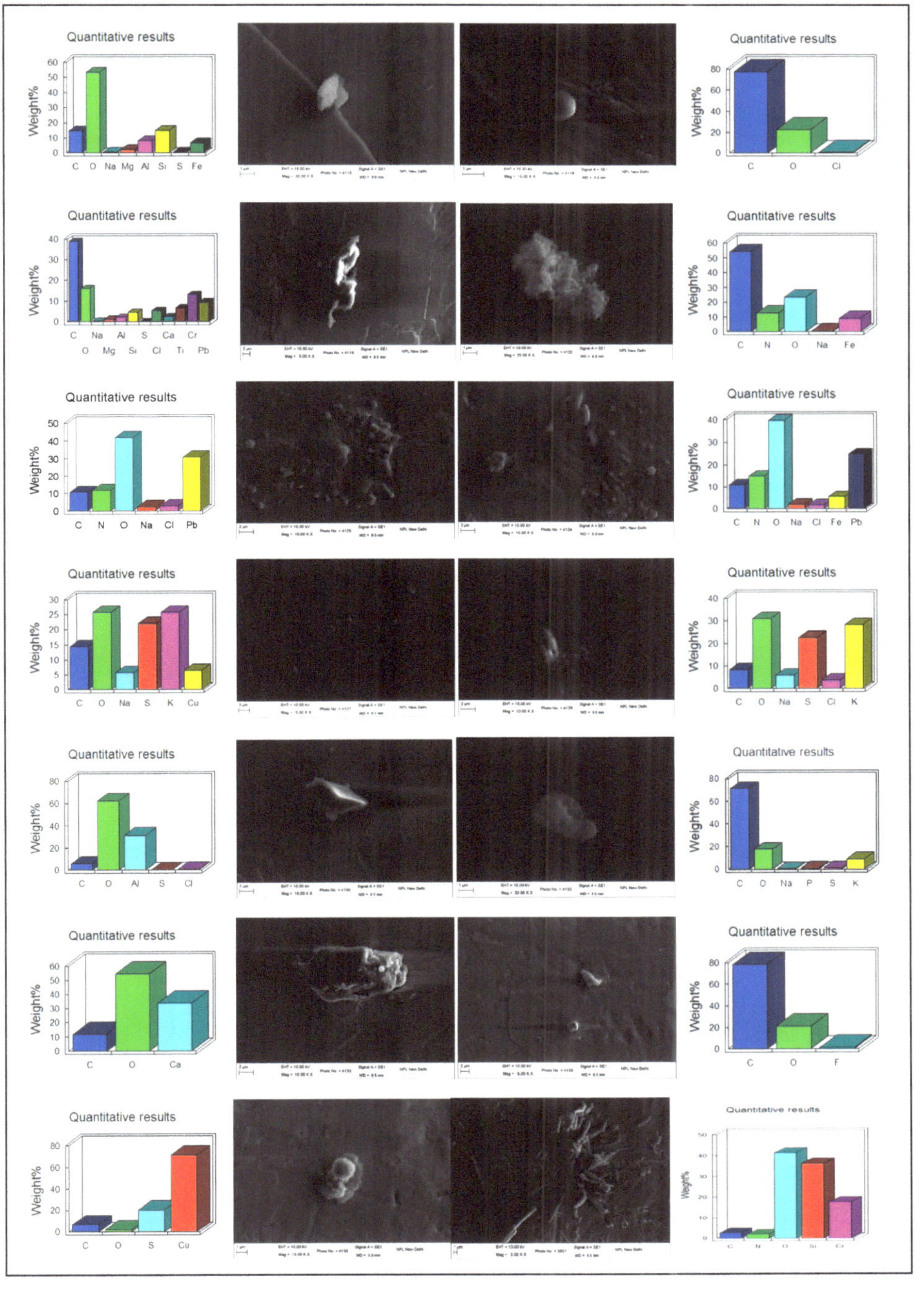

**Figure 5.** *Cont.*

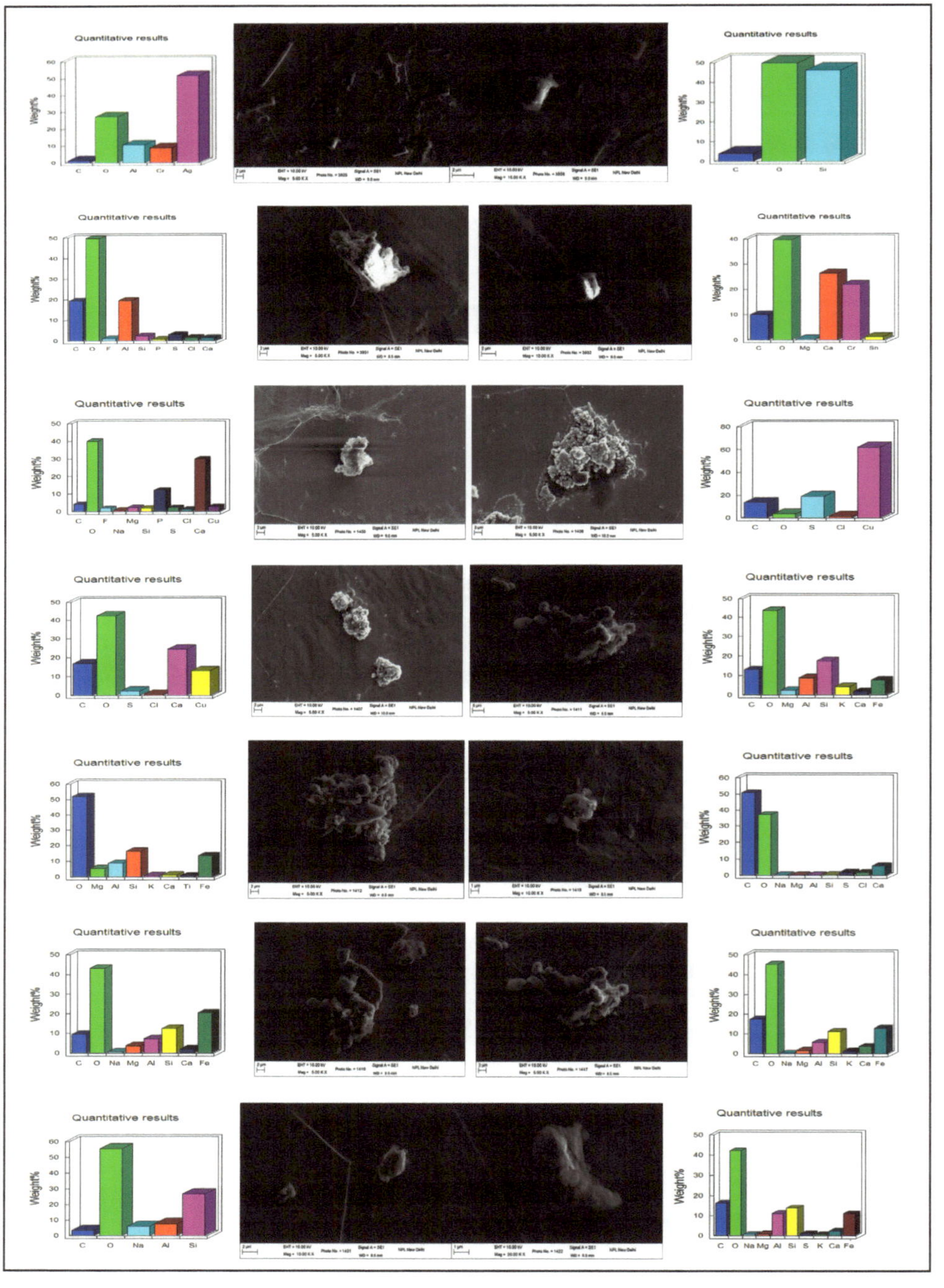


**Figure 5.** Morphology and chemical composition of individual particles during Smog Event-2017.

**Table 6.** $PM_{2.5}$ deposited mass rate (μg/min) for different age groups during pre-, during, and post-Smog-2017.

| | Pre-smog | Age Groups | | | | | | | | |
|---|---|---|---|---|---|---|---|---|---|---|
| Sr. No. | Deposition Potential | 3-Month | 21-Month | 28-Month | 3-Year | 8-Year | 14-Year | 18-Year | 21-Year | 30-Year |
| 1 | Head deposited mass rate (μg/min) | 0.04 | 0.07 | 0.09 | 0.10 | 0.16 | 0.20 | 0.35 | 0.34 | 0.44 |
| 2 | TB deposited mass rate (μg/min) | 0.02 | 0.04 | 0.02 | 0.02 | 0.03 | 0.05 | 0.04 | 0.04 | 0.06 |
| 3 | Pulmonary deposited mass rate (μg/min) | 0.04 | 0.08 | 0.07 | 0.10 | 0.24 | 0.26 | 0.23 | 0.26 | 0.18 |
| 4 | Total deposited mass rate (μg/min) | 0.10 | 0.19 | 0.18 | 0.22 | 0.43 | 0.51 | 0.62 | 0.64 | 0.67 |
| | Smog | | | | | | | | | |
| 5 | Head deposited mass rate (μg/min) | 0.20 | 0.41 | 0.55 | 0.59 | 0.92 | 1.18 | 2.02 | 1.97 | 2.51 |
| 6 | TB deposited mass rate (μg/min) | 0.11 | 0.23 | 0.11 | 0.11 | 0.19 | 0.27 | 0.24 | 0.22 | 0.34 |
| 7 | Pulmonary deposited mass rate (μg/min) | 0.26 | 0.48 | 0.40 | 0.59 | 1.36 | 1.49 | 1.34 | 1.47 | 1.05 |
| 8 | Total deposited mass rate (μg/min) | 0.56 | 1.12 | 1.05 | 1.29 | 2.47 | 2.93 | 3.60 | 3.67 | 3.90 |
| | Post-smog | | | | | | | | | |
| 9 | Head deposited mass rate (μg/min) | 0.07 | 0.14 | 0.18 | 0.19 | 0.30 | 0.39 | 0.66 | 0.65 | 0.82 |
| 10 | TB deposited mass rate (μg/min) | 0.03 | 0.08 | 0.04 | 0.04 | 0.06 | 0.09 | 0.08 | 0.07 | 0.11 |
| 11 | Pulmonary deposited mass rate (μg/min) | 0.08 | 0.16 | 0.13 | 0.19 | 0.45 | 0.49 | 0.44 | 0.48 | 0.34 |
| 12 | Total deposited mass rate (μg/min) | 0.18 | 0.37 | 0.35 | 0.42 | 0.81 | 0.96 | 1.18 | 1.20 | 1.28 |

The variations in $PM_{2.5}$ deposition fractions for different age groups are provided in Table 5. Deposition fractions in the pulmonary region were found highest, followed by the head region for the 3-month, 21-month, 8-year, and 14-year age groups, which signify more deposition of fine particles in the alveolar region for these age groups (Table 5). Deposition fractions in the head region were found highest, followed by pulmonary region for the 28-month, 3-year, 18-year, 21-year, and 30-year age groups, which signify more deposition of $PM_{2.5}$ particles in the outer respiratory tract region than the alveolar region for these age groups (Table 5). Lowest deposition fractions were found in TB region for all age groups (Table 5). This signifies that babies (3-month and 21-month) and growing age children (8-year and 14-year) are more susceptible to fine particles deposition in inner parts of human lungs, e.g., alveolar region with potential health effects. The values of deposition fraction in our study are found to be in accordance with another study carried out in Chennai city, India, as shown in comparative Table 5 [29].

The MPPD model outputs suggested that on pre-smog day, lower deposited mass rate and mass flux for $PM_{2.5}$ were observed, which drastically increased during smog conditions and later decreased on the last day of post-smog (15 November 2017) (Tables 6 and 7). Increase/decrease in deposited mass rate and mass flux were observed due to respective increase/decrease in $PM_{2.5}$ concentrations. Similar to the deposition fraction, deposited mass rate (μg/min) was found to be highest for the pulmonary region followed by the head region in specific age groups (3-month, 21-month, 8-year, and 14-year), whereas found highest for the head region followed by pulmonary region for remainder of the age groups studied (28-month, 3-year, 18-year, 21-year, and 30-year) (Table 6). Lowest deposited mass rate was found in TB region for all age groups during pre-, during, and post-Smog Event-2017 but with different values (Table 6). In addition, much higher values of deposited mass rate (μg/min) and mass flux (μg/min/m$^2$) during smog conditions for all age groups showed the higher extent of $PM_{2.5}$ deposition during pollution episodes in Delhi (Tables 6 and 7). In the present study (Delhi, India), the highest total $PM_{2.5}$ deposited mass rates (μg/min) in different age groups were reported during the smog event as 0.10 (3-month), 0.19 (21-month), 0.18 (28-month), 0.22 (3-year), 0.43 (8-year), 0.51 (14-year), 0.62 (18-year), 0.64 (21-year), and 0.67 (30-year) (Table 6). Among all age groups, the highest total deposited mass rate (μg/min) was observed for 30-year-old followed by 21-year-old adults throughout pre-, during, and post-Smog Event-2017 (Table 6). The highest value of total deposited mass rate for 30-year-old adults during pre-smog event was observed as 0.67 μg/min, which drastically increased up to ~6 times as 3.90 μg/min during smog event, which shows quite higher deposition of $PM_{2.5}$ during episodic cases such as smog (Table 6). This higher rate of PM deposition may cause blocking of the upper respiratory tract or head region when individual particles with higher surface area are inhaled and get deposited because in this study, highest deposited mass rates are found for the head region for specific age groups (28-month, 3-year, 18-year, 21-year, and 30-year). On the other hand, smaller particles may reach the alveoli and bloodstreams and become deposited in internal organs of our body such as mitochondria, cytoplasm, and cytoplasm-bound vesicles [25] because in this study, highest deposited mass rate are found for pulmonary region for some age groups (3-month, 21-month, 8-year, and 14-year). When deposited inside cell organelles, PM causes oxidative stress, inflammation, mitochondria-induced apoptosis, and can interfere with intracellular proteins, organelles, and DNA present inside the cell [69]. The present values of deposited mass rate during smog were found to be quite a lot higher than the previously reported MPPD model outputs in India. For example, highest $PM_{2.5}$ deposited mass rate (μg/hr) in Chennai, India, were found as $4.7 \times 10^{-6}$ (3-month), $1.2 \times 10^{-5}$ (28-month), $1.4 \times 10^{-5}$ (3-year), $2.8 \times 10^{-5}$ (8-year), $4.3 \times 10^{-5}$ (9-year), $4.8 \times 10^{-5}$ (14-year), $4.7 \times 10^{-5}$ (18-year), and $6.0 \times 10^{-5}$ (21-year) [29]. Highest $PM_{2.5}$ deposited mass rate (μg/hr) in Dehradun, India, were found as $9.1 \times 10^{-3}$ (3-month), 0.09 (21-month), 0.03 (3-year), 0.08 (9-year), 0.09 (18-year), and 0.12 (21-year) [28].

**Table 7.** Mass flux maps (μg/min/m$^2$) for different age groups during pre-, during, and post-Smog-2017.

| Age Groups | Normal Lung Geometry | Pre-smog | Smog | Post-smog |
| --- | --- | --- | --- | --- |
| 3-month |  | 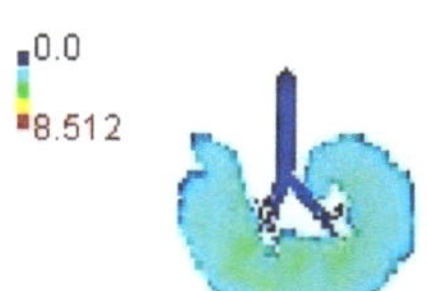  | 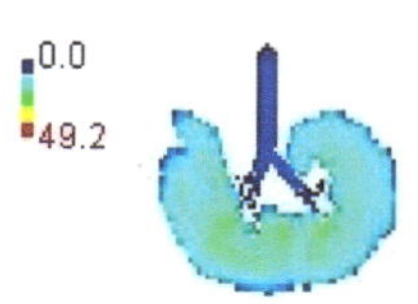  | 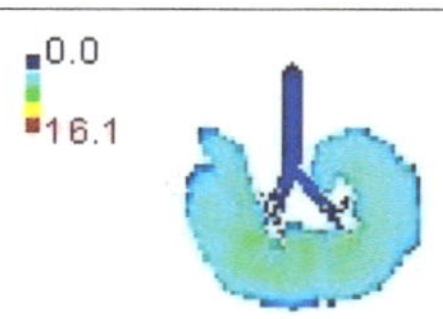  |
| 21-month | 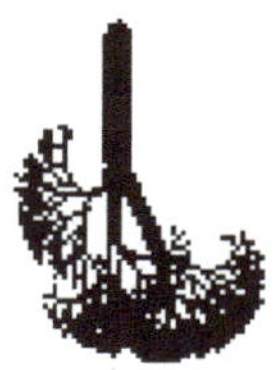 | 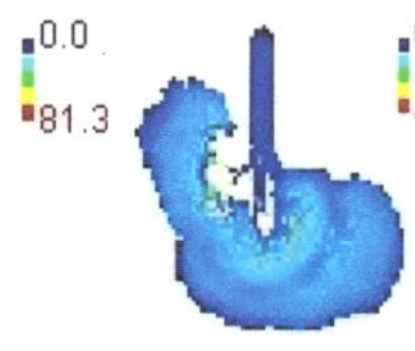  | 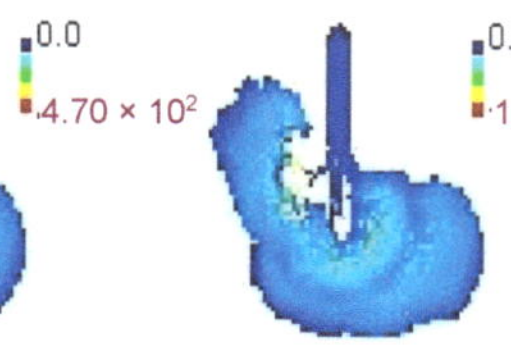  | 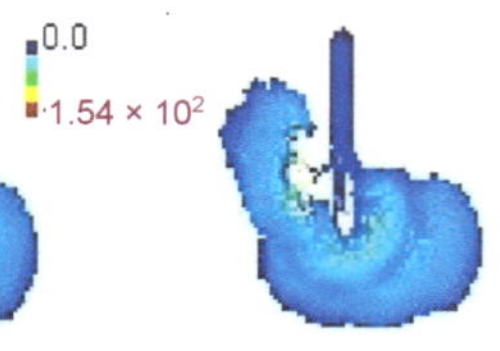  |

**Table 7.** *Cont.*

| Age Groups | Normal Lung Geometry | Pre-smog | Smog | Post-smog |
|---|---|---|---|---|
| 28-month | 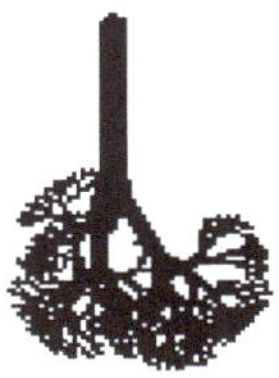 | 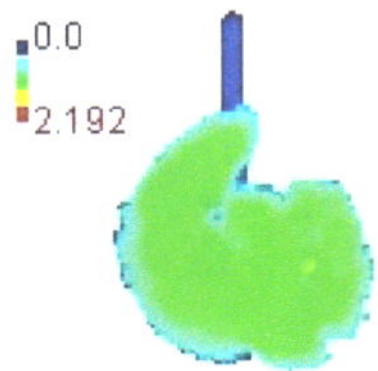  | 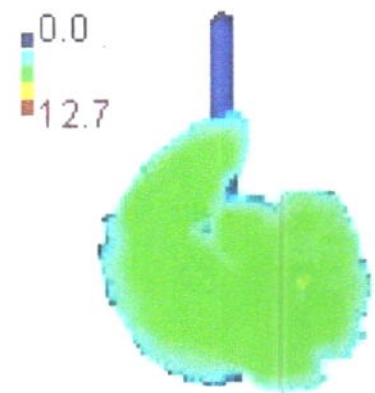  | 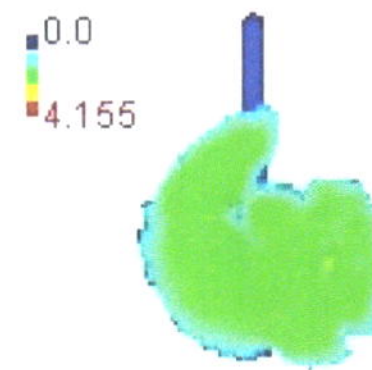  |
| 3-year | 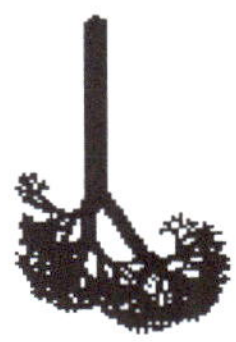 | 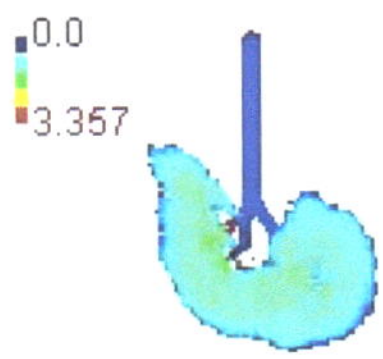  | 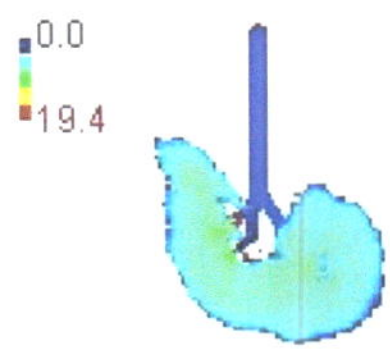  | 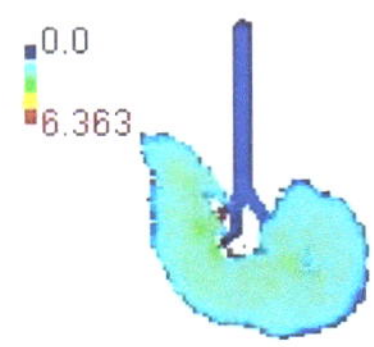  |
| 8-year | 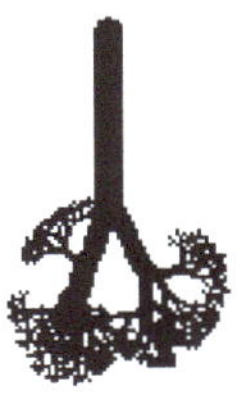 | 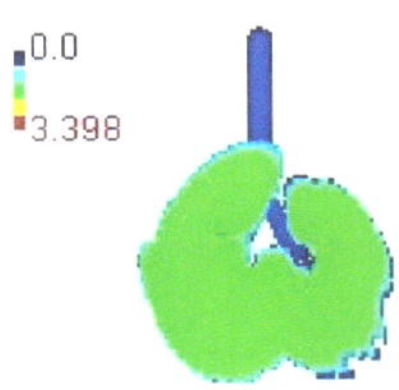  | 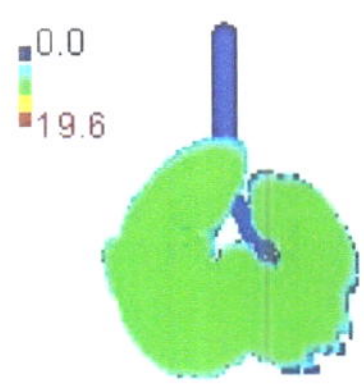  | 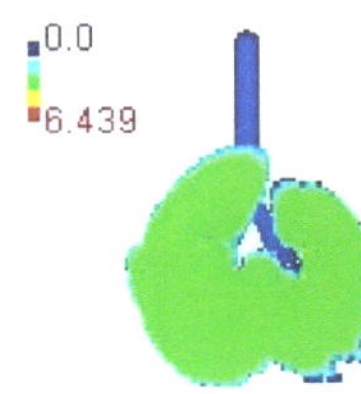  |
| 14-year | 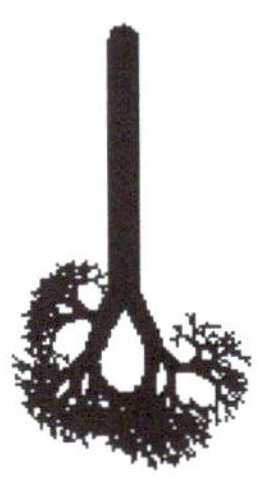 | 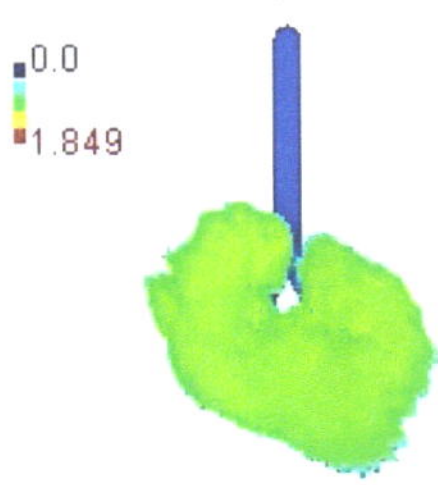  | 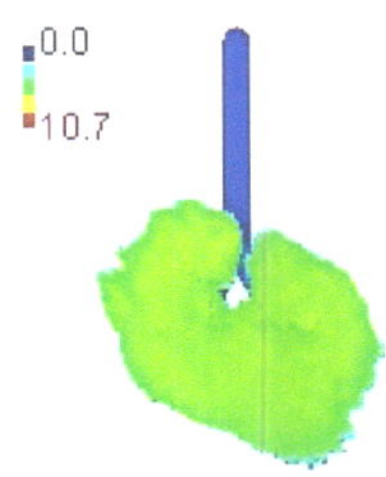  | 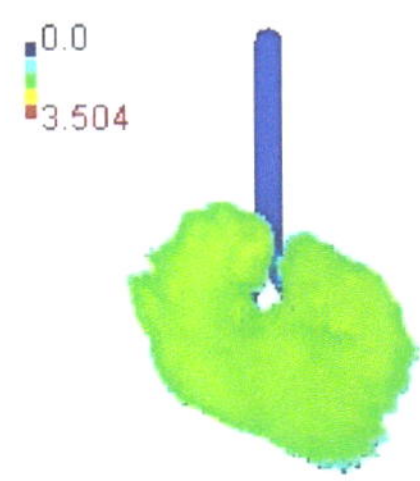  |
| 18-year | 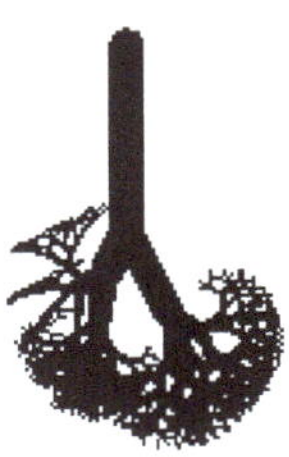 | 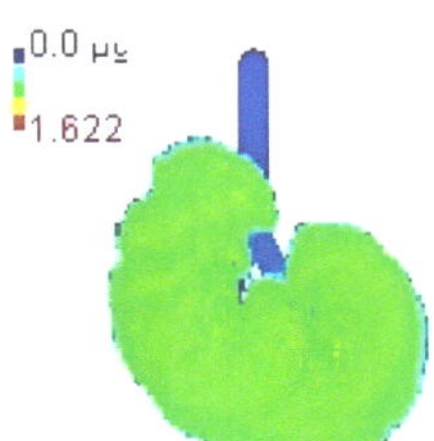  | 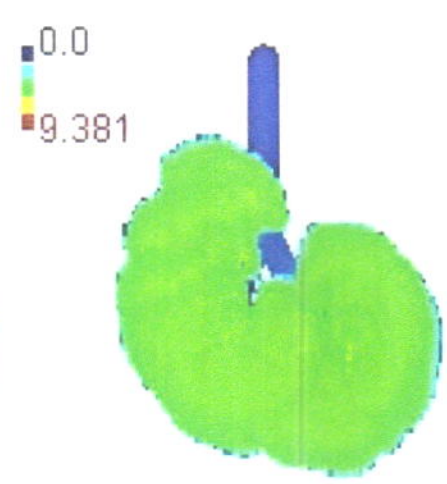  | 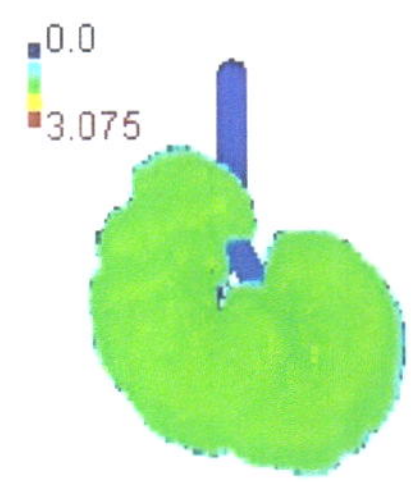  |

**Table 7.** *Cont.*

| Age Groups | Normal Lung Geometry | Pre-smog | Smog | Post-smog |
|---|---|---|---|---|
| 21-year |  | 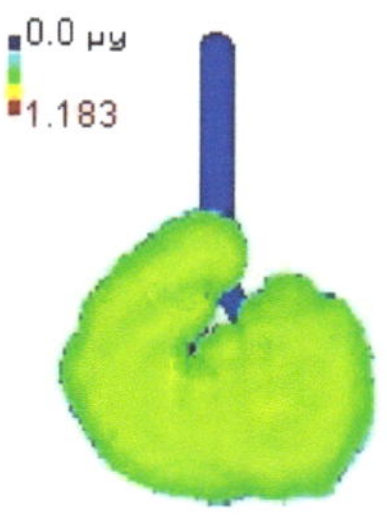  | 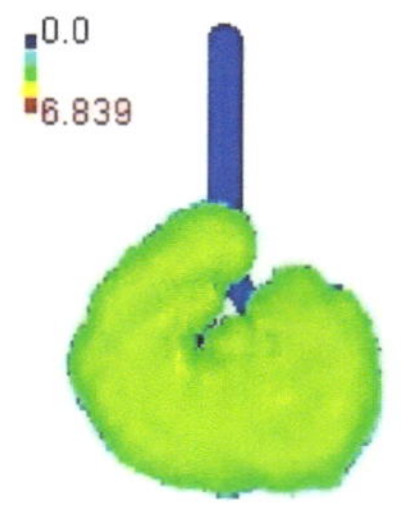  | 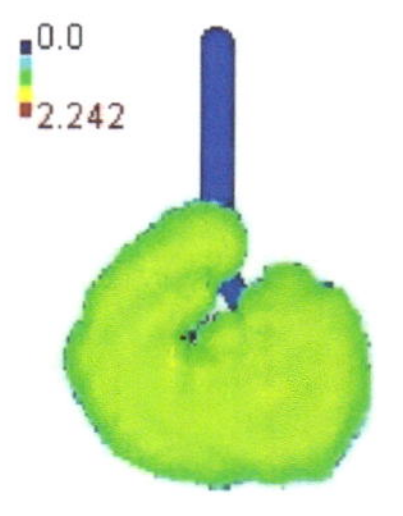  |
| 30-year | 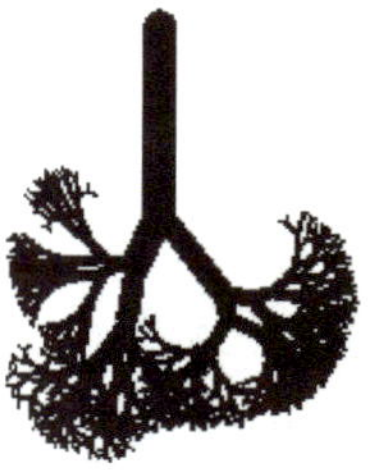 | 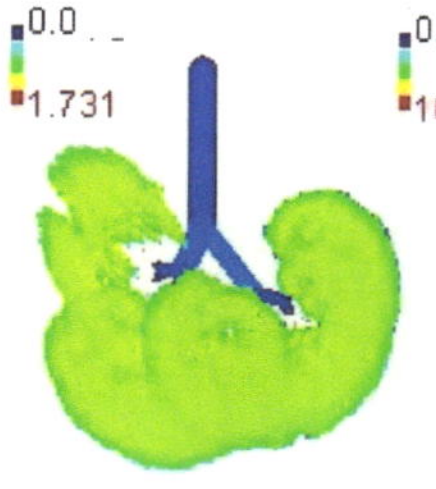  | 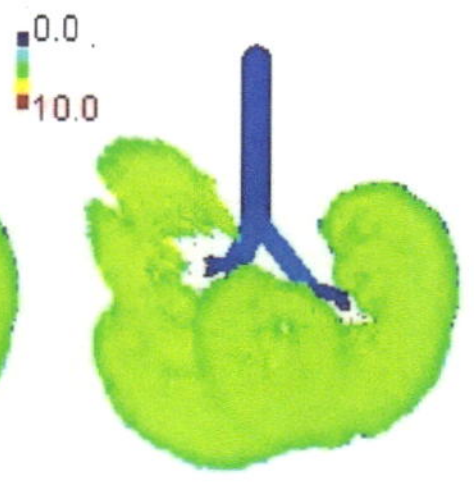  | 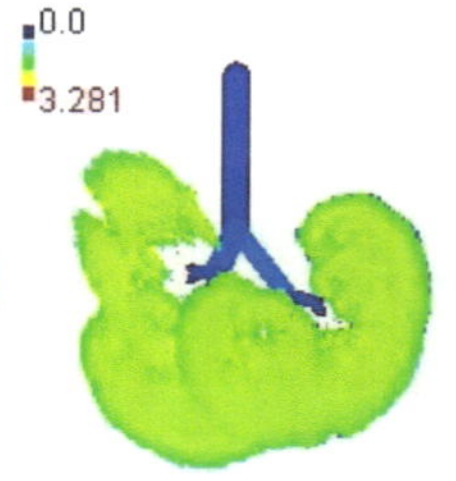  |

Mass flux maps are shown for different age groups pre-, during, and post-Smog Event-2017 in Table 7. Highest mass flux (µg/min/m$^2$) values were reported for 21-month-old, followed by 3-month-old children, whereas lowest values were reported for 21-year-old adults, which highlights that $PM_{2.5}$ exposure effects in terms of mass flux are higher for children than for adults (Table 7). This is due mainly to the fact that 21-month-old and 3-month-old babies have lowest lung surface area, whereas 21-year-old adults have largest lung surface area among all the age groups studied (Table 7). Highest mass flux during smog conditions were reported for 21-month-olds as $4.7 \times 10^2$ µg/min/m$^2$, followed by 3-month-olds as 49.2 µg/min/m$^2$,whereas lowest as 6.8 µg/min/m$^2$ for 21-year-old adults, which signify ~69 times and ~7 times more mass flux deposition in 21-month-old children, compared with 21-year-old adults (Table 6). This showed newborn baby and young child groups are more vulnerable than adults to $PM_{2.5}$ mass flux, especially under episodic cases such as smog, which raises serious concern for pollution management strategies to avoid pollution episodes. Highest $PM_{2.5}$ mass flux (µg/h/m$^2$) values in Chennai city, India, were reported as 0.12 (3-month), 0.05 (28-month), 0.08 (3-year), 0.11 (8-year), 0.43 (9-year), 0.06 (14-year), 0.06 (18-year), and 0.05 (21-years) (Table 6) [29].

### *3.4. Variations in Average Daily Dose of $PM_{2.5}$ Associated Elements during Smog Event-2017*

Variations in average daily dose (ADD; in µg/kg/day) of 18 $PM_{2.5}$-associated elements for children and adults on different days of sampling during Smog Event-2017 are shown in Tables 8 and 9. Higher ADD were found for children than adults during smog conditions and lowest on pre-smog day, showing more elemental deposition in children lungs during smog events than adults (Tables 8 and 9). During smog, highest ADD (µg/kg/day) was reported for different elements such as S ($1.81 \times 10^{-7}$), K ($2.01 \times 10^{-8}$), Ca ($1.17 \times 10^{-8}$), Cr ($3.27\times 10^{-8}$), Fe ($2.65 \times 10^{-8}$), Zn ($1.79\times 10^{-8}$), and Si ($2.37 \times 10^{-8}$) for adults (Table 8). For children, highest ADD (µg/kg/day) during smog was observed as S ($5.54 \times 10^{-7}$), Al ($2.81 \times 10^{-8}$), Cl ($2.28 \times 10^{-8}$), K ($6.15 \times 10^{-8}$), Ca ($3.57 \times 10^{-8}$), Cr ($1.0 \times 10^{-7}$), Fe ($8.09 \times 10^{-8}$), Zn ($5.48 \times 10^{-8}$), Pb ($1.71 \times 10^{-8}$), and Si ($7.24 \times 10^{-8}$) (Table 9). These highest ADD values during smog conditions show potential health effects of these elements after lung deposition in children and adults (Tables 8 and 9). ADD for potential toxic elements were reported in the order of Cr > Fe > Zn > Pb > Cu > Mn > Ni with their highest ADD concentrations during Smog Event-2017 having potential toxicity after lung deposition (Tables 8 and 9). In addition, values of ADD for toxic elements for children during post-smog condition (15 November 2017) were found comparable to those of ADD for adults during smog conditions. This signify that even after the smog episode was over in Delhi, children were still vulnerable to the post-effects of episodic cases such as smog in terms of $PM_{2.5}$ lung deposition and associated health effects. The ADD for children is found to be higher than ADD for adults in other studies, which shows more exposure of children to the elements than adults [70]. According to a study conducted in Delhi, in the case of human exposure to metals, individuals were found exposed most to Fe, Zn, and Co and least exposed to Cd, Cr, and Pb during normal (non-episodic) days [33].

**Table 8.** Average daily dose (μg/kg/day) of different elements present in $PM_{2.5}$ for adults during different days of sampling.

| Adult | 01 November 2017 | 08 November 2017 | 09 November 2017 | 10 November 2017 | 11 November 2017 | 13 November 2017 | 14 November 2017 | 15 November 2017 |
|---|---|---|---|---|---|---|---|---|
| Na (μg/kg/day) | $3.46 \times 10^{-10}$ | $1.96 \times 10^{-9}$ | $1.31 \times 10^{-9}$ | $7.31 \times 10^{-10}$ | $1.27 \times 10^{-9}$ | $1.00 \times 10^{-9}$ | $6.54 \times 10^{-10}$ | $6.54 \times 10^{-10}$ |
| Mg (μg/kg/day) | $3.46 \times 10^{-10}$ | $2.06 \times 10^{-9}$ | $1.37 \times 10^{-9}$ | $7.69 \times 10^{-10}$ | $1.31 \times 10^{-9}$ | $1.06 \times 10^{-9}$ | $6.93 \times 10^{-10}$ | $6.73 \times 10^{-10}$ |
| Al (μg/kg/day) | $1.6 \times 10^{-9}$ | $9.19 \times 10^{-9}$ | $6.08 \times 10^{-9}$ | $3.46 \times 10^{-9}$ | $5.89 \times 10^{-9}$ | $4.69 \times 10^{-9}$ | $3.06 \times 10^{-9}$ | $3.02 \times 10^{-9}$ |
| P (μg/kg/day) | $3.65 \times 10^{-10}$ | $2.14 \times 10^{-9}$ | $1.40 \times 10^{-9}$ | $8.08 \times 10^{-10}$ | $1.37 \times 10^{-9}$ | $1.10 \times 10^{-9}$ | $7.12 \times 10^{-10}$ | $6.93 \times 10^{-10}$ |
| S (μg/kg/day) | $3.13 \times 10^{-8}$ | $1.81 \times 10^{-7}$ | $1.20 \times 10^{-7}$ | $6.81 \times 10^{-8}$ | $1.16 \times 10^{-7}$ | $9.25 \times 10^{-8}$ | $6.03 \times 10^{-8}$ | $5.96 \times 10^{-8}$ |
| Cl (μg/kg/day) | $1.29 \times 10^{-9}$ | $7.46 \times 10^{-9}$ | $4.92 \times 10^{-9}$ | $2.81 \times 10^{-9}$ | $4.79 \times 10^{-9}$ | $3.81 \times 10^{-9}$ | $2.48 \times 10^{-9}$ | $2.44 \times 10^{-9}$ |
| K (μg/kg/day) | - | $2.01 \times 10^{-8}$ | $1.33 \times 10^{-8}$ | $7.56 \times 10^{-9}$ | $1.29 \times 10^{-8}$ | $1.03 \times 10^{-8}$ | $6.69 \times 10^{-9}$ | $3.48 \times 10^{-9}$ |
| Ca (μg/kg/day) | $2.02 \times 10^{-9}$ | $1.17 \times 10^{-8}$ | $7.73 \times 10^{-9}$ | $4.39 \times 10^{-9}$ | $7.48 \times 10^{-9}$ | $5.96 \times 10^{-9}$ | $3.89 \times 10^{-9}$ | $3.85 \times 10^{-9}$ |
| Cr (μg/kg/day) | $5.66 \times 10^{-9}$ | $3.27 \times 10^{-8}$ | $2.16 \times 10^{-8}$ | $1.23 \times 10^{-8}$ | $2.10 \times 10^{-8}$ | $1.67 \times 10^{-8}$ | $1.09 \times 10^{-8}$ | $1.08 \times 10^{-8}$ |
| Fe (μg/kg/day) | $4.58 \times 10^{-9}$ | $2.65 \times 10^{-8}$ | $1.75 \times 10^{-8}$ | $9.95 \times 10^{-9}$ | $1.70 \times 10^{-8}$ | $1.35 \times 10^{-8}$ | $8.81 \times 10^{-9}$ | $8.69 \times 10^{-9}$ |
| Ni (μg/kg/day) | $2.31 \times 10^{-10}$ | $1.33 \times 10^{-9}$ | $8.85 \times 10^{-10}$ | $5.00 \times 10^{-10}$ | $8.46 \times 10^{-10}$ | $6.73 \times 10^{-10}$ | $4.42 \times 10^{-10}$ | $4.42 \times 10^{-10}$ |
| Zn (μg/kg/day) | $3.1 \times 10^{-9}$ | $1.79 \times 10^{-8}$ | $1.18 \times 10^{-8}$ | $6.73 \times 10^{-9}$ | $1.15 \times 10^{-8}$ | $9.14 \times 10^{-9}$ | $5.96 \times 10^{-9}$ | $5.89 \times 10^{-9}$ |
| Mn (μg/kg/day) | $3.08 \times 10^{-10}$ | $1.75 \times 10^{-9}$ | $1.15 \times 10^{-9}$ | $6.54 \times 10^{-10}$ | $1.12 \times 10^{-9}$ | $8.85 \times 10^{-10}$ | $5.77 \times 10^{-10}$ | $5.77 \times 10^{-10}$ |
| Pb (μg/kg/day) | $9.62 \times 10^{-10}$ | $5.58 \times 10^{-9}$ | $3.69 \times 10^{-9}$ | $2.10 \times 10^{-9}$ | $3.58 \times 10^{-9}$ | $2.85 \times 10^{-9}$ | $1.87 \times 10^{-9}$ | $1.83 \times 10^{-9}$ |
| Cu (μg/kg/day) | $3.85 \times 10^{-10}$ | $2.21 \times 10^{-9}$ | $1.46 \times 10^{-9}$ | $8.27 \times 10^{-10}$ | $1.42 \times 10^{-9}$ | $1.14 \times 10^{-9}$ | $7.31 \times 10^{-10}$ | $7.31 \times 10^{-10}$ |
| Br (μg/kg/day) | - | $1.62 \times 10^{-9}$ | $1.08 \times 10^{-9}$ | $6.16 \times 10^{-10}$ | $1.04 \times 10^{-9}$ | $8.27 \times 10^{-10}$ | $5.39 \times 10^{-10}$ | $5.39 \times 10^{-10}$ |
| Si (μg/kg/day) | $4.1 \times 10^{-9}$ | $2.37 \times 10^{-8}$ | $1.56 \times 10^{-8}$ | $8.89 \times 10^{-9}$ | $1.52 \times 10^{-8}$ | $1.21 \times 10^{-8}$ | $7.89 \times 10^{-9}$ | $7.77 \times 10^{-9}$ |
| Ti (μg/kg/day) | $3.46 \times 10^{-10}$ | $2.02 \times 10^{-9}$ | $1.33 \times 10^{-9}$ | $7.50 \times 10^{-10}$ | $1.29 \times 10^{-9}$ | $1.04 \times 10^{-9}$ | $6.73 \times 10^{-10}$ | $6.73 \times 10^{-10}$ |

**Table 9.** Average daily dose (μg/kg/day) of different elements present in $PM_{2.5}$ for children during different days of sampling.

| Children | 01 November 2017 | 08 November 2017 | 09 November 2017 | 10 November 2017 | 11 November 2017 | 13 November 2017 | 14 November 2017 | 15 November 2017 |
|---|---|---|---|---|---|---|---|---|
| Na (μg/kg/day) | $1.06 \times 10^{-9}$ | $6.00 \times 10^{-9}$ | $4.00 \times 10^{-9}$ | $2.24 \times 10^{-9}$ | $3.88 \times 10^{-9}$ | $3.06 \times 10^{-9}$ | $2.00 \times 10^{-9}$ | $2.00 \times 10^{-9}$ |
| Mg (μg/kg/day) | $1.06 \times 10^{-9}$ | $6.29 \times 10^{-9}$ | $4.18 \times 10^{-9}$ | $2.35 \times 10^{-9}$ | $4.00 \times 10^{-9}$ | $3.24 \times 10^{-9}$ | $2.12 \times 10^{-9}$ | $2.06 \times 10^{-9}$ |
| Al (μg/kg/day) | $4.88 \times 10^{-9}$ | $2.81 \times 10^{-8}$ | $1.86 \times 10^{-8}$ | $1.06 \times 10^{-8}$ | $1.80 \times 10^{-8}$ | $1.44 \times 10^{-8}$ | $9.35 \times 10^{-9}$ | $9.24 \times 10^{-9}$ |
| P (μg/kg/day) | $1.12 \times 10^{-9}$ | $6.53 \times 10^{-9}$ | $4.29 \times 10^{-9}$ | $2.47 \times 10^{-9}$ | $4.18 \times 10^{-9}$ | $3.35 \times 10^{-9}$ | $2.18 \times 10^{-9}$ | $2.12 \times 10^{-9}$ |
| S (μg/kg/day) | $9.58 \times 10^{-8}$ | $5.54 \times 10^{-7}$ | $3.66 \times 10^{-7}$ | $2.08 \times 10^{-7}$ | $3.55 \times 10^{-7}$ | $2.83 \times 10^{-7}$ | $1.85 \times 10^{-7}$ | $1.82 \times 10^{-7}$ |
| Cl (μg/kg/day) | $3.94 \times 10^{-9}$ | $2.28 \times 10^{-8}$ | $1.51 \times 10^{-8}$ | $8.59 \times 10^{-9}$ | $1.47 \times 10^{-8}$ | $1.17 \times 10^{-8}$ | $7.59 \times 10^{-9}$ | $7.47 \times 10^{-9}$ |

**Table 9.** *Cont.*

| Children | 01 November 2017 | 08 November 2017 | 09 November 2017 | 10 November 2017 | 11 November 2017 | 13 November 2017 | 14 November 2017 | 15 November 2017 |
|---|---|---|---|---|---|---|---|---|
| K (μg/kg/day) | - | $6.15 \times 10^{-8}$ | $4.07 \times 10^{-8}$ | $2.31 \times 10^{-8}$ | $3.94 \times 10^{-8}$ | $3.14 \times 10^{-8}$ | $2.05 \times 10^{-8}$ | $1.07 \times 10^{-8}$ |
| Ca (μg/kg/day) | $6.18 \times 10^{-9}$ | $3.57 \times 10^{-8}$ | $2.37 \times 10^{-8}$ | $1.34 \times 10^{-8}$ | $2.29 \times 10^{-8}$ | $1.82 \times 10^{-8}$ | $1.19 \times 10^{-8}$ | $1.18 \times 10^{-8}$ |
| Cr (μg/kg/day) | $1.73 \times 10^{-8}$ | $1.00 \times 10^{-7}$ | $6.62 \times 10^{-8}$ | $3.76 \times 10^{-8}$ | $6.42 \times 10^{-8}$ | $5.11 \times 10^{-8}$ | $3.34 \times 10^{-8}$ | $3.29 \times 10^{-8}$ |
| Fe (μg/kg/day) | $1.4 \times 10^{-8}$ | $8.09 \times 10^{-8}$ | $5.35 \times 10^{-8}$ | $3.04 \times 10^{-8}$ | $5.19 \times 10^{-8}$ | $4.14 \times 10^{-8}$ | $2.69 \times 10^{-8}$ | $2.66 \times 10^{-8}$ |
| Ni (μg/kg/day) | $7.06 \times 10^{-10}$ | $4.06 \times 10^{-9}$ | $2.71 \times 10^{-9}$ | $1.53 \times 10^{-9}$ | $2.59 \times 10^{-9}$ | $2.06 \times 10^{-9}$ | $1.35 \times 10^{-9}$ | $1.35 \times 10^{-9}$ |
| Zn (μg/kg/day) | $9.47 \times 10^{-9}$ | $5.48 \times 10^{-8}$ | $3.62 \times 10^{-8}$ | $2.06 \times 10^{-8}$ | $3.51 \times 10^{-8}$ | $2.79 \times 10^{-8}$ | $1.82 \times 10^{-8}$ | $1.80 \times 10^{-8}$ |
| Mn (μg/kg/day) | $9.41 \times 10^{-10}$ | $5.35 \times 10^{-9}$ | $3.53 \times 10^{-9}$ | $2.00 \times 10^{-9}$ | $3.41 \times 10^{-9}$ | $2.71 \times 10^{-9}$ | $1.77 \times 10^{-9}$ | $1.77 \times 10^{-9}$ |
| Pb (μg/kg/day) | $2.94 \times 10^{-9}$ | $1.71 \times 10^{-8}$ | $1.13 \times 10^{-8}$ | $6.41 \times 10^{-9}$ | $1.09 \times 10^{-8}$ | $8.71 \times 10^{-9}$ | $5.71 \times 10^{-9}$ | $5.59 \times 10^{-9}$ |
| Cu (μg/kg/day) | $1.18 \times 10^{-9}$ | $6.76 \times 10^{-9}$ | $4.47 \times 10^{-9}$ | $2.53 \times 10^{-9}$ | $4.35 \times 10^{-9}$ | $3.47 \times 10^{-9}$ | $2.24 \times 10^{-9}$ | $2.24 \times 10^{-9}$ |
| Br (μg/kg/day) | - | $4.94 \times 10^{-9}$ | $3.29 \times 10^{-9}$ | $1.88 \times 10^{-9}$ | $3.18 \times 10^{-9}$ | $2.53 \times 10^{-9}$ | $1.65 \times 10^{-9}$ | $1.65 \times 10^{-9}$ |
| Si (μg/kg/day) | $1.25 \times 10^{-8}$ | $7.24 \times 10^{-8}$ | $4.78 \times 10^{-8}$ | $2.72 \times 10^{-8}$ | $4.64 \times 10^{-8}$ | $3.69 \times 10^{-8}$ | $2.41 \times 10^{-8}$ | $2.38 \times 10^{-8}$ |
| Ti (μg/kg/day) | $1.06 \times 10^{-9}$ | $6.18 \times 10^{-9}$ | $4.06 \times 10^{-9}$ | $2.29 \times 10^{-9}$ | $3.94 \times 10^{-9}$ | $3.18 \times 10^{-9}$ | $2.06 \times 10^{-9}$ | $2.06 \times 10^{-9}$ |

## 4. Conclusions

In the present study, lower concentrations of $PM_{2.5}$, elemental composition, and organic functional groups were observed during pre-smog conditions which drastically increased during Smog Event-2017 in Delhi. Pre-smog $PM_{2.5}$ concentration was reported as 124.0 μg/m$^3$ which drastically increased up to 717.2 μg/m$^3$ (~6 times higher) during the studied smog episode. The meteorological conditions, such as lower temperature, low wind speed, and higher RH played a crucial role in $PM_{2.5}$ accumulation, thereby, reducing the air visibility to a greater extent in Delhi during the smog. During smog conditions, significantly higher concentration of elements such as N, S, Cl, K, Cr, Fe, Zn, Pb, Cu, and Br and the presence of dominant organic functional groups in $PM_{2.5}$ confirmed biomass-burning signatures along with local emission sources in Delhi. Non-spherical-shaped individual particles dominated, followed by some spherical/nearly spherical-shaped particles with round, flaky structures, solid and irregular-shaped particles, carbon fractals, sticky aggregates, sharp-edged particles, and rod-shaped and rectangular particles during Smog Event-2017.

The MPPD model outputs suggested that during Smog Event-2017 much higher rate of $PM_{2.5}$ deposition potential (mass rate and mass flux) were reported than that of pre- or post-smog days. The highest deposition fraction and mass rate (μg/min) were reported for the alveolar region of the lungs, followed by the head region in some age groups of babies and children, whereas, for the head region, followed by alveolar region in the remainder of the age groups, including adults. This shows that mostly the baby and child groups were affected by particle deposition in the alveolar region, causing adverse health effects, whereas in adults and other age groups, most of the particle deposition took place in the head region, affecting upper respiratory tract more than the alveolar region. Mass flux (μg/min/m$^2$) in 3-month-old babies was found to be highest, followed by 21-month-old children, which shows babies and children were the most vulnerable to $PM_{2.5}$ mass flux during the smog period due to lower lung surface area. In addition, ADD (μg/kg/h) for $PM_{2.5}$-associated elements were found higher for children than adults. ADD for elements with potential toxicity, such as Cr, Fe, Zn, Pb, Cu, Mn, and Ni, were also found higher for children during smog conditions followed by the post-smog period, showing potential adverse effects of these toxic elements to children, even after the smog event was over. On the basis of this study, it can be suggested that there is an urgent need for atmospheric scientists, meteorologists, and policy makers to contemplate more stringent mitigation policies to combat pollution episodes such as smog to minimize the adverse health effects to the sensitive population, such as babies and children, in Delhi.

**Author Contributions:** S.F.: conceptualization, investigation, data curation, writing—original draft; S.K.M.: supervision, resources, writing—review and editing; A.A.: data curation, writing—review and editing; A.P.D.: writing—review and editing. All authors have read and agreed to the published version of the manuscript.

**Funding:** This research received no external funding.

**Institutional Review Board Statement:** The study did not require ethical approval as it does not include human beings as subjects.

**Informed Consent Statement:** Not applicable.

**Acknowledgments:** Authors would like to acknowledge the Director, CSIR-NPL, New Delhi for providing support during this research.

**Conflicts of Interest:** The authors declare no conflict of interest.

## Abbreviations

| | |
|---|---|
| MPPD | Multiple path particle dosimetry |
| PM | Particulate matter |
| $PM_{2.5}$ | Particulate matter with aerodynamic diameter less than 2.5 |
| IGP | Indo-Gangetic Plains |
| COPD | Chronic Obstructive Pulmonary Disease |
| NCT | National Capital Territory |
| CSIR-NPL | CSIR-National Physical Laboratory |
| PTFE | Polytetrafluoroethylene |
| SEM | Scanning Electron Microscopy |
| EDS | Energy Dispersive X-ray spectroscopy |
| WD-XRF | Wavelength Dispersive-X-ray Fluorescence |
| OP-FTIR | Open Path-Fourier Transform Infrared Spectroscopy |
| MIR | Mid-infrared |
| MCT | Mercury Cadmium Telluride |
| IMD | India Meteorological Department |
| HRT | Human respiratory tract |
| TB | Trachea-bronchial |
| P | Pulmonary |
| FRC | Functional residual capacity |
| URT | Upper respiratory tract |
| USEPA | United States Environment Protection Agency |
| ADD | Average daily dose |
| WHO | World Health Organization |
| NAAQS | National Ambient air Quality Standards |
| RH | Relative humidity |
| % | Percentage |
| NIST | National Institute of Standards and Technology |

## References

1. de Leeuw, F.A.A.M.; Moussiopoulos, N.; Sahm, P.; Bartonova, A. Urban air quality in larger conurbations in the European Union. *Environ. Model. Softw.* **2001**, *16*, 399–414. [CrossRef]
2. Angyal1, A.; Ferenczi, Z.; Manousakas, M.; Furu1, E.; Szoboszlai1, Z.; Török, Z.; Papp, E.; Szikszai1, Z.; Kertész, Z. Source identification of fine and coarse aerosol during smog episodes in Debrecen, Hungary. *Air Qual. Atmos. Health* **2021**, *14*, 1017–1032. [CrossRef]
3. Mikušaka, P.; Křůmal, K.; Večeřa, Z. Characterization of organic compounds in the $PM_{2.5}$ aerosols in winter in an industrial urban area. *Atmos. Environ.* **2015**, *105*, 97–108. [CrossRef]
4. Kanawade, V.P.; Srivastava, A.K.; Ram, K.; Asmi, E.; Vakkari, V.; Soni, V.K.; Varaprasad, V.; Sarangi, C. What caused severe air pollution episode of November 2016 in New Delhi? *Atmos. Environ.* **2020**, *222*, 117125. [CrossRef]
5. Sawlani, R.; Agnihotri, R.; Sharma, C.; Patra, P.K.; Dimri, A.P.; Ram, K.; Verma, R.L. The severe Delhi SMOG of 2016: A case of delayed crop residue burning, coincident firecracker emissions, and atypical meteorology. *Atmos. Pollut. Res.* **2019**, *10*, 868–879. [CrossRef]

6. Badarinath, K.V.S.; Kharol, S.K.; Sharma, A.R.; Krishna Prasad, V. Analysis of aerosol and carbon monoxide characteristics over Arabian Sea during crop residue burning period in the Indo-Gangetic Plains using multi-satellite remote sensing datasets. *J. Atmos. Sol. Terr. Phy.* **2009**, *71*, 1267–1276. [CrossRef]
7. Chowdhury, S.; Dey, S.; Guttikunda, S.; Pillarisetti, A.; Smith, K.R.; Di Girolamo, L. Indian annual ambient air quality standard is achievable by completely mitigating emissions from household sources. *Proc. Natl. Acad. Sci. USA* **2019**, *116*, 10711–10716. [CrossRef]
8. Reinmuth-Selzle, K.; Kampf, C.J.; Lucas, K.; Lang-Yona, N.; Fröhlich-Nowoisky, J.; Shiraiwa, M.; Lakey, P.S.J.; Lai, S.; Liu, F.; Kunert, A.T.; et al. Air Pollution and Climate Change Effects on Allergies in the Anthropocene: Abundance, Interaction, and Modification of Allergens and Adjuvants. *Environ. Sci. Technol.* **2017**, *51*, 4119–4141. [CrossRef]
9. Subramanian, M. Can Delhi save itself from its toxic air? *Nature* **2016**, *534*, 166–169. [CrossRef]
10. National Academy of Agricultural Sciences (NAAS). 2017. Available online: naas.org.in (accessed on 8 September 2022).
11. Li, W.; Shao, L.; Zhang, D.; Ro, C.-U.; Hu, M.; Bi, X.; Geng, H.; Matsuki, A.; Niu, H.; Chen, J. A review of single aerosol particle studies in the atmosphere of East Asia: Morphology, mixing state, source, and heterogeneous reactions. *J. Clean. Prod.* **2016**, *112*, 1330–1349. [CrossRef]
12. Jacobson, M.C.; Hansson, H.C.; Noone, K.J.; Charlson, R.J. Organic atmospheric aerosols: Review and state of the science. *Rev. Geophys.* **2000**, *38*, 267–294. [CrossRef]
13. Pöschl, U. Atmospheric Aerosols: Composition, Transformation, Climate and Health Effects. *Angew. Chem. Int. Ed.* **2005**, *44*, 7520–7540. [CrossRef] [PubMed]
14. Baklanov, A.; Grimmond, C.S.B.; Carlson, D.; Terblanche, D.; Tang, X.; Bouchet, V.; Lee, B.; Langendijk, G.; Kolli, R.K.; Hovsepyan, A. From urban meteorology, climate and environment research to integrated city services. *Urban Clim.* **2018**, *23*, 330–341. [CrossRef]
15. Zhang, J.; Liu, L.; Wang, Y.; Ren, Y.; Wang, X.; Shi, Z.; Zhang, D.; Che, H.; Zhao, H.; Liu, Y.; et al. Chemical composition, source, and process of urban aerosols during winter haze formation in Northeast China. *Environ. Pollut.* **2017**, *231*, 357–366. [CrossRef] [PubMed]
16. González-Flecha, B. Oxidant mechanisms in response to ambient air particles. *Mol. Aspects Med.* **2004**, *25*, 169–182. [CrossRef]
17. Aust, A.E.; Ball, J.C.; Hu, A.A.; Lighty, J.S.; Smith, K.R.; Straccia, A.M.; Veranth, J.M.; Young, W.C. Particle characteristics responsible for effects on human lung epithelial cells. *Res. Rep. Health Eff. Inst.* **2002**, *11*, 67–76.
18. Gilli, G.; Traversi, D.; Rovere, R.; Pignata, C.; Schilirò, T. Chemical characteristics and mutagenic activity of $PM_{10}$ in Torino, a Northern Italian City. *Sci. Total Environ.* **2007**, *385*, 97–107. [CrossRef]
19. Ostro, B.; Hu, J.; Goldberg, D.; Reynolds, P.; Hertz, A.; Bernstein, L.; Kleeman, M.J. Associations of mortality with long-term exposures to fine and ultrafine particles, species and sources: Results from the California Teachers Study Cohort. *Environ. Health Perspect.* **2015**, *123*, 549–556. [CrossRef]
20. Available online: unep.org (accessed on 8 September 2022).
21. Campbell, A.; Oldham, M.; Becaria, A.; Bondy, S.C.; Meacher, D.; Sioutas, C.; Misra, C.; Mendez, L.B.; Kleinman, M. Particulate Matter in polluted air may increase biomarkers of inflammation in mouse brain. *Neurotoxicology* **2006**, *26*, 133–140. [CrossRef]
22. Fayad, M.A.; Tsolakis, A.; Martos, F.J.; Bogarra, M.; Lefort, I.; Dearn, K.D. Investigation the effect of fuel injection strategies on combustion and morphology characteristics of PM in modern diesel engine operated with oxygenate fuel blending. *Therm. Sci. Eng. Prog.* **2022**, *35*, 101456. [CrossRef]
23. Valavanidis, A.; Fiotakis, K.; Vlachogianni, T. Airborne particulate matter and human health: Toxicological assessment and importance of size and composition of particles for oxidative damage and carcinogenic mechanisms. *J. Environ. Sci. Health C Environ. Carcinog. Ecotoxicol. Rev.* **2008**, *26*, 339–362. [CrossRef] [PubMed]
24. Löndahl, J.; Pagels, J.; Swietlicki, E.; Zhou, J.; Ketzel, M.; Massling, A.; Bohgard, M. A set-up for field studies of respiratory tract deposition of fine and ultrafine particles in humans. *J. Aerosol Sci.* **2006**, *37*, 1152–1163. [CrossRef]
25. Muenchen, H.Z. *Particulate Air Pollution: Exposure to Ultrafine Particles Influences Cardiac Function*; Science Daily; German Research Centre for Environmental Health: Neuherberg, Germany, 2015.
26. Pražnikar, Z.J.; Pražnikar, J. The Effects of Particulate Matter Air Pollution on Respiratory Health and on the Cardiovascular System. *Slov. J. Public Health* **2011**, *51*, 190–199. [CrossRef]
27. Peters, A.; Veronesi, B.; Calderòn-Garciduenas, L.; Gehr, P.; Chen, L.C.; Geiser, M.; Reed, W.; Rothen-Rutishauser, B.; Schürch, S.; Schulz, H. Translocation and potential neurological effects of fine and ultrafine particles a critical update. *Part. Fibre Toxicol.* **2006**, *3*, 13. [CrossRef] [PubMed]
28. Madhwal, S.; Prabhu, V.; Sundriyal, S.; Shridhar, V. Distribution, characterization and health risk assessment of size fractionated bioaerosols at an open landfill site in Dehradun, India. *Atmos. Pollut. Res.* **2020**, *11*, 156–169. [CrossRef]
29. Manojkumar, N.; Srimuruganandam, B.; Nagendra, S.M.S. Application of multiple-path particle dosimetry model for quantifying age specified deposition of particulate matter in human airway. *Ecotoxicol. Environ. Saf.* **2019**, *168*, 241–248. [CrossRef]
30. Lv, H.; Li, H.; Qiu, Z.; Zhang, F.; Song, J. Assessment of pedestrian exposure and deposition of $PM_{10}$, $PM_{2.5}$ and ultrafine particles at an urban roadside: A case study of Xi'an, China. *Atmos. Pollut. Res.* **2021**, *12*, 112–121. [CrossRef]
31. Izhar, S.; Goel, A.; Chakraborty, A.; Gupta, T. Annual trends in occurrence of submicron particles in ambient air and health risk posed by particle bound metals. *Chemosphere* **2016**, *146*, 582–590. [CrossRef]

32. Jena, S.; Singh, G. Human health risk assessment of airborne trace elements in Dhanbad, India. *Atmos. Pollut. Res.* **2017**, *8*, 490–502. [CrossRef]
33. Kushwaha, R.; Lal, H.; Srivastava, A.; Jain, V.K. Human exposure to particulate matter and their risk assessment over Delhi, India. *Natl. Acad. Sci. Lett.* **2012**, *35*, 497–504. [CrossRef]
34. Pothirat, C.; Chaiwong, W.; Liwsrisakun, C.; Bumroongkit, C.; Deesomchok, A.; Theerakittikul, T.; Limsukon, A.; Tajaroenmuang, P.; Phetsuk, N. Influence of Particulate Matter during Seasonal Smog on Quality of Life and Lung Function in Patients with Chronic Obstructive Pulmonary Disease. *Int. J. Environ. Res. Public Health* **2019**, *16*, 106. [CrossRef] [PubMed]
35. Carabali, G.; Villanueva-Macias, J.; Ladino, L.A.; Álvarez-Ospina, H.; Raga, G.B.; Andraca-Ayala, G.; Miranda, J.; Grutter, M.; Silva, M.M.; Riveros-Rosas, D. Characterization of aerosol particles during a high pollution episode over Mexico City. *Sci. Rep.* **2021**, *11*, 22533. [CrossRef]
36. Yazdani, A.; Dudani, N.; Takahama, S.; Bertrand, A.; Prévôt, A.S.H.; El Haddad, I.; Dillner, A.M. Characterization of primary and aged wood burning and coal combustion organic aerosols in an environmental chamber and its implications for atmospheric aerosols. *Atmos. Chem. Phys.* **2021**, *21*, 10273–10293. [CrossRef]
37. Gautam, R.; Patel, P.N.; Singh, M.K.; Liu, T.; Mickley, L.J.; Jethva, H.; DeFries, R.S. Extreme smog challenge of India intensified by increasing lower tropospheric stability. *EarthArXiv* **2021**. [CrossRef]
38. Sati, A.P.; Mohan, M. Analysis of air pollution during a severe smog episode of November 2012 and the Diwali Festival over Delhi, India. *Int. J. Remote Sens.* **2014**, *35*, 6940–6954. [CrossRef]
39. Available online: citypopulation.de/en/india/cities/delhi/ (accessed on 9 September 2022).
40. Goel, V.; Mishra, S.K.; Pal, P.; Ahlawat, A.; Vijayan, N.; Jain, S.; Sharma, C. Influence of chemical aging on physico-chemical properties of mineral dust particles: A case study of 2016 dust storms over Delhi. *Environ. Pollut.* **2020**, *267*, 115338. [CrossRef] [PubMed]
41. Available online: earth.google.com (accessed on 9 September 2022).
42. Fatima, S.; Sehgal, A.; Mishra, S.K.; Mina, U.; Goel, V.; Vijayan, N.; Tawale, J.S.; Kothari, R.; Ahlawat, A.; Sharma, C. Particle composition and morphology over urban environment (New Delhi): Plausible effects on wheat leaves. *Environ. Res.* **2021**, *202*, 111552.
43. Available online: ready.noaa.gov/hypub-bin/trajtype.pl (accessed on 9 September 2022).
44. Available online: firms.modaps.eosdis.nasa.gov (accessed on 10 September 2022).
45. Available online: hindustantimes.com (accessed on 10 September 2022).
46. Available online: ara.com/mppd/ (accessed on 10 September 2022).
47. Goel, V.; Mishra, S.K.; Lodhi, N.; Singh, S.; Ahlawat, A.; Gupta, B.; Das, R.M.; Kotnala, R.K. Physico-chemical characterization of individual Antarctic particles: Implications to aerosol optics. *Atmos. Environ.* **2018**, *192*, 173–181.
48. Sarangi, B.; Aggarwal, S.G.; Sinha, D.; Gupta, P.K. Aerosol effective density measurement using scanning mobility particle sizer and quartz crystal microbalance with the estimation of involved uncertainty. *Atmos. Meas. Technol.* **2016**, *9*, 859–875. [CrossRef]
49. USEPA. Guidance for evaluating the oral bioavailability of metals in soils for use in human health risk assessment. *OSWER* **2007**, *9285*, 7–80.
50. Jimenez, J.L.; Canagaratna, M.R.; Donahue, N.M.; Prevot, A.S.H.; Zhang, Q.; Kroll, J.H.; DeCarlo, P.F.; Allan, J.D.; Coe, H.; Ng, N.L.; et al. Evolution of organic aerosols in the atmosphere. *Science* **2009**, *326*, 1525–1529. [CrossRef] [PubMed]
51. Zhang, X.; Wu, Y.; Gu, B. Characterization of haze episodes and factors contributing to their formation using a panel model. *Chemosphere* **2016**, *149*, 320–327. [CrossRef]
52. Janhäll, S. Review on urban vegetation and particle air pollution–Deposition and dispersion. *Atmos. Environ.* **2015**, *105*, 130–137. [CrossRef]
53. Available online: ppcb.punjab.gov.in/en (accessed on 10 September 2022).
54. Minguillón, M.C.; Querol, X.; Baltensperger, U.; Prévôt, A.S.H. Fine and coarse PM composition and sources in rural and urban Switzerland: Local or regional pollution? *Sci. Total Environ.* **2012**, *427–428*, 191–202. [CrossRef] [PubMed]
55. Moffet, R.C.; Desyaterik, Y.; Hopkins, R.J.; Tivanki, A.V.; Gilles, M.K.; Wang, Y.; Shutthanandan, V.; Molina, L.T.; Abraham, R.G.; Johnsin, K.S.; et al. Characterization of aerosols containing Zn, Pb, and Cl from an industrial region of Mexico City. *Environ. Sci. Technol.* **2008**, *42*, 7091–7097. [CrossRef]
56. Li, W.; Sun, J.; Xu, L.; Shi, Z.; Riemer, N.; Sun, Y.; Fu, P.; Zhang, J.; Lin, Y.; Wang, X.; et al. A conceptual framework for mixing structures in individual aerosol particles. *J. Geophys. Res. Atmos.* **2016**, *121*, 13784–13798. [CrossRef]
57. Huang, K.; Zhuang, G.; Lin, Y.; Fu, J.S.; Wang, Q.; Liu, T.; Zhang, R.; Jiang, Y.; Deng, C.; Fu, Q.; et al. Typical types and formation mechanisms of haze in an Eastern Asia Megacity, Shanghai. *Atmos. Chem. Phys.* **2012**, *12*, 105–124. [CrossRef]
58. Cao, G.; Yan, Y.; Zou, X.; Zhu, R.; Ouyang, F. Applications of infrared spectroscopy in analysis of organic aerosols. *Spectr. Anal. Rev.* **2018**, *6*, 12–32. [CrossRef]
59. Liu, S.; Ahlm, L.; Day, D.A.; Russell, L.M.; Zhao, Y.; Gentner, D.R.; Weber, R.J.; Goldstein, A.H.; Jaoui, M.; Offenberg, J.H.; et al. Secondary organic aerosol formation from fossil fuel sources contribute majority of summertime organic mass at Bakersfield. *J. Geophys. Res. Atmos.* **2012**, *117*. [CrossRef]
60. Schade, G.W.; Crutzen, P.J. Emission of aliphatic amines from animal husbandry and their reactions: Potential source of $N_2O$ and HCN. *J. Atmos. Chem.* **1995**, *22*, 319–346. [CrossRef]

61. Gilman, J.B.; Lerner, B.M.; Kuster, W.C.; Goldan, P.D.; Warneke, C.; Veres, P.R.; Roberts, J.M.; De Gouw, J.A.; Burling, I.R.; Yokelson, R.J. Biomass burning emissions and potential air quality impacts of volatile organic compounds and other trace gases from fuels common in the US. *Atmos. Chem. Phys.* **2015**, *15*, 13915–13938. [CrossRef]
62. Liu, A.; Wang, H.; Cui, Y.; Shen, L.; Yin, Y.; Wu, Z.; Guo, S.; Shi, S.; Chen, K.; Zhu, B.; et al. Characteristics of Aerosol during a Severe Haze-Fog Episode in the Yangtze River Delta: Particle Size Distribution, Chemical Composition, and Optical Properties. *Atmosphere* **2020**, *11*, 56. [CrossRef]
63. Zhai, Y.; Liu, X.; Chen, H.; Xu, B.; Zhu, L.; Li, C.; Zeng, G. Source identification and potential ecological risk assessment of heavy metals in $PM_{2.5}$ from Changsha. *Sci. Total Environ.* **2014**, *493*, 109–115. [CrossRef] [PubMed]
64. Bhardwaj, P.; Singh, B.P.; Pandey, A.K.; Jain, V.K.; Kumar, K. Characterization and Morphological Analysis of Summer and Wintertime $PM_{2.5}$ Aerosols over Urban-Rural Locations in Delhi-NCR. *Int. J. Appl. Environ. Sci.* **2017**, *12*, 1009–1030.
65. Mishra, S.K.; Agnihotri, R.; Yadav, P.K.; Singh, S.; Prasad, M.V.S.N.; Praveen, P.S.; Tawale, J.S.; Mishra, N.D.; Arya, B.C.; Sharma, C. Morphology of atmospheric particles over Semi-Arid region (Jaipur, Rajasthan) of India: Implications for optical properties. *Aerosol Air Qual. Res.* **2015**, *15*, 974–984. [CrossRef]
66. Mishra, S.K.; Saha, N.; Singh, S.; Sharma, C.; Prasad, M.V.S.N.; Gautam, S.; Misra, A.; Gaur, A.; Bhattu, D.; Ghosh, S.; et al. Morphology, mineralogy and mixing of individual atmospheric particles over Kanpur (IGP): Relevance of homogeneous equivalent sphere approximation in radiative models. *MAPAN* **2017**, *32*, 229–241. [CrossRef]
67. Xiong, C.; Friedlander, S.K. Morphological properties of atmospheric aerosol aggregates. *Proc. Natl. Acad. Sci. USA* **2001**, *98*, 11851–11856. [CrossRef]
68. China, S.; Mazzoleni, C.; Gorkowski, K.; Aiken, A.C.; Dubey, M.K. Morphology and mixing state of individual freshly emitted wildfire carbonaceous particles. *Nat. Commun.* **2013**, *4*, 2122. [CrossRef]
69. Møller, P.; Folkmann, J.K.; Forchhammer, L.; Bräuner, E.V.; Danielsen, P.H.; Risom, L.; Loft, S. Air pollution, oxidative damage to DNA, and carcinogenesis. *Cancer Lett.* **2008**, *266*, 84–97. [CrossRef]
70. Xu, X.; Lu, X.; Han, X.; Zhao, N. Ecological and health risk assessment of metal in re-suspended particles of urban street dust from an industrial city in China. *Curr. Sci.* **2015**, *108*, 72–79.

International Journal of
*Environmental Research and Public Health*

MDPI

*Article*

# Spatial Distribution of Air Pollution, Hotspots and Sources in an Urban-Industrial Area in the Lisbon Metropolitan Area, Portugal—A Biomonitoring Approach

**Leonor Abecasis [1], Carla A. Gamelas [1,2], Ana Rita Justino [1], Isabel Dionísio [1], Nuno Canha [1,3,*], Zsofia Kertesz [4] and Susana Marta Almeida [1]**

[1] Centro de Ciências e Tecnologias Nucleares, Instituto Superior Técnico, Universidade de Lisboa, Estrada Nacional 10, 2695-066 Bobadela, Portugal; maria.leonor.rente.abecasis@tecnico.ulisboa.pt (L.A.); carla.gamelas@ctn.tecnico.ulisboa.pt (C.A.G.); ana.justino@tecnico.ulisboa.pt (A.R.J.); dionisio@ctn.tecnico.ulisboa.pt (I.D.); smarta@ctn.tecnico.ulisboa.pt (S.M.A.)
[2] ESTSetúbal/IPS and CINEA, IPS Campus, Polytechnic Institute of Setúbal, 2914-508 Setúbal, Portugal
[3] CESAM—Centre for Environmental and Marine Studies, Department of Environment and Planning, University of Aveiro, 3810-193 Aveiro, Portugal
[4] Laboratory for Heritage Science, Institute for Nuclear Research, H-4026 Debrecen, Hungary; zsofi@atomki.hu
* Correspondence: nunocanha@ctn.tecnico.ulisboa.pt

**Abstract:** This study aimed to understand the influence of industries (including steelworks, lime factories, and industry of metal waste management and treatment) on the air quality of the urban-industrial area of Seixal (Portugal), where the local population has often expressed concerns regarding the air quality. The adopted strategy was based on biomonitoring of air pollution using transplanted lichens distributed over a grid to cover the study area. Moreover, the study was conducted during the first period of national lockdown due to COVID-19, whereas local industries kept their normal working schedule. Using a set of different statistical analysis approaches (such as enrichment and contamination factors, Spearman correlations, and evaluation of spatial patterns) to the chemical content of the exposed transplanted lichens, it was possible to assess hotspots of air pollution and to identify five sources affecting the local air quality: (i) a soil source of natural origin (based on Al, Si, and Ti), (ii) a soil source of natural and anthropogenic origins (based on Fe and Mg), (iii) a source from the local industrial activity, namely steelworks (based on Co, Cr, Mn, Pb, and Zn); (iv) a source from the road traffic (based on Cr, Cu, and Zn), and (v) a source of biomass burning (based on Br and K). The impact of the industries located in the study area on the local air quality was identified (namely, the steelworks), confirming the concerns of the local population. This valuable information is essential to improve future planning and optimize the assessment of particulate matter levels by reference methods, which will allow a quantitative analysis of the issue, based on national and European legislation, and to define the quantitative contribution of pollution sources and to design target mitigation measures to improve local air quality.

**Keywords:** air pollution; biomonitoring; transplanted lichens; spatial analysis; urban-industrial area; steelworks; source apportionment



**Citation:** Abecasis, L.; Gamelas, C.A.; Justino, A.R.; Dionísio, I.; Canha, N.; Kertesz, Z.; Almeida, S.M. Spatial Distribution of Air Pollution, Hotspots and Sources in an Urban-Industrial Area in the Lisbon Metropolitan Area, Portugal—A Biomonitoring Approach. *Int. J. Environ. Res. Public Health* **2022**, *19*, 1364. https://doi.org/10.3390/ijerph19031364

Academic Editor: Tareq Hussein

Received: 23 November 2021
Accepted: 21 January 2022
Published: 26 January 2022



## 1. Introduction

Industrial emissions have an important contribution to particulate matter (PM) in urban-industrial areas [1,2]. To solve environmental problems in this type of areas it is crucial to understand and identify prevailing emission sources to promote targeted and successful mitigation measures.

It is known that particulate matter may have adverse health effects, since PM may contain potentially toxic elements (PTEs) [3–5]. Thus, the determination of the chemical composition of PM aims not only to determine the sources of the PM sampled at receptor sites but also to identify its potential health hazards.

In the last years, the episodic deposition of airborne particulates onto homes and properties was a common complaint from the residents of Aldeia de Paio Pires-Seixal (municipality of Seixal of the Lisbon metropolitan area), an urban-industrial area in Portugal [6]. To answer the concerns raised by the population, the local Council of Seixal promoted a set of actions, such as the assessment of the chemical composition of the settled dust to determine its sources and potential health hazards [7]. In this previous study, the settled dust was collected in January 2019, characterized by micro-PIXE, and the influence of the steel industries was identified due to the content of Fe, Cr, and Mn, along with a minor traffic influence.

PM emissions from steelworks are a complex mixture of stationary and diffuse emissions, associated with the process and with operations such as stocking and transportation of raw materials and slags [1,8,9]. Atmospheric PM from iron and steel industries have high concentrations of As, Cd, Cr, Cu, Fe, Mn, Ni, Se, V, and Zn [10–12]. Moreover, steelworks are often close to other industries and traffic zones, making it difficult to distinguish between the contributions of the processes.

Traditionally, air pollution studies are performed through instrumental techniques that are limited to a small number of sampling stations [13]. Biomonitoring offers several advantages over standard sampling methods since biomonitors can be used in vast areas with many monitoring points while requiring little maintenance and are low cost [14]. Biomonitors can be native organisms in the ecosystem or they can be collected in an unpolluted site and transplanted to the area of interest [15,16].

Lichens have the ability to accumulate elements, in correlation with atmospheric levels. In fact, since lichens have no roots, they are dependent on the atmosphere for the uptake of water and mineral substances [17]. Lichens are extensively used in biomonitoring studies [13,18–22] and for spatial mapping of air contaminants [14,23,24], allowing the identification of pollution hotspots.

Continuing the set of actions promoted by the local Council of Seixal to understand the population exposure and the associated health risks of the settled dust events, the present study was conducted for the Council of Seixal and it aimed to assess the spatial distribution of air pollution, to identify its hotspots and its sources, by the use of the biomonitoring technique of air pollution using transplanted lichens in the Aldeia de Paio Pires-Seixal. The elemental characterization of the transplanted lichens was performed to assess the potential pollution sources, through the analysis of their chemical tracers. The Micro-X-ray Fluorescence technique was applied since it has been previously used to quantify metals in lichens and has several advantages: it is nondestructive, presents no need for sample digestion, and accuracy and reproducibility are at least equal to spectroscopic methods [25].

## 2. Materials and Methods

### 2.1. Study Area

This study was carried out in the parish of União das Freguesias do Seixal, Arrentela e Aldeia de Paio Pires (UFSAAPP), in the municipality of Seixal (Portugal), which is located in the peninsula of Setúbal and it is a part of the Lisbon Metropolitan Area (Portugal), next to the Nature Reserve of the Tagus Estuary (Figure 1). The municipality of Seixal is one of the most densely populated municipalities in Portugal, with 167,294 inhabitants in 95.5 $Km^2$ [26] (PORDATA, 2021).

Besides including a significant network of highways and national roads (A2-IP7, A33, and EN10), the study area (UFSAAPP) comprises small and medium-sized industries and an industrial park where the following facilities are located (as shown in Figure 1): industry A–that is a steelwork that manufactures galvanized sheet metal (with Cr passivation) and cold rolled sheet, with an installed capacity of 800,000 tons per year [27]; industry B–that is steelwork with an installed capacity of 215 tons per hour for the production of steel, where the main processes consist of an Electric Arc Furnace for steelmaking and hot rolling [28]; industry C–that it is a lime factory, which manufactures lime by calcination of limestone in a coke kiln [29]; and industry D that is a company focused on metal waste (such as iron and aluminum) management and treatment (recovered iron is a raw material for industry B).

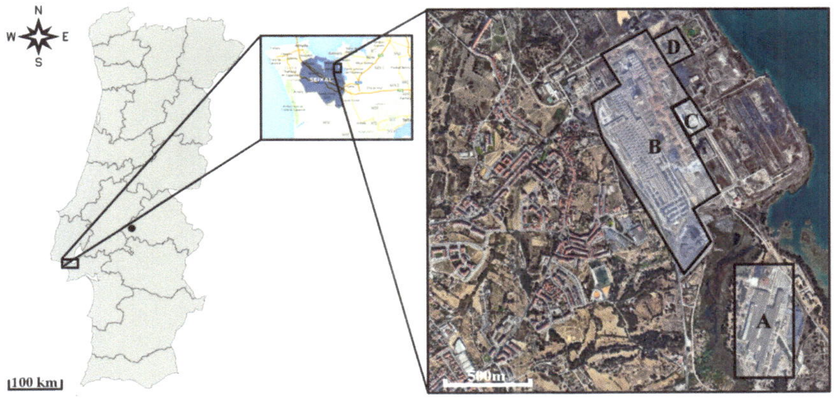


**Figure 1.** Location of the study area: (**left**) Framework of the study area (black rectangle) and the clean area (black dot, Montargil) in Portugal mainland; (**center**) Seixal municipality location; (**right**) location of industries A, B, C and D within the study area (geographical data obtained from Google Earth Pro [30]).

## 2.2. Transplantation and Sampling

Samples of the lichen *Flavoparmelia caperata* (L.) Hale were collected from olive trees at about 1.5 m above the soil, in Montargil (39°03′24″ N, 8°10′36″ W) (as shown in Figure 1), on 22 January 2020. Montargil is a rural area considered clean from an air pollution point of view [31]. Lichens were collected using powderless gloves and stored temporarily in paper envelopes to be transported to the laboratory. In the laboratory, four lichen samples were separated randomly, as reference base levels. Lichens were placed in nylon mesh bags and exposed in the study area, fixed to an appropriate substrate using nylon string, at about 1.8 m above the soil (Figure 2). The exposure period of the samples was approximately four and a half months (total of 137 days), from 1 February to 17 June 2020.

**Figure 2.** Exposed transplanted *Flavoparmelia caperata* lichens in nylon bags (**right**) that were placed in the study area (**left**).

A georeferenced grid of 4.55 km × 6.82 km, with 77 cells of 650 m × 620 m, was drawn for the lichens exposure, between the coordinates −9.11, 38.65 and −9.05, 38.59 (upper left and lower right corner of the grid), corresponding to the extremes of UFSAAPP (Figure 3).

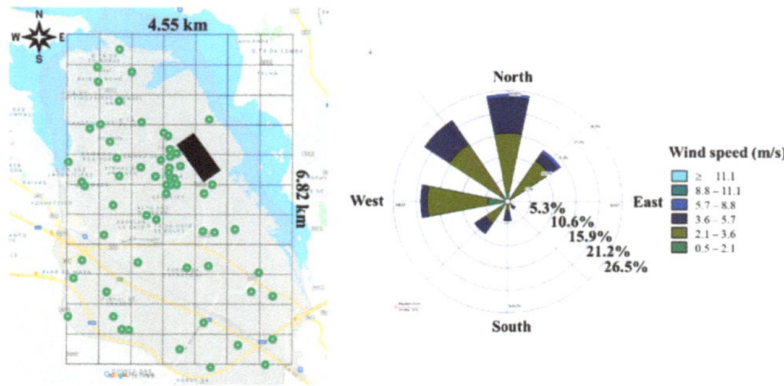


**Figure 3.** Grid of the spatial distribution of the transplanted and retrieved lichens after the exposure period (green dots) in the study area (**left**), with identification of the industrial area described before (black rectangle) (geographical data obtained from Google Earth Pro [30]), and (**right**) wind rose of the prevailing winds monitored in the weather station described in Section 2.3 (data obtained from the Portuguese Institute of Sea and Atmosphere, IPMA), during the exposure period.

Lichens samples were distributed over the study area based on two strategies: (1) participation of the local population by exposing the transplanted lichens in outdoor of their dwellings, and (2) to have good coverage of the study area, lichen samples were exposed in selected sites to cover the established grid cells that were not covered by the strategy 1. For the population engagement, a public presentation session was held on 1 February 2020, to explain the study and to identify volunteers. A total of 88 lichen samples were exposed but only 63 could be retrieved at the end of the exposure period and subsequently analyzed (retrieved lichen samples are shown in Figure 3, left).

### 2.3. Meteorological Data

Meteorological data for the exposure period were obtained from the Barreiro-Lavradio weather station next to the study area (38°40′28″ N, 9°2′51″ W, it is the nearest weather station, and it is located at 5.5 km from the study area) and it was supplied by the Portuguese Institute of Sea and Atmosphere (Instituto Português do Mar e da Atmosfera-IPMA).

During the exposure period, the average temperature in the study area was 16.9 ± 3.2 °C (ranging from 9.7 to 25.1 °C) and the average relative humidity was 80 ± 8% (ranging from 63 to 98%), with a total of 36 rainy days registered.

The wind rose for the exposure period is presented in Figure 3, where a predominance of winds from North (N), Northwest (NW), and West (W) was registered.

### 2.4. Chemical Analysis

At the laboratory, lichens were cleaned and cleared of extraneous material (such as dust, bark remaining, leaf debris, fungus contamination, and degraded material), and rinsed for 5 s with demineralized water, freeze-dried, and ground into powder in a ball mill with PTFE capsules under liquid nitrogen, for homogenization [32]. Pellets with an average thickness of 12 mm, were then prepared for elemental characterization by micro-X-ray Fluorescence [33]. For each lichen sample, a total of three pellet replicates were done. Figure S1 provides an overview of the procedure for the preparation of pellets.

Micro-X-ray Fluorescence (micro-XRF) analysis was conducted in the Laboratory for Heritage Science (Debrecen, Hungary), ATOMKI, Debrecen, Hungary using a Bruker M4 Tornado micro-XRF equipment (Bruker Corporation, Billerica, MA, USA). All measurements were made in a vacuum (~20 mbar). Rh excitation source set to 50 kV with 300 μA current was applied. The beam was focused down to 25 μm using polycapillary optics. To set the optimal measurement conditions, different filter combinations were applied on selected samples: (1) no filter, (2) 12.5 μm Al, (3) 100 μm Al + 25 μm Ti [34]. Test measurements to detect elements like Cd, Sn, Sb were carried out with W (50 kV, 700 μA) excitation source. Since Cd, Sn, Sb were under detection limit even with the W tube, an Rh excitation source was applied for the measurements. To achieve the result for the widest possible elemental range (Na-U), measurements were performed without a filter.

X-ray spectra were collected simultaneously with 2 detectors (XFlash® SDD with 30 $mm^2$ active surfaces, Be window). Spectra and maps of ~6 mm × 7 mm areas were acquired to get the best estimate of the average composition of the samples and to eliminate the possible fluctuations in concentrations due to sampling inhomogeneity. For each lichen sample, only one of the three replicate pellets was measured. Regarding the quality control of the measurements, randomly selected pellets were re-measured 2 more times on different days with the same settings. In the case of 5 random pellets, both sides were measured, and in the case of 6 samples, all of their replicate prepared pellets were analyzed (three replicates per sample). In all cases, the resulting concentration data were within 2 sigmas.

For the quantification, the fundamental parameter (FA) method [35] was applied using the MQuant built-in software [36]. The composition of cellulose $(C_6H_{10}O_5)_n$ was set as an "unknown" matrix.

The elemental composition of exposed and unexposed biomonitors was assessed for a total of 20 elements (Al, As, Br, Ca, Co, Cr, Cu, Fe, K, Mg, Mn, Pb, Rb, S, Se, Si, Sr, Ti, Zn, and Zr).

### 2.5. Statistical Analysis

Statistical analysis was performed using STATISTICA software version 13. The variables in the data set exhibited a non-normal distribution, and so the Mann-Whitney U nonparametric statistics was applied [37], at a significance level of 0.05, for independent groups, to suggest whether samples come from the same population or not. One Sample Wilcoxon nonparametric test was also applied, at a significance level of 0.05, to indicate if there is a significant difference between the median of a sample group and an hypothesized value [37]. Spearman correlations [37] were used to understand the associations between parameters.

The enrichment factor (EF) is used for identifying the crustal and non-crustal origin of elements [38], and it is applied to particulate matter collected in filters [14,39], plants/lichens [32,40], and sediments [41]. The enrichment factor of the element was determined for each sample, using the reference values of soil composition (Table S1) [42] and taking Si as the crustal reference element [43], according to Equation (1):

$$EF_X = \frac{\left(\frac{[X]}{[Si]}\right)_{Lichen}}{\left(\frac{[X]}{[Si]}\right)_{Soil}} \tag{1}$$

where $\left(\frac{[X]}{[Si]}\right)_{Lichen}$ is the ratio between the concentrations of element $X$ and $Si$ in the lichen and $\left(\frac{[X]}{[Si]}\right)_{Soil}$ is the ratio between the reference concentrations of the element $X$ and $Si$ in *Soil*. To account for the local variation in the soil composition in the EF calculations, elements with EF values between 1 and 10 are considered to be of a crustal origin [32], while elements with EF values greater than 10 were considered to be enriched by other sources rather than the crust [44].

To compare the mean elemental concentrations assessed in the exposed lichens with the unexposed lichens, the calculation of the Contamination Factor (CF) was conducted, where CF for a specific element is the ratio between the exposed and the unexposed lichen [45]. CF is also known as Exposure to Control ratio, which has been divided in five classes [46], namely (i) 0–0.25: severe loss, (ii) 0.25–0.75: loss, (iii) 0.75–1.25: normal, (iv) 1.25–1.75: accumulation, and (v) >1.75: severe accumulation.

Geostatistical modeling maps were built using ArcGIS 10.1 software (ESRI, 2020, Redlands, CA, USA), for studying the geospatial distribution of the elemental concentrations in the study area. The mapping was performed using the IDW (Inverse Distance Weighted) tool, which has been applied in similar studies [45]. IDW calculates the cell value for the unmeasured location, averaging the sampled data around each processing cell; the closer a measured point is to the center of the prediction cell, the more weight it will have [47]. In the overall equation for IDW, $v_0$ represents the estimated value at point 0 (Equation (2)), $v_i$ the value in the known point I, $d_i$ is the distance to point I, $S$ is the number of known points applied in the estimate, and $k$ is assumed to equal to 2. A set of 12 neighboring samples was chosen to represent the spatial variation of the elements under study.

$$v_0 = \frac{\sum_{i=1}^{S} v_i \left(\frac{1}{d_i^k}\right)}{\sum_{i=1}^{S} \left(\frac{1}{d_i^k}\right)} \quad (2)$$

The range between the maximum and minimum concentrations obtained for each element was subdivided into five bands, corresponding to 0–20, 20–40, 40–60, 60–80, and 80–100 percentiles.

## 3. Results and Discussion

### 3.1. Elemental Characterisation

Table 1 presents the mean elemental concentrations in unexposed lichens and in the transplanted lichens after the exposure period. The major elements assessed in the exposed lichens were (by decreasing order): Ca (22.6%) > Fe (1.18%) > K (0.79%) > Si (0.62%) > Al (0.26%) > S (0.16%) > Mg (0.12%) > Ti (0.12%), with the remaining elements contributing with less than 0.10% to the total mass of the lichen samples (namely, Zn > Sr > Mn > Br > Cu > Pb > Zr > Se > Rb > Co > Cr > As).

**Table 1.** Mean elemental mass fractions (in mg·Kg$^{-1}$, dry weight) in unexposed and exposed lichens samples and comparison with values from other studies.

| | **Present Study** | | | | | | **(Godinho et al., 2009) [48]** | **(Pacheco et al., 2008) [49]** |
|---|---|---|---|---|---|---|---|---|
| | **Unexposed Lichens** | | | **Exposed Lichens** | | | **Industrial Area, Sines [1]** | **Industrial Area, Sines [1]** |
| Element | Mean ± SD | Min | Max | Mean ± SD | Min | Max | Mean ± SD | Mean ± SD |
| Al | 2310 ± 190 | 2140 | 2550 | 2590 ± 450 | 1770 | 3790 | 1040 ± 146 | 3170 |
| As | 58.5 ± 24.8 | 36.0 | 88.0 | 30.0 ± 28.4 | 1 | 125 | 0.33 ± 0.11 | 1.33 |
| Br | 192 ± 17 | 170 | 211 | 180 ± 16 | 143 | 221 | 14.6 ± 1.8 | 13.7 |
| Ca | 208,000 ± 23,000 | 180,000 | 232,000 | 226,000 ± 15,000 | 187,000 | 257,000 | 8890 ± 1690 | 4390 |
| Co | 47.0 ± 10.5 | 32.0 | 56.0 | 54.9 ± 16.2 | 24 | 130 | 0.68 ± 0.16 | 1.33 |
| Cr | 7.33 ± 0.58 | 7.00 | 8.00 | 52.1 ± 90.8 | 3 | 607 | 6.4 ± 0.4 | 7.1 |
| Cu | 140 ± 16 | 122 | 158 | 163 ± 90 | 102 | 654 | n.d. | n.d. |
| Fe | 9860 ± 1480 | 8430 | 11,700 | 11,800 ± 3800 | 6500 | 30,600 | 651 ± 137 | 3190 |
| K | 8910 ± 690 | 8200 | 9710 | 7900 ± 1120 | 4410 | 11,750 | 3877 ± 737 | 4170 |
| Mg | 1130 ± 90 | 1020 | 1220 | 1230 ± 270 | 830 | 2670 | 992 ± 79 | 1320 |
| Mn | 274 ± 33 | 246 | 314 | 406 ± 297 | 225 | 2303 | 24.0 ± 0.5 | 28.2 |
| Pb | 101 ± 15 | 82 | 117 | 130 ± 111 | 81 | 967 | n.d. | n.d. |
| Rb | 69.8 ± 17.0 | 48.0 | 86.0 | 69.7 ± 15.1 | 45 | 142 | 4.4 ± 1.3 | 23.4 |
| S | 1390 ± 120 | 1280 | 1560 | 1580 ± 410 | 1070 | 3950 | n.d. | n.d. |
| Se | 77.5 ± 6.0 | 58.0 | 86.0 | 74.7 ± 5.7 | 58 | 86 | n.d. | 0.49 |
| Si | 5460 ± 370 | 5010 | 5790 | 6240 ± 1280 | 3810 | 11,220 | n.d. | n.d. |
| Sr | 553 ± 36 | 508 | 586 | 565 ± 74 | 415 | 737 | n.d. | n.d. |
| Ti | 1010 ± 100 | 910 | 1140 | 1150 ± 230 | 690 | 1840 | 102 ± 1 | 274 |
| Zn | 347 ± 16 | 333 | 365 | 608 ± 454 | 280 | 2270 | 140 ± 34 | 134 |
| Zr | 83.8 ± 17.1 | 59.0 | 98.0 | 78.5 ± 19.7 | 43 | 156 | n.d. | n.d. |

[1] after 4-month exposure; SD—standard deviation; n.d.—not determined.

Compared with studies reported in the literature that used the same lichen species in industrial environments, it is possible to conclude that the concentrations found in our study area were always higher than in Sines (Portugal), an industrial area, with heavy chemical industry, a coal power plant and a deep water harbor [48,49].

It is important to highlight that the exposure period of the transplanted lichens in the present study was between February and June 2020, a period when the COVID-19 pandemic situation reached Europe. In fact, due to the global COVID-19 pandemic situation, on 19 March 2020, Portugal decreed the state of national emergency, which included mandatory confinement, movement restriction of citizens, and closure of non-essential business. As a consequence of the nationwide lockdown that occurred from 19 March to 31 May 2020, concentrations in mainland Portugal presented a mean reduction of 41% for $NO_2$ and 18% for $PM_{10}$, considering the 20 air quality monitoring stations analyzed [50], following a tendency observed all over the world [51].

In the present study area, a drastic decrease of concentrations of $PM_{10}$ and $NO_2$ was registered in April 2020 (−40.3% and −44.0%, respectively), compared to the previous six years (a significant difference, with $p$-value = 0.000) [52]. In May 2020, a lower but still significant decrease of $PM_{10}$ (−16.7%, $p$-value = 0.000) and $NO_2$ (−17.4%, $p$-value = 0.048) was also registered. These reductions were accentuated again in June 2020 for both pollutants: $PM_{10}$ (−38.0%, $p$-value = 0.000) and $NO_2$ (−29.0%, $p$-value = 0.000). Despite the reduction of emissions from anthropogenic activities (such as traffic) due to the national lockdown (which reflects in pollutants, such as $PM_{10}$ and $NO_2$), the industries (namely, steelworks) that exist in the study area never stopped operating during this period. Therefore, the present biomonitoring study has the potential to assess and evaluate the contribution of these specific industries since they were the main sources of air pollution in the area during the exposure period of the transplanted lichens.

### *3.2. Enrichment and Contamination Factors*

Figure 4 presents the mean EFs assessed for each element in the exposed lichens.

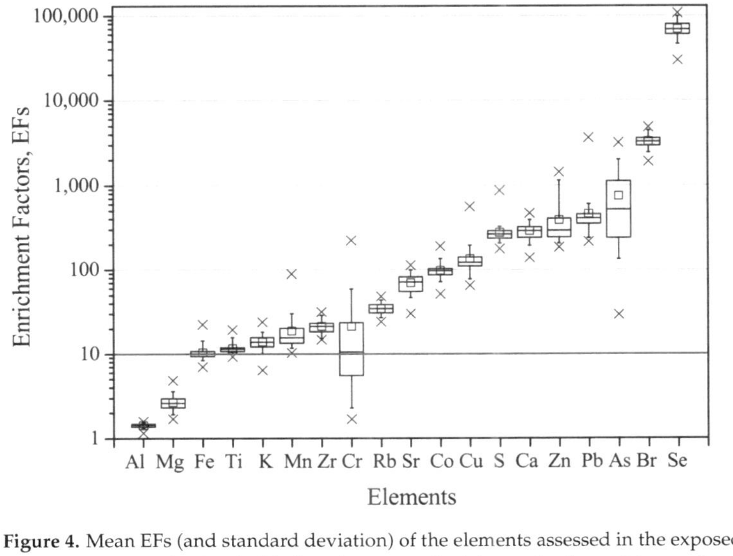


**Figure 4.** Mean EFs (and standard deviation) of the elements assessed in the exposed lichens. The red line represents the threshold of 10, the accepted minimum for the enrichment from a non-crustal source. Regarding the box plot, the square represents the mean, upper and lower times sign (×) shows the maximum and minimum values, the box provides the 25 percentile, the median and the 75% percentile, and the lower and upper minus sign (−) shows the 5 and 95 percentiles, respectively.

The elements Al and Mg were found to have a predominantly crustal origin (with EFs always below 10), while the remaining elements showed the contribution of non-crustal emissions to their levels. The elements Fe, Ti, and K, typically associated with a soil source [43], presented EFs slightly above 10 (with means of 10.5 ± 2.3, 11.7 ± 1.6, and 14.0 ± 2.9, respectively), indicating also anthropogenic contributions to their levels. A mix of different anthropogenic sources may have contributed to lichens enrichments, given the urban-industrial location of the study area.

Figure 5 provides the contamination factors regarding all the studied elements and it is observed a severe accumulation for Cr (CF = 7.11 ± 0.25) and an accumulation for Zn (CF = 1.75 ± 0.43), Mn (CF = 1.48 ± 0.50) and Pb (CF = 1.29 ± 0.67). However, a significant difference between exposed and unexposed elemental contents of lichens was only found for Cr ($p$-value = 0.034) and Zn ($p$-value = 0.022), but not for Mn ($p$-value = 0.064) neither Pb ($p$-value = 0.190). However, a significant difference between the medians of elemental contents of exposed and unexposed lichens was found for all the elements, except for Cu and Sr, as ascertained by the One Sample Wilcoxon test (at 0.05 significance level). Arsenic showed a loss (CF = 0.51 ± 2.02), with significant difference between exposed and unexposed levels ($p$-value = 0.031), as well as K (CF = 0.89 ± 0.18, $p$-value = 0.033). There is evidence that metals (both mineral and PTEs) may be lost due to leaching under certain meteorological conditions [48].

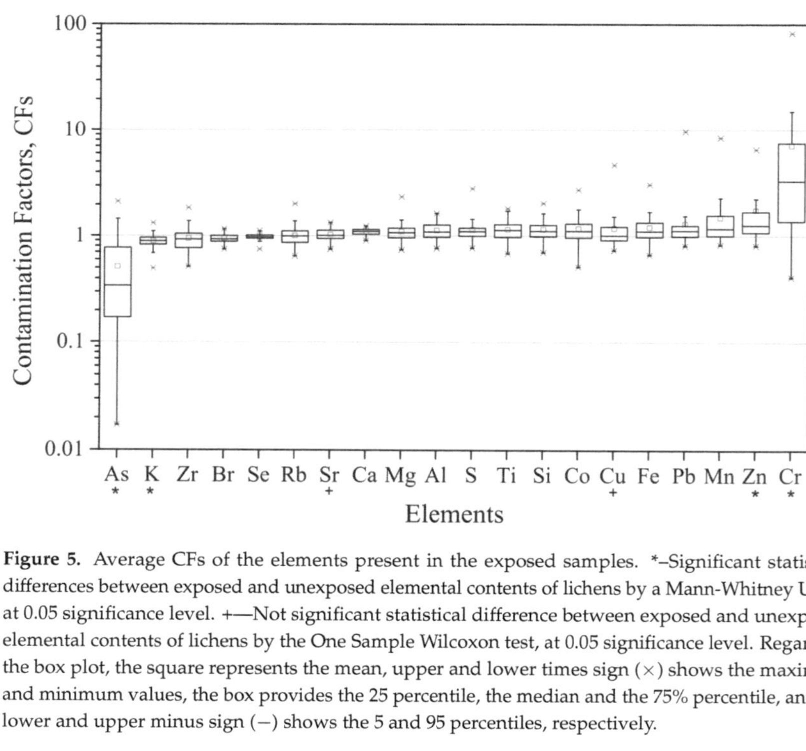


**Figure 5.** Average CFs of the elements present in the exposed samples. *–Significant statistical differences between exposed and unexposed elemental contents of lichens by a Mann-Whitney U test, at 0.05 significance level. +—Not significant statistical difference between exposed and unexposed elemental contents of lichens by the One Sample Wilcoxon test, at 0.05 significance level. Regarding the box plot, the square represents the mean, upper and lower times sign (×) shows the maximum and minimum values, the box provides the 25 percentile, the median and the 75% percentile, and the lower and upper minus sign (−) shows the 5 and 95 percentiles, respectively.

The CF ratios found in the present study are within the same intervals of loss/accumulation when compared to CF ratios for lichens exposed in the urban outdoor environment [31], except for Cr and Zn, which presented higher CF ratios in the present study.

### *3.3. Identification of Emission Sources*

#### 3.3.1. Spearman Correlations

Table 2 presents the positive significant Spearman correlations between the elements assessed in the exposed lichens, where values in bold represent the strongest correlations. The possible emission sources which contribute to the atmospheric PM in the study area can be qualitatively identified from the correlation matrix.

**Table 2.** Spearman correlations between elements in the exposed lichens, significant at the 0.05 level. Values in bold represent strong correlations (correlation coefficient above 0.70), while weak and medium correlations refer to correlation coefficients between 0.30–0.49 and 0.50–0.70, respectively.

| Elements | Al | As | Br | Ca | Co | Cr | Cu | Fe | K | Mg | Mn | Pb | Rb | S | Se | Si | Ti | Zn | Zr |
|---|---|---|---|---|---|---|---|---|---|---|---|---|---|---|---|---|---|---|---|
| Al | | | 0.33 | | **0.71** | 0.62 | | **0.87** | 0.29 | 0.50 | 0.56 | | 0.66 | 0.56 | | **0.97** | **0.84** | 0.42 | 0.67 |
| As | | | | | | | | | | | | | | | | | | | 0.52 |
| Br | | | | | | | 0.43 | 0.32 | 0.54 | 0.42 | 0.25 | | 0.46 | 0.49 | 0.35 | 0.39 | 0.37 | | 0.38 |
| Ca | | | | | | | | | | | | | | | | | | | |
| Co | | | | | | 0.64 | 0.27 | **0.86** | 0.28 | 0.41 | 0.59 | 0.34 | 0.62 | 0.49 | | **0.72** | 0.64 | 0.44 | 0.60 |
| Cr | | | | | | | 0.54 | **0.81** | | 0.39 | **0.92** | 0.37 | 0.32 | 0.62 | | 0.62 | 0.64 | **0.73** | 0.42 |
| Cu | | | | | | | | 0.39 | | 0.35 | 0.57 | 0.32 | | 0.52 | | 0.32 | 0.37 | 0.50 | 0.34 |
| Fe | | | | | | | | | | 0.51 | **0.77** | 0.38 | 0.67 | 0.62 | | **0.87** | **0.84** | 0.60 | 0.65 |
| K | | | | | | | | | | 0.44 | | | 0.61 | 0.36 | | 0.35 | 0.33 | | 0.34 |
| Mg | | | | | | | | | | | 0.44 | | 0.46 | 0.48 | | 0.48 | 0.50 | 0.29 | 0.57 |
| Mn | | | | | | | | | | | | 0.41 | 0.33 | 0.62 | | 0.58 | 0.62 | **0.81** | 0.43 |
| Pb | | | | | | | | | | | | | | 0.27 | | | | 0.48 | |
| Rb | | | | | | | | | | | | | | 0.48 | | 0.66 | **0.73** | 0.27 | 0.58 |
| S | | | | | | | | | | | | | | | | 0.61 | 0.61 | 0.58 | 0.41 |
| Si | | | | | | | | | | | | | | | | | **0.84** | 0.42 | 0.68 |
| Ti | | | | | | | | | | | | | | | | | | 0.55 | 0.66 |
| Zn | | | | | | | | | | | | | | | | | | | 0.29 |

Given the significant positive correlations between Al-Si (r = 0.97), Al-Fe (r = 0.87), Al-Ti (r = 0.84), Si-Fe (r = 0.87), Si-Ti (r = 0.84), Ti-Fe (r = 0.84), it is concluded that these elements come from a same source, namely soil, since Al, Si, and Ti are typical soil elements [53,54]. In fact, Al, Fe, Si, and trace elements, such as Rb, are associated with feldspars, quartz, micas and Ti is associated with the titanite silicate [43].

In South European regions, such as Portugal, it is known that atmospheric PM can have a contribution to dust transport episodes from the interior of the Iberian Peninsula and the Saharan desert [55]. It was already found that Fe could reach high concentrations during these episodes of long-transportation of dust [56]. In fact, during the exposure period of the lichens, some Saharan dust events were identified (e.g., from 18 to 21 March 2020) [50].

An important amount of Al, Si, Mg, and Ti was also found in sinter plant emissions (fugitive and emissions coming from the cooling area) [1], consisting of internally mixed aluminosilicates/metallic particles, suggesting also an association between these elements and sinter plant emissions [12,53]. However, the steelworks installed in the study area do not include sintering in their processes, since scrap is used as raw material and not iron ore, and, therefore, this specific source is excluded in the study area.

Strong correlations were found between Fe-Mn (r = 0.77), Fe-Cr (r = 0.81), Cr-Mn (r = 0.92), Cr-Zn (r = 0.73), and Mn-Zn (r = 0.81) in the exposed lichens, which may indicate the existence of a common emission source, assigned to the steelworks in the study area, considering that previous studies also found strong correlations between Fe-Mn-Zn near iron and steel industry [8,57]. Typically, Fe, Cr, and Mn are considered tracers of the iron and steel industry [43]. Therefore, in the present study, despite a natural origin (soil), Fe also has the contribution of this industrial activity. This influence of different sources on Fe was found in other studies [58].

In fact, especially when scrap is employed as a raw material in the steel industry, levels of Cr and Mn can be higher [1]. Zn is also considered a tracer of the steel industry [43,59], being especially related to the use of galvanized scrap [60].

Finally, the association of Pb with the steel industry has been reported by several authors [1,61]. In the present study, significant but weak correlations between Pb-Fe (r = 0.38), Pb-Cr (r = 0.37), Pb-Mn (r = 0.41), and Pb-Zn (r = 0.48) were found, which may suggest some association of Pb emissions with the steelworks, but also the existence of other sources.

A high correlation was found for Fe and Co (r = 0.86), which suggests that Co probably is originated from the steelworks, as already found in previous studies [8].

The Electric Arc Furnace (EAF) steelmaking plants, as the one installed in the study area, typically present fugitive emissions from raw materials and slag handling and storage piles, in addition to the emission from the furnace and rolling [8].

According to the Best Available Techniques (BAT) for Iron and Steel Production [62], emissions from EAF steelmaking plants include the following metals, after abatement: Hg, Pb, Cr, Ni, Zn, Cd, and Cu (where Cr and Ni are, naturally, higher in the production of stainless steel). Typical emissions have concentrations of 0.5–50 mg dust/$Nm^3$. Emissions from the secondary metallurgy (ladle metallurgy, ingot and continuous casting, oxygen blow unit) include the following metals, after abatement: Pb, Co, Ni, Se, Te, Sb, Cr, Cu, Mn, V, Sn [62].

The primary constituents of the slag produced in the EAF process are Ca, Fe, Al, Mg, Mn, Si, and Cr, where calcium and magnesium oxides are added as fluxing agents [63]. In the present study, no correlation was found between Ca-Fe, Ca-Al, Ca-Mg, Ca-Mn, Ca-Si, or Ca-Cr, and, therefore, EAF slag is not identified as a probable source.

A moderate correlation has been found between Cu-Cr (r = 0.54) and Cu-Zn (r = 0.50), which may suggest multiple emission sources. These elements are usually associated with emissions from road traffic, namely through tire (Cr, Zn) and brake wear (Zn, Cu, Ba) and motor oil (Zn) [38,43,64]. Moderate correlations between Cu, Cd, and Zn were also found regarding car traffic [65].

USEPA-SPECIATE and SPECIEUROPE databases were used to identify emission sources, selecting source profiles from the steel industry (general and specific processes) and traffic (break and tire wearing, highway vehicles) [54]. Additionally, a comparison was also done with the elemental ratios determined in the settled dust collected in the study area in 2019 [7]. Figure 6A shows a visible high correlation between Fe and Mn concentrations assessed in the exposed lichens ($r^2$ = 0.80) and a notable association to the steelmaking processes according to the databases of source profiles. It is also verified some similarity between the Fe/Mn ratio in the exposed lichens and the settled dust collected previously in the study area (11.5 and 7.0, respectively).

Figure 6B shows a very high correlation between Cr and Mn measured in lichens ($r_2$ = 0.97) and a notable similarity between the Cr/Mn ratio in the exposed lichens and the settled dust (0.30 and 0.36, respectively), supporting that both elements are associated with the steelworks emissions.

Furthermore, Figure 6C,D evidence the association of Cr, Zn, and Cu to traffic emissions. Concurrently it is verified that Cr/Zn ratios in the exposed lichens and in the settled dust (0.1 and 3.6, respectively) are substantially different, suggesting that Zn emissions in the study area are substantially affected by traffic and not only by the steelworks.

### 3.3.2. Concentrations versus Distance to Steelworks

The mean elemental concentrations were calculated for the lichens exposed in a circle with center in industry B, which is the biggest steelwork, and with 1 km radius ("(0–1)"), and in consecutive concentric crowns of 1 or 2 km radius, namely, "(1–2)", "(2–3)", "(3–4)" and "(4–6)", where the edges of each interval are the distance (in km) from the established center in industry B. Figure 7 shows the mean concentrations of selected elements regarding the distance from industry B. It should be mentioned that there is a distance of around 2 km from both steelworks in the study area (namely, industries A and B), and thus both facilities are included (and are expected to directly influence the samples) in the 2 km radius (namely the samples located in "(0–1)" and "(1–2)").

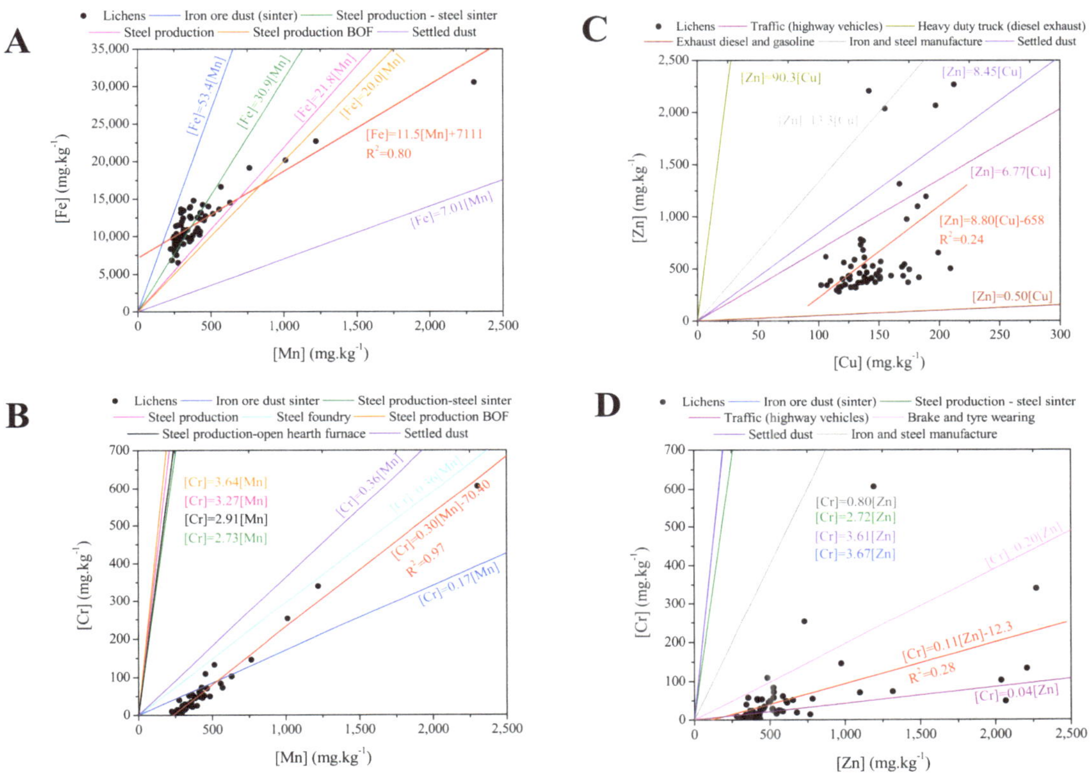

**Figure 6.** Relation between elements in the exposed lichens, in the settled dust and according to USEPA-SPECIATE and SPECIEUROPE profiles for iron and steel industry and traffic: (**A**) Mn vs. Fe, (**B**) Mn vs. Cr, (**C**) Cu vs. Zn, and (**D**) Zn vs. Cr. The red line presents the linear regression regarding the data of the exposed lichens.

As expected, elements associated with the steel industry (Fe, Cr, Mn, Zn, and Pb) registered the highest concentrations in the lichens exposed near the steelworks, within a 1 or 2 km radius from industry B, and with concentrations decreasing with distance, as shown by Figure 7. Mean Zn levels showed a peak at 4 km from industry B, and this may reflect the traffic influence since it is in the distance range of the nearby highway.

Comparing the mean concentrations of exposed lichens in the 1 km radius circle, to the control sample exposed at a 6580 m distance from industry B, levels decreased more sharply for Cr (6.1 times), than for Zn (2.5), Pb (1.9), Mn (1.7), and Fe (1.5).

The same analysis regarding the variability of EFs with the distance to industry B was also conducted, as shown by Figure 8. The same trend was observed, namely, the EFs of Mn and Fe decrease on a regular basis with the distance; in the case of Zn and Cr, there is an increase in the EF at a distance of 3–4 km from the industry B, consistent with the traffic highway influence.

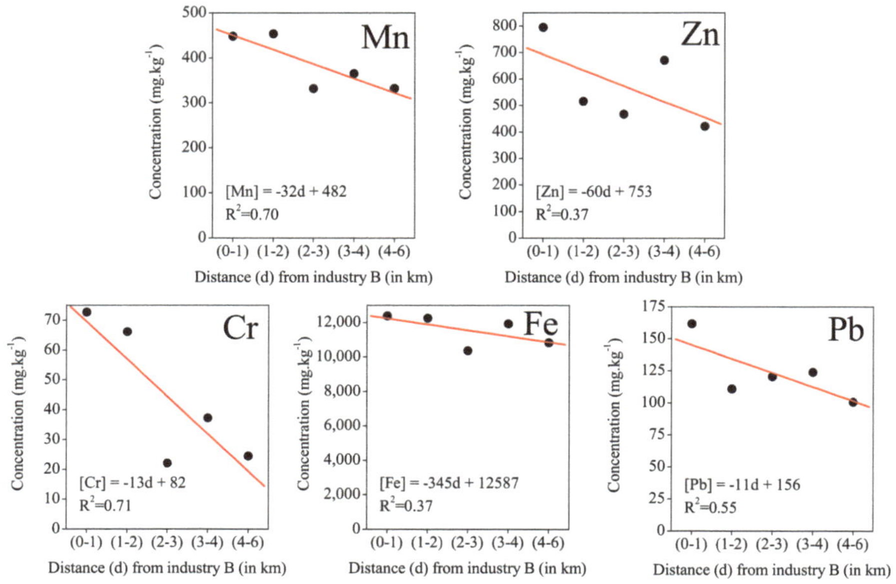


**Figure 7.** Mean elemental concentration versus distance to the steelworks.

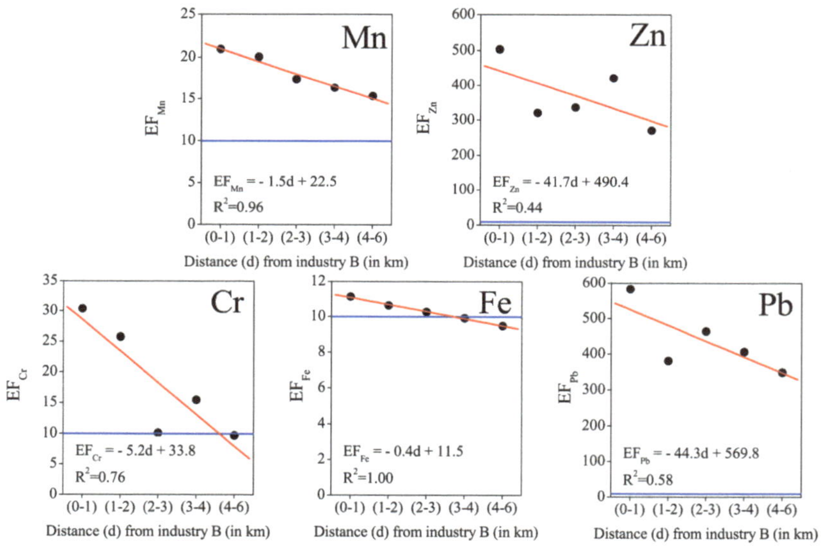


**Figure 8.** Mean Enrichment Factors (EFs) versus distance to the steelworks. The blue line represents the threshold of 10, the accepted minimum for the enrichment from a non-crustal source.

The ratio between the EFs for samples exposed at a distance less than and more than 1 km from the steelworks' area is shown in Figure 9. This ratio is higher for Cr (1.77), Zn (1.50), and Pb (1.45), i.e., close to steelworks there was an increase in EF for these elements, as well as for Mn and Fe. Nevertheless, only the EFs of the elements Cr ($p$ = 0.015), Mn ($p$ = 0.035), and Fe ($p$ = 0.031) presented significant differences, between the samples located at less than and more than 1 km from the steelworks.

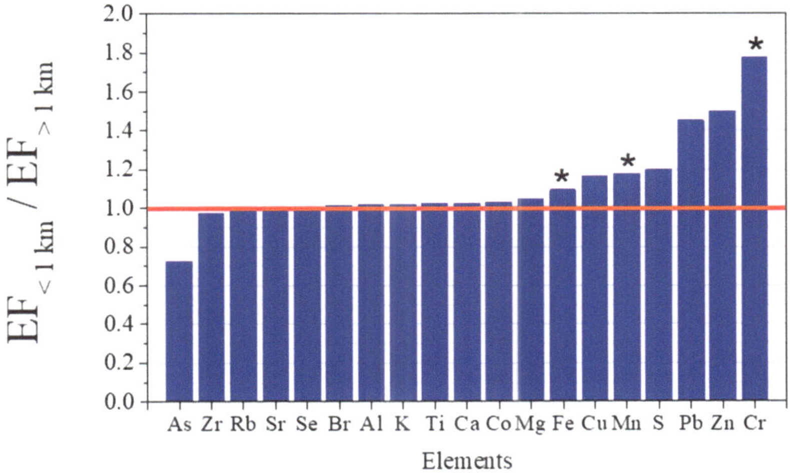


**Figure 9.** Ratio between EF in samples located at less than and more than 1 km from steelworks. * Significant statistical difference between the EF in the two groups by a Mann-Whitney U test, at 0.05 significance level.

3.3.3. Spatial Distribution Patterns

As a final step in the identification of emission sources, the spatial distribution of elemental concentrations in the exposed lichens was mapped (Figures 10–12). The similarities of the spatial distributions obtained between elements and the correlations found above, support the grouping of elements by types of emission source: natural origin from the soil (Al, Si, and Ti—Figure 10, top); a mixture of soil with natural and anthropogenic origins (Fe and Mg—Figure 10, bottom); anthropogenic resulting from the industrial activity, namely steelworks (Co, Cr, Mn, Pb, and Zn—Figure 11); anthropogenic from road traffic (Cu and Zn—Figure 12, top) and biomass burning (Br and K—Figure 12, bottom).

There is a close resemblance of the spatial distributions of soil origin elements Al, Si, and Ti, as expected from the high correlation coefficients obtained for these elements. The relatively high concentrations of these elements nearby the steelworks is certainly due to the resuspension of particles of geological origin, by the wind and heavy duty traffic [66].

In the studied area, Fe and Mg concentrations may result from the natural emission of soil and the activity of the steelworks, which may provide these elements from fugitive emissions. This hypothesis is supported by the similarity of the spatial distribution of these elements compared to Co, Cr, Mn, Cr, and Zn, associated with the steelworks. Regarding Pb, an additional hot spot can be observed in the north of the study area, but no industry could be identified. Additionally, the spatial distribution of Cu evidences high concentrations near roads due to the influence of traffic [43].

As the wind blew predominantly from the north (N), northwest (NW), and west (W) (Figure 3) during the exposure period, the highest concentrations of the elements related to the steelworks would be expected downwind of this emission source, that is in the south-southeast sector of the sampling grid, as is indeed apparent from the spatial distribution in Figure 11.

The S enrichment in the industrial area may be associated with metal smelting [43] and, in fact, good correlations were obtained above for S-Fe ($r = 0.62$), S-Cr ($r = 0.62$), and S-Mn ($r = 0.62$); but may also derive from local emissions of SOx related to coal/coke burning [43], since there is a coke kiln at a lime company located in the steelworks park-industry C. The significant correlation S-Cu ($r = 0.52$) also suggests some association with traffic emissions [12], which may be due to heavy-duty traffic in the steelworks area (Figure 12).

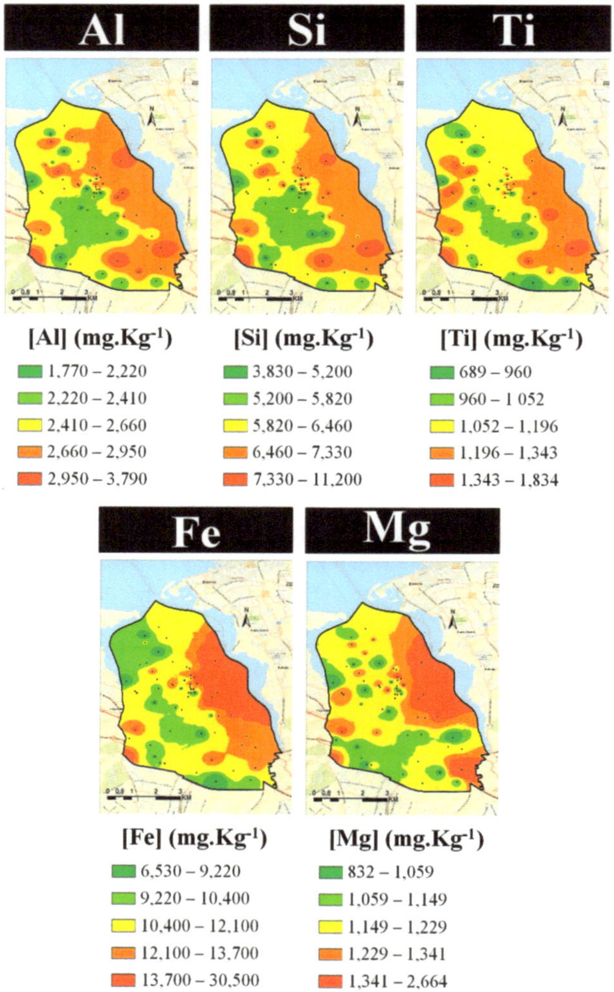


**Figure 10.** Spatial distribution of soil-related elements in the study area: (**top**) natural origin and (**bottom**) a mixture of natural and anthropogenic origins.

Elements Br and K present a spatial distribution with some similarities (and a significant correlation, r = 0.54), indicating a common anthropogenic source, such as biomass burning [43]. This is in accordance with the analysis of the national inventory of emissions for the municipality of Seixal, which identified residential combustion as one of the main emission sources in the region, concerning both $PM_{10}$ and $PM_{2.5}$, besides industry and road transport [67].

Figure S2 provides the spatial distribution in the study area of the chemical elements that were not associated with an identified source, such as Ca. Lime and calcium carbide are mineral additions used in the production process of industry B [28], and industry C is a factory of lime products and derivatives, with both industries located in the steelworks park. Nevertheless, high levels of Ca were not obtained in the steelworks area, but in the southeast of the grid (Figure S2). Since there is no cement industry or quarry in the study area, Ca enrichment is probably due to the physiological characteristics of the lichens [32], as granules with calcium oxalate may occur in lichens due to environmental factors [68].

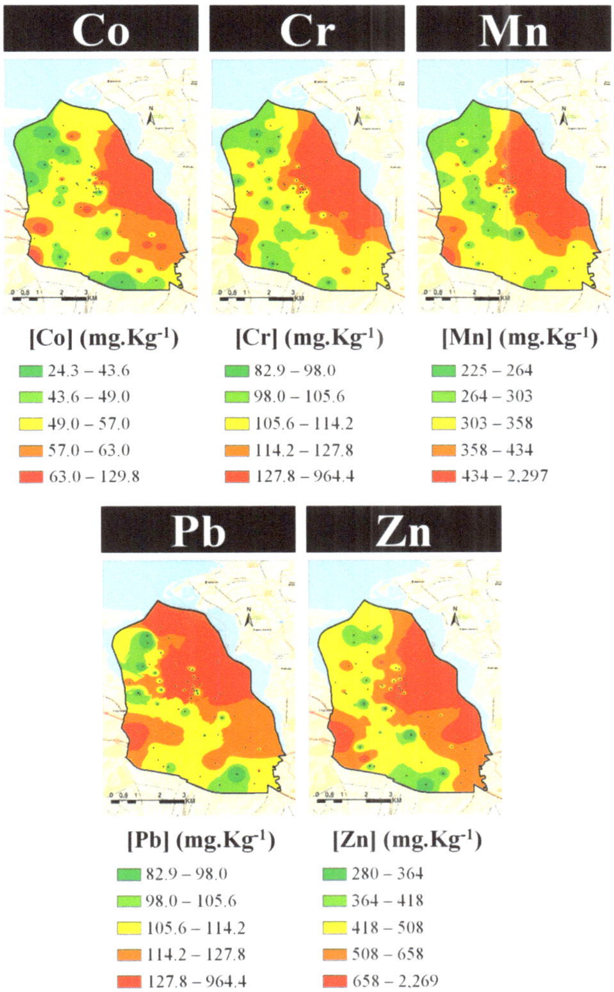


**Figure 11.** Spatial distribution of elements related with anthropogenic origin, namely industry.

The spatial distribution of the elements discussed previously is in accordance with the Air Quality Map of Seixal, according to which the highest values of particulate matter ($PM_{10}$ and $PM_{2.5}$) were registered in the industrial areas and main roads of the municipality [67]. The authors concluded that the air quality index regarding $PM_{10}$ is "average" generically throughout the municipality, except in the industrial area where it is "weak" or "bad"; for $PM_{2.5}$, the air quality index varies from generically "weak" to "very bad" in the industrial area.

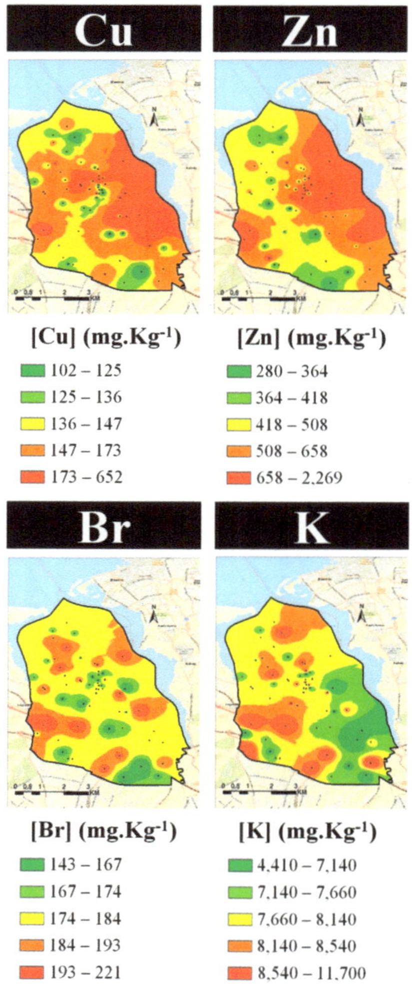


**Figure 12.** Spatial distribution of elements related to an anthropogenic origin, namely from road traffic (**top**) and biomass burning (**bottom**).

## 4. Conclusions

The present study aimed to perform a biomonitoring study of air pollution using lichens in an urban-industrial area where the local population has expressed several concerns regarding air quality in recent years. The methodology applied in this study allowed to involve the local population and to obtain a spatial distribution of the chemical elements absorbed by the lichens during a period where the activities in the area were mainly the industries, due to the COVID-19 national lockdown.

Using a set of different analysis approaches to the chemical content of the exposed transplanted lichens in the study area, it was possible to identify hotspots of air pollution in the area (taking advantage of the possibility to assess the spatial variability due to the use of a grid of biomonitors) and to define the potential sources affecting the local air quality.

A total of five different sources were identified as contributing to the local air quality: (i) a soil source of natural origin (based on Al, Si, and Ti), (ii) a soil source of natural and anthropogenic origins (based on Fe and Mg), (iii) a source from the local industrial activity, namely steelworks (based on Co, Cr, Mn, Pb, and Zn); (iv) a source from the road traffic (based on Cr, Cu, and Zn), and (v) a source of biomass burning (based on Br and K).

The impact of the industries located in the study area on the local air quality was identified (namely, the steelworks). To identify the quantitative contribution of such sources and to design targeted mitigation measures, future efforts should be conducted to assess particulate matter levels by reference methods, which will allow a more comprehensive and quantitative analysis of the issue, taking into account the national and European legislation.

Despite biomonitoring being a technique known for decades, this study showed again the potentialities and advantages of such an approach (used as a complementary strategy to reference methodologies), providing spatial information of air pollution. This information is crucial to define future strategies, based on knowledge: understand exposure levels of the population to air pollution, their potential health impacts, and to define mitigation measures to improve air quality.

**Supplementary Materials:** The following supporting information can be downloaded at: https://www.mdpi.com/article/10.3390/ijerph19031364/s1, Figure S1: Procedure for preparation of pellets: (**A**) used ball mill RETSCH, (**B**) homogeneised powder of lichen sample, (**C**) used Pelletiser SPECAC, and (**D**) pellet of a lichen sample. Figure S2. Spatial distribution of elements without a specific identified source. Table S1. Reference values of soil composition (in mg·Kg$^{-1}$) defined by Mason and Moore [42], used for the calculation of the Enrichment factors.

**Author Contributions:** L.A., C.A.G. and A.R.J. were responsible for the management of volunteers; L.A., C.A.G., A.R.J., N.C. and S.M.A. were responsible for the methodology of biomonitoring using transplanted lichens from a clean area (regarding exposure and collection of samples); L.A., C.A.G., A.R.J. and I.D. performed the experimental work at the laboratory regarding the treatment of samples and their preparation for further analysis; Z.K. performed the chemical analysis of samples; L.A. and C.A.G. performed the data analysis, including statistical, and wrote the original draft. N.C. and S.M.A. contributed to the review and editing of the original draft. S.M.A. was also responsible for the coordination and management of the project. All authors have read and agreed to the published version of the manuscript.

**Funding:** N. Canha acknowledges the funding by national funds through FCT-Fundação para a Ciência e Tecnologia, I.P. (Portugal) for his IST-ID contract (IST-ID/098/2018). The FCT support is also acknowledged by C$^2$TN/IST authors (UIDB/04349/2020+UIDP/04349/2020) and by CESAM author (UIDB/50017/2020+UIDP/50017/2020).

**Institutional Review Board Statement:** Not applicable.

**Informed Consent Statement:** Not applicable.

**Data Availability Statement:** The datasets generated and analysed during the current study are not publicly available but are available from the corresponding author on reasonable request.

**Acknowledgments:** Acknowledgements are due to Câmara Municipal do Seixal, for funding and all the support given in the sampling process. Instituto Português do Mar e da Atmosfera (IPMA) is also gratefully acknowledged for providing the meteorological data.

**Conflicts of Interest:** The authors declare no conflict of interest.

## References

1. Hleis, D.; Fernández-Olmo, I.; Ledoux, F.; Kfoury, A.; Courcot, L.; Desmonts, T.; Courcot, D. Chemical profile identification of fugitive and confined particle emissions from an integrated iron and steelmaking plant. *J. Hazard. Mater.* **2013**, *250–251*, 246–255. [CrossRef] [PubMed]
2. Karagulian, F.; Belis, C.A.; Dora, C.F.C.; Prüss-Ustün, A.M.; Bonjour, S.; Adair-Rohani, H.; Amann, M. Contributions to cities' ambient particulate matter (PM): A systematic review of local source contributions at global level. *Atmos. Environ.* **2015**, *120*, 475–483. [CrossRef]
3. Harrison, R.M.; Yin, J. Particulate matter in the atmosphere: Which particle properties are important for its effects on health? *Sci. Total Environ.* **2000**, *249*, 85–101. [CrossRef]

4. Loomis, D.; Grosse, Y.; Lauby-Secretan, B.; El Ghissassi, F.; Bouvard, V.; Benbrahim-Tallaa, L.; Guha, N.; Baan, R.; Mattock, H.; Straif, K. The carcinogenicity of outdoor air pollution. *Lancet Oncol.* **2013**, *14*, 1262–1263. [CrossRef]
5. Pope, C.A., III; Dockery, D.W. Health Effects of Fine Particulate Air Pollution: Lines that Connect. *J. Air Waste Manag. Assoc.* **2012**, *2247*, 709–742. [CrossRef] [PubMed]
6. Chaíça, I. Poeira negra que cobre Paio Pires não é nociva para a saúde. Origem Permanece Desconhecida. *Público*, 11 May 2019. Available online: https://www.publico.pt/2019/05/11/local/noticia/poeira-negra-cobre-paio-pires-nao-inalavel-sao-precisos-estudos-saber-onde-vem-1872369(accessed on 9 January 2022).
7. Justino, A.R.R.; Canha, N.; Gamelas, C.; Coutinho, J.T.T.; Kertesz, Z.; Almeida, S.M.M. Contribution of micro-PIXE to the characterization of settled dust events in an urban area affected by industrial activities. *J. Radioanal. Nucl. Chem.* **2019**, *322*, 1953–1964. [CrossRef]
8. Mohiuddin, K.; Strezov, V.; Nelson, P.F.; Stelcer, E. Characterisation of trace metals in atmospheric particles in the vicinity of iron and steelmaking industries in Australia. *Atmos. Environ.* **2014**, *83*, 72–79. [CrossRef]
9. Silva, A.V.; Oliveira, C.M.; Canha, N.; Miranda, A.I.; Almeida, S.M. Long-Term Assessment of Air Quality and Identification of Aerosol Sources at Setúbal, Portugal. *Int. J. Environ. Res. Public Health* **2020**, *17*, 5447. [CrossRef]
10. Machemer, S.D. Characterization of Airborne and Bulk Particulate from Iron and Steel Manufacturing Facilities. *Environ. Sci. Technol.* **2004**, *38*, 381–389. [CrossRef]
11. Querol, X.; Viana, M.; Alastuey, A.; Amato, F.; Moreno, T.; Castillo, S.; Pey, J.; de la Rosa, J.; Sánchez de la Campa, A.; Artíñano, B.; et al. Source origin of trace elements in PM from regional background, urban and industrial sites of Spain. *Atmos. Environ.* **2007**, *41*, 7219–7231. [CrossRef]
12. Taiwo, A.M.; Beddows, D.C.S.; Calzolai, G.; Harrison, R.M.; Lucarelli, F.; Nava, S.; Shi, Z.; Valli, G.; Vecchi, R. Receptor modelling of airborne particulate matter in the vicinity of a major steelworks site. *Sci. Total Environ.* **2014**, *490*, 488–500. [CrossRef] [PubMed]
13. Almeida, S.M.; Lage, J.; Freitas, M.D.C.; Pedro, A.I.; Ribeiro, T.; Silva, A.V.; Canha, N.; Almeida-Silva, M.; Sitoe, T.; Dionisio, I.; et al. Integration of biomonitoring and instrumental techniques to assess the air quality in an industrial area located in the coastal of central Asturias, Spain. *J. Toxicol. Environ. Health Part A Curr. Issues* **2012**, *75*, 1392–1403. [CrossRef] [PubMed]
14. Lage, J.; Almeida, S.M.; Reis, M.A.; Chaves, P.C.; Ribeiro, T.; Garcia, S.; Faria, J.P.; Fernández, B.G.; Wolterbeek, H.T. Levels and Spatial Distribution of Airborne Chemical Elements in a Heavy Industrial Area Located in the North of Spain. *J. Toxicol. Environ. Health Part A Curr. Issues* **2014**, *77*, 856–866. [CrossRef] [PubMed]
15. Fränzle, O. Chapter 2 Bioindicators and environmental stress assessment. *Trace Met. Other Contam. Environ.* **2003**, *6*, 41–84. [CrossRef]
16. Markert, B.A.; Breure, A.M.; Zechmeister, H.G. Chapter 1 Definitions, strategies and principles for bioindication/biomonitoring of the environment. *Trace Met. Other Contam. Environ.* **2003**, *6*, 3–39. [CrossRef]
17. Gür, F.; Yaprak, G. Biomonitoring of metals in the vicinity of Soma coal-fired power plant in western Anatolia, Turkey using the epiphytic lichen, Xanthoria parietina. *J. Environ. Sci. Health Part A Toxic/Hazard. Subst. Environ. Eng.* **2011**, *46*, 1503–1511. [CrossRef]
18. Bermejo-Orduna, R.; McBride, J.R.; Shiraishi, K.; Elustondo, D.; Lasheras, E.; Santamaría, J.M. Biomonitoring of traffic-related nitrogen pollution using *Letharia vulpina* (L.) Hue in the Sierra Nevada, California. *Sci. Total Environ.* **2014**, *490*, 205–212. [CrossRef]
19. Canha, N.; Freitas, M.D.C.; Almeida, S.M. Contribution of short irradiation instrumental neutron activation analysis to assess air pollution at indoor and outdoor environments using transplanted lichens. *J. Radioanal. Nucl. Chem.* **2019**, *320*, 129–137. [CrossRef]
20. Conti, M.E.; Cecchetti, G. Biological monitoring: Lichens as bioindicators of air pollution assessment—A review. *Environ. Pollut.* **2001**, *114*, 471–492. [CrossRef]
21. Garty, J. Biomonitoring atmospheric heavy metals with lichens: Theory and application. *CRC. Crit. Rev. Plant Sci.* **2001**, *20*, 309–371. [CrossRef]
22. Leonardo, L.; Damatto, S.R.; Gios, B.R.; Mazzilli, B.P. Lichen specie *Canoparmelia texana* as bioindicator of environmental impact from the phosphate fertilizer industry of São Paulo, Brazil. *J. Radioanal. Nucl. Chem.* **2014**, *299*, 1935–1941. [CrossRef]
23. Bozkurt, Z. Determination of airborne trace elements in an urban area using lichens as biomonitor. *Environ. Monit. Assess.* **2017**, *189*, 573. [CrossRef] [PubMed]
24. Cruz, A.M.J.; Freitas, M.D.C.; Canha, N.; Verburg, T.G.; Almeida, S.M.; Wolterbeek, H.T. Spatial mapping of the city of Lisbon using biomonitors. *Int. J. Environ. Health* **2012**, *6*, 1. [CrossRef]
25. Bontempi, E.; Bertuzzi, R.; Ferretti, E.; Zucca, M.; Apostoli, P.; Tenini, S.; Depero, L.E. Micro X-ray fluorescence as a potential technique to monitor in-situ air pollution. *Microchim. Acta* **2008**, *161*, 301–305. [CrossRef]
26. PORDATA Base de Dados Portugal Contemporâneo. Available online: http://www.pordata.pt/Municipios/Ambiente+de+Consulta/Tabela (accessed on 22 September 2021).
27. APA—Agência Portuguesa do Ambiente. *Licença Ambiental Lusosider Aços Planos, S.A.*; APA: Amadora, Portugal, 2008.
28. APA—Agência Portuguesa do Ambiente. *Licença Ambiental LA no 658_1.1_2017—SN Seixal_Siderurgia Nacional, S.A.*; APA: Amadora, Portugal, 2017.
29. APA—Agência Portuguesa do Ambiente. *Licença Ambiental Microlime—Produtos de Cal e Derivados, S.A.*; APA: Amadora, Portugal, 2011.
30. Google Earth Pro 2021, Version 7.3.4.8248. Available online: https://www.google.com/intl/pt-PT/earth/versions/#earth-pro (accessed on 22 September 2021).

31. Canha, N.; Almeida, S.M.; Freitas, M.C.; Wolterbeek, H.T. Indoor and outdoor biomonitoring using lichens at urban and rural primary schools. *J. Toxicol. Environ. Health Part A Curr. Issues* **2014**, *77*, 900–915. [CrossRef]
32. Canha, N.; Almeida-Silva, M.; Freitas, M.C.; Almeida, S.M.; Wolterbeek, H.T. Lichens as biomonitors at indoor environments of primary schools. *J. Radioanal. Nucl. Chem.* **2012**, *291*, 123–128. [CrossRef]
33. Nečemer, M.; Kump, P.; Ščančar, J.; Jaćimović, R.; Simčič, J.; Pelicon, P.; Budnar, M.; Jeran, Z.; Pongrac, P.; Regvar, M.; et al. Application of X-ray fluorescence analytical techniques in phytoremediation and plant biology studies. *Spectrochim. Acta Part B At. Spectrosc.* **2008**, *63*, 1240–1247. [CrossRef]
34. Pessanha, S.; Samouco, A.; Adão, R.; Carvalho, M.L.; Santos, J.P.; Amaro, P. Detection limits evaluation of a portable energy dispersive X-ray fluorescence setup using different filter combinations. *X-ray Spectrom.* **2017**, *46*, 102–106. [CrossRef]
35. Sitko, R.; Zawisz, B. Quantification in X-Ray Fluorescence Spectrometry. In *X-ray Spectroscopy*; InTech: Nappanee, IN, USA, 2012; pp. 137–162. [CrossRef]
36. Carvalho, P.M.S.; Pessanha, S.; Machado, J.; Silva, A.L.; Veloso, J.; Casal, D.; Pais, D.; Santos, J.P. Energy dispersive X-ray fluorescence quantitative analysis of biological samples with the external standard method. *Spectrochim. Acta Part B At. Spectrosc.* **2020**, *174*, 105991. [CrossRef]
37. Miller, J.N.; Miller, J.C.; Miller, R.D. *Statistics and Chemometrics for Analytical Chemistry*, 7th ed.; Pearson: London, UK, 2018; ISBN 9781292186719.
38. Belis, C.A.; Karagulian, F.; Larsen, B.R.; Hopke, P.K. Critical review and meta-analysis of ambient particulate matter source apportionment using receptor models in Europe. *Atmos. Environ.* **2013**, *69*, 94–108. [CrossRef]
39. Mohd Tahir, N.; Poh, S.C.; Suratman, S.; Ariffin, M.M.; Shazali, N.A.M.; Yunus, K. Determination of trace metals in airborne particulate matter of Kuala Terengganu, Malaysia. *Bull. Environ. Contam. Toxicol.* **2009**, *83*, 199–203. [CrossRef] [PubMed]
40. Canha, N.; Freitas, M.C.; Anawar, H.M.; Dionísio, I.; Dung, H.M.; Pinto-Gomes, C.; Bettencourt, A. Characterization and phytoremediation of abandoned contaminated mining area in Portugal by INAA. *J. Radioanal. Nucl. Chem.* **2010**, *286*, 577–582. [CrossRef]
41. Pekey, H. The distribution and sources of heavy metals in Izmit Bay surface sediments affected by a polluted stream. *Mar. Pollut. Bull.* **2006**, *52*, 1197–1208. [CrossRef] [PubMed]
42. Mason, B.; Moore, C.B. *Principles of Geochemistry*; Wiley: New York, NY, USA, 1982.
43. Calvo, A.I.; Alves, C.; Castro, A.; Pont, V.; Vicente, A.M.; Fraile, R. Research on aerosol sources and chemical composition: Past, current and emerging issues. *Atmos. Res.* **2013**, *120–121*, 1–28. [CrossRef]
44. Almeida-Silva, M.; Canha, N.; Vogado, F.; Baptista, P.C.; Faria, A.V.; Faria, T.; Coutinho, J.T.; Alves, C.; Almeida, S.M. Assessment of particulate matter levels and sources in a street canyon at Loures, Portugal—A case study of the REMEDIO project. *Atmos. Pollut. Res.* **2020**, *11*, 1857–1869. [CrossRef]
45. Fernández, J.A.; Carballeira, A. A comparison of indigenous mosses and topsoils for use in monitoring atmospheric heavy metal deposition in Galicia (northwest Spain). *Environ. Pollut.* **2001**, *114*, 431–441. [CrossRef]
46. Frati, L.; Brunialti, G.; Loppi, S. Problems related to lichen transplants to monitor trace element deposition in repeated surveys: A case study from central Italy. *J. Atmos. Chem.* **2005**, *52*, 221–230. [CrossRef]
47. Watson, D.F.; Philip, G.M. A Refinement of Inverse Distance Weighted Interpolation. *Geoprocessing* **1985**, *2*, 315–327.
48. Godinho, R.M.; Verburg, T.G.; Freitas, M.C.; Wolterbeek, H.T. Accumulation of trace elements in the peripheral and central parts of two species of epiphytic lichens transplanted to a polluted site in Portugal. *Environ. Pollut.* **2009**, *157*, 102–109. [CrossRef]
49. Pacheco, A.M.G.; Freitas, M.C.; Baptista, M.S.; Vasconcelos, M.T.S.D.; Cabral, J.P. Elemental levels in tree-bark and epiphytic-lichen transplants at a mixed environment in mainland Portugal, and comparisons with an in situ lichen. *Environ. Pollut.* **2008**, *151*, 326–333. [CrossRef]
50. Gama, C.; Relvas, H.; Lopes, M.; Monteiro, A. The impact of COVID-19 on air quality levels in Portugal: A way to assess traffic contribution. *Environ. Res.* **2021**, *193*. [CrossRef] [PubMed]
51. Liu, F.; Wang, M.; Zheng, M. Effects of COVID-19 lockdown on global air quality and health. *Sci. Total Environ.* **2021**, *755*, 142533. [CrossRef] [PubMed]
52. Gamelas, C.; Abecasis, L.; Canha, N.; Almeida, S.M. The Impact of COVID-19 Confinement Measures on the Air Quality in an Urban-Industrial Area of Portugal. *Atmosphere* **2021**, *12*, 1097. [CrossRef]
53. Almeida, S.M.M.; Lage, J.; Fernández, B.; Garcia, S.; Reis, M.A.A.; Chaves, P.C.C. Chemical characterization of atmospheric particles and source apportionment in the vicinity of a steelmaking industry. *Sci. Total Environ.* **2015**, *521–522*, 411–420. [CrossRef] [PubMed]
54. Lage, J.; Wolterbeek, H.T.; Reis, M.A.; Chaves, P.C.; Garcia, S.; Almeida, S.M. Source apportionment by positive matrix factorization on elemental concentration obtained in PM10 and biomonitors collected in the vicinities of a steelworks. *J. Radioanal. Nucl. Chem.* **2016**, *309*, 397–404. [CrossRef]
55. Rodríguez, S.; Querol, X.; Alastuey, A.; Kallos, G.; Kakaliagou, O. Saharan dust contributions to PM10 and TSP levels in Southern and Eastern Spain. *Atmos. Environ.* **2001**, *35*, 2433–2447. [CrossRef]
56. Almeida, S.M.; Freitas, M.C.; Pio, C.A. Neutron activation analysis for identification of African mineral dust transport. *J. Radioanal. Nucl. Chem.* **2008**, *276*, 161–165. [CrossRef]

57. Dai, Q.L.; Bi, X.H.; Wu, J.H.; Zhang, Y.F.; Wang, J.; Xu, H.; Yao, L.; Jiao, L.; Feng, Y.C. Characterization and source identification of heavy metals in ambient PM10 and PM2.5 in an integrated Iron and Steel industry zone compared with a background site. *Aerosol Air Qual. Res.* **2015**, *15*, 875–887. [CrossRef]
58. Mazzei, F.; Lucarelli, F.; Marenco, F.; Nava, S.; Prati, P.; Valli, G.; Vecchi, R. Elemental composition and source apportionment of particulate matter near a steel plant in Genoa (Italy). *Nucl. Instrum. Meth. B* **2006**, *249*, 548–551. [CrossRef]
59. Sammut, M.L.; Noack, Y.; Rose, J. Zinc speciation in steel plant atmospheric emissions: A multi-technical approach. *J. Geochem. Explor.* **2006**, *88*, 239–242. [CrossRef]
60. Timonen, K.L.; Wiikinkoski, T.; Ruuskanen, A.R. Source contributions to PM2.5 particles in the urban air of a town situated close to a steel works. *Atmos. Environ.* **2003**, *37*, 1013–1022. [CrossRef]
61. Tsai, J.H.; Lin, K.H.; Chen, C.Y.; Ding, J.Y.; Choa, C.G.; Chiang, H.L. Chemical constituents in particulate emissions from an integrated iron and steel facility. *J. Hazard. Mater.* **2007**, *147*, 111–119. [CrossRef]
62. Remus, R.; Aguado Monsonet, M.; Roudier, S. *DSL Best Available Techniques (BAT) Reference Document:for:Iron and Steel Production:Industrial Emissions Directive 2010/75/EU:(Integrated Pollution Prevention and Control)*; Joint Research Centre: Luxembourg, 2012.
63. Proctor, D.M.; Fehling, K.A.; Shay, E.C.; Wittenborn, J.L.; Green, J.J.; Avent, C.; Bigham, R.D.; Connolly, M.; Lee, B.; Shepker, T.O.; et al. Physical and chemical characteristics of blast furnace, basic oxygen furnace, and electric arc furnace steel industry slags. *Environ. Sci. Technol.* **2000**, *34*, 1576–1582. [CrossRef]
64. Almeida-Silva, M.; Canha, N.; Freitas, M.C.; Dung, H.M.; Dionísio, I. Air pollution at an urban traffic tunnel in Lisbon, Portugal-an INAA study. *Appl. Radiat. Isot.* **2011**, *69*, 1586–1591. [CrossRef] [PubMed]
65. Zgłobicki, W.; Telecka, M.; Skupiński, S.; Pasierbińska, A.; Kozieł, M. Assessment of heavy metal contamination levels of street dust in the city of Lublin, E Poland. *Environ. Earth Sci.* **2018**, *77*, 774. [CrossRef]
66. Hernández-Mena, L.; Murillo-Tovar, M.; Ramírez-Muñíz, M.; Colunga-Urbina, E.; De La Garza-Rodríguez, I.; Saldarriaga-Noreña, H. Enrichment factor and profiles of elemental composition of PM2.5 in the city of Guadalajara, Mexico. *Bull. Environ. Contam. Toxicol.* **2011**, *87*, 545–549. [CrossRef]
67. *Universidade de Aveiro—Departamento de Ambiente e Ordenamento, Carta de Qualidade do Ar para o Município do Seixal*; University of Aveiro: Aveiro, Portugal, 2019.
68. Osyczka, P.; Boroń, P.; Lenart-Boroń, A.; Rola, K. Modifications in the structure of the lichen Cladonia thallus in the aftermath of habitat contamination and implications for its heavy-metal accumulation capacity. *Environ. Sci. Pollut. Res.* **2018**, *25*, 1950–1961. [CrossRef]

International Journal of
*Environmental Research and Public Health*

MDPI

*Article*

# The Role of Portable Air Purifiers and Effective Ventilation in Improving Indoor Air Quality in University Classrooms

**Mohammad Aldekheel [1,2], Abdulmalik Altuwayjiri [3], Ramin Tohidi [1], Vahid Jalali Farahani [1] and Constantinos Sioutas [1,*]**

[1] Department of Civil and Environmental Engineering, University of Southern California, Los Angeles, CA 90089, USA
[2] Department of Civil Engineering, Kuwait University, P.O. Box 5969, Kuwait City 13060, Kuwait
[3] Department of Civil and Environmental Engineering, College of Engineering, Majmaah University, Al-Majmaah 11952, Saudi Arabia
* Correspondence: sioutas@usc.edu; Tel.: +1-213-740-6134; Fax: +1-213-744-1426

**Abstract:** In this study we investigated the effectiveness of air purifiers and in-line filters in ventilation systems working simultaneously inside various classrooms at the University of Southern California (USC) main campus. We conducted real-time measurements of particle mass (PM), particle number (PN), and carbon dioxide ($CO_2$) concentrations in nine classrooms from September 2021 to January 2022. The measurement campaign was carried out with different configurations of the purifier (i.e., different flow rates) while the ventilation system was continuously working. Our results showed that the ventilation systems in the classrooms were adequate in providing sufficient outdoor air to dilute indoor $CO_2$ concentrations due to the high air exchange rates (2.63–8.63 $h^{-1}$). The particle penetration coefficients (P) of the investigated classrooms were very low for PM (<0.2) and PN (<0.1), with the exception of one classroom, corroborating the effectiveness of in-line filters in the ventilation systems. Additionally, the results showed that the efficiency of the air purifier exceeded 95% in capturing ultrafine and coarse particles and ranged between 82–88% for particles in the accumulation range (0.3–2 µm). The findings of this study underline the effectiveness of air purifiers and ventilation systems equipped with efficient in-line filters in substantially reducing indoor air pollution.

**Keywords:** air pollution; indoor air quality; particulate matter (PM); ventilation; ultrafine particles; portable air purifier

check for updates

**Citation:** Aldekheel, M.; Altuwayjiri, A.; Tohidi, R.; Jalali Farahani, V.; Sioutas, C. The Role of Portable Air Purifiers and Effective Ventilation in Improving Indoor Air Quality in University Classrooms. *Int. J. Environ. Res. Public Health* **2022**, *19*, 14558. https://doi.org/10.3390/ijerph192114558

Academic Editors: Nuno Canha, Marta Almeida, Evangelia Diapouli and Paul B. Tchounwou

Received: 5 September 2022
Accepted: 3 November 2022
Published: 6 November 2022

**Publisher's Note:** MDPI stays neutral with regard to jurisdictional claims in published maps and institutional affiliations.



## 1. Introduction

Improving indoor air quality using air filtration technologies is essential since people spend most of their time in closed environments [1–4]. Indoor air pollution leads to adverse health outcomes and almost 3.8 million premature deaths annually [5]. Occupants' exposure to indoor particle pollutants can cause a number of adverse health effects, including respiratory illnesses, lung cancer, strokes, heart failure, asthma, and eye problems [6–10]. In addition to the health drawbacks, indoor pollution in working environments can lead to fatigue and a decline in focus and productivity [11].

In addition to ambient particles infiltrating indoors and particles emitted by indoor sources, human-generated particles (i.e., airborne aerosol particles released by the exhaled breath of humans) are major routes of airborne transmission of bacteria and viruses, including SARS-CoV-2, especially in confined environments with high population density, such as classrooms [12–15]. The exhaled particles generally have an aerodynamic diameter of less than 1 µm, mostly in the range of 0.1 to 0.5 µm [16–18]. The larger exhaled droplets settle on the ground within seconds due to the gravitational force and/or evaporate to smaller particles in a few seconds [19]. The smaller particles remain suspended in the indoor environment for several hours and can be carried by air currents as far as several meters from their source [15,20]. These smaller particles have a greater capacity to increase the infection potential than large particles since they can travel longer distances [21].

Given the health impacts caused by indoor airborne pollutants, employing air purification means in indoor spaces is essential for decreasing pollutant concentrations [15,18]. There are two main methods used to enhance indoor air quality and remove indoor particle pollutants, including in-line filters in ventilation systems and portable air purifiers [22]. Portable air purification units have been widely used in recent years due to their efficient removal of indoor pollutants [23–25]. They have been placed in approximately 30% of private residential buildings in developed countries, and a steady growth in the use of these cleaning devices is expected in the upcoming years [1,24]. The existence of in-line filtration in mechanical ventilation systems reduces the infiltration of outdoor particles to indoor spaces to a certain level depending on the filter's characteristics, filter efficiency, and particle size [26–28]. The effectiveness of these in-line filters in capturing ambient particles is assessed by the penetration coefficient (P), which describes the fraction of outdoor particles that successfully penetrate the building into the indoor environment [26,29,30]. Moreover, the adequacy of the ventilation systems in bringing sufficient outdoor air to the indoor environment is assessed by the air exchange rate value [31,32]. Air exchange rate (*AER*) is the rate at which indoor air is entirely replaced by outdoor air in a specific closed environment (e.g., classrooms). The replacement of indoor air with outdoor air occurs by various means, such as natural ventilation (e.g., doors and windows) and forced ventilation (e.g., mechanical ventilation systems). Indoor air quality can be improved by increasing the air exchange rate, since allowing more air to enter the space will dilute indoor pollution, except in cases where outdoor pollution is substantially high [31,33] in which the outdoor air needs to be purified by some sort of in-line filter.

The main objective of this study was to explore the effectiveness of air purifiers and mechanical ventilation systems equipped with in-line filters in removing indoor airborne particles originating from outdoor and indoor sources in university classrooms. Several studies supported the effective work of the air purifier inside classrooms in improving indoor air quality and mitigating the transmission of bacteria and viruses [34–37]. However, this study provided additional insights by examining the performance of both air purifiers and in-line filters in the ventilation systems working simultaneously inside various university classrooms with different characteristics. In addition, we investigated the adequacy of the ventilation systems in bringing sufficient outdoor air to the indoor environment. The findings of this work provide significant insights for public health officials, especially in educational institutions, to implement air pollution control systems and enhance the quality of air in indoor environments.

## 2. Methods

### 2.1. Measurement Sites and Protocol

The measurements were conducted inside classrooms in the University of Southern California (USC) main campus area over a 5-month period from September 2021 to January 2022. Table 1 shows the details of the selected classrooms, including volume, floor area, and the total number of students. These classrooms were solely dependent on forced ventilation (i.e., mechanical ventilation systems) as the means of air exchange between outdoor and indoor environments. Natural ventilation was minimized in all classrooms by closing all doors and windows. The indoor monitoring was performed in two separate campaigns; the first campaign was held in all selected classrooms with students attending classes, while the second campaign was conducted in an empty classroom (i.e., classroom 3). In the first measurement campaign, indoor air quality measurements were conducted in three phases in 9 classrooms located in 7 different buildings; each phase had a different setting of an air purifier (Model Trio Plus™, Field Controls, Kinston, NC, USA) equipped with H13 HEPA filters. The first phase was carried out without the presence of the portable air purifier to evaluate the effectiveness of ventilation systems equipped with in-line filters in reducing indoor pollutant levels without the interference of additional air-cleaning devices. In the second phase, we conducted the measurements while both the classroom's ventilation system and the air purifier (with flow rate of 267 $m^3/h$) were active. In the third phase, the air purifier was set at the highest possible flow rate (i.e., 748 $m^3/h$) while the ventilation system was operating simultaneously. We performed real-time measurements for indoor and outdoor $PM_{2.5}$ mass concentrations (PM), particle number concentrations (PN), and $CO_2$ levels during active 2-hour lectures in the presence of students. It should be noted that strong indoor particle generation sources (e.g., chalkboard dust and cleaning activities) were not present in the classrooms during the lectures. Pollutants' monitoring in each classroom started 15 min before the beginning of the lecture and continued until 15 min after the end of the lecture. On the same day, we also monitored the outdoor pollutant concentrations for 15 min before and after the lecture to ensure that the outdoor concentration had not changed considerably while the lecture was ongoing. For each classroom, we repeated the previous procedure on three different days by changing the configuration of the purifier according to the three phases discussed earlier. Moreover, the location of the monitoring devices in the classrooms could affect the readings of the indoor pollutant concentrations. Therefore, we investigated the spatial variance in the pollutant concentrations by placing the monitoring devices in the middle and corners of the classrooms, the results of which are shown in Figure 1 for two different classrooms as an example (the rest of the classrooms yielded similar results). The results indicate overall spatial homogeneity for PM, PN, and $CO_2$ concentrations inside the classrooms. This observation shows that the particle and gas pollutants are well mixed due to the overall high air exchange rates in the classrooms; thus, the location of the measuring devices in different spots within the classroom should not result in notable differences between the readings. According to the findings of Küpper et al. (2019) [22] regarding the possible spatial variance in the purifier's removal efficacy, changing the location of the purifier will provide almost identical removal efficiencies and lead to the same distribution of clean air in the space. Therefore, we positioned the purifier in a fixed location (i.e., the center of the classroom) during the entire campaign.

**Table 1.** Characteristics of the investigated classrooms.

| No. | Classroom | Building Name | Number of Students | Room Height (m) | Floor Area ($m^2$) | Volume ($m^3$) |
|---|---|---|---|---|---|---|
| 1 | OHE136 | Olin Hall (OHE) | 21 | 3.05 | 86.12 | 262.50 |
| 2 | RTH 105 | Tutor Hall (RTH) | 7 | 3.05 | 97.73 | 297.89 |
| 3 | SGM 226 | Seeley G. Mudd Building (SGM) | 8 | 3.05 | 63.64 | 193.97 |
| 4 | GFS 221 | Grace Ford Salvatori Hall (GFS) | 20 | 3.05 | 36.42 | 111.00 |
| 5 | GFS 205 | Grace Ford Salvatori Hall (GFS) | 12 | 3.05 | 36.51 | 111.29 |
| 6 | KAP159 | Kaprielian Hall (KAP) | 20 | 3.05 | 37.63 | 114.68 |
| 7 | OHE 120 | Olin Hall (OHE) | 6 | 3.05 | 56.49 | 172.17 |
| 8 | KDC 236 | Glorya Kaufman International Dance Center (KDC) | 26 | 3.05 | 89.00 | 271.28 |
| 9 | THH 118 | Taper Hall (THH) | 22 | 3.05 | 76.83 | 234.18 |

The second measurement campaign was carried out in classroom 3 in the presence of an indoor pollution source (i.e., sodium chloride aerosols) to simulate exhaled particles of humans. Particles can be generated by humans through various activities, including breathing, speaking, coughing, and sneezing. The particle size that is generally produced by breathing ranges between 0.1 and 1.0 μm [17,38–40]. On the other hand, coughing, sneezing, and speaking generate larger particles compared to breathing; these particles are typically larger than 5 μm and will either settle gravitationally or evaporate to smaller particles (<1 μm) [19,34,41,42]. To corroborate the use of sodium chloride (NaCl) as a representative for human exhaled particles, we measured the size distribution of NaCl particles by means of an optical particle sizer (OPS) (Model 3330, TSI, Shoreview, MN, USA) and a scanning mobility particle sizer (SMPS) (Model 3936, TSI, Shoreview, MN, USA). At first, we prepared a suspension by dissolving 50 mg of sodium chloride in 100 mL of ultrapure Milli-Q water to reach a concentration of 500 µg/mL. The suspension was sonicated in an ultrasonic bath for 30 min to achieve a homogenous solution. NaCl suspension was subsequently aerosolized using a commercially available nebulizer (Model 11310 HOPE$^{TM}$ nebulizer, B&B Medical Technologies, Carlsbad, CA, USA) that was connected to a compressor pump (Model VP0625-V1014-P2-0511, Medo Inc., Roselle, IL, USA) equipped with a HEPA capsule (Model 12144, Pall Corporation, USA) to supply compressed filtered air to the nebulizer. The nebulizer was connected to both the SMPS and OPS by a clear vinyl tube to obtain the number-based particle size distribution. The particle size distribution is shown in Figure S1 and indicates that NaCl particles mostly fall in the range of 0.071 to 1.13 µm with a peak at 0.51 µm, which supports the use of NaCl as a representative of the particles generated by humans. A number of previously published studies used NaCl as the aerosol test agent to assess the effectiveness of air filtration means [18,43–46]. In addition, the National Institute for Occupational Safety and Health (NIOSH) considered NaCl as a standard test aerosol for measuring the effectiveness of respiratory protective equipment (e.g., N95 masks) [47]. Following the same setup and sample preparation discussed earlier, NaCl suspension was aerosolized in classroom 3 to act as an indoor source of aerosols.

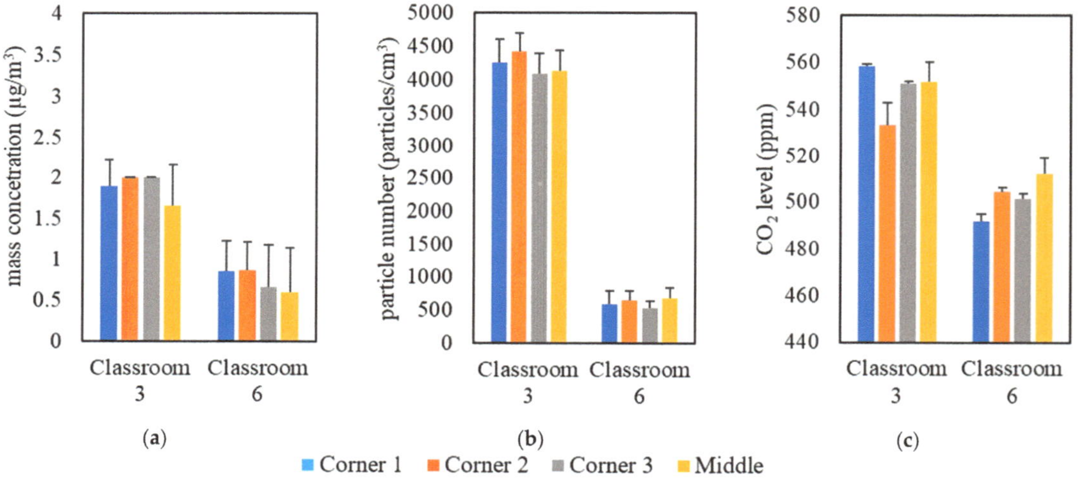


**Figure 1.** Spatial variability in classrooms 3 and 6 based on (**a**) PM, (**b**) PN, and (**c**) $CO_2$. The error bars indicate standard deviations of values measured in a single day.

## 2.2. Instrumentation

Various air quality monitors were used in this study to measure different pollutant concentrations. We employed the DiSCmini nanoparticle counter (Model Testo DiSCmini, Testo, West Chester, PA, USA) in our experiments to measure particle number concentrations. The TSI DustTrak monitor (Model 8520, TSI, Shoreview, MN, USA) was used to obtain real-time $PM_{2.5}$ particle mass concentrations. In addition, we employed a Q-track device (Model 8551, TSI, Shoreview, MN, USA) to measure indoor and outdoor $CO_2$ levels. One of the objectives of the second measurement campaign in the empty classroom was to estimate the purifier's efficiency for each particle size. This was done using the optical particle sizer (OPS) (Model 3330, TSI, Shoreview, MN, USA) to obtain particle concentrations and size distributions. The OPS measures particle sizes from 0.3 to 10 µm, which are particles in the coarse and accumulation size ranges. Further details about the calibration of the monitoring instruments are available in the supplementary material.

### 2.3. Additional Calculations

#### 2.3.1. Indoor Particle Penetration (P) in the Set of Classrooms

The indoor particle penetration (P) was calculated based on the steady-state approach, which is similar to that of Chao et al. (2003) [48]. Treating the classroom as a closed system allows for the application of the mass balance equation. Equation (1) illustrates the mass balance applied in the tested classrooms:

$$\frac{dC_{in}}{dt} = \frac{S}{V} + C_{out}AER\ (P) - C_{in}AER - C_{in}k \tag{1}$$

where $dC_{in}/dt$ is the change of indoor particle concentration with time, $S$ represents the indoor particle production rate, $V$ is the volume of the classroom ($m^3$), $k$ is the deposition rate of particles ($h^{-1}$), and $C_{in}$ and $C_{out}$ are the indoor and outdoor particle concentrations, respectively. The indoor particle production rate in Equation (1) was neglected (i.e., $S = 0$) since there was no indoor source for particles in the studied classrooms during the active lectures. The presence of students inside the classrooms did not result in noticeable increases in the indoor particle concentrations since the particle emission rate by humans is negligible compared to the particles infiltrating from outdoor sources [49–51]. The indoor particle concentration in the classrooms reached a steady-state condition after 5–8 min from the beginning of the lecture (i.e., $dC_{in}/dt = 0$), which means Equation (1) can be rearranged to the following equation:

$$P = \frac{C_{in}(AER + k)}{C_{out}AER} \tag{2}$$

The above expression has been widely used for the calculation of effective penetration or the steady-state indoor concentration ($C_{in}$) in numerous previous studies [52–54]. The calculation of particle penetration indoors was carried out in the first phase of measurements when the air purifier was switched off. The particle penetration should not be affected by using the air purifier in the second and third phases of measurements. However, to show the agreement of penetration coefficients in the three phases, the following equation was used when the air purifier was active:

$$P = \frac{C_{in}\left(AER + k + \frac{CADR}{V}\right)}{C_{out}\ AER} \tag{3}$$

where $CADR$ is the clean air delivery rate of the purifier ($m^3/h$), and $V$ is the volume of the classroom ($m^3$). Although the operation of an air purifier does not affect the penetration coefficient, it significantly affects the indoor-to-outdoor ratio. Equation (3) demonstrates that the addition of the $(CADR/V)$ term will decrease the $(C_{in}/C_{out})$ ratio. Moreover, increasing the purifier's flow rate leads to a further reduction in the indoor-to-outdoor ratio.

The penetration coefficient in the studied classrooms was used as a metric for assessing the effectiveness of the in-line filters of the ventilation systems in capturing particles penetrating the building from the outdoor space. The air handling units in all tested classrooms, except classroom 3, were equipped with a dual direction 12-inch MERV 14 filter with a fiberglass media (Model Aerostar FP Mini-Pleat, Filtration Group, Santa Rosa, CA, USA). MERV 14 efficiently filters the outside air and the air returning from the indoor space. The air handling unit of classroom 3 had a 2-inch MERV 13 filter with a synthetic air media (Model Aerostar Green Pleat, Filtration Group, Santa Rosa, CA, USA). According to the manual of Aerostar filters, MERV 13 filters have lower particle removal efficiency than MERV 14 filters.

#### 2.3.2. Air Purifier's Efficiency in Classroom 3

The second measurement campaign consisted of three stages leading to the determination of the purifier's efficiency. In the first stage, the background indoor pollutant concentrations were measured without operating the pollution source and the purifier. The second stage started when the indoor pollutant generator was switched on until a stabilized particle concentration was reached. In the third and last stage, the indoor pollutant source was switched off, and the air purifier was switched on. The purpose of the third stage was to investigate the particle decay rate ($K_{purifier}$) in the presence of the air purifier. The experiment was repeated three times by changing the third-stage scenario. In the first scenario, the purifier was switched off in order to measure the natural decay rate of particles ($K_{natural}$) when the ventilation system was only switched on. In the second scenario, the purifier was operated at a flow rate of 267 $m^3/h$ to obtain the particle decay rate ($K_{purifier\ (low)}$). The last scenario was conducted while operating the purifier at a flow rate of 748 $m^3/h$ to measure the decay rate at the purifier's maximum fan speed ($K_{purifier\ (high)}$). In order to calculate the particle decay rate after switching off the NaCl source, we treated the classroom as a closed system and applied the mass balance equation below:

$$\frac{dC_{in}}{dt} = C_{out}AER\ (P) - C_{in}(K) \tag{4}$$

where $dC_{in}/dt$ is the change of the indoor particle concentration with time, $K$ is the particle decay rate ($h^{-1}$), $AER$ is the air exchange rate ($h^{-1}$), $P$ is the particle penetration coefficient, and $C_{in}$ and $C_{out}$ are the indoor and outdoor particle concentrations, respectively. Based on the integration of Equation (4), the general equation for the indoor concentration is expressed as follows:

$$C_{in(t)} = \frac{C_{out}AER\ (P)}{K}\left(1 - e^{-(K)t}\right) + C_{in(o)}e^{-(K)t} \tag{5}$$

where $C_{in(t)}$ is the concentration of the particles at time $t$ and $C_{in(o)}$ represents the concentration of the particles at time 0. In order to analyze the decay of the particles (i.e., reduction in particle concentration) with time, we subtracted the concentration of the particles continuously infiltrating indoors (i.e., the first term of Equation (5)) from the measured concentrations during the decay period. This allowed us to use the exponential equation below to obtain the decay rate of the particles:

$$C_{in(t)} = C_{in(o)}e^{-(K)\ t} \tag{6}$$

The particle decay rate is a function of the air exchange rate, particle deposition rate, and particle removal rate by the purifier. Thus, Equations (7) and (8) were used to express the decay rate in the natural condition (i.e., without the purifier) and in the presence of the purifier, respectively:

$$K_{Natural} = AER + k \tag{7}$$

$$K_{Purifier} = AER + k + \eta\frac{Q}{V} \tag{8}$$

where $AER$ is the air exchange rate ($h^{-1}$), $k$ is the particle deposition rate ($h^{-1}$), $\eta$ represents the purifier efficiency, $Q$ is the air volumetric flow rate of the purifier ($m^3/h$), and $V$ represents the volume of the classroom ($m^3$). By combining Equations (7) and (8), we can calculate the purifier's efficiency, as shown in Equation (9):

$$\eta = \frac{(K_{Purifier} - K_{Natural})_V}{Q} \tag{9}$$

The decay in the particle mass and number concentrations was plotted as a function of time after switching off the aerosol source. Decay curves were obtained for a range of particle sizes (0.3–10 μm) to estimate the purifier's efficiency in removing different particle sizes. In addition, the purifier removal efficiency for ultrafine particles was estimated using PN data obtained from the DiSCmini since it mainly detected particles with diameters below 0.3 μm.

## 3. Results and Discussion

### 3.1. Indoor Monitoring of PM, PN, and $CO_2$ Concentrations in the Set of Classrooms

#### 3.1.1. Indoor $CO_2$ Levels

Figure 2 demonstrates the actual air exchange rates (*AER*) in the selected classrooms, which were measured and showed a very good agreement with the *AER* received from the USC facilities and the management department as shown in Figure S2. The detailed methodology of calculating the *AER* inside the classrooms is described in the supplementary material. *AER* is the metric for assessing the adequacy of the applied ventilation (i.e., mechanical ventilation system) in bringing in sufficient outdoor air and diluting indoor $CO_2$ concentrations. However, high air exchange rates will also increase the infiltration of outdoor particulate pollutants, especially if the ventilation system operates without an in-line filtration system that removes a fraction of outdoor particle pollutants [26,31]. As shown in Figure 2, the classrooms' *AER* values ranged from 2.63 $h^{-1}$ (Classroom 7) to 8.63 $h^{-1}$ (Classroom 4). The American Society of Heating, Refrigerating and Air-Conditioning Engineers (ASHRAE) standard 62.1 (2016) recommended a minimum ventilation rate of 7.5 L/sec (27 $m^3$/h) per person in closed environments [55]. Figure 2 shows the alignment of the *AER* values with the ASHRAE's recommendation in all investigated classrooms. Therefore, these *AER* values indicate sufficient outdoor-to-indoor air circulation and adequate ventilation.

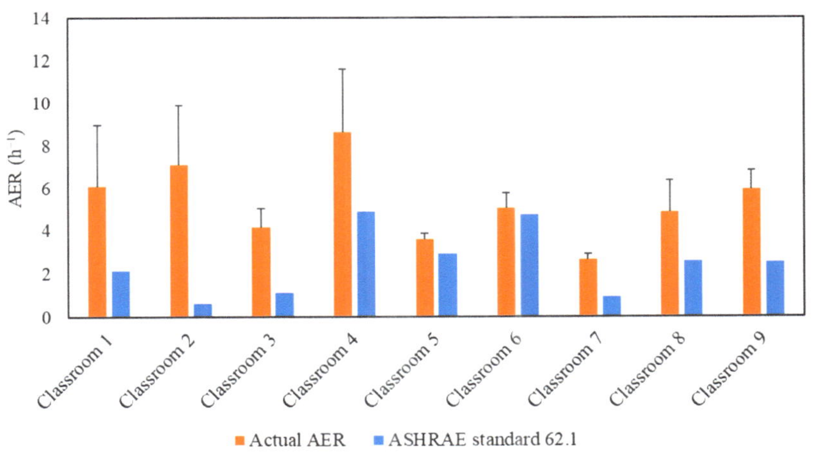


**Figure 2.** Air exchange rate (*AER*) values for the tested classrooms. The error bars indicate standard deviations of the values measured on three different days.

After approximately 10 min from the beginning of each lecture, the indoor $CO_2$ concentration reached a well-mixed steady-state condition when the production of $CO_2$ by the students was equal to the losses of $CO_2$ due to air circulation in the ventilation system. The ASHRAE standard 62.1 (2016) recommended that the indoor steady-state $CO_2$ concentration should not exceed the outdoor $CO_2$ level by more than 700 ppm [55]. Figure 3 presents the average outdoor and indoor $CO_2$ concentrations during the three phases of measurements in the studied classrooms. The comparable indoor $CO_2$ levels in the three phases confirm that the indoor $CO_2$ concentrations are not affected by the use of air purifiers since the latter remove particulate and not gaseous air pollutants. Elevated concentrations of $CO_2$ can impact productivity [56–58] and lead to headaches, tiredness [59,60], and sick building syndrome (SBS) symptoms (e.g., difficulty in concentration, dizziness) [61–64]. According to the recommended indoor $CO_2$ level by ASHRAE (not exceeding the outdoor level by 700 ppm) and the measured outdoor $CO_2$ level (400–500 ppm), the indoor $CO_2$ levels in the tested classrooms should not exceed 1100–1200 ppm. This is consistent with our measurements inside the classrooms which showed values ranging between

500 ppm and 900 ppm. This observation corroborates that the ventilation systems in all the tested classrooms are adequate and provide sufficient outdoor air to dilute indoor $CO_2$ concentrations as a result of the generally high air exchange rates (2.63–8.63 $h^{-1}$) in each classroom [32,65,66].

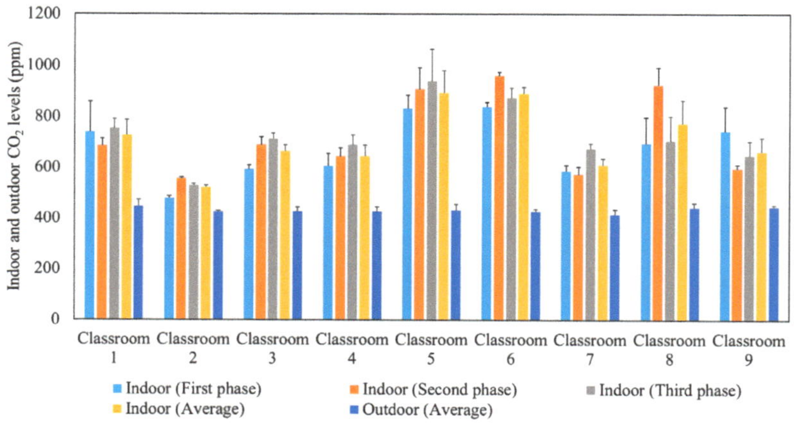


**Figure 3.** Average outdoor and indoor $CO_2$ levels during the three phases in the studied classrooms. The error bars indicate standard deviations of values measured in a single day.

### 3.1.2. Particle Mass and Number Concentrations and Indoor-to-Outdoor Ratios inside the Classrooms

Table 2 summarizes the ambient, indoor, and indoor-to-outdoor (I/O) ratios of $PM_{2.5}$ mass concentrations and particle number concentrations in the occupied classrooms for the first, second, and third measurement phases. During the first phase, classroom 3 exhibited the highest indoor $PM_{2.5}$ mass concentration (8.62 $\mu g/m^3$), followed by classroom 9 with an indoor mass concentration of 2.43 $\mu g/m^3$. The indoor mass concentration during the three phases does not accurately reflect the effectiveness of the air purification unit in reducing indoor pollution because the ambient pollution has a significant influence on the indoor concentration. For example, classroom 9 showed a higher PM indoor concentration (2.43 $\mu g/m^3$) compared to classroom 8 (0.95 $\mu g/m^3$) in the first phase, while the corresponding outdoor levels were 21.67 and 5.69 $\mu g/m^3$, respectively. Therefore, we used the indoor-to-outdoor ratio as a metric for measuring the effectiveness of ventilation and air purifiers in reducing indoor pollutant levels. Excluding classroom 3, all classrooms had PM and PN I/O ratios below 0.2 in the first phase without using the purification unit. This observation indicates that the ambient PM and PN were initially reduced by 80% or more in most classrooms by just the in-line filters of the ventilation system. In classroom 3, the ambient PM and PN concentrations in the first phase were reduced by 56 % and 65%, respectively. The low I/O values in the first phase did not allow for a proper investigation of the purifier's efficiency in removing particles in the subsequent phases. For example, the PN I/O ratio in classroom 4 decreased from 0.05 in the first phase to 0.04 in the third phase when the purifier was operated at the maximum volumetric flow rate (748 $m^3/h$). Starting with a low I/O value did not allow the purifier to reduce the I/O ratio substantially and, more importantly, the indoor PM levels approached the limit of detection of the DustTrak, such as classrooms 2 and 7.

**Table 2.** Indoor, ambient, and indoor-to-outdoor (I/O) ratios of PM and PN in the three measurement phases. LOD refers to the limit of detection of the employed instrument.

| | **$PM_{2.5}$ Mass Concentration (PM) (μg/m$^3$)** | | | | | | | | |
|---|---|---|---|---|---|---|---|---|---|
| | **First Phase** | | | **Second Phase** | | | **Third Phase** | | |
| | **Indoor** | **Outdoor** | **I/O** | **Indoor** | **Outdoor** | **I/O** | **Indoor** | **Outdoor** | **I/O** |
| Classroom 1 | 1.19 | 7.90 | 0.15 | 0.07 | 2.04 | 0.03 | 0.21 | 9.00 | 0.02 |
| Classroom 2 | 0.31 | 8.2 | 0.04 | 0.48 | 46.01 | 0.01 | <LOD | 24.33 | NA |
| Classroom 3 | 8.62 | 19.55 | 0.44 | 1.25 | 3.04 | 0.41 | 10.97 | 42.39 | 0.26 |
| Classroom 4 | 1.29 | 10.60 | 0.12 | 0.61 | 9.56 | 0.06 | 2.03 | 28.22 | 0.07 |
| Classroom 5 | 1.24 | 13.44 | 0.09 | 1.04 | 10.31 | 0.10 | 0.73 | 6.60 | 0.11 |
| Classroom 6 | 1.00 | 5.63 | 0.18 | 0.15 | 2.75 | 0.05 | 0.77 | 22.93 | 0.03 |
| Classroom 7 | 0.27 | 7.83 | 0.03 | <LOD | 4.08 | NA | <LOD | 11.67 | NA |
| Classroom 8 | 0.95 | 5.69 | 0.17 | 2.99 | 33.91 | 0.09 | 2.20 | 60.25 | 0.04 |
| Classroom 9 | 2.43 | 21.67 | 0.11 | 1.92 | 18.71 | 0.10 | 0.87 | 11.85 | 0.07 |
| | **Particle Number Concentration (PN) (particles/cm$^3$)** | | | | | | | | |
| | **First Phase** | | | **Second Phase** | | | **Third Phase** | | |
| | **Indoor** | **Outdoor** | **I/O** | **Indoor** | **Outdoor** | **I/O** | **Indoor** | **Outdoor** | **I/O** |
| Classroom 1 | 207.4 | 8523.5 | 0.02 | 90.6 | 4798.5 | 0.02 | 84.0 | 2972.8 | 0.03 |
| Classroom 2 | 113.73 | 2290.07 | 0.05 | 165.87 | 5285.31 | 0.03 | 42.76 | 3672.92 | 0.01 |
| Classroom 3 | 2328.7 | 6693.6 | 0.35 | 1800.0 | 5557.3 | 0.32 | 1917.8 | 7050.2 | 0.27 |
| Classroom 4 | 258.6 | 5537.9 | 0.05 | 306.2 | 6704.9 | 0.05 | 253.8 | 6944.3 | 0.04 |
| Classroom 5 | 345.6 | 8453.6 | 0.04 | 151.0 | 8414.1 | 0.02 | 363.1 | 17704.5 | 0.02 |
| Classroom 6 | 1389.0 | 14338.6 | 0.10 | 299.1 | 12158.4 | 0.02 | 89.4 | 10312.7 | 0.01 |
| Classroom 7 | 76.36 | 6215.27 | 0.01 | 52.33 | 4425.42 | 0.01 | 53.41 | 5573.16 | 0.01 |
| Classroom 8 | 388.0 | 13783.6 | 0.03 | 220.0 | 7050.2 | 0.03 | 90.4 | 5807.8 | 0.02 |
| Classroom 9 | 1100.2 | 17763.5 | 0.06 | 673.5 | 14862.2 | 0.05 | 543.5 | 15294.1 | 0.04 |

The effective indoor penetration was measured for each classroom to assess the effectiveness of the in-line filtration in the air handling units. Figure 4 shows the penetration coefficients for PM and PN during each phase, as well as the average values throughout all three phases. Unlike the I/O ratio, the penetration coefficient values are independent of the purifier as corroborated by the comparable values in the three phases. The penetration coefficients for PM were higher than PN as the latter primarily consists of ultrafine particles (i.e., size < 0.3 μm), which are easier to remove by filters due to their diffusivity. The P values in the majority of classrooms were low for both PM (<0.2) and PN (<0.1), which can be attributed to the presence of efficient in-line filters (i.e., MERV 14) in the ventilation systems of almost all classrooms. Higher penetration coefficient values for PM (0.51) and PN (0.45) were observed in classroom 3 due to the less efficient in-line filter (i.e., MERV 13) used in its mechanical ventilation system. Based on the penetration values in classroom 3, the in-line filtration system could only reduce ambient PM and PN by approximately 49% and 55%, respectively. Therefore, we selected classroom 3 to conduct our experiments for the second measurement campaign.

### *3.2. Indoor Monitoring of PM, PN, and $CO_2$ Concentrations in Classroom 3 in the Presence of Indoor Particle Pollution Source*

Real-time monitoring of PM, PN, and $CO_2$ was conducted in the presence of an aerosol-generating source emitting sodium chloride in classroom 3. As discussed earlier, classroom 3 was selected for the second measurement campaign due to its higher penetration coefficient compared to the other classrooms. The measured indoor $CO_2$ level in classroom 3 was constant during the three stages due to the absence of indoor $CO_2$ sources (e.g., students). Indoor $CO_2$ levels were not affected by the generation of aerosols or the change of the purifier setting, as we would expect; however, PN and PM concentrations were heavily affected.

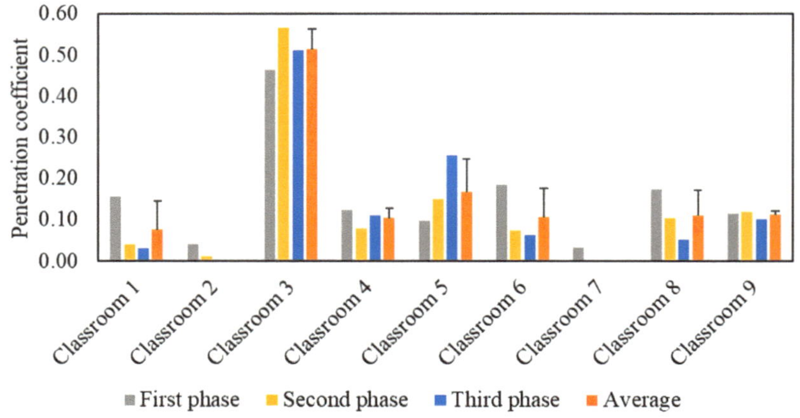


(**a**)

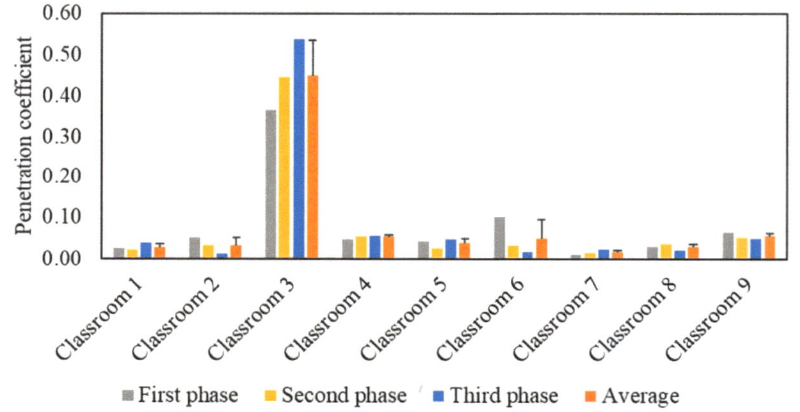


(**b**)

**Figure 4.** Particle indoor penetration based on: (**a**) $PM_{2.5}$ mass concentration and (**b**) particle number concentration. The error bars indicate standard deviations of the values measured on three different days.

#### 3.2.1. PM and PN Decay Rates with and without the Use of Air Purifier at Different Volumetric Flow Rates

Figure S3 presents the real-time measurements of PM and PN concentrations during the three stages: (i) background condition, (ii) NaCl indoor source is switched on, (iii) purifier is switched on and the source is switched off. It also shows the PM and PN measurements for different configurations of the third stage (i.e., without the purifier, the purifier at a low flow rate of 267 $m^3/hr$, and the purifier at a high flow rate of 748 $m^3/h$). Figure 5 shows the PM and PN particle decay curves in classroom 3, which were obtained and analyzed based on the third-stage data. The natural decay rates of the particles without the application of the air purifier were in the range of 3.9 to 4.8 $h^{-1}$, where the K values were 4.74 $h^{-1}$ and 3.95 $h^{-1}$ for PM and PN, respectively. When the purifier was switched on at a low flow rate (267 $m^3/h$), the decay rates increased to 5.0–5.3 $h^{-1}$, with K values of 5.09 $h^{-1}$ for PM and 5.26 $h^{-1}$ for PN. Operating the purifier at the maximum air flow rate (748 $m^3/h$) resulted in a significant increase in the particle decay rates (6.5–6.7 $h^{-1}$), with decay values of 6.70 $h^{-1}$ and 6.58 $h^{-1}$ for PM and PN, respectively. The theoretical values of the decay rates were calculated using Equations (7) and (8). According to Long et al. (2001) [54], the deposition rate is dependent on the particle size and ranges between 0.10–0.25 $h^{-1}$ for $PM_{2.5}$ particles. Table 3 shows a good agreement between the theoretical and experimental decay rates for PM.

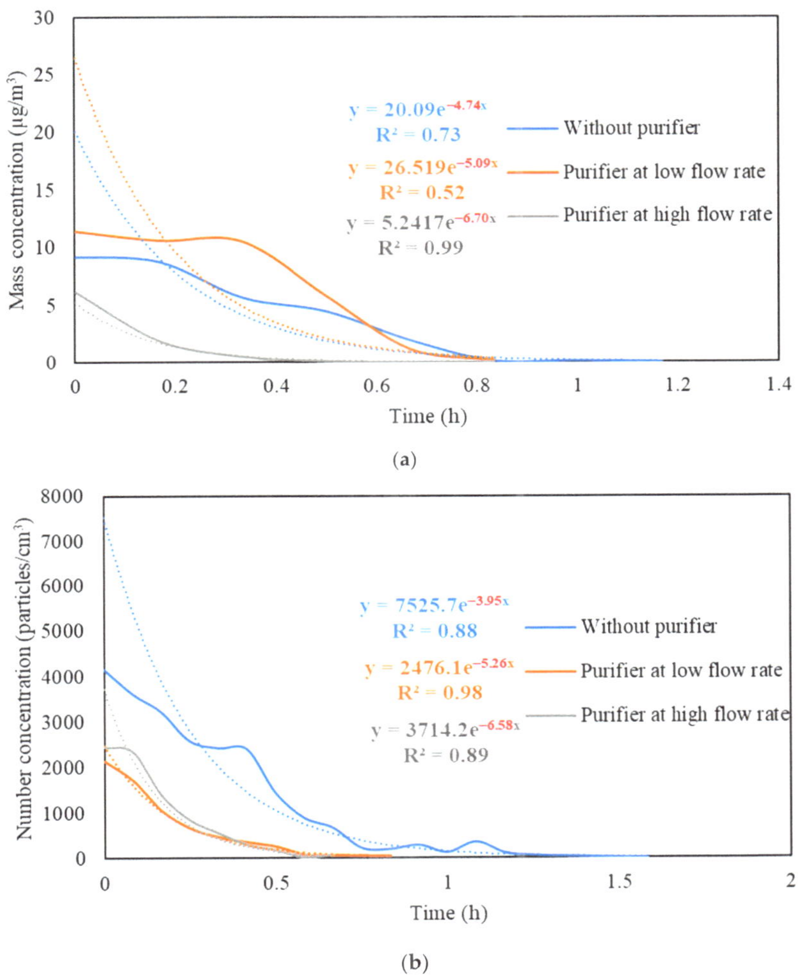


**Figure 5.** Decay rate curves with and without the use of purifier at different flow rates based on (**a**) PM and (**b**) PN in classroom 3. The dotted curves represent exponential trendlines.

**Table 3.** Theoretical versus experimental decay rates for particle mass (PM) with and without the use of purifier at different settings in classroom 3.

| | Theoretical K ($h^{-1}$) | Experimental K ($h^{-1}$) |
|---|---|---|
| Without purifier ($K_{natural}$) | 4.32 | 4.74 ± 0.15 |
| Purifier at low setting ($K_{purifier\ (low)}$) | 5.22 | 5.09 ± 0.13 |
| Purifier at high setting ($K_{purifier\ (high)}$) | 6.90 | 6.70 ± 0.33 |

The quick reduction in particle concentrations clearly demonstrates the effectiveness of the air purifier. In the first stage, the initial PM and PN concentrations at the beginning of the decay period reached a 50% reduction after 35–40 min when only mechanical ventilation was on. Using the purifier at a low flow rate of 267 $m^3/h$ and a high flow rate of 748 $m^3/h$ reduced the particle number concentrations by 50% after 25–30 min and 10–15 min, respectively. According to Szabadi et al. (2022) [18], operating the purifier at the maximum flow rate caused a 50% reduction in the particle number concentration after 20 min of switching off the aerosol source, which is consistent with our study. Lower decay rates will result in longer particle residence times indoors and, if these aerosols contain viruses (e.g., SARS-CoV-2), the probability of transmission and infection will increase [35,36,44,67]. Zuraimi et al. (2011) [44] reported that using an air purifier at its maximum fan setting reduced the residence time of coughing and sneezing particles from 4–6 h to 30–40 min. All the aforementioned studies support the use of an air purifier at the maximum flow rate to increase the particle decay, which will decrease the risk of viruses' transmission in case an infectious person is present in the classroom.

#### 3.2.2. Removal Efficiency as a Function of Particle Size

The measurements of the particle number concentration at the purifier's flow rate of 267 $m^3/h$ were used to determine the purifier's removal efficiency as a function of particle size in classroom 3. Figure 6 presents different particle decay rates based on various particle size ranges. As shown in the figure, the increase in particle size is associated with a higher value of particle decay rate. The efficiencies for each particle size range were calculated and shown in Figure 7. The particle removal efficiencies of the purifier for the size ranges (0.3–0.5 μm), (0.5–1 μm), (1–2 μm), (2–5 μm), and (5–10 μm) were 82.8%, 85.3%, 87.7%, 95.0%, and 99.4%, respectively. Higher efficiencies were achieved for coarse particles, which indicates the efficient performance of HEPA filters in capturing coarse particles. HEPA filters are less efficient in removing particles in the accumulation mode (0.3–2 μm), with removal efficiencies between 82% and 88%.

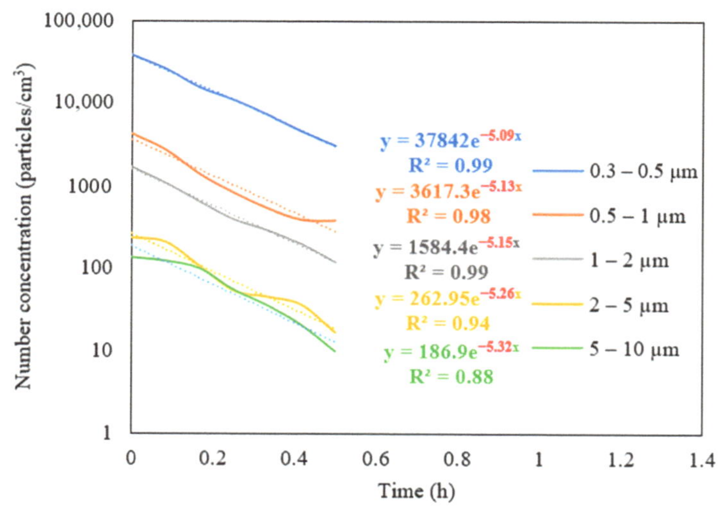


**Figure 6.** Decay rate curves as a function of particle size in classroom 3 after switching off the NaCl source and operating the purifier at a flow rate of 267 $m^3/h$. The dotted curves represent exponential trendlines.

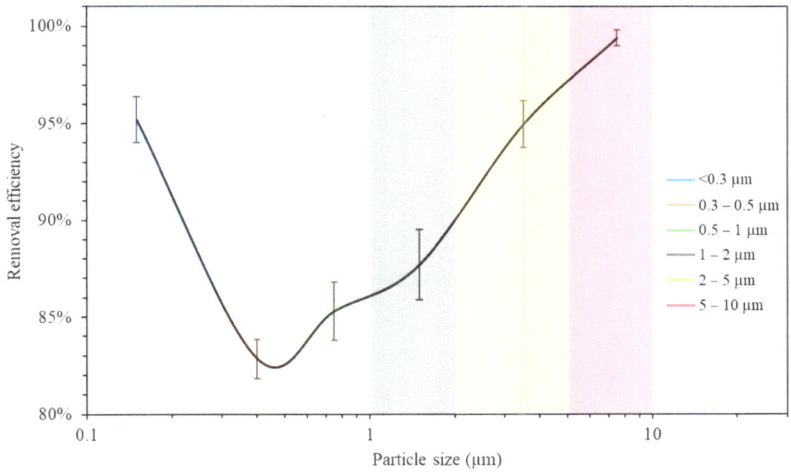


**Figure 7.** Air purifier removal efficiency as a function of particle size. The error bars indicate standard deviations of values measured in a single day. Values in x-axis are the mid-point diameters of each particle size range.

The measurements of the particle number concentration in Figure 5b were used to assess the removal efficiency of ultrafine particles since PN data were dominated by particles with a size range of less than 0.3 μm. The decay rate at the purifier's flow rate of 267 $m^3/h$ was calculated as 5.26 $h^{-1}$, corresponding to a removal efficiency value of 95.2%. These results lead to the conclusion that the air purifier equipped with HEPA filters is more efficient in removing both ultrafine particles (<0.3 μm) and coarse particles (2–10 μm). However, particles in the intermediate size range (0.3–2 μm) were somewhat less efficiently removed compared to those in the coarse and ultrafine ranges, although the removal efficiency even in that particle range was between 82 and 88%. These results are consistent with various previously published studies and are a result of the fact that smaller particles are easily removed by filters due to their high diffusivity, and larger particles are primarily removed because of their high interception and inertia impaction [68,69].

## 4. Summary and Conclusions

This work investigated the effectiveness of air purifiers working in conjunction with in-line filters of mechanical ventilation systems inside different classrooms and their role in improving air quality and capturing pollutants originating from both indoor and outdoor sources. The mechanical ventilation systems in all classrooms, except one, were equipped with 12-inch MERV 14 filters that significantly reduced ambient PM and PN concentrations by more than 80%. The less efficient in-line filter (MERV 13) in the ventilation system of classroom 3 reduced ambient PM and PN by 49% and 55%, respectively. The indoor $CO_2$ levels in the analyzed classrooms (500–900 ppm) were below the ASHRAE 62.1 standard, indicating adequate ventilation and sufficient outdoor-to-indoor air circulation due to the high air exchange rates (2.63–8.63 $h^{-1}$). Moreover, operating the purifier at the maximum flow rate (748 $m^3/h$) in classroom 3 resulted in increasing the particle decay rate from 3.9–4.8 $h^{-1}$ (without the purifier) to 6.5–6.7 $h^{-1}$, corresponding to a 50% reduction in indoor PM and PN after 10–15 min of switching off the aerosol source. The efficiency of the HEPA air purifier exceeded 95% in capturing ultrafine and coarse particles and ranged between 82–88% for particles in the accumulation range. This study highlighted the significance of mitigating indoor pollution in closed environments, especially in densely seated classrooms where the infection risk of viruses' transmission is high. The findings of this study recommend the use of HEPA air purifiers in closed environments, especially when the ventilation system is not equipped with an efficient in-line filter.

**Supplementary Materials:** The following supporting information can be downloaded at: https://www.mdpi.com/article/10.3390/ijerph192114558/s1. Figure S1: Number-based size distribution curve of NaCl particles; Figure S2: Actual *AER* measured in the selected classrooms in comparison with the *AER* received from USC facilities and management (FM) department; Figure S3: PM and PN measurements in classroom 3 (a) without using purifier, (b) with purifier at low flow rate (267 $m^3$/h), and (c) with purifier at maximum flow rate (748 $m^3$/h). References [70,71] are cited in Supplementary Materials.

**Author Contributions:** Conceptualization, M.A. and C.S.; Data curation, M.A., A.A., R.T. and V.J.F.; Methodology, M.A., A.A., R.T. and V.J.F.; Project administration, C.S.; Supervision, C.S.; Validation, V.J.F.; Visualization, R.T.; Writing—original draft, M.A.; Writing—review & editing, A.A., R.T., V.J.F. and C.S. All authors have read and agreed to the published version of the manuscript.

**Funding:** This research was funded by the dean's office at USC Viterbi School of Engineering (Internal grant).

**Institutional Review Board Statement:** Not applicable.

**Informed Consent Statement:** Informed consent was obtained from all subjects involved in the study.

**Data Availability Statement:** The data that support the findings of this study are available on request from the corresponding author.

**Acknowledgments:** The authors would like to acknowledge the Ph.D. fellowship awards from the University of Southern California (USC) and Kuwait University. They would also wish to thank the dean's office at USC Viterbi School of Engineering for the internal grant to support this study.

**Conflicts of Interest:** The authors declare no conflict of interest.

## References

1. Ma, H.; Shen, H.; Shui, T.; Li, Q.; Zhou, L. Experimental Study on Ultrafine Particle Removal Performance of Portable Air Cleaners with Different Filters in an Office Room. *Int. J. Environ. Res. Public Health* **2016**, *13*, 102. [CrossRef] [PubMed]
2. Cheek, E.; Guercio, V.; Shrubsole, C.; Dimitroulopoulou, S. Portable Air Purification: Review of Impacts on Indoor Air Quality and Health. *Sci. Total Environ.* **2021**, *766*, 142585. [CrossRef] [PubMed]
3. Cooper, E.; Wang, Y.; Stamp, S.; Burman, E.; Mumovic, D. Use of Portable Air Purifiers in Homes: Operating Behavior, Effect on Indoor PM2.5 and Perceived Indoor Air Quality. *Build. Environ.* **2021**, *191*, 107621. [CrossRef]
4. Pacitto, A.; Amato, F.; Moreno, T.; Pandolfi, M.; Fonseca, A.; Mazaheri, M.; Stabile, L.; Buonanno, G.; Querol, X. Effect of Ventilation Strategies and Air Purifiers on the Children's Exposure to Airborne Particles and Gaseous Pollutants in School Gyms. *Sci. Total Environ.* **2020**, *712*, 135673. [CrossRef] [PubMed]
5. World Health Organization (WHO). Household Air Pollution and Health. Available online: https://www.who.int/health-topics/air-pollution (accessed on 19 March 2022).
6. Polichetti, G.; Cocco, S.; Spinali, A.; Trimarco, V.; Nunziata, A. Effects of Particulate Matter (PM10, PM2.5 and PM1) on the Cardiovascular System. *Toxicology* **2009**, *261*, 1–8. [CrossRef] [PubMed]
7. Anderson, J.O.; Thundiyil, J.G.; Stolbach, A. Clearing the Air: A Review of the Effects of Particulate Matter Air Pollution on Human Health. *J. Med. Toxicol.* **2012**, *8*, 166–175. [CrossRef]
8. Oberdörster, G. Toxicology of Ultrafine Particles: In Vivo Studies. *Philos. Trans. R. Soc. A Math. Phys. Eng. Sci.* **2000**, *358*, 2719–2740. [CrossRef]
9. Hoskins, J.A. Health Effects Due to Indoor Air Pollution. *Indoor Built Environ.* **2003**, *12*, 427–433. [CrossRef]
10. Perez-Padilla, R.; Schilmann, A.; Riojas-Rodriguez, H.; Murray, J.F. Respiratory Health Effects of Indoor Air Pollution. *Int. J. Tuberc. Lung Dis.* **2010**, *14*, 1079–1086.
11. Chang, T.Y.; Zivin, J.G.; Gross, T.; Neidell, M. The Effect of Pollution on Worker Productivity: Evidence from Call Center Workers in China. *Am. Econ. J. Appl. Econ.* **2019**, *11*, 151–172. [CrossRef]
12. Jayaweera, M.; Perera, H.; Gunawardana, B.; Manatunge, J. Transmission of COVID-19 Virus by Droplets and Aerosols: A Critical Review on the Unresolved Dichotomy. *Environ. Res.* **2020**, *188*, 109819. [CrossRef] [PubMed]
13. Morawska, L.; Cao, J. Airborne Transmission of SARS-CoV-2: The World Should Face the Reality. *Environ. Int.* **2020**, *139*, 105730. [CrossRef] [PubMed]
14. Kähler, C.J.; Fuchs, T.; Hain, R. Can Mobile Indoor Air Cleaners Effectively Reduce an Indirect Risk of SARS-CoV-2 Infection by Aerosols? *MMWR* **2020**, *70*, 972–976.

15. Piscitelli, P.; Miani, A.; Setti, L.; de Gennaro, G.; Rodo, X.; Artinano, B.; Vara, E.; Rancan, L.; Arias, J.; Passarini, F.; et al. The Role of Outdoor and Indoor Air Quality in the Spread of SARS-CoV-2: Overview and Recommendations by the Research Group on COVID-19 and Particulate Matter (RESCOP Commission). *Environ. Res.* **2022**, *211*, 113038. [CrossRef]
16. Schwarz, K.; Biller, H.; Windt, H.; Koch, W.; Hohlfeld, J.M. Characterization of Exhaled Particles from the Human Lungs in Airway Obstruction. *J. Aerosol Med. Pulm. Drug Deliv.* **2015**, *28*, 52–58. [CrossRef]
17. Scheuch, G. Breathing Is Enough: For the Spread of Influenza Virus and SARS-CoV-2 by Breathing Only. *J. Aerosol Med. Pulm. Drug Deliv.* **2020**, *33*, 230–234. [CrossRef]
18. Szabadi, J.; Meyer, J.; Lehmann, M.; Dittler, A. Simultaneous Temporal, Spatial and Size-Resolved Measurements of Aerosol Particles in Closed Indoor Environments Applying Mobile Filters in Various Use-Cases. *J. Aerosol Sci.* **2022**, *160*, 105906. [CrossRef]
19. Lindsley, W.G.; Noti, J.D.; Blachere, F.M.; Szalajda, J.V.; Beezhold, D.H. Efficacy of Face Shields against Cough Aerosol Droplets from a Cough Simulator. *J. Occup. Environ. Hyg.* **2014**, *11*, 509–518. [CrossRef]
20. Xie, X.; Li, Y.; Chwang, A.T.Y.; Ho, P.L.; Seto, W.H. How Far Droplets Can Move in Indoor Environments-Revisiting the Wells Evaporation-Falling Curve. *Indoor Air* **2007**, *17*, 211–225. [CrossRef]
21. Fennelly, K.P. Particle Sizes of Infectious Aerosols: Implications for Infection Control. *Lancet Respir. Med.* **2020**, *8*, 914–924. [CrossRef]
22. Küpper, M.; Asbach, C.; Schneiderwind, U.; Finger, H.; Spiegelhoff, D.; Schumacher, S. Testing of an Indoor Air Cleaner for Particulate Pollutants under Realistic Conditions in an Office Room. *Aerosol Air Qual. Res.* **2019**, *19*, 1655. [CrossRef]
23. Sultan, Z.M.; Nilsson, G.J.; Magee, R.J. Removal of Ultrafine Particles in Indoor Air: Performance of Various Portable Air Cleaner Technologies. *HVAC R Res.* **2011**, *17*, 513–525. [CrossRef]
24. Shaughnessy, R.J.; Sextro, R.G. What Is an Effective Portable Air Cleaning Device? A Review. *J. Occup. Environ. Hyg.* **2006**, *3*, 169–181. [CrossRef]
25. Kim, H.J.; Han, B.; Kim, Y.J.; Jeong, C.S.; Lee, S.H. A Simple and Efficient Method for Evaluating Air-Cleaning Performance against Airborne Allergen Particles. *Build. Environ.* **2013**, *60*, 272–279. [CrossRef]
26. van der Zee, S.C.; Strak, M.; Dijkema, M.B.A.; Brunekreef, B.; Janssen, N.A.H. The Impact of Particle Filtration on Indoor Air Quality in a Classroom near a Highway. *Indoor Air* **2017**, *27*, 291–302. [CrossRef] [PubMed]
27. Sublett, J.L. Effectiveness of Air Filters and Air Cleaners in Allergic Respiratory Diseases: A Review of the Recent Literature. *Curr. Allergy Asthma Rep.* **2011**, *11*, 395–402. [CrossRef]
28. Batterman, S.; Godwin, C.; Jia, C. Long Duration Tests of Room Air Filters in Cigarette Smokers' Homes. *Environ. Sci. Technol.* **2005**, *39*, 7260–7268. [CrossRef]
29. Barn, P.; Larson, T.; Noullett, M.; Kennedy, S.; Copes, R.; Brauer, M. Infiltration of Forest Fire and Residential Wood Smoke: An Evaluation of Air Cleaner Effectiveness. *J. Expo. Sci. Environ. Epidemiol.* **2008**, *18*, 503–511. [CrossRef]
30. Lin, C.C.; Peng, C.K. Characterization of Indoor PM10, PM2.5, and Ultrafine Particles in Elementary School Classrooms: A Review. *Environ. Eng. Sci.* **2010**, *27*, 915–922. [CrossRef]
31. Guo, H.; Morawska, L.; He, C.; Gilbert, D. Impact of Ventilation Scenario on Air Exchange Rates and on Indoor Particle Number Concentrations in an Air-Conditioned Classroom. *Atmos. Environ.* **2008**, *42*, 757–768. [CrossRef]
32. You, Y.; Niu, C.; Zhou, J.; Liu, Y.; Bai, Z.; Zhang, J.; He, F.; Zhang, N. Measurement of Air Exchange Rates in Different Indoor Environments Using Continuous $CO_2$ Sensors. *J. Environ. Sci.* **2012**, *24*, 657–664. [CrossRef]
33. Li, Y.; Chen, Z. A Balance-Point Method for Assessing the Effect of Natural Ventilation on Indoor Particle Concentrations. *Atmos. Environ.* **2003**, *37*, 4277–4285. [CrossRef]
34. Zhao, B.; Liu, Y.; Chen, C. Air Purifiers: A Supplementary Measure to Remove Airborne SARS-CoV-2. *Build. Environ.* **2020**, *177*, 106918. [CrossRef] [PubMed]
35. Curtius, J.; Granzin, M.; Schrod, J. Testing Mobile Air Purifiers in a School Classroom: Reducing the Airborne Transmission Risk for SARS-CoV-2. *Aerosol Sci. Technol.* **2021**, *55*, 586–599. [CrossRef]
36. Burgmann, S.; Janoske, U. Transmission and Reduction of Aerosols in Classrooms Using Air Purifier Systems. *Phys. Fluids* **2021**, *33*, 033321. [CrossRef]
37. Liu, D.T.; Phillips, K.M.; Speth, M.M.; Besser, G.; Mueller, C.A.; Sedaghat, A.R. Portable HEPA Purifiers to Eliminate Airborne SARS-CoV-2: A Systematic Review. *Otolaryngol.-Head Neck Surg.* **2022**, *166*, 615–622. [CrossRef]
38. Holmgren, H.; Ljungström, E.; Almstrand, A.-C.; Bake, B.; Olin, A.-C. Size Distribution of Exhaled Particles in the Range from 0.01 to 2.0 μm. *J. Aerosol Sci.* **2010**, *41*, 439–446. [CrossRef]
39. Johnson, G.R.; Morawska, L. The Mechanism of Breath Aerosol Formation. *J. Aerosol Med. Pulm. Drug Deliv.* **2009**, *22*, 229–237. [CrossRef]
40. Fabian, P.; McDevitt, J.J.; DeHaan, W.H.; Fung, R.O.P.; Cowling, B.J.; Chan, K.H.; Leung, G.M.; Milton, D.K. Influenza Virus in Human Exhaled Breath: An Observational Study. *PLoS ONE* **2008**, *3*, e2691. [CrossRef]
41. Han, Z.Y.; Weng, W.G.; Huang, Q.Y. Characterizations of Particle Size Distribution of the Droplets Exhaled by Sneeze. *J. R. Soc. Interface* **2013**, *10*, 20130560. [CrossRef]
42. Chao, C.Y.H.; Wan, M.P.; Morawska, L.; Johnson, G.R.; Ristovski, Z.D.; Hargreaves, M.; Mengersen, K.; Corbett, S.; Li, Y.; Xie, X.; et al. Characterization of Expiration Air Jets and Droplet Size Distributions Immediately at the Mouth Opening. *J. Aerosol Sci.* **2009**, *40*, 122–133. [CrossRef] [PubMed]

43. Elsaid, A.M.; Ahmed, M.S. Indoor Air Quality Strategies for Air-Conditioning and Ventilation Systems with the Spread of the Global Coronavirus (COVID-19) Epidemic: Improvements and Recommendations. *Environ. Res.* **2021**, *199*, 111314. [CrossRef] [PubMed]
44. Zuraimi, M.S.; Nilsson, G.J.; Magee, R.J. Removing Indoor Particles Using Portable Air Cleaners: Implications for Residential Infection Transmission. *Build. Environ.* **2011**, *46*, 2512–2519. [CrossRef]
45. Heim, M.; Mullins, B.J.; Wild, M.; Meyer, J.; Kasper, G. Filtration Efficiency of Aerosol Particles below 20 Nanometers. *Aerosol Sci. Technol.* **2005**, *39*, 782–789. [CrossRef]
46. Edwards, N.J.; Colder, B.; Sullivan, J.; Naramore, L. A Practical Approach to Indoor Air Quality for Municipal Public Health and Safety. *Open J. Polit. Sci.* **2021**, *11*, 176–191. [CrossRef]
47. National Institute for Occupational Safety and Health (NIOSH). *Determination of Particulate Filter Efficiency Level for N95 Series Filters against Solid Particulates for Non-Powered, Air-Purifying Respirators Standard Testing Procedure (STP)*; National Personal Protective Technology Laboratory: Washington, DC, USA, 2019.
48. Chao, C.Y.H.; Wan, M.P.; Cheng, E.C.K. Penetration Coefficient and Deposition Rate as a Function of Particle Size in Non-Smoking Naturally Ventilated Residences. *Atmos. Environ.* **2003**, *37*, 4233–4241. [CrossRef]
49. Morawska, L.; Johnson, G.R.; Ristovski, Z.D.; Hargreaves, M.; Mengersen, K.; Corbett, S.; Chao, C.Y.H.; Li, Y.; Katoshevski, D. Size Distribution and Sites of Origin of Droplets Expelled from the Human Respiratory Tract during Expiratory Activities. *J. Aerosol Sci.* **2009**, *40*, 256–269. [CrossRef]
50. Alsved, M.; Matamis, A.; Bohlin, R.; Richter, M.; Bengtsson, P.-E.; Fraenkel, C.-J.; Medstrand, P.; Löndahl, J. Exhaled Respiratory Particles during Singing and Talking. *Aerosol Sci. Technol.* **2020**, *54*, 1245–1248. [CrossRef]
51. Asadi, S.; Wexler, A.S.; Cappa, C.D.; Barreda, S.; Bouvier, N.M.; Ristenpart, W.D. Effect of Voicing and Articulation Manner on Aerosol Particle Emission during Human Speech. *PLoS ONE* **2020**, *15*, e0227699. [CrossRef]
52. Koutrakis, P.; Briggs, S.; Leaderer, B. Source Apportionment of Indoor Aerosols in Suffolk and Onondaga Counties, New York. *Environ. Sci. Technol.* **1992**, *26*, 521–527. [CrossRef]
53. Wallace, L. Indoor Particles: A Review. *J. Air Waste Manag. Assoc.* **1996**, *46*, 98–126. [CrossRef] [PubMed]
54. Long, C.M.; Suh, H.H.; Catalano, P.J.; Koutrakis, P. Using Time- and Size-Resolved Particulate Data to Quantify Indoor Penetration and Deposition Behavior. *Environ. Sci. Technol.* **2001**, *35*, 2089–2099. [CrossRef] [PubMed]
55. American Society of Heating Refrigerating and Air Conditioning (ASHRAE). Engineers Standard 62.1: Ventilation for Acceptable Indoor Air Quality. Available online: https://www.ashrae.org/technical-resources/standards-and-guidelines/read-only-versions-of-ashrae-standards (accessed on 23 April 2022).
56. MacNaughton, P.; Spengler, J.; Vallarino, J.; Santanam, S.; Satish, U.; Allen, J. Environmental Perceptions and Health before and after Relocation to a Green Building. *Build. Environ.* **2016**, *104*, 138–144. [CrossRef] [PubMed]
57. Allen, J.G.; MacNaughton, P.; Satish, U.; Santanam, S.; Vallarino, J.; Spengler, J.D. Associations of Cognitive Function Scores with Carbon Dioxide, Ventilation, and Volatile Organic Compound Exposures in Office Workers: A Controlled Exposure Study of Green and Conventional Office Environments. *Environ. Health Perspect.* **2016**, *124*, 805–812. [CrossRef]
58. Simoni, M.; Annesi-Maesano, I.; Sigsgaard, T.; Norback, D.; Wieslander, G.; Nystad, W.; Cancianie, M.; Sestini, P.; Viegi, G. School Air Quality Related to Dry Cough, Rhinitis and Nasal Patency in Children. *Eur. Respir. J.* **2010**, *35*, 742–749. [CrossRef]
59. Myhvold, A.; Olsen, E.; Lauridsen, O. Indoor Environment in Schools: Pupils Health & Performance in Regard to $CO_2$ Concentrations. *Indoor Air* **1996**, *96*, 369–374.
60. Norbäck, D.; Nordström, K.; Zhao, Z. Carbon Dioxide ($CO_2$) Demand-Controlled Ventilation in University Computer Classrooms and Possible Effects on Headache, Fatigue and Perceived Indoor Environment: An Intervention Study. *Int. Arch. Occup. Environ. Health* **2013**, *86*, 199–209. [CrossRef]
61. Vehviläinen, T.; Lindholm, H.; Rintamäki, H.; Pääkkönen, R.; Hirvonen, A.; Niemi, O.; Vinha, J. High Indoor $CO_2$ Concentrations in an Office Environment Increases the Transcutaneous $CO_2$ Level and Sleepiness during Cognitive Work. *J. Occup. Environ. Hyg.* **2016**, *13*, 19–29. [CrossRef]
62. Shriram, S.; Ramamurthy, K.; Ramakrishnan, S. Effect of Occupant-Induced Indoor $CO_2$ Concentration and Bioeffluents on Human Physiology Using a Spirometric Test. *Build. Environ.* **2019**, *149*, 58–67. [CrossRef]
63. Kim, J.; Hong, T.; Kong, M.; Jeong, K. Building Occupants' Psycho-Physiological Response to Indoor Climate and $CO_2$ Concentration Changes in Office Buildings. *Build. Environ.* **2020**, *169*, 106596. [CrossRef]
64. Azuma, K.; Kagi, N.; Yanagi, U.; Osawa, H. Effects of Low-Level Inhalation Exposure to Carbon Dioxide in Indoor Environments: A Short Review on Human Health and Psychomotor Performance. *Environ. Int.* **2018**, *121*, 51–56. [CrossRef] [PubMed]
65. Hou, Y.; Liu, J.; Li, J. Investigation of Indoor Air Quality in Primary School Classrooms. *Procedia Eng.* **2015**, *121*, 830–837. [CrossRef]
66. di Gilio, A.; Palmisani, J.; Pulimeno, M.; Cerino, F.; Cacace, M.; Miani, A.; de Gennaro, G. $CO_2$ Concentration Monitoring inside Educational Buildings as a Strategic Tool to Reduce the Risk of Sars-CoV-2 Airborne Transmission. *Environ. Res.* **2021**, *202*, 111560. [CrossRef]
67. Zhai, Z.; Li, H.; Bahl, R.; Trace, K. Application of Portable Air Purifiers for Mitigating COVID-19 in Large Public Spaces. *Buildings* **2021**, *11*, 329. [CrossRef]
68. Lowther, S.D.; Deng, W.; Fang, Z.; Booker, D.; Whyatt, D.J.; Wild, O.; Wang, X.; Jones, K.C. How Efficiently Can HEPA Purifiers Remove Priority Fine and Ultrafine Particles from Indoor Air? *Environ. Int.* **2020**, *144*, 106001. [CrossRef] [PubMed]

69. Christopherson, D.A.; Yao, W.C.; Lu, M.; Vijayakumar, R.; Sedaghat, A.R. High-Efficiency Particulate Air Filters in the Era of COVID-19: Function and Efficacy. *Otolaryngol.-Head Neck Surg.* **2020**, *163*, 1153–1155. [CrossRef]
70. Fruin, S.A.; Hudda, N.; Sioutas, C.; Delfino, R.J. Predictive Model for Vehicle Air Exchange Rates Based on a Large, Representative Sample. *Environ. Sci. Technol.* **2011**, *45*, 3569–3575. [CrossRef]
71. Hudda, N.; Kostenidou, E.; Sioutas, C.; Delfino, R.J.; Fruin, S.A. Vehicle and Driving Characteristics That Influence In-Cabin Particle Number Concentrations. *Environ. Sci. Technol.* **2011**, *45*, 8691–8697. [CrossRef]

International Journal of
*Environmental Research and Public Health*

MDPI

*Article*

# Modeling Primary Emissions of Chemicals from Liquid Products Applied on Indoor Surfaces

**Wenjuan Wei [1,*], John C. Little [2], Mélanie Nicolas [1], Olivier Ramalho [1] and Corinne Mandin [1]**

[1] Scientific and Technical Center for Building (CSTB), Health and Comfort Department, French Indoor Air Quality Observatory (OQAI), University of Paris-Est, CEDEX 2, 77447 Marne la Vallée, France
[2] Department of Civil and Environmental Engineering, Virginia Tech, Blacksburg, VA 24060, USA
* Correspondence: wenjuan.wei@cstb.fr

**Abstract:** Liquid products applied on material surfaces and human skin, including many household cleaning products and personal care products, can lead to intermittent emissions of chemicals and peak concentrations in indoor air. The existing case-based models do not allow inter-comparison of different use scenarios and emission mechanisms. In this context, the present work developed a mechanistic model based on mass transfer theories, which allowed emissions into the air from the liquid product to be characterized. It also allowed for diffusion into the applied surface during product use and re-emission from the applied surface after the depletion of the liquid product. The model was validated using literature data on chemical emissions following floor cleaning and personal care product use. A sensitivity analysis of the model was then conducted. The percentage of the chemical mass emitted from the liquid to the air varied from 45% (applied on porous material) to 99% (applied on human skin), and the rest was absorbed into the applied material/skin. The peak gas-phase concentration, the time to reach the peak concentration, and the percentage of the liquid-to-air emission depended significantly on the chemical's octanol/gas and material/gas partition coefficients and the diffusion coefficient of the chemical in the applied material/skin.

**Keywords:** volatile organic compounds; consumer exposure; household cleaning products; personal care products; intermittent source

check for updates

**Citation:** Wei, W.; Little, J.C.; Nicolas, M.; Ramalho, O.; Mandin, C. Modeling Primary Emissions of Chemicals from Liquid Products Applied on Indoor Surfaces. *Int. J. Environ. Res. Public Health* **2022**, *19*, 10122. https://doi.org/10.3390/ijerph191610122

Academic Editor: Paul B. Tchounwou

Received: 29 July 2022
Accepted: 12 August 2022
Published: 16 August 2022



## 1. Introduction

Chemicals in indoor environments, including aldehydes, volatile organic compounds (VOCs), and semi-volatile organic compounds (SVOCs), come from various sources. Common indoor sources include (1) porous solid/soft materials, such as building materials, furniture, and children's toys; (2) applied and sprayed liquids, such as surface cleaning products, air fresheners, body lotion, and household pesticides; and (3) heating and combustion products, such as incense, gas stoves, and candles [1]. According to the emission characteristics, these indoor sources can be classified into (1) continuous regular sources, such as furniture, which have constant source strength; (2) continuous irregular sources, such as wet paints, which have variable source strength during curing; (3) intermittent regular sources, such as gas stoves, which have a regular time-pattern of use; and (4) intermittent irregular sources, such as household cleaning products, which have variable time-pattern of use [2].

Most of the studies in the past decades focused on characterizing chemical emissions from continuous sources. They include many materials that can emit chemicals continuously to the indoor environment, from when the source materials are placed indoors to when they are removed, resulting in long-term emissions. A pioneering mechanistic model was developed in 1994 to predict the diffusion of VOCs contained in carpets through the source material to a ventilated environment [3]. The model has been improved and extended, forming a series of models that address chemical emissions from continuous regular and irregular sources. The improvements include (1) extending the model to address emissions from other materials, such as multi-layer dry porous materials [4–7] and paint film coating that transforms from wet to dry states [8–13]; (2) extending the model to address the emissions of other chemicals, such as formaldehyde [14–16] and SVOCs [17–23]; (3) considering other mechanisms for chemical emission and transport, such as convective mass transfer in the boundary layer adjacent to the source material [24,25], VOC adsorption on sink surfaces [26–31], and SVOC partitioning between the different phases [32–35]; and (4) considering the influence of environmental factors, such as temperature and relative humidity, on the emission rate [36–39].

Whereas continuous sources lead to long-term emissions, intermittent sources can cause peak chemical concentrations in the environment when they are used, leading to intense short-term occupant exposure [2]. For intermittent sources, fewer emission models have been developed compared to continuous sources. A possible reason is that the emission mechanism of an intermittent source is dependent on the form of the source, e.g., spray, film, and combustion; the use scenario, i.e., the chemical mass of the emission source that is used, the dimension of the emission source, and the emission period; and the associated environment, e.g., office, living room and bathroom [40]. Among indoor intermittent sources, liquid products applied on material surfaces or human skin for cleaning or personal care purposes can be characterized by surface emission [40]. Three types of models have been developed, i.e., models based on convective mass transfer, diffusion, and evaporation. Models based on convective mass transfer aim to characterize the emission of liquid products during product use [41,42]. Models based on diffusion aim to characterize the re-emission of adsorbed products from permeable surfaces after product use [43–45]. Models based on evaporation aim to provide a simplified estimation of the average emission rate [46–50]. Despite the different scope of applications, the case-based models do not allow inter-comparison of different use scenarios and emission mechanisms, and there is a lack of a comprehensive understanding of the emission profiles and exposure characteristics for these liquid products.

Since intermittent chemical sources include household cleaning and personal care products that are commonly used in indoor environments and on human skin, chemicals emitted from these products contribute to the aggregated exposure of occupants to indoor pollutants via multiple pathways. Due to the lack of a holistic modeling framework to address the mass transfer of indoor intermittent sources, the objective of the present work was to develop a mechanistic model to characterize different chemical transfer phenomena following the use of liquid products on material surfaces and human skin. The present work focused on primary emissions of the initial product ingredients. The decomposition of some product ingredients may lead to the formation of other VOCs, such as formaldehyde formation from bronopol decomposition [51]. However, the formation of decomposed chemicals is not discussed in the present work. Once the product ingredients are emitted into the air, they can react with oxidants leading to emissions of secondary chemicals [52]. This chemical reaction process can be characterized using indoor chemistry models [53], but the formation of secondary chemicals after emission is not discussed in the present work.

## 2. Materials and Methods

### 2.1. Model Development

The model considers two emission stages for products used on surfaces (Figure 1), i.e., when a product liquid layer exists (stage 1) and after the product liquid layer is depleted (stage 2). During product use (stage 1), a product liquid layer exists on the applied surface, and the model considers three zones, i.e., the product liquid layer as the chemical emission source, the surface material below the liquid where the product ingredients can diffuse through if it is permeable, and the air above the liquid where the chemicals emitted to the air can be characterized by convective mass transfer. After the product application (stage 2), the liquid layer is depleted progressively (the thickness of the liquid layer reduces to zero), and the model considers two zones, i.e., the surface material and the air. If the surface material is permeable and contains adsorbed product ingredients, it re-emits the chemicals to the air as the emission source. If the surface material is impermeable, no more emission source exists in stage 2. Mathematical equations based on diffusion and convective mass transfer theories to characterize the two emission stages are as follows.

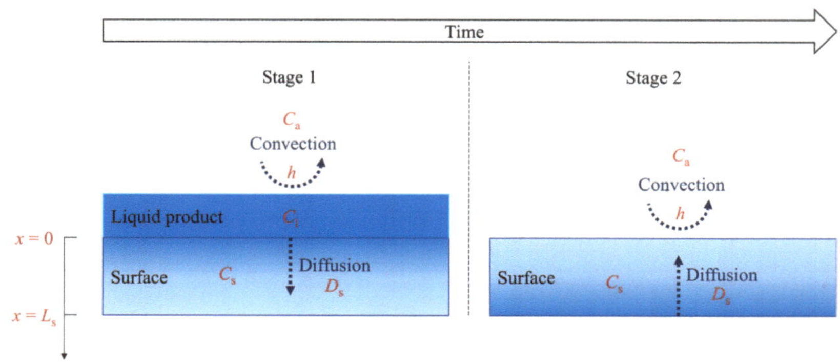


**Figure 1.** Schematic representation of the model to predict the primary emission of chemicals associated with the use of a liquid product.

The chemical concentration in the surface material or skin is:

$$\frac{\partial C_s}{\partial t} = D_s \frac{\partial^2 C_s}{\partial x^2} \tag{1}$$

where $C_s$ is the chemical concentration in the material ($\mu g/m^3$), $t$ is time (s), $D_s$ is the chemical diffusion coefficient in the material ($m^2/s$), and $x$ is the distance below the surface of the material (m). The surface material or skin is considered to be clean before product use, thus, the initial condition is:

$$C_s = 0, \text{ if } t = 0 \tag{2}$$

The current model does not consider chemical degradation in the human body if the product is applied to the skin, thus the boundary conditions are:

$$C_s = \frac{C_l}{K_{ls}}, \text{ if } x = 0, \text{ for stage 1} \tag{3}$$

$$D_s \frac{\partial C_s}{\partial x} = h\left(\frac{C_s}{K_{sa}} - C_a\right), \text{ if } x = 0, \text{ for stage 2} \tag{4}$$

$$\frac{\partial C_s}{\partial x} = 0, \text{ if } x = L_s \tag{5}$$

where $C_l$ is the concentration of the target chemical in the product liquid ($\mu g/m^3$), $L_s$ is the thickness of the material (m), $K_{ls}$ is the liquid/material partition coefficient (-), $K_{sa}$ is the material/air partition coefficient (-), and $h$ is the convective mass transfer coefficient in the air (m/s).

Due to the diversity of product formulae that are often not accessible, the model cannot consider the emission rate for each product ingredient. Therefore, the mass percentage of the target chemical in the product liquid is considered to be constant, so that the concentration of the target chemical ($C_l$) in the liquid is constant before the liquid is depleted. This is a mathematical simplification of the multi-ingredient emission process, nevertheless, complete knowledge of the product formulae should be acquired to conduct more complex modeling of multi-ingredient emissions. The chemical mass in the liquid is:

$$m_l = m_0 - \int_0^t \left[ hA\left(\frac{C_l}{K_{oa}} - C_a\right) + \int_0^{L_s} A\frac{\partial C_s}{\partial t}\mathrm{d}x \right]\mathrm{d}t \quad (6)$$

where $m_l$ is the chemical mass in the liquid (μg), $m_0$ is the initial chemical mass in the liquid (μg), $K_{oa}$ is chemical octanol/air partition coefficient (-), $A$ is the surface area ($m^2$), $C_a$ is the chemical concentration in the air ($\mu g/m^3$). The chemical concentration in the air is:

$$\frac{\mathrm{d}C_a}{\mathrm{d}t} = Q(C_{out} - C_a) + hA\left(\frac{C_l}{K_{oa}} - C_a\right),\ \text{for stage 1} \quad (7)$$

$$V\frac{\mathrm{d}C_a}{\mathrm{d}t} = Q(C_{out} - C_a) + hA\left(\frac{C_s}{K_{sa}} - C_a\right),\ \text{for stage 2} \quad (8)$$

where $V$ is the air volume ($m^3$), $Q$ is the air flow rate in the indoor environment ($m^3/s$), $C_{out}$ is the chemical concentration in the outdoor air that enters the environment ($\mu g/m^3$), and $K_{sa}$ is the material/air partition coefficient (-). The indoor air is considered to be clean before product use, thus, the initial condition is:

$$C_a = 0,\ \text{if } t = 0 \quad (9)$$

The model implies the following assumptions for its current development: (1) the surface material or skin and the indoor air are considered to be clean before product use, thus the current model should be used independently and is not to be coupled with other emission models; (2) chemical diffusion into and out of the surface material or skin is considered a one-dimensional diffusion driven by the concentration gradient of the chemical; (3) chemical degradation in the human body is neglected if the product is applied to the skin, thus it may underestimate the chemical mass that enters the skin in stage 1 and overestimate the chemical mass that exists the skin in stage 2; and (4) the mass percentage of the target chemical in the product liquid is constant during the liquid emission process, thus the prediction error in the emission profile depends on the product formulae and may be significant if the product contains ingredients that have diverse volatilities. The four assumptions could be relaxed, but this would lead to models that are more complex mathematically, requiring detailed information on product formulation and human physiological parameters.

Numerical solutions of the differential equations were obtained by the finite-difference method applying the central difference approximation. After conducting simulations with different numbers of layers, the material or skin was divided into 10 layers along its thickness, thus the spacing is 1/10 of the thickness because more simulation layers did not improve the results. The time step for the calculation varied with materials and chemicals and was changed for each calculation to ensure that the results would converge.

The model requires 13 input parameters (Table 1). Nine parameters are associated with indoor/outdoor environments, i.e., the chemical's initial concentration in the material/skin, chemical concentration in the liquid, material/skin thickness and area, liquid mass applied, convective mass transfer coefficient, indoor volume, indoor air flow rate, and the chemical's initial concentrations in indoor and outdoor air. Only four parameters are associated with chemical and material properties, i.e., the chemical's diffusion coefficient in the material/skin, the chemical's octanol/gas partition coefficient, the chemical's material/gas partition coefficient, and the chemical's liquid/material partition coefficient, which needs to be obtained from theoretical or experimental studies.

### 2.2. Model Validation Using Measured Data

The model was validated using experimental data from two studies obtained from the literature. The first study measured acetic acid (a common ingredient in cleaning products) concentration profiles in indoor air in a house following the use of an all-purpose cleaner on hardwood flooring [46]. The study conducted measurements for only 6 min following product use, thus the results were used to validate only stage 1 of the present model. The second study measured decamethylcyclopentasiloxane (D5, a common ingredient in cosmetic products) concentration profiles in indoor air in a university classroom following the use of personal care products on human skin [45]. The study determined the initial D5 concentration in human skin after the depletion of the product liquid, thus the results were used to validate only stage 2 of the present model. The two stages of the model were validated using experimental data from two different studies because no study was found to measure both liquid emission and surface re-emission of the product ingredients. The input data for the model prediction were obtained mainly from the two studies and are shown in Table 1. The chemical's octanol/gas partition coefficient was obtained using the EPI Suite calculator developed by the U.S. Environmental Protection Agency [54].

**Table 1.** Input data for modeling indoor concentrations associated with product use.

| Parameter | Acetic Acid Emission from Floor Cleaning [46] | Decamethylcyclopentasiloxane (D5) Emission from Human Skin [45] |
|---|---|---|
| Diffusion coefficient in the material/skin ($m^2/s$) | $5.05 \times 10^{-10}$ [55] | $1.46 \times 10^{-16}$ [45] |
| Initial chemical concentration in the material/skin ($\mu g/m^3$) | 0 (Assumed) | $8.80 \times 10^{10}$ [45] |
| Octanol/gas partition coefficient, $K_{oa}$ (-) | $1.66 \times 10^5$ (Calculated using EPI Suite) | $8.57 \times 10^6$ (Calculated using EPI Suite) |
| Material/gas partition coefficient, $K_{sa}$ (-) | $6.27 \times 10^2$ [55] | $3.27 \times 10^4$ [45] |
| Liquid/material partition coefficient, $K_{ls}$ (-) | $2.65 \times 10^2$ (Estimated as $K_{oa}/K_{sa}$) | $2.65 \times 10^2$ (Estimated as $K_{oa}/K_{sa}$) |
| Chemical concentration in the liquid ($\mu g/m^3$) | $4.2 \times 10^{10}$ [46] | Not needed for the model in stage 2 |
| Material/skin thickness | 1.9 cm [55] | 1 μm [45] |
| Material/skin area ($m^2$) | $5.6 \times 10^{-1}$ [46] | 2.28 (24 students) [45] |
| Liquid mass applied (μg) | $2.52 \times 10^6$ [46] | Not needed for the model in stage 2 |
| Convective mass transfer coefficient (m/s) | $9 \times 10^{-4}$ [45] | $9 \times 10^{-4}$ [45] |
| Indoor volume ($m^3$) | $2.45 \times 10^1$ [46] | 670 [45] |
| Air flow rate ($m^3/s$) | $8.33 \times 10^{-3}$ [46] | 0.93 [45] |
| Initial chemical concentration in indoor air ($\mu g/m^3$) | 0 [46] | 0 [45] |

## 3. Results and Discussion

### 3.1. Predicted Indoor Concentrations Following Product Use

Following the use of all-purpose cleaner on hardwood flooring [46] and personal care products [45] respectively in two indoor environments, the gas-phase concentrations of acetic acid and D5 were predicted using the present model and compared to the measured data in the literature. Figure 2a shows the predicted gas-phase concentration profile of acetic acid in indoor air during the entire emission process. The acetic acid in the air reached the peak gas-phase concentration 5 h after the use of 60 mL of a cleaning product containing 2.52 g of acetic acid on a hardwood floor of 0.56 $m^2$ in a room of 24.5 $m^3$ with an air change rate of 0.5 $h^{-1}$. During this liquid emission period (stage 1), 90.7% of the initial acetic acid mass was emitted from the liquid to the air. After the liquid depletion, the acetic acid diffused into the wood was re-emitted into the air. The wood re-emission period (stage 2) lasted 5 h, and 9.3% of the initial acetic mass was re-emitted from the floor. The predicted acetic acid concentrations were compared with the measured concentrations for 6 min after the use of the product, and the differences were within a factor of two. A possible explanation of the difference is that the measurement study did not provide data on the diffusion coefficient and partition coefficient of acetic acid in the hardwood floor, as well as the thickness of the floor. These three input parameters were obtained from another emission measurement of acetic acid from wood furniture [55], and the differences in the material properties may lead to bias in the prediction.

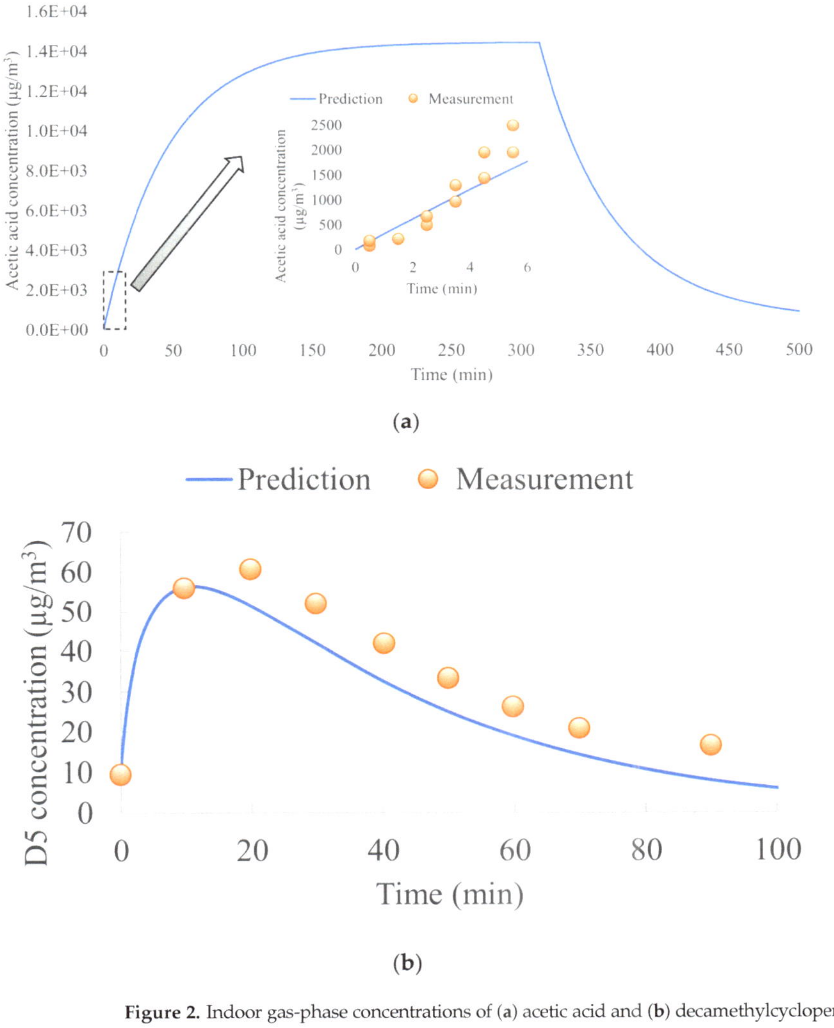


**Figure 2.** Indoor gas-phase concentrations of (**a**) acetic acid and (**b**) decamethylcyclopentasiloxane (D5).

Figure 2b shows the comparison of the predicted and measured gas-phase concentrations of D5 in the air of a classroom of 670 $m^3$ with an air change rate of 5 $h^{-1}$. Since the D5 was re-emitted from human skin according to the measurement study, the prediction did not consider liquid emission. The D5 in the air reached the peak gas-phase concentration 10 min after emission according to the predictions, and 20 min according to the measurements. Then the D5 concentrations decreased for 1 h to reach the background concentration in the classroom. The predicted and measured concentrations were well correlated, and the differences were less than 30% for each sampling point.

Measurements of the two above-mentioned studies in the literature allowed the present model to be partially validated. Due to the lack of existing experimental data measuring both liquid-to-air and liquid-to-surface mass transfers in stage 1 and surface-to-air emission in stage 2, full validation of the present model can be challenging. The lack of experimental data and the need to design a complete experimental study in the future highlights the necessity of conducting a sensitivity analysis at the current stage to characterize the influence of the chemical and surface material properties on the emission profile.

### 3.2. Sensitivity Analysis of Parameters Influencing the Emission

The present model requires 13 input parameters, as shown in Table 1. Depending on whether the liquid product was used on a wood floor or human skin, significant differences were observed in five parameters, i.e., the chemical's diffusion coefficient in the material/skin, the chemical's octanol/gas, and material/gas partition coefficients, the material/skin thickness, and the indoor air flow rate (Table 1). A sensitivity analysis of these 5 parameters was conducted to study their influences on the emission profile.

#### 3.2.1. Chemical Diffusion Coefficient in the Material/Skin

Figure 3 shows the indoor gas-phase concentration profiles of acetic acid, for which the chemical's diffusion coefficient in the material/skin varied between $5.05 \times 10^{-7}$ and $5.05 \times 10^{-16}$ $m^2/s$ (the orders of magnitude can cover indoor materials and skin), and the other input parameters had the same values as in the study of acetic acid emission from floor cleaning (Table 1). The sensitivity analysis of the diffusion coefficient aimed to characterize the influence of the material's permeability on the emission profile. During the liquid emission period (stage 1), the concentration profiles were identical regardless of the material's diffusion coefficient until the product liquid was depleted and the peak gas-phase concentration was reached. The time to reach the peak concentration depended on the diffusion coefficient and varied between 159 and 346 min (Figure 3a). The increased diffusion coefficient resulted in a higher rate of chemical diffusion through the material, thus faster depletion of the chemical in the liquid product. The peak concentration value in the gas phase was hardly influenced by the diffusion coefficient since the chemical emission from the liquid to the air during stage 1 was due to airflow on the surface of the product liquid. The percentage of the total chemical mass emitted from the product liquid to the air was also dependent on the diffusion coefficient and varied between 46.4% and 99.9% (Figure 3b). The rest of the chemical mass was absorbed into the material/skin and was re-emitted from the material/skin to the air during stage 2. When the diffusion coefficient was less than $10^{-11}$ $m^2/s$, the percentage of the liquid-to-air emission reached 99.7% and was hardly influenced by the decreased diffusion coefficient. Since many building materials are porous and have diffusion coefficients higher than $10^{-11}$ $m^2/s$, a non-neglectable proportion of the chemical applied on indoor material surfaces can be absorbed into the materials and re-emitted to the air after the liquid depletion. However, when the product liquid is applied to human skin, most of the chemical is emitted into the air from the liquid, as human skin has low permeability and low diffusion coefficients between $10^{-14}$ and $10^{-16}$ $m^2/s$ [42]. During the material re-emission period (stage 2), materials with high diffusion coefficients contained more absorbed chemical mass compared to materials with low diffusion coefficients, leading to a higher material re-emission and a slower decrease of the indoor gas-phase concentrations.

#### 3.2.2. Chemical's Octanol/Gas and Material/Gas Partition Coefficients

Figure 4 shows the indoor chemical gas-phase concentration profiles, for which the chemical's octanol/gas partition coefficient varied between $1.66 \times 10^3$ and $1.66 \times 10^7$, and the material/gas partition coefficient varied between 6.27 and $6.27 \times 10^4$. The two partition coefficients had large ranges to cover volatile chemicals and were varied together to maintain a constant ratio so that the liquid/material partition coefficient remained unchanged for the analysis. The other input parameters had the same values as in the study of acetic acid emission from floor cleaning (Table 1). The two partition coefficients can significantly affect the peak gas-phase concentrations, the time to reach the peak concentration, and the percentage of the chemical mass emitted from the product liquid to the air (Figure 4). This is because chemicals with low partition coefficients are more volatile than those with high partition coefficients and tend to be emitted from the liquid to the air. During the liquid emission period (stage 1), decreased octanol/gas and material/gas partition coefficients led to a shorter time for the liquid to be depleted (from 4.5 days to 3 min) and higher chemical peak concentrations in the gas phase (from $1.44 \times 10^2$ to $9.92 \times 10^4$ $\mu g/m^3$) (Figure 4a). The percentage of the chemical mass emitted from the product liquid to the air decreased with higher octanol/gas and material/gas partition coefficients and varied between 46.7% and 99.6% (Figure 4b). During the material re-emission period (stage 2), materials with higher material/gas partition coefficients contained more absorbed chemical mass compared to materials with low material/gas partition coefficients, leading to a higher material re-emission and a slower decrease of the indoor gas-phase concentrations.

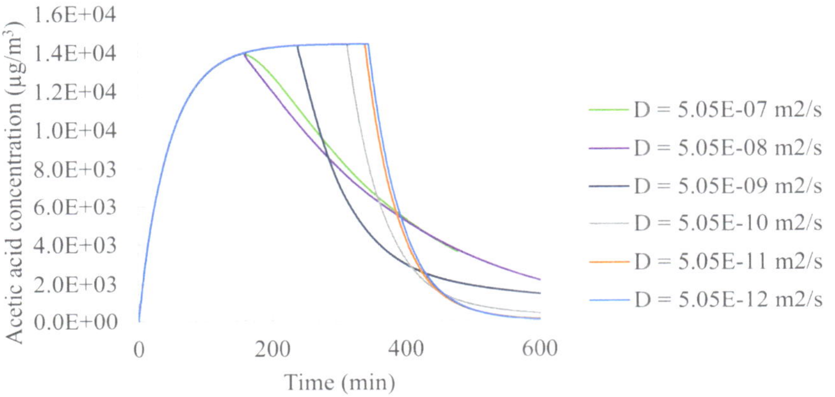


(a)

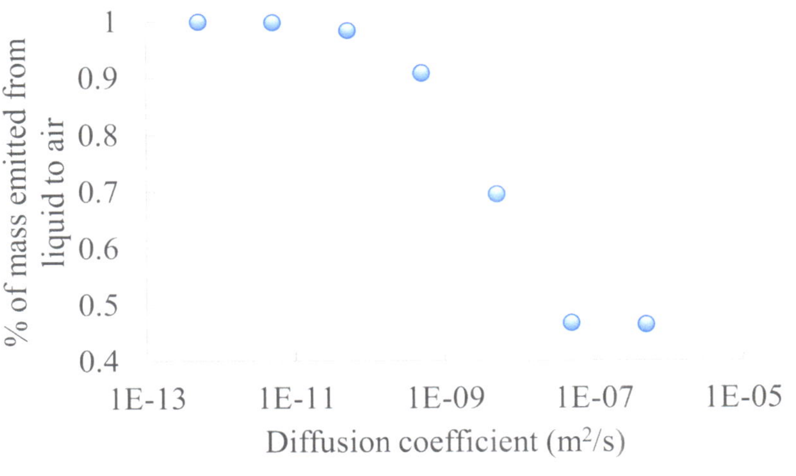


(b)

**Figure 3.** Influence of diffusion coefficient ($D$) on the indoor gas-phase concentration of acetic acid: (**a**) concentration profile and (**b**) fraction of mass emitted from the product liquid to air.

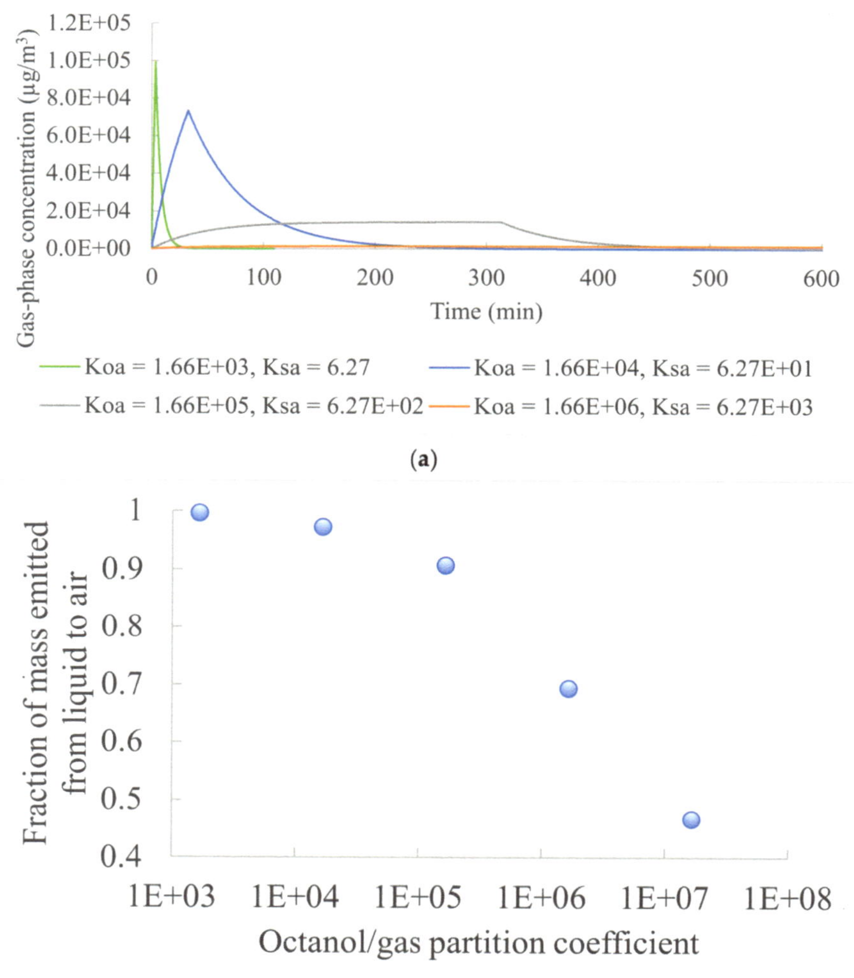


(**b**)

**Figure 4.** Influence of octanol/gas and material/gas partition coefficients ($K_{oa}$ and $K_{sa}$) on the chemical indoor gas-phase concentration: (**a**) concentration profile and (**b**) fraction of mass emitted from the product liquid to air.

### 3.2.3. Material/Skin Thickness

The material/skin thickness varied between 1.90 × $10^3$ and 1.90 × $10^{-1}$ mm for the sensitivity analysis, and the other input parameters had the same values as in the study of acetic acid emission from floor cleaning (Table 1). Changes in the material/skin thickness by 5 orders of magnitude led to less than 1% difference in the peak gas-phase concentrations, less than 30 min difference in the time to reach the peak concentrations, and less than 10% difference in the percentage of the chemical mass emitted from the product liquid to the air. Therefore, the material/skin thickness had much less influence on the liquid-to-air emission and liquid-to-material diffusion processes, compared to the chemical diffusion and partition coefficients.

### 3.2.4. Indoor Air Flow Rate

Figure 5 shows the indoor gas-phase concentration profiles of acetic acid, for which the indoor air flow rate varied between $8.33 \times 10^{-5}$ and $8.33 \times 10^{-1}$ $m^3/s$ (0.0122–122 air change per hour), and the other input parameters had the same values as in Table 1. The indoor air flow rate can significantly affect the peak gas-phase concentration value by orders of magnitude and change the time to reach the peak concentrations by hundreds of minutes. Changes in the airflow rate by 5 orders of magnitude led to less than a 2% difference in the percentage of the total chemical mass emitted from the product liquid to the air.

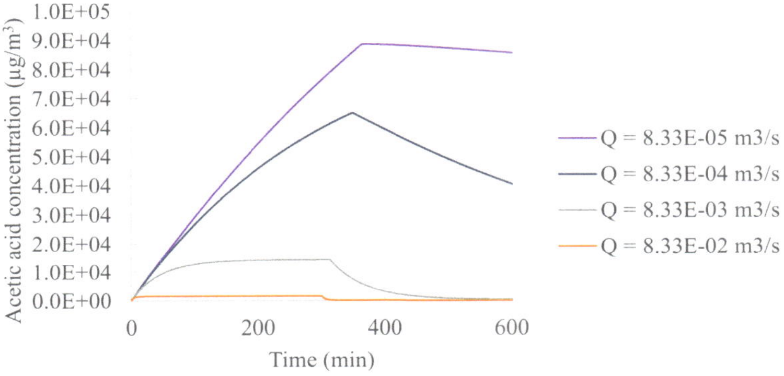


**Figure 5.** Influence of the airflow rate (Q) on the chemical indoor gas-phase concentration.

## 4. Conclusions

A mechanistic model was developed to characterize the chemical emission profiles following the use of liquid products on indoor surfaces and human skin. The model characterizes the liquid-to-air emission and the liquid-to-surface diffusion during the product use, as well as the surface-to-air emission after the depletion of the product liquid. The multiple chemical emission phenomena considered in the model allow it to address occupants' exposure via multiple pathways for indoor intermittent sources applied on surfaces, including household cleaning and personal care products. The simulated indoor gas-phase chemical concentrations agreed with the measured indoor concentrations of acetic acid following the use of a cleaning product on hardwood flooring, as well as the measured indoor concentrations of D5 due to the use of personal care products on human skin.

Following the product use, the chemical's emission profile depended significantly on the chemical's diffusion coefficient in the material/skin, the chemical's octanol/gas and material/gas partition coefficients, and the indoor air flow rate. The partition coefficients affected the peak gas-phase concentration by orders of magnitude and the time to reach the peak concentration by hours.

The percentage of the chemical mass emitted from the product liquid to the air increased with decreasing octanol/gas partition coefficients and was higher than 90% for volatile chemicals whose octanol/gas partition coefficients were lower than $10^{-5}$. The increased diffusion coefficient reduced the time to reach the peak gas-phase concentration by hundreds of minutes but had a minor influence on the peak concentration value. The percentage of the chemical mass emitted from the product liquid to the air increased with decreasing diffusion coefficient and was higher than 90% if the product liquid was applied on materials whose diffusion coefficients were lower than $10^{-10}$ $m^2/s$, e.g., human skin. The gas-phase concentrations also decreased with increasing air flow rate by orders of magnitude. The results suggest that when volatile chemicals are applied on surfaces with low permeability, most of the chemicals may be emitted from the product liquid to the air as a primary emission associated with the product use. When chemicals with low volatility are applied on surfaces with high permeability, the material absorption can be significant, and the emission can last long after the depletion of the product liquid. The results have highlighted that the use of cleaning and personal care products can result in exposure via inhalation and dermal contact, and the percentage of the contribution of the two exposure pathways is dependent on the chemical and skin properties.

**Author Contributions:** Conceptualization, W.W.; methodology, W.W. and J.C.L.; investigation, W.W. and J.C.L.; writing—original draft preparation, W.W. and J.C.L.; writing—review and editing, M.N., C.M. and O.R.; project administration, C.M. and O.R.; funding acquisition, C.M. and O.R. All authors have read and agreed to the published version of the manuscript.

**Funding:** This work was funded by the CSTB research program (Grant SEC21).

**Conflicts of Interest:** The authors declare no conflict of interest.

## References


1. Eichler, C.M.A.; Hubal, E.A.C.; Xu, Y.; Cao, J.; Bi, C.; Weschler, C.J.; Salthammer, T.; Morrison, G.C.; Koivisto, A.J.; Zhang, Y.; et al. Assessing Human Exposure to SVOCs in Materials, Products, and Articles: A Modular Mechanistic Framework. *Environ. Sci. Technol.* **2021**, *55*, 25–43. [CrossRef] [PubMed]
2. Seifert, B.; Ullrich, D. Methodologies for evaluating sources of volatile organic chemicals (VOC) in homes. *Atmos. Environ.* **1987**, *21*, 395–404. [CrossRef]
3. Little, J.C.; Hodgson, A.T.; Gadgil, A.J. Modeling emissions of volatile organic compounds from new carpets. *Atmos. Environ.* **1994**, *28*, 227–234. [CrossRef]
4. Wang, X.; Zhang, Y. General analytical mass transfer model for VOC emissions from multi-layer dry building materials with internal chemical reactions. *Chin. Sci. Bull.* **2011**, *56*, 222–228. [CrossRef]
5. Hu, H.P.; Zhang, Y.P.; Wang, X.K.; Little, J.C. An analytical mass transfer model for predicting VOC emissions from multi-layered building materials with convective surfaces on both sides. *Int. J. Heat Mass Transf.* **2007**, *50*, 2069–2077. [CrossRef]
6. Yuan, H.; Little, J.C.; Marand, E.; Liu, Z. Using fugacity to predict volatile emissions from layered materials with a clay/polymer diffusion barrier. *Atmos. Environ.* **2007**, *41*, 9300–9308. [CrossRef]
7. Haghighat, F.; Huang, H. Integrated IAQ model for prediction of VOC emissions from building material. *Build. Environ.* **2003**, *38*, 1007–1017. [CrossRef]
8. Li, F.; Niu, J.; Zhang, L. A physically-based model for prediction of VOCs emissions from paint applied to an absorptive substrate. *Build. Environ.* **2006**, *41*, 1317–1325. [CrossRef]
9. Yang, X.; Chen, Q.; Zeng, J.; Zhang, J.; Shaw, C. A mass transfer model for simulating volatile organic compound emissions from 'wet' coating materials applied to absorptive substrates. *Int. J. Heat Mass Transf.* **2001**, *44*, 1803–1815. [CrossRef]
10. Zhou, X.; Gao, Z.; Wang, X.; Wang, F. Mathematical model for characterizing the full process of volatile organic compound emissions from paint film coating on porous substrates. *Build. Environ.* **2020**, *182*, 107062. [CrossRef]
11. Chang, Y.-M.; Hu, W.-H.; Fang, W.-B.; Chen, S.-S.; Chang, C.-T.; Ching, H.-W. A Study on Dynamic Volatile Organic Compound Emission Characterization of Water-Based Paints. *J. Air Waste Manag. Assoc.* **2011**, *61*, 35–45. [CrossRef] [PubMed]
12. Deng, B.; Zhang, B.; Qiu, Y. Analytical solution of VOCs emission from wet materials with variable thickness. *Build. Environ.* **2016**, *104*, 145–151. [CrossRef]
13. Altinkaya, S.A. Predicting emission characteristics of volatile organic compounds from wet surface coatings. *Chem. Eng. J.* **2009**, *155*, 586–593. [CrossRef]
14. Xiong, J.; Zhang, Y.; Huang, S. Characterisation of VOC and Formaldehyde Emission from Building Materials in a Static Environmental Chamber: Model Development and Application. *Indoor Built Environ.* **2011**, *20*, 217–225. [CrossRef]
15. Bourdin, D.; Mocho, P.; Desauziers, V.; Plaisance, H. Formaldehyde emission behavior of building materials: On-site measurements and modeling approach to predict indoor air pollution. *J. Hazard. Mater.* **2014**, *280*, 164–173. [CrossRef]
16. Wei, W.; Xiong, J.; Zhao, W.; Zhang, Y. A framework and experimental study of an improved VOC/formaldehyde emission reference for environmental chamber tests. *Atmos. Environ.* **2014**, *82*, 327–334. [CrossRef]
17. Cao, J.; Zhang, X.; Zhang, Y. Predicting Dermal Exposure to Gas-Phase Semivolatile Organic Compounds (SVOCs): A Further Study of SVOC Mass Transfer between Clothing and Skin Surface Lipids. *Environ. Sci. Technol.* **2018**, *52*, 4676–4683. [CrossRef]
18. Xu, Y.; Liu, Z.; Park, J.; Clausen, P.A.; Benning, J.L.; Little, J.C. Measuring and Predicting the Emission Rate of Phthalate Plasticizer from Vinyl Flooring in a Specially-Designed Chamber. *Environ. Sci. Technol.* **2012**, *46*, 12534–12541. [CrossRef]
19. Xu, Y.; Little, J.C. Predicting Emissions of SVOCs from Polymeric Materials and Their Interaction with Airborne Particles. *Environ. Sci. Technol.* **2006**, *40*, 456–461. [CrossRef]
20. Xu, Y.; Cohen Hubal, E.A.; Clausen, P.A.; Little, J.C. Predicting Residential Exposure to Phthalate Plasticizer Emitted from Vinyl Flooring: A Mechanistic Analysis. *Environ. Sci. Technol.* **2009**, *43*, 2374–2380. [CrossRef]

21. Bi, C.; Wang, X.; Li, H.; Li, X.; Xu, Y. Direct Transfer of Phthalate and Alternative Plasticizers from Indoor Source Products to Dust: Laboratory Measurements and Predictive Modeling. *Environ. Sci. Technol.* **2021**, *55*, 341–351. [CrossRef]
22. Liang, Y.; Bi, C.; Wang, X.; Xu, Y. A general mechanistic model for predicting the fate and transport of phthalates in indoor environments. *Indoor Air* **2019**, *29*, 55–69. [CrossRef] [PubMed]
23. Wei, W.; Ramalho, O.; Mandin, C. A long-term dynamic model for predicting the concentration of semivolatile organic compounds in indoor environments: Application to phthalates. *Build. Environ.* **2019**, *148*, 11–19. [CrossRef]
24. Deng, B.; Kim, C.N. An analytical model for VOCs emission from dry building materials. *Atmos. Environ.* **2004**, *38*, 1173–1180. [CrossRef]
25. Xu, Y.; Zhang, Y. An improved mass transfer based model for analyzing VOC emissions from building materials. *Atmos. Environ.* **2003**, *37*, 2497–2505. [CrossRef]
26. Kang, D.H.; Choi, D.H.; Seong, Y.-B.; Yeo, M.S.; Kim, K.W. A numerical simulation of VOC emission and sorption behaviors of adhesive-bonded materials under floor heating condition. *Build. Environ.* **2013**, *68*, 193–201. [CrossRef]
27. Deng, B.; Zhang, H.; Wu, J. Modeling VOCs emission/sorption with variable operating parameters and general boundary conditions. *Environ. Pollut.* **2021**, *271*, 116315. [CrossRef] [PubMed]
28. Zhu, L.; Deng, B.; Guo, Y. A unified model for VOCs emission/sorption from/on building materials with and without ventilation. *Int. J. Heat Mass Transf.* **2013**, *67*, 734–740. [CrossRef]
29. Lee, C.-S.; Haghighat, F.; Ghaly, W.S. A study on VOC source and sink behavior in porous building materials—Analytical model development and assessment. *Indoor Air* **2005**, *15*, 183–196. [CrossRef]
30. Xiong, J.; Liu, C.; Zhang, Y. A general analytical model for formaldehyde and VOC emission/sorption in single-layer building materials and its application in determining the characteristic parameters. *Atmos. Environ.* **2012**, *47*, 288–294. [CrossRef]
31. Yang, X.; Chen, Q.; Zhang, J.S.; An, Y.; Zeng, J.; Shaw, C.Y. A mass transfer model for simulating VOC sorption on building materials. *Atmos. Environ.* **2001**, *35*, 1291–1299. [CrossRef]
32. Zhou, X.; Luo, C.; Wang, X.; Liu, K. Simplified methods for characterizing dynamic sorption of SVOCs on viscous particles. *Atmos. Environ.* **2020**, *241*, 117796. [CrossRef]
33. Wu, Y.; Eichler, C.M.A.; Leng, W.; Cox, S.S.; Marr, L.C.; Little, J.C. Adsorption of Phthalates on Impervious Indoor Surfaces. *Environ. Sci. Technol.* **2017**, *51*, 2907–2913. [CrossRef]
34. Eichler, C.M.A.; Cao, J.; Isaacman-VanWertz, G.; Little, J.C. Modeling the formation and growth of organic films on indoor surfaces. *Indoor Air* **2019**, *29*, 17–29. [CrossRef]
35. Liu, C.; Shi, S.; Weschler, C.; Zhao, B.; Zhang, Y. Analysis of the Dynamic Interaction Between SVOCs and Airborne Particles. *Aerosol Sci. Technol.* **2013**, *47*, 125–136. [CrossRef]
36. Wei, W.; Mandin, C.; Ramalho, O. Influence of indoor environmental factors on mass transfer parameters and concentrations of semi-volatile organic compounds. *Chemosphere* **2018**, *195*, 223–235. [CrossRef]
37. Zhang, Y.; Luo, X.; Wang, X.; Qian, K.; Zhao, R. Influence of temperature on formaldehyde emission parameters of dry building materials. *Atmos. Environ.* **2007**, *41*, 3203–3216. [CrossRef]
38. Xiong, J.; Zhang, P.; Huang, S.; Zhang, Y. Comprehensive influence of environmental factors on the emission rate of formaldehyde and VOCs in building materials: Correlation development and exposure assessment. *Environ. Res.* **2016**, *151*, 734–741. [CrossRef] [PubMed]
39. Haghighat, F.; De Bellis, L. Material emission rates: Literature review, and the impact of indoor air temperature and relative humidity. *Build. Environ.* **1998**, *33*, 261–277. [CrossRef]
40. Wei, W.; Little, J.C.; Ramalho, O.; Mandin, C. Predicting chemical emissions from household cleaning and personal care products: A review. *Build. Environ.* **2022**, *207*, 108483. [CrossRef]
41. Earnest, C.M.; Corsi, R.L. Inhalation exposure to cleaning products: Application of a two-zone model. *J. Occup. Environ. Hyg.* **2013**, *10*, 328–335. [CrossRef] [PubMed]
42. McCready, D.; Fontaine, D. Refining ConsExpo Evaporation and Human Exposure Calculations for REACH. *Hum. Ecol. Risk Assess. Int. J.* **2010**, *16*, 783–800. [CrossRef]
43. Teixeira, M.A.; Rodríguez, O.; Rodrigues, A.E. Diffusion and performance of fragranced products: Prediction and validation. *AIChE J.* **2013**, *59*, 3943–3957. [CrossRef]
44. Kasting, G.B.; Miller, M.A. Kinetics of finite dose absorption through skin 2: Volatile compounds. *J. Pharm. Sci.* **2006**, *95*, 268–280. [CrossRef]
45. Yang, T.; Xiong, J.; Tang, X.; Misztal, P.K. Predicting Indoor Emissions of Cyclic Volatile Methylsiloxanes from the Use of Personal Care Products by University Students. *Environ. Sci. Technol.* **2018**, *52*, 14208–14215. [CrossRef] [PubMed]
46. Arnold, S.; Ramachandran, G.; Kaup, H.; Servadio, J. Estimating the time-varying generation rate of acetic acid from an all-purpose floor cleaner. *J. Expo. Sci. Environ. Epidemiol.* **2020**, *30*, 374–382. [CrossRef]
47. Kasting, G.B.; Saiyasombati, P. A physico-chemical properties based model for estimating evaporation and absorption rates of perfumes from skin. *Int. J. Cosmet. Sci.* **2001**, *23*, 49–58. [CrossRef]
48. Saiyasombati, P.; Kasting, G.B. Two-stage kinetic analysis of fragrance evaporation and absorption from skin. *Int. J. Cosmet. Sci.* **2003**, *25*, 235–243. [CrossRef]
49. Ernstoff, A.S.; Fantke, P.; Csiszar, S.A.; Henderson, A.D.; Chung, S.; Jolliet, O. Multi-pathway exposure modeling of chemicals in cosmetics with application to shampoo. *Environ. Int.* **2016**, *92–93*, 87–96. [CrossRef]

50. Mackay, D.; Van Wesenbeeck, I. Correlation of chemical evaporation rate with vapor pressure. *Environ. Sci. Technol.* **2014**, *48*, 10259–10263. [CrossRef]
51. Kajimura, K.; Tagami, T.; Yamamoto, T.; Iwagami, S. The Release of Formaldehyde upon Decomposition of 2-Bromo-2-nitropropan-1, 3-diol (Bronopol). *J. Health Sci.* **2008**, *54*, 488–492. [CrossRef]
52. Nazaroff, W.W.; Weschler, C.J. Cleaning products and air fresheners: Exposure to primary and secondary air pollutants. *Atmos. Environ.* **2004**, *38*, 2841–2865. [CrossRef]
53. Yeoman, A.M.; Shaw, M.; Carslaw, N.; Murrells, T.; Passant, N.; Lewis, A.C. Simplified speciation and atmospheric volatile organic compound emission rates from non-aerosol personal care products. *Indoor Air* **2020**, *30*, 459–472. [CrossRef] [PubMed]
54. U.S. EPA. Download EPI SuiteTM—Estimation Program Interface v4.11. 2012. Available online: https://www.epa.gov/tsca-screening-tools/download-epi-suitetm-estimation-program-interface-v411 (accessed on 28 July 2022).
55. Yao, Y.; Xiong, J.; Liu, W.; Mo, J.; Zhang, Y. Determination of the equivalent emission parameters of wood-based furniture by applying C-history method. *Atmos. Environ.* **2011**, *45*, 5602–5611. [CrossRef]

International Journal of
*Environmental Research and Public Health*

MDPI

*Article*

# Tracing of Heavy Metals Embedded in Indoor Dust Particles from the Industrial City of Asaluyeh, South of Iran

**Mahsa Tashakor [1], Reza Dahmardeh Behrooz [2,*], Seyed Reza Asvad [3] and Dimitris G. Kaskaoutis [4,5,*]**

1 School of Geology, College of Science, University of Tehran, Tehran 14155-6455, Iran; mahsa.tashakor@ut.ac.ir
2 Department of Environmental Sciences, Faculty of Natural Resources, University of Zabol, Zabol 98615-538, Iran
3 Department of Environment, Faculty of Natural Resources and Marine Science, Tarbiat Modares University, Tehran 14115-111, Iran; reza.asvad@gmail.com
4 Institute for Environmental Research and Sustainable Development, National Observatory of Athens, Palaia Penteli, 15236 Athens, Greece
5 Environmental Chemical Processes Laboratory, Department of Chemistry, University of Crete, 70013 Heraklion, Greece
* Correspondence: dahmardehbehrooz@uoz.ac.ir or dahmardeh_behrooz@yahoo.com (R.D.B.); dkask@noa.gr (D.G.K.)

**Abstract:** Assessment of indoor air quality is especially important, since people spend substantial amounts of time indoors, either at home or at work. This study analyzes concentrations of selected heavy metals in 40 indoor dust samples obtained from houses in the highly-industrialized Asaluyeh city, south Iran in spring and summer seasons (20 samples each). Furthermore, the health risk due to exposure to indoor air pollution is investigated for both children and adults, in a city with several oil refineries and petrochemical industries. The chemical analysis revealed that in both seasons the concentrations of heavy metals followed the order of Cr > Ni > Pb > As > Co > Cd. A significant difference was observed in the concentrations of potential toxic elements (PTEs) such as Cr, As and Ni, since the mean (±stdev) summer levels were at 60.2 ± 9.1 mg $kg^{-1}$, 5.6 ± 2.7 mg $kg^{-1}$ and 16.4 ± 1.9 mg $kg^{-1}$, respectively, while the concentrations were significantly lower in spring (17.6 ± 9.7 mg $kg^{-1}$, 3.0 ± 1.7 mg $kg^{-1}$ and 13.5 ± 2.4 mg $kg^{-1}$ for Cr, As and Ni, respectively). Although the hazard index (HI) values, which denote the possibility of non-carcinogenic risk due to exposure to household heavy metals, were generally low for both children and adults (HI < 1), the carcinogenic risks of arsenic and chromium were found to be above the safe limit of $1 \times 10^{-4}$ for children through the ingestion pathway, indicating a high cancer risk due to household dust in Asaluyeh, especially in summer.

**Keywords:** potential toxic elements; household dust; petrochemicals; health risk; Asaluyeh; Persian Gulf



**Citation:** Tashakor, M.; Behrooz, R.D.; Asvad, S.R.; Kaskaoutis, D.G. Tracing of Heavy Metals Embedded in Indoor Dust Particles from the Industrial City of Asaluyeh, South of Iran. *Int. J. Environ. Res. Public Health* **2022**, *19*, 7905. https://doi.org/10.3390/ijerph19137905

Academic Editors: Nuno Canha, Marta Almeida and Evangelia Diapouli

Received: 18 May 2022
Accepted: 24 June 2022
Published: 28 June 2022



## 1. Introduction

Particulate matter (PM) is considered a major indicator of air quality, composed of various particles of different origin, size, chemical composition and toxicity [1]. In the urban agglomerations and industrialized areas, fossil-fuel combustion from vehicles and industries is the main source of PM pollution, which is escalated by emissions from residential heating and cooking, road dust resuspension and long-range transported aerosols [2–4]. Especially in urban areas largely affected by dust storms, PM concentrations can be highly increased during such episodes causing cardiovascular and respiratory diseases, asthma, lung cancer, etc. [5,6]. Among PM substances, various chemical compounds, such as polycyclic aromatic hydrocarbons (PAHs) and heavy metal(loid)s (e.g., Arsenic, As; Cadmium, Cd; Chromium, Cr; Lead, Pb; Nickel, Ni; Zinc, Zn), can adhere to dust surfaces, with deleterious effects on human health and ecosystems [7–9]. According to the International

Agency for Research on Cancer (IARC), these substances are considered carcinogenic, and the World Health Organization (WHO) has defined annual limits for As (6 ng $m^{-3}$), Cd (5 ng $m^{-3}$), Ni (20 ng $m^{-3}$) and Pb (0.5 ng $m^{-3}$), in a way to control concentrations and protect human health.

People spend substantial amounts of time indoors, either in homes, offices or at schools, and for entertainment; this time was estimated at about 88% of the daily life for adults and 71–79% for children [10–12]. Therefore, indoor air quality is especially important for human health, while it depends on outdoor air pollution and human activities indoors such as cooking, cleaning and smoking [13–15]. Indoor air pollution is an issue concerning all people due to its deleterious effects on human health, thus attracting the interest of many researchers [16,17]. Similar to outdoor ambient conditions, household aerosols may be composed of a heterogeneous mixture of organic and inorganic compounds, including PAHs and heavy metals [18–20]. Furthermore, indoor dust particle may have considerable range in size, thus causing different effects on the human respiratory system, with the smaller and more toxic elements being more hazardous, as they can reach deep into the alveoli [21]. Other studies have shown that escalated outdoor PM concentrations from anthropogenic or natural sources (i.e., wildfires) negatively affect the indoor air quality, with indoor concentrations being even higher than outdoors [22]. Heavy metals and potentially toxic elements (PTEs) may originate from various sources in urban areas such as combustion emissions from vehicles, industrial zones, power plants, refineries and municipal waste disposal [23–25]. Heavy metals can enter the human organism via inhalation, corresponding to the airborne dust fraction, via ingestion due to contamination in food and via dermal contact [26,27].

Iran has been facing serious air pollution issues during the last decades due to rapid industrialization, urbanization and an increase in energy demands [28], which contribute highly to a background turbid atmosphere due to natural aerosols such as dust [29]. The Iranian economy is largely based on combustion of fossil fuels for energy production, exports of oil, natural gas and petrochemical products, which further deteriorate the urban environments [30,31]. The industrial sector is a major source of ambient air pollution, which may also highly contribute to indoor air quality, although this effect has not been well quantified [32]. Asaluyeh, located on the northern shore of the Persian Gulf, is one of the most highly industrialized cities in Iran, due to the operation of several oil refineries, petrochemical industries and the Pars Special Energy Economic Zone (PSEEZ). The gas and petrochemical industries release a large amount of pollutants such as volatile organic compounds (VOCs) and heavy metals [33,34]. Previous study in Asaluyeh examined the concentrations of PTEs and mineralogy of 43 street dust samples collected from industrial and urban areas in summer [35]. Statistical analyses revealed the main sources of trace elements to be the resuspended dust, traffic emissions and industrial sources. Another study analyzed the concentrations of heavy metals (Cu, Zn, Pb, Co, Cr, Fe, Ni and Mn) in dust deposited on surface and palm tree leaves at industrial, urban and non-urban areas in Bushehr and Asaluyeh, and assessed the contamination levels and possible sources [36], while a recent study evaluated the Hg concentrations of indoor dust samples in Asaluyeh against those in Ahvaz, southwest Iran and Zabol, southeast Iran [37].

As a measure of warning people for the deleterious effects of excessive indoor pollution, studies analyzing the dispersed pollutants and heavy metals indoors are very important, and also help in mitigation strategies for their reduction [17,38]. Although extensive analysis has been performed on PTEs in soil and airborne outdoor dust in Asaluyeh county [39–41], there are limited data related to heavy metal concentrations indoors. This study analyzes dust samples obtained indoors in the spring and summer seasons in Asaluyeh city, south Iran. The main objectives are (i) to determine the levels and distribution of selected heavy metal(oid)s of indoor dust in a highly-industrialized area, (ii) to compare the concentrations between different homes, areas and seasons, (iii) to estimate the health risk (carcinogenic and non-carcinogenic) posed by exposure to PTEs for children and adults via inhalation, ingestion and dermal contact pathways for the first time in Asaluyeh.

## 2. Material and Methods

### 2.1. Heavy Metal Concentrations

Asaluyeh county, with an area of about 30,000 hectares, is situated in the Bushehr province, southwest Iran, between 27°14′–30°16′ N, and 50°6′–52°58′ E, limited by the Persian Gulf in the south and the Zagros Mountains in the north (Figure 1). The area experiences a hot and arid climate with annual temperatures ranging between 5 °C and 50 °C, relative humidity between 50% and 88% and average annual rainfall of 100 mm [36]. The prevailing wind direction is from northwest to southeast (Figure 1). The region has experienced remarkable industrial and economic activity due to the establishment of the Pars Special Energy Economic Zone (PSEEZ) in 1998 near Asaluyeh city, which covers a total area of 14,000 ha. This zone includes the world's second-largest natural gas reserve and the largest oil and gas resources in Iran (16 gas processing plants and 15 petrochemical complexes), with a standard capacity for natural gas of $3.4 \times 10^{12}$ $m^3$, releasing a large amount of pollutants into the atmosphere [42–44]. The area also experiences a high frequency of dust storms, especially during summer [45]. The increased industrialization and fossil-fuel combustion escalated the levels of fine particulate matter, organic compounds and heavy metals in Asaluyeh, and studies examined the health risk of exposure to ambient conditions for local residents and people working in the industrial zones [32,44].

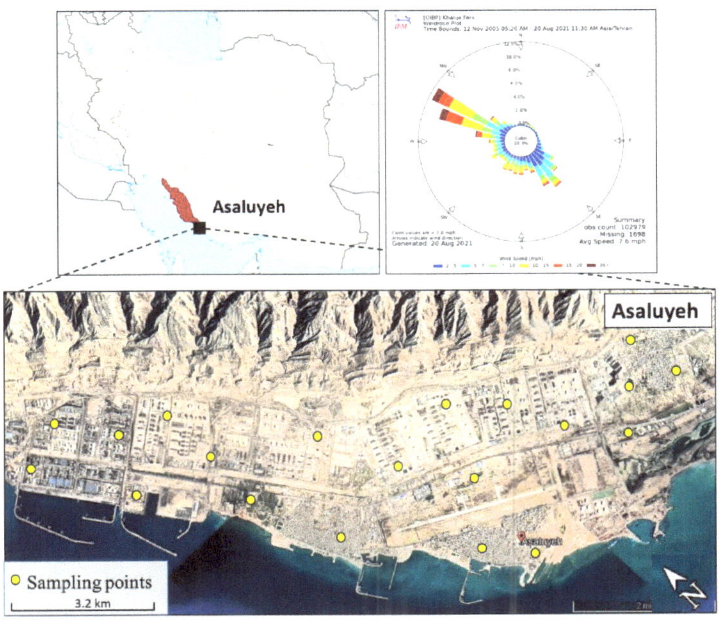


**Figure 1.** District distribution map of the Bushehr province in south Iran and the Asaluyeh county (**top left**) and the locations of the sampling points (houses) in Asaluyeh (**bottom**), along with wind rose diagram during the sampling days (**top right**).

### 2.2. Indoor Sampling Measurements

In this study, we analyzed the concentrations of selected heavy metals via chemical analysis of 40 samples obtained indoors in spring and summer 2021 (20 samples of household dust in each season). These samples were collected at random district zones in Asaluyeh city, as shown in Figure 1. The surveyed residences were private dwelling units selected based on the willingness of landlords to participate in this study. Dust samples were collected from vacuum cleaner bags, using volunteers living in governmental houses in Asaluyeh city. In all the sampling days, the meteorological conditions were fair without strong winds or dust storms.

All the houses had an area of 90 $m^2$, with two bedrooms and were 15 years old. All were built in the same way by the government for employees working in the South Pars zone. There were two to four people living in, and all the houses had only one air conditioner in the kitchen. Vacuum sampling of house dust was performed with a vacuum cleaner 9 times per season (three times per month). Sampling was performed (in three bags) and each bag contained 30 to 50 g of soil. All three bags were passed through a sieve after the sample was obtained. The soil that passed through the sieve was kept in plastic bags until the sample was transported to the laboratory and stored in the refrigerator.

Sampling of all parts of the house (bedrooms, living room, kitchen, etc.) was performed. The last cleaning time at each house was not accurately known, but it was assumed to occur within one week from the sampling time and likely contributes to the range of the heavy metal concentrations between the houses. About 30 to 50 g of soil was generally collected in each vacuum cleaner bag. All of these houses have the same architectural form and equipment in terms of construction. Dust samples were limited only to homes with non-smoking residents in order to eliminate the bias from smoke contamination in the concentrations of trace metals. Furthermore, during the sampling period in spring and summer, due to very hot conditions (35–40 °C), the domestic cooling via air conditions was at its maximum.

The content in the vacuum cleaner compartment was placed in plastic resealable bags, labeled and returned to the laboratory where they were kept in the fridge at 4 °C until chemical analysis. For chemical preparation and digestion of the samples, the following materials and tools were used: (i) 63-micron mesh sieve, (ii) PTFE Teflon tubes for digestion of samples, (iii) nitric acid 65%, (iv) hydrofluoric acid 40%, (v) perchloric acid 70% suprapure with very high purity, all provided by Merck, Germany. Polyethylene funnel, Whatman 42 filter paper and 25 mL bellows for filtering digested samples were also used. The dust samples were initially dried in an oven at 105 °C for 24 h and then sieved [17,46]. After drying the samples, approximately 0.25 g of soft sifted dust was weighed with a digital microbalance with an accuracy of 0.0001 g, and using polyethylene pipes for digestion and 7 mL of a mixture of $HClO_4$-HF-$HNO_3$ acids in 1:2:4 ratios, the samples were placed on a hot plate until the white vapor rose and the digestion process was complete. After steaming and drying the acid, the residue was dissolved in 2% nitric acid to a volume of 25 mL. Control samples were also prepared by repeating all digestion steps without the presence of samples [47]. Inductively Coupled Plasma Mass Spectrometry (ICP-MS; 7800 Series) was used to determine the concentrations of selected heavy metal(loid)s (As, Cd, Cr, Co, Ni and Pb). In order to increase the accuracy of the test and reduce the error rate, the quality control method was performed as follows: (i) all containers used in sampling, digestion and storage of samples were pre-soaked in dilute nitric acid (20%) for 24 h, then washed with distilled water and dried; (ii) during digestion, a blank sample was prepared for each group of samples and analyzed against other samples; the calibration curve was drawn using a blank sample and four standard samples and its accuracy was then confirmed using a control solution and a standard sample close to the middle concentration range (approximately once every 10 samples); (iii) the controls for all stages of the work should not differ by more than 2% from the initial curve; for every 15 samples, one sample was re-measured randomly and a standard sample was also selected and re-measured to ensure that the device was working properly. Soil standards (CRM: NIST 2710) for elements were also used to evaluate the accuracy of the measurement method and the recovery percentage. The recovery for the studied elements was 79% to 115%. Moreover, the concentration of elements in the control samples was between 0.004 and 1 $\mu g\ mL^{-1}$, which was much lower than the amount of these contaminants in the dust samples. Replicate soil standards (CRM:

NIST 2710) and dust sample measurements (repeat five times) had a relative standard deviation (RSD) < 5%.

The geochemical results were analyzed for descriptive statistics using Microsoft Excel and SPSS 23.0 software. The Shapiro–Wilk test was carried out to control the normality assumption of data and revealed non-normal distribution ($p$ value $\leq 0.05$) for all analyzed elements. Based on this result, the non-parametric Mann–Whitney U test was used to examine the difference in the concentrations of elements for indoor dust between spring and summer samples.

*2.3. Health Risk Assessment*

Similar to many studies examining the exposure to ambient outdoor dust, the health effects of PTEs in indoor dust were considered via the three major pathways of chemical ingestion, inhalation and dermal contact. The health risk assessment model introduced by the U.S. Environmental Protection Agency was applied to estimate the cancer and non-cancer risks associated with heavy metals in household dust samples in Asaluyeh. This model takes into account three major pathways of adult and children exposure to heavy metals: (i) intake by direct ingestion of dust particles, (ii) intake through mouth and nose breathing of resuspended particles (inhalation), (iii) intake via absorption from skin adhered dust particles (dermal contact). The average daily intakes of heavy metals (mg $kg^{-1}$ BW $d^{-1}$) received through the ingestion ($ADI_{ing}$), inhalation ($ADI_{inh}$) and dermal contact ($ADI_{drm}$) pathways are calculated using the below formulas [3]:

$$ADI_{ing} = C_s \times \frac{IngR \times EF \times ED}{BW \times AT} \times 10^{-6} \quad (1)$$

$$ADI_{inh} = C_s \times \frac{InhR \times EF \times ED}{PEF \times BW \times AT} \quad (2)$$

$$ADI_{drm} = C_s \times \frac{ESA \times SAF \times DAF \times EF \times ED}{BW \times AT} \times 10^{-6} \quad (3)$$

The hazard quotient (HQ) represents the non-carcinogenic risk of a single element and was calculated by dividing the average daily intake (ADI) for each element and exposure pathway to a specific reference dose (Equation (4)). The overall non-carcinogenic hazards caused by multiple metals are accounted for by the hazard index (HI) [48] and calculated by summing the HQ values of the measured metals (Equation (5)).

$$HQ_e = \frac{ADI_e}{RfD} \quad (4)$$

$$HI = \sum HQ_e = \sum \frac{ADI_e}{RfD} \quad (5)$$

where RfD stands for the reference dose (mg/kg/day), established by the USEPA, as an estimation of the maximum allowable rate, and e represents the route of exposure (ingestion, inhalation, dermal contact). The occurrence of non-carcinogenic effects is more likely when HI value is greater than 1, whereas a value of HI lower than 1 indicates no significant risk.

The carcinogenic or cancer risk (CR) for each metal indicates the probability of developing cancer over a person's lifetime due to exposure to that pollutant [24], and according to USEPA [49], the acceptable limits are in the range of $1 \times 10^{-6}$ to $1 \times 10^{-4}$. TCR is the sum of obtained carcinogenic risks from various exposure pathways. CR and TCR were calculated using the following equations (Equations (6) and (7)):

$$CR_e = ADI_e \times SF_e \quad (6)$$

$$TCR = \sum CR_e \quad (7)$$

$SF_e$ represents the carcinogenic slope factor from Regional Screening Levels for each pathway. The CR related to skin contact was calculated only for As, because the dermal slope factor has not been established for the rest of the examined elements. Ingestion and dermal contact slope factors for Cd and Ni were also not provided, while Co is not considered as a carcinogenic element, and therefore, no CR values were estimated. The CR values surpassing $1.00 \times 10^{-4}$ are considered to be unacceptable with the potential for causing cancer during the individual's lifetime, while the values below $1.00 \times 10^{-6}$ are considered to be safe, without causing significant health effects. The values between $1.00 \times 10^{-6}$ and $1.00 \times 10^{-4}$ indicate tolerable risk [50].

## 3. Results

### *3.1. Heavy Metal Concentrations*

The present study focuses on analyzing the concentrations of arsenic, cadmium, cobalt, chromium, nickel and lead, which are among the most dangerous types of trace element pollutants [51]. Metalloids such as arsenic (As) often fall into the category of heavy metals due to their similarity in chemical properties and environmental behavior [52]. Figure 2 shows the concentrations of the 6 examined heavy metals of household dust in Asaluyeh during spring and summer in a box–whisker chart view.

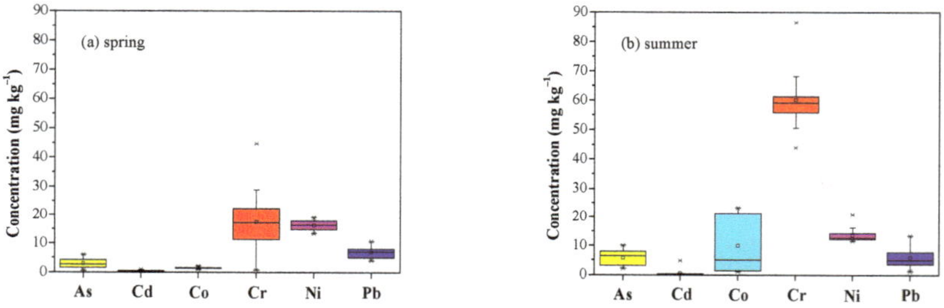


**Figure 2.** Box plots of the concentrations of heavy metals from indoor dust samples in Asaluyeh in spring (**a**) and summer (**b**). The bottom and top of the box are the first and third quartiles, respectively, while the mean is denoted by circle and the median by line within the boxes. Whiskers (the vertical lines) are the 1.5 interquartile ranges of the lower and upper quartiles.

In both seasons, the highest concentrations occurred for Cr, followed by Ni, while in spring, concentrations of these elements were similar, although Cr exhibited a wider variability. In summer, Co concentrations were highly variable, while the mean was similar to that of Ni. For the collected household dust in summer, Cr exhibited the highest median concentration (59.3 mg $kg^{-1}$), followed by Ni (12.6 mg $kg^{-1}$), As (6.41 mg $kg^{-1}$), Co (6.25 mg $kg^{-1}$), Pb (4.91 mg $kg^{-1}$) and Cd (0.24 mg $kg^{-1}$). This pattern is slightly different in the spring collected samples, so that after Cr (17.3 mg $kg^{-1}$) and Ni (16.5 mg $kg^{-1}$), Pb exhibited higher concentrations (7 mg $kg^{-1}$) than As (2.76 mg $kg^{-1}$), Co (1.55 mg $kg^{-1}$) and Cd (0.49 mg $kg^{-1}$). Amongst the analyzed elements, Cr and Cd revealed the highest and the lowest concentrations, respectively, considering all available dust samples. Although Pb has been stopped as an additive in petrol and gasoline in Iran, Pb concentrations are likely due to the industrial sector, but also due to traffic emissions, as also shown in European urban sites [24,53].

The influence of seasonal effect on the concentrations of trace elements was statistically evaluated by applying the Mann–Whitney U test. As can be seen in Figure 3, concentrations of As, Cd, Co, Cr and Ni significantly differ between spring and summer ($p$ value < 0.01). This difference was not identified for Pb, which presented a $p$ value larger than 0.05 ($p$ value: 0.08). It should be noted that the concentration distributions for heavy metals were far from normality (Figure 3), so small changes in initial assumptions or data could lead to significant changes in $p$ value. Especially for Cd, removing one outlier measurement from the summer dataset could significantly increase the $p$ value.

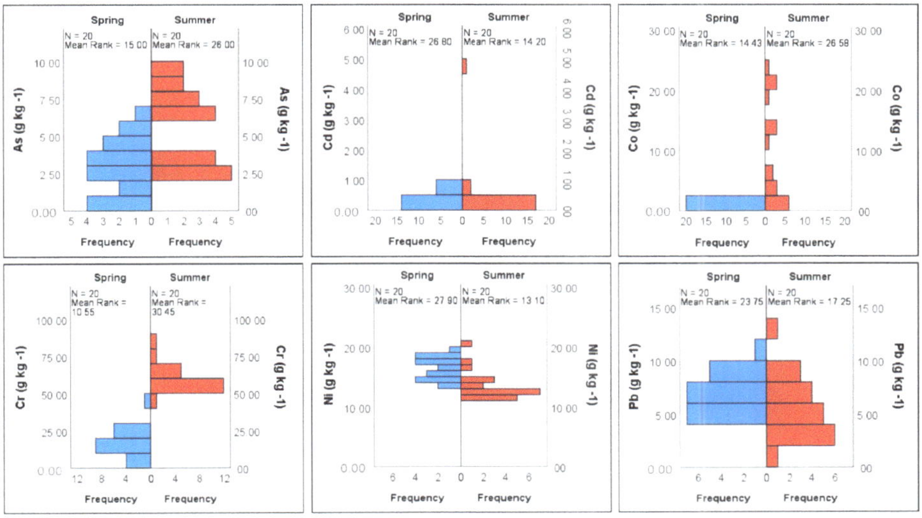


| | As | Cd | Co | Cr | Ni | Pb |
|---|---|---|---|---|---|---|
| Mann–Whitney U | 90.000 | 74.000 | 78.500 | 1.000 | 52.000 | 135.000 |
| Wilcoxon W | 300.000 | 284.000 | 288.500 | 211.000 | 262.000 | 345.000 |
| Z | -2.976 | -3.415 | -3.288 | -5.383 | -4.004 | -1.758 |
| Asymp. Sig. [(2-tailed Sig.)] | 0.003 | 0.001 | 0.001 | 0.000 | 0.000 | 0.079 |
| Exact Sig. [2*(1-tailled Sig.)] | 0.002 | 0.000419 | 0.001 | $2.90 \times 10^{-11}$ | 0.000021 | 0.081 |

The significance level is 0.050
Null hypothesis: The distribution of As, Cd, Co, Cr, Ni and Pb is the same across categories of Difference.
The null hypothesis is rejected for As, Cd, Co, Cr and Ni and retained for Pb.

**Figure 3.** Histograms of the heavy metal concentrations and statistical results of the Mann–Whitney U test for examining the difference of the concentrations between spring and summer samples.

Indoor air quality is highly affected by outdoor air conditions (infiltration of aerosols and pollutants) and indoor human activities, while the fine particles such as carbonaceous aerosols and heavy metals from urban/industrial emissions have higher infiltration rates than coarse desert-dust particles due to their smaller sizes [54,55]. The current results showed a general slight variation in PTE concentrations between the sampling homes in both seasons, except for Cr (in both seasons) and Co (in summer) where a larger variability was observed (Figure 2). In spring, the ratios between the highest and lowest concentrations were for Cr (~55), followed by As (~9), Cd (~4), Pb (~3), Co (~2) and Ni (~1). In summer, this ratio was Cd (~32), Co (~21), Pb (~11), As (~5), Cr and Ni (~2). In general, variation between the concentration of elements can be attributed to different sources of household dust particles, as well as different sources of metals introduced to each house [56]. Furthermore, the correlations between the PTEs were generally low, supporting the large heterogeneity in the indoor air pollution sources between the houses, while moderate correlations were found only between As and Cd (r = 0.31) and Cr (r = 0.51).

Descriptive statistics of heavy metal concentrations (in mg $kg^{-1}$) for the analyzed indoor dust in Asaluyeh are shown in Table 1. Compared to the respective concentrations presented in studies of indoor dust samples from other regions in the world (Table 1), the obtained median values in this study, both from spring and summer samples, are either in the same range or significantly lower. Median concentrations of heavy metals in Asaluyeh were also below the limits defined for average global soils [57], except for As, which seems to be marginally higher in summer. Analysis of Asaluyeh surface sediments demonstrated a noticeable amount of arsenic released by diagenesis [58], while high concentrations of As in sediments from the Persian Gulf have also been mentioned in previous studies [59,60]. Previous work showed that Cr, Ni and Fe concentrations in street dust were lower in the industrial area of Asaluyeh compared to urban and non-urban areas in Bushehr [36]. Furthermore, the concentrations of Co, Cr and Ni can also be attributed to geogenic parameters, such as atmospheric inputs and the weathering of crustal materials [58]. The non-normal distribution of elements indicates the influence of multiple factors on the composition of indoor dust particles. Apart from the industrial and traffic emissions, increased dust activity and influence from Shamal dust storms increase the outside dust during summer, which may highly affect the concentrations of PTEs in indoor air samples.

**Table 1.** Total concentrations of heavy metals (mg kg$^{-1}$, dry weight) from indoor dust samples in Asaluyeh and other urban areas, and for average global soils.

| Locations | | As | Cd | Co | Cr | Ni | Pb | Reference |
|---|---|---|---|---|---|---|---|---|
| Asaluyeh (Spring) | Min | 0.65 | 0.25 | 0.99 | 0.81 | 13.4 | 4.05 | This study |
| | Max | 6.02 | 0.98 | 2.17 | 44.7 | 19.1 | 10.6 | |
| | Mean | 2.98 | 0.49 | 1.54 | 17.6 | 16.4 | 6.87 | |
| | Med | 2.76 | 0.49 | 1.55 | 17.3 | 16.5 | 7.00 | |
| | Std. Dev | 1.66 | 0.18 | 0.29 | 9.66 | 1.85 | 1.80 | |
| | Skew | 0.19 | 1.16 | 0.44 | 0.91 | −0.13 | 0.32 | |
| | Kurt | −0.98 | 1.92 | 0.84 | 2.21 | −1.44 | -0.43 | |
| Asaluyeh (Summer) | Min | 2.10 | 0.15 | 1.12 | 43.9 | 11.4 | 1.24 | This study |
| | Max | 9.96 | 4.80 | 23.22 | 86.5 | 20.9 | 13.4 | |
| | Mean | 5.62 | 0.50 | 9.83 | 60.2 | 13.5 | 5.79 | |
| | Med | 6.41 | 0.24 | 6.25 | 59.3 | 12.6 | 4.91 | |
| | Std. Dev | 2.68 | 1.02 | 8.17 | 9.09 | 2.37 | 2.89 | |
| | Skew | 0.06 | 4.38 | 0.48 | 1.20 | 1.98 | 0.92 | |
| | Kurt | −1.63 | 19.4 | −1.37 | 3.01 | 4.27 | 1.02 | |
| Ahvaz, Iran | | n.a | 0.25–0.65 | 5.8–11.8 | 10–26 | 5–20 | 39.6–124 | Neisi, et al. [61] |
| Neyshabur, Iran | | n.a | 0.5–12.9 | 1.3–21.4 | 28.1–190 | 24.7–162 | 13.7–5345 | Naimabadi, et al. [62] |
| Giza and Cairo, Egypt | | n.a | 2.23 | n.a | 68.1 | 39.2 | 222 | Hassan [63] |
| Istanbul, Turkey | | n.a | 0.8 | 5 | 54.9 | 263 | 28.1 | Kurt-Karakus [17] |
| Ilorin, North central Nigeria | | 0.08 | 0.12 | 3.35 | 1.92 | 1.35 | 5.55 | Abdulraheem, et al. [64] |
| Chengdu, China | | n.a | 2.37 | n.a | 82.7 | 52.6 | 123 | Cheng, et al. [65] |
| Kwun Tong, China | | n.a | 39 | n.a | n.a | n.a | 308 | Tong and Lam [66] |
| Selangor, Malaysia | | n.a | 190 | n.a | n.a | 830 | 850 | Latif, et al. [18] |
| Warsaw, Poland | | n.a | n.a | n.a | 90 | 30 | 124 | Lisiewicz, et al. [67] |
| United Kingdom | | n.a | 1.3 | n.a | n.a. | 56.5 | 181 | Turner and Simmonds [68] |
| Toronto, Canada | | n.a | 1.7 | n.a | 42 | 23 | 36 | Al Hejami, et al. [56] |
| Ottawa, Canada | | 4.1 | 4.3 | 8.8 | 69 | 52 | 222 | Rasmussen, et al. [69] |
| Tokyo and Hiroshima, Japan | | n.a | 1.02 | n.a | 67.8 | 59.6 | 57.9 | Yoshinaga, et al. [70] |
| Christchurch, New Zealand | | n.a | 5.2 | n.a | n.a | n.a | 724 | Kim and Fergusson [71] |
| Sydney, Australia | | n.a | 1.6 | n.a | 65 | 15 | 76 | Chattopadhyay, et al. [72] |
| Average World Soils | | 6.83 | 0.41 | | 60 | 18 | 27 | Kabata-Pendias [57] |

The arid surroundings in Asaluyeh may highly affect the pollution levels, thus contributing to an already degraded urban environment due to refineries, petrochemical industries and large fossil-fuel combustion [32]. Furthermore, the high frequency of calcium carbonate in street dust in Asaluyeh can cause high arsenic immobilization [32]. This suggests that a large fraction of As, Cr, Fe and Ni is from local soil sources, while Pb is attributed to anthropogenic emissions. On the other hand, fine dust particles are more efficient in carrying heavy metals due to their higher specific surface area, and presence of clay minerals and organic matter in the soil [73]. A previous study in Asaluyeh showed that the mean concentrations of heavy metals in outdoor dust and soil samples decreased in the order of Fe (~11 g kg$^{-1}$) > Mn (365 mg kg$^{-1}$) > Zn (170 mg kg$^{-1}$) > Ni (79 mg kg$^{-1}$) > Pb (68.1 mg kg$^{-1}$) > Cu (54 mg kg$^{-1}$) > Cr (35.7 mg kg$^{-1}$) > Co (11.7 mg kg$^{-1}$) [36], which presented notable differences from the decreasing order of the heavy-metal concentrations of household dust, indicating large influence from human intervention and/or significant limitation of outdoor dust indoors. Similar results regarding the heavy metal concentrations from street dust samples in Asaluyeh were reported by Abbasi et al. [36]. This large difference between outdoor and indoor concentrations of heavy metals should be examined in more detail with new sampling strategies of concurrent outdoor and indoor measurements.

Although traffic is not considered as a major influential factor in Asaluyeh, proximity to the main roads, especially those driving to the PARS industrial zone, as well as to oil refineries and petrochemical plants may influence the variations in heavy-metal concentrations between the houses [74,75]. However, differences in human activities indoors is likely the most influential factor for the changes between concentrations of heavy metals in Asaluyeh. Compared to indoor heavy metal concentrations in other urban/industrialized areas around the world (Table 1), it is seen that the petrochemical industries in Asaluyeh have a small effect on indoor air quality. Higher heavy metal concentrations occur in cities larger than Asaluyeh (i.e., Istanbul, Cairo, Sydney, Toronto, Tokyo), where the traffic effect is especially high. Low traffic emissions in Asaluyeh is also a reason for the much lower PTEs concentrations in this city compared to other urban/industrial areas in Iran [32,36,37]. Continuous monitoring of indoor air pollution in Asaluyeh using gravimetric analysis of airborne $PM_{2.5}$ samples, which is more health relevant to the citizens, is necessary and will allow apportionment of the sources affecting indoor air pollution.

Analysis of indoor air quality in two churches in Faisalabad, Pakistan revealed high concentrations of $PM_{2.5}$ ($69 \pm 28$ μg $m^{-3}$), $CO_2$ ($1459 \pm 714$ ppm), $NO_2$ ($216 \pm 37$ ppm) and $SO_2$ ($125 \pm 65$ ppm), attributed to a crowded population and poor ventilation systems, as well as to high outdoor pollution [76]. Coal and biomass combustion was identified as the highest contributing source to measured polycyclic aromatic hydrocarbons (PAHs) in indoor dust in Kocaeli, Turkey [77]. On the other hand, indoor black carbon (BC) concentrations at households in rural areas across the Ganges valley, north India, were about 6–7 times higher when traditional cookstoves were used instead of liquefied petroleum gas for cooking, thus highlighting the importance of using new clean technologies for reducing indoor air pollution [78]. Recently, the lockdown intervention due to the COVID-19 pandemic significantly reduced the outdoor air pollution, but the longer stay and working at home resulted in an increase in indoor PM concentrations [79,80].

A previous study in Denver, Colorado during wildfire seasons revealed that indoor BC levels were about 2.4 times higher at homes that kept windows open for more than 12 h a day than homes with closed windows, while similar features were observed for $PM_{2.5}$ and $NO_x$ [22]. This indicates that different times of natural ventilation in homes by opening windows may highly differentiate the PM and PTE levels. Beyond open windows and doors, outdoor air pollutants can also enter indoors through mechanical air-ventilation systems and unintentional openings in the buildings. The construction rules for naturally ventilated buildings, the materials and the building air tightness are important factors determining the indoor air pollution levels [81,82]. Furthermore, the number of residents in each house may be an important determining factor for the levels of heavy metals indoors [17], although other studies have reported higher concentrations of Fe and Zn in houses with only 1–2 residents [72]. Another reason for the variation in heavy-metal concentrations between different houses may be the operation of air conditioning systems, which was found to be associated with increased levels of Cu, Cr, Cd and Mn in Istanbul [17]. On the other hand, recent studies have shown that paint walls are a significant source of heavy metals, since green is mostly associated with higher concentrations of Cu, purple with Zn and Pb and yellow with Cd, Cu, Pb and Zn [17,66,72]. Consequently, the time passed from the last wall painting and the quality of the paints used, may be a regulatory factor for the levels of heavy metals and the variability in concentrations between the houses.

### *3.2. Health Risk Assessment*

#### 3.2.1. Average Daily Intake of HEAVY Metals from Indoor Dust

The average daily intakes of heavy metals through the ingestion ($ADI_{ing}$), inhalation ($ADI_{inh}$) and dermal contact ($ADI_{drm}$) pathways in Asaluyeh, were calculated using the Equations (1)–(3). The results showed that $ADI_{ing}$ was found to be the main pathway of heavy metal intake from indoor dust followed by skin contact and inhalation. The results showed that the average daily intake of all of the investigated heavy metals are higher in children than adults. The highest $ADI_{ing}$ values were found for Cr ($7.70 \times 10^{-4}$ for summer and $2.24 \times 10^{-4}$ for spring) and Ni ($2.09 \times 10^{-4}$ for spring and $1.73 \times 10^{-4}$ for summer) in children (Table 2). Similarly, Cr and Ni showed the highest degree of average daily intake for ingestion in adults ($5.33 \times 10^{-5}$ and $2.04 \times 10^{-5}$, respectively). Regarding the minimum $ADI_{ing}$, Cd revealed the lowest probability of intake by adults and children. The $ADI_{drm}$ values were in the order of $10^{-6}$–$10^{-7}$ for children and $10^{-7}$–$10^{-8}$ for adults, while $ADI_{inh}$ values were two orders of magnitudes lower (Table 2).

**Table 2.** Reference dose values of heavy metals for the ingestion, inhalation and dermal contact pathways used in this study.

| | Population | $ADI_{ing}$ | | $ADI_{inh}$ | | $ADD_{der}$ | |
|---|---|---|---|---|---|---|---|
| | | Spring | Summer | Spring | Summer | Spring | Summer |
| As | Adult | $4.08 \times 10^{-6}$ | $7.70 \times 10^{-6}$ | $6.00 \times 10^{-10}$ | $1.13 \times 10^{-9}$ | $1.63 \times 10^{-8}$ | $3.07 \times 10^{-8}$ |
| | Children | $3.81 \times 10^{-5}$ | $7.18 \times 10^{-5}$ | $1.06 \times 10^{-9}$ | $2.01 \times 10^{-9}$ | $1.07 \times 10^{-7}$ | $2.01 \times 10^{-7}$ |
| Cd | Adult | $6.75 \times 10^{-7}$ | $6.89 \times 10^{-7}$ | $9.93 \times 10^{-11}$ | $1.01 \times 10^{-10}$ | $2.69 \times 10^{-9}$ | $2.75 \times 10^{-9}$ |
| | Children | $6.30 \times 10^{-6}$ | $6.43 \times 10^{-6}$ | $1.76 \times 10^{-10}$ | $1.80 \times 10^{-10}$ | $1.76 \times 10^{-8}$ | $1.80 \times 10^{-8}$ |
| Co | Adult | $2.11 \times 10^{-6}$ | $1.35 \times 10^{-5}$ | $3.10 \times 10^{-10}$ | $1.98 \times 10^{-9}$ | $8.40 \times 10^{-9}$ | $5.37 \times 10^{-8}$ |
| | Children | $1.97 \times 10^{-5}$ | $1.26 \times 10^{-4}$ | $5.49 \times 10^{-10}$ | $3.51 \times 10^{-9}$ | $5.50 \times 10^{-8}$ | $3.52 \times 10^{-7}$ |
| Cr | Adult | $2.40 \times 10^{-5}$ | $8.25 \times 10^{-5}$ | $3.54 \times 10^{-9}$ | $1.21 \times 10^{-8}$ | $9.59 \times 10^{-8}$ | $3.29 \times 10^{-7}$ |
| | Children | $2.24 \times 10^{-4}$ | $7.70 \times 10^{-4}$ | $6.27 \times 10^{-9}$ | $2.15 \times 10^{-8}$ | $6.28 \times 10^{-7}$ | $2.16 \times 10^{-6}$ |
| Ni | Adult | $2.24 \times 10^{-5}$ | $1.86 \times 10^{-5}$ | $3.30 \times 10^{-9}$ | $2.73 \times 10^{-9}$ | $8.95 \times 10^{-8}$ | $7.40 \times 10^{-8}$ |
| | Children | $2.09 \times 10^{-4}$ | $1.73 \times 10^{-4}$ | $5.85 \times 10^{-9}$ | $4.84 \times 10^{-9}$ | $5.86 \times 10^{-7}$ | $4.85 \times 10^{-7}$ |
| Pb | Adult | $9.42 \times 10^{-6}$ | $7.93 \times 10^{-6}$ | $1.38 \times 10^{-9}$ | $1.17 \times 10^{-9}$ | $3.76 \times 10^{-8}$ | $3.16 \times 10^{-8}$ |
| | Children | $8.79 \times 10^{-5}$ | $7.40 \times 10^{-5}$ | $2.46 \times 10^{-9}$ | $2.07 \times 10^{-9}$ | $2.46 \times 10^{-7}$ | $2.07 \times 10^{-7}$ |

### 3.2.2. Non-Carcinogenic Risk Assessment

Regarding the non-carcinogenic risk, the estimates showed that the ingestion of dust particles was found to be the predominant pathway triggering non-carcinogenic risk in both adults and children, while inhalation was the lowest harmful pathway for all examined heavy metals except for Co (Figure 4).

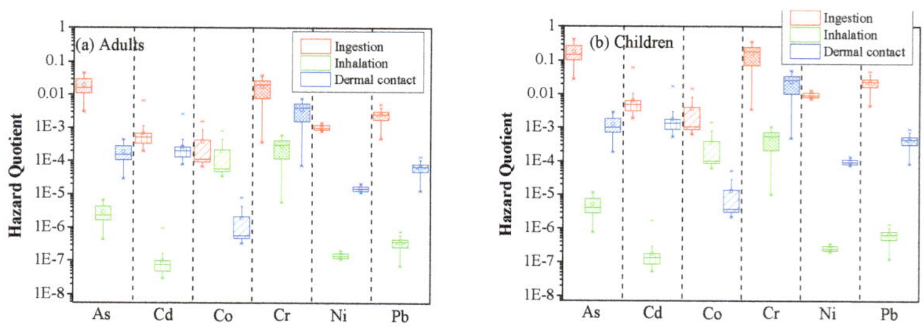


**Figure 4.** Hazard Quotient (HQ) values for non-carcinogenic health risk of selected heavy metals in household dust in Asaluyeh for adults (**a**) and children (**b**), and for the three pathways (ingestion, inhalation and dermal contact).

The larger HQ values for children compared to adults revealed their higher vulnerability to heavy metals [27,83,84]. The highest $HQ_{ing}$ was obtained for As in summer dust ($4.25 \times 10^{-1}$ for children and $4.55 \times 10^{-2}$ for adults). However, As has still low potential to cause non-carcinogenic risk ($0.1 < HQ < 1$ for children and $0.01 < HQ < 0.1$ for adults). The obtained $HQ_{ing}$ values were also relatively high for Cr, revealing an average of $3.95 \times 10^{-2}$ for adults and $3.69 \times 10^{-1}$ for children. Regarding the $HQ_{inh}$ and $HQ_{drm}$, Cr was also recognized as the most hazardous element for indoor dust samples in Asaluyeh (Figure 4). The possible non-carcinogenic risk in case of long exposure through inhalation and dermal contact was the least for Cd and Co, respectively. The average HQ values for children showed to be one order of magnitude higher compared to adults.

The overall cumulative non-carcinogenic risks of heavy metals from household dust was found in the safe allowable limit of HI $\leq 1$ for both sub-population groups (Figure 5). Due to lower pollution levels in spring, the exposure to PTEs reduced, and consequently, the risk to human health was lower via all the examined pathways. The median HI values decreased in the order of As > Cr > Pb > Ni > Cd > Co for the spring and Cr > As > Pb > Ni > Co > Cd for the summer samples. These orders are similar for adults and children though the trend of higher levels for children than adults was consistent for all of the elements, indicating that children are at higher risk by PTEs during their lifetime [85].

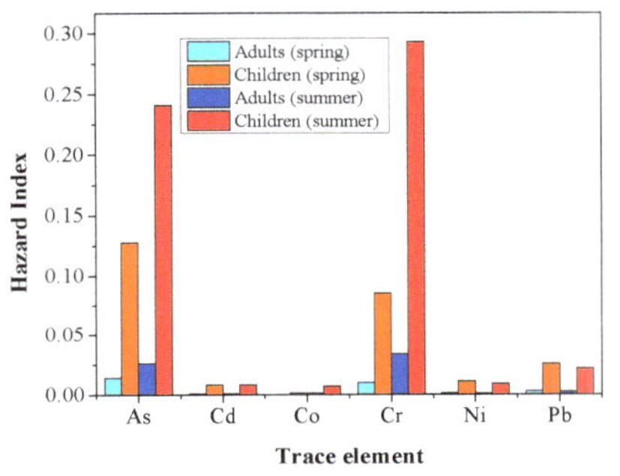


**Figure 5.** Hazard index (HI) values of selected heavy metals for non-carcinogenic health risk in Asaluyeh for adults and children.

### 3.2.3. Carcinogenic Risk Assessment

Regarding the carcinogenic health risk (CR), ingestion of indoor dust particles was again estimated to be the foremost pathway threatening the health of Asaluyeh residents (Figure 6). The CR values for children were significantly higher than those for adults, while the main factors affecting this difference are the higher ingestion rate and the lower body weight of children [3]. The carcinogenic risks associated with oral intake of As and Cr were found to be within the tolerable thresholds for adults and at high risk ($>10^{-4}$) for children. The cancer effect associated with the inhalation pathway was found to be insignificant for both adults and children ($CR_{inh} < 10^{-6}$), except for Cr in children, while CR values above $10^{-6}$ were also found for Pb via ingestion and for As via dermal contact, but only for children (Figure 6).

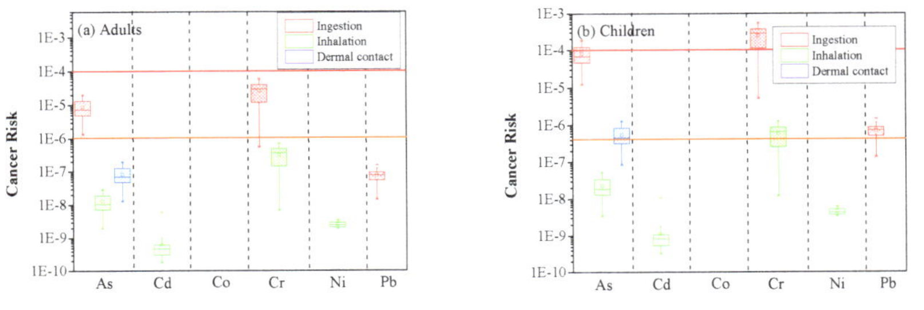


**Figure 6.** Cancer risk (CR) values of selected heavy metals in household dust in Asaluyeh for adults (**a**) and children (**b**), and for the three pathways (ingestion, inhalation and dermal contact). The orange and red lines denote the lower and upper safe limits, respectively.

The results of total cancer risk (TCR) suggested that As and Cr are the elements of major concern, with the potential for causing a lifetime risk of developing cancer (Figure 7). The TCR values for Cr ranged from $5.20 \times 10^{-6}$ to $5.55 \times 10^{-4}$ for children and $5.63 \times 10^{-7}$

to $6.00 \times 10^{-5}$ for adults, representing a moderate-to-high risk. Similar results were obtained for As, with the TCR values ranging from $1.26 \times 10^{-5}$ to $1.92 \times 10^{-4}$ for children and $1.35 \times 10^{-6}$ to $2.07 \times 10^{-5}$ for adults. Hence, the mean TCR values for children exceeded the safe threshold level of $10^{-4}$ for Cr in spring and summer, while the TCR for As was close to this level ($10^{-4}$). For adults, TCR values for Cr and As were within the tolerable limits, while the carcinogenic risks posed by Cd, Ni and Pb were assessed to be very low (CR < $10^{-6}$) in both age groups. Men et al. [4] also reported higher CR for As, Cr and Ni in road dust in Beijing. Spring and summer dust samples followed the same trend of health risk; however, the TCR appeared to be higher in summer dust due to much higher concentrations of heavy metals (Figure 2).

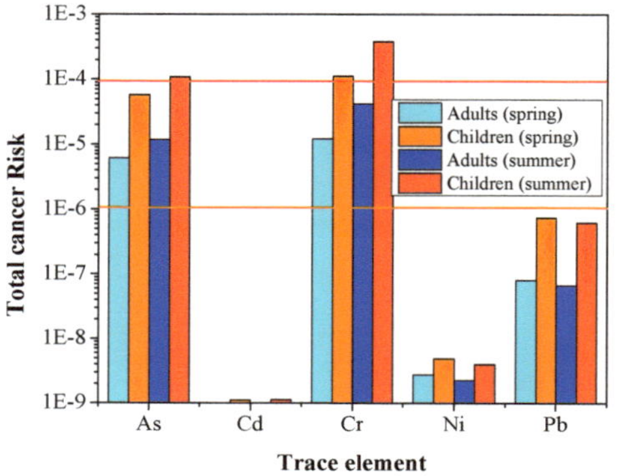


**Figure 7.** Total cancer risk values of selected heavy metals in Asaluyeh for adults and children. The orange and red lines denote the lower and upper safe limits, respectively.

## 4. Discussion

In residential areas located close to large industrial sectors, excessive accumulation of heavy metals in street and airborne dust can cause chronic respiratory diseases for local inhabitants through ingestion, inhalation and dermal contact routes [86,87]. Household dust plays a dual role in health exposure to PTEs, combining the transfer of outdoor dust and pollutants indoors [88,89]. The current results revealed that household dust in Asaluyeh industrial city can cause a cancer risk to the local population, mainly in children, through exposure to As and Cr, while ingestion was the riskiest pathway. Children are at the highest risk of exposure to PTEs, due to their physiological and behavioral characteristics (physical activity, playing in playgrounds, etc.) [90,91]. Furthermore, children are also more sensitive than adults to indoor air pollution, especially for the inhalation and ingestion pathways, as they breathe in a higher volume of polluted air relative to their body weight and have more frequent hand-to-mouth activities [92].

Heavy metals are of high concern for health-related issues due to their toxicity, bio-accumulation in human body and low degradation potential [93,94]. Cd and Pb were characteristic tracers of traffic emissions from engines and brake wear, while As, Cr and Ni are highly released from fuel combustion, among other sources [95,96]. Arsenic may cause deleterious effects on the human organism by negatively affecting the digestive, cardiovascular and DNA systems [97], and this is especially important in Asaluyeh due to increased CR values of As. Furthermore, Pb can negatively affect the kidneys, the soil pH and its absorption capacity, while high quantities of Pb can cause neurological disorders [8,98,99], although the indoor concentrations of Pb in Asaluyeh were not at such alarming levels. Exposure to excess Ni levels can be associated with increased carcinogenic risk and heart attack, as it can be bio-accumulated in the bones, liver, kidneys and aorta [100]. However, it should be noted that CR values for each element were mostly sensitive to the ingestion rate and skin adherence factor, which control the human health risk assessment [3].

Studies examining the health risk for indoor dust are few compared to those related to outdoor ambient or street dust. Similar to the current results, the HI values for eight PTEs in indoor dust in Istanbul were below the safe limit of 1 for both children and adults, indicating less potential health risk [17]. Ingestion was found to be the major pathway of exposure to indoor dust for non-cancer effects in homes and offices in Istanbul, followed by dermal contact and inhalation [17]. Our results showed highest CR for Cr in Asaluyeh, especially in summer, which was above the safe limit for children (Figure 5). The CR values of Cr for children in indoor dust in Istanbul was estimated at $3.7 \times 10^{-5}$, $3.4 \times 10^{-10}$ and $3.9 \times 10^{-5}$ for ingestion, inhalation and dermal contact, respectively, which were within the acceptable limits of EPA. The respective CR values for adults were $2.7 \times 10^{-5}$, $1.3 \times 10^{-9}$ and $1.1 \times 10^{-4}$, indicating significant risk for dermal contact [17]. On the other hand, a study in Vilnius, Lithuania showed that despite the much lower fine $PM_1$ concentration indoors than outdoors ($8.8 \pm 2.7$ μg $m^{-3}$ against $43.2 \pm 22.3$ μg $m^{-3}$) during forest fire events, the highly acidic indoor $PM_1$ could have harmful health effects [101], thus highlighting the importance of aerosol PH in health-related studies. Furthermore, it should be noted that apart from the PTEs that are examined in this work, many other components of PM, such as VOCs, formaldehyde and carbon monoxide (CO), may negatively affect air quality and human health indoors [102,103].

Previous estimates of the health risk based on street dust samples in Asaluyeh revealed HI values below 1 and less health risk to residents, with ingestion being the dominant pathway for both children and adult [32], in accordance with the current results. It is worth mentioning that although the examined PTEs do not seem to cause non-carcinogenic risk for adults (HI < 1), the population of Asaluyeh is predicted to be exposed to morbidity and mortality risk due to long-term exposure to fine particulate matter generated by intense dust storms over the adjacent arid and semi-arid regions [104]. So, the combination of dust and heavy metals from oil refineries and petrochemical industries deteriorates air quality, leading to unhealthy conditions at least for sensitive population groups such as children and elderly. A recent study regarding the health effects of indoor mercury (Hg) in Asaluyeh showed moderate-to-low health risk and lower levels than those observed in the larger city of Ahvaz, southwest Iran [37]. A recent study in Zabol, southeast Iran [84] reported high non-carcinogenic and carcinogenic risks (HI > 1; CRs > $10^{-4}$) for children and adults due to exposure to outdoor dust. In developing countries of south Asia, despite awareness of the strong relationship between environmental degradation and human health, diseases and deaths related to atmospheric pollution have been increasing during the last decades [105,106].

Although widely used, the EPA model for health risk assessment is subjected to some biases, since the input variables (i.e., concentrations of heavy metals, ingestion rates, body weight, skin adherence factor) may not accurately represent the population groups at any time, and may vary among different people and even for the same person [3]. Evaluation of these uncertainties and sensitivity of each input variable to the health risk assessment were not considered in most studies, while Men et al. [3] provided such sensitivity analysis based on Monte Carlo simulations. They found that the cancer risk was more sensitive to heavy metal concentrations rather than other input variables, while the skin adherence factor and the ingestion rate were the most sensitive input variables for CR assessment.

## 5. Mitigation Strategies and Future Projections

In the hot environment of the Middle East, operation of cooling devices for avoiding opening windows under highly polluted atmospheres would be a good strategy to protect outdoor pollution from coming indoors, and this is recommended by public authorities during dust storms. This practice is also necessary during pollution events caused by accumulation of pollutants due to excess emissions and/or unfavorable meteorological conditions, such as calms, shallow boundary layer and temperature inversions. On the other hand, air filtering systems could be especially efficient in removing fine particulate from indoor air [101,107]. For new energy-conserving buildings in Asaluyeh, special treatment should be given to the ventilation options and in reducing exchanges between outdoor and indoor air in a way to limit health risk [108–110]. Therefore, improving air tightness of a building is an important pathway to limit the incoming of outdoor air pollutants, while mechanical ventilation with efficient air purification systems, such as HEPA, should be followed for highly polluted conditions [14,15,111].

This study provided first results concerning heavy metal concentrations and potential health risks due to exposure to indoor dust in Asaluyeh. Indoor and outdoor sampling will be continued in Asaluyeh city to assess possible changes in PTE concentrations and the effect of petrochemical industries and meteorology. More samples and chemical analysis will also allow for the determination of pollution sources in the city via statistical techniques, in order to evaluate the contribution of each source and to propose solutions. On the other hand, continuous sampling and assessment of the pollution sources and health risk will allow for warning strategies for the local population to be aware of the effects of short and long-term exposure to PTEs. Future research in Asaluyeh county in a way to prevent human health from exposure to petrochemicals and PTEs should emphasize (i) low-cost sensors that can provide real-time measurements of PM concentrations [112], (ii) analysis of PAHs and other organic compounds such as formaldehyde [113], both in outdoor and indoor air, (iii) dissemination of the pollution levels to local population via apps in smart phones or in public walls. Furthermore, more studies are recommended to be undertaken on spatial distribution patterns and source identification of PTEs in Asaluyeh and other highly-industrialized areas along the Persian Gulf due to highly turbid atmospheres from the combination of traffic and industrial emissions and dust storms. Continuous recordings of hospital admissions and statistics for morbidity and mortality in Asaluyeh county will further help in associating pollution levels and potential health risk assessment with hospitalization and morbidity rates.

## 6. Conclusions

This study analyzed the concentrations of selected heavy metal(oid)s (As, Cd, Co, Cr, Ni and Pb) of household dust in the industrial city of Asaluyeh in south Iran. The key objectives were to assess the indoor air quality and health risk for children and adults due to exposure to heavy metals via three pathways, i.e., ingestion, inhalation and dermal contact. The dust samples were collected at houses in Asaluyeh during spring and summer (20 samples in each season) and analyzed for heavy metals through Inductively Coupled Plasma Mass Spectrometry (ICP-MS).

The mean concentrations of heavy metals in the analyzed household dust samples followed the order of Cr > Ni > Pb > As > Co > Cd in both seasons, while significant differences were observed from spring to summer, with a large increase in mean Cr concentrations (60.3 mg $kg^{-1}$ against 17.6 mg $kg^{-1}$ in spring), as well as in As (5.62 mg $kg^{-1}$ vs. 2.98 mg $kg^{-1}$) and Co (9.85 vs. 1.54 mg $kg^{-1}$).

Estimates of the health risk for the local residents due to exposure to household dust revealed generally low and acceptable levels for non-carcinogenic effects, since HI values were below 1 for both children and adults. The highest health risk was through the ingestion pathway, followed by dermal contact and inhalation, while As and Cr were the most hazardous elements for human health. Regarding the cancer risk (CR), the ingestion pathway for children presented values above the safe threshold of $10^{-4}$, while for adults the CR for ingestion was in the order of $10^{-5}$. Cr and As exhibited the highest CR values, while As through dermal contact, Cr through inhalation and Pb through ingestion also presented tolerable CR values for children, but lower than $10^{-6}$ for adults. Although in most of the cases the HI and CR values were within the acceptable limits of EPA (HI < 1; CR: < $10^{-6}$), the bioaccumulation of heavy metals to the tissues and other organs of the human body can cause chronic deleterious effects that could not be overlooked, especially in highly industrialized areas with several oil refineries and petrochemical industries.

**Author Contributions:** Conceptualization, M.T., R.D.B. and D.G.K.; methodology, M.T., R.D.B. and S.R.A.; formal analysis, R.D.B., M.T.; data curation, R.D.B., M.T. and S.R.A.; writing—original draft preparation, M.T.; writing—review and editing, R.D.B., M.T. and D.G.K.; supervision, D.G.K. All authors have read and agreed to the published version of the manuscript.

**Funding:** This article was supported by the University of Zabol (Iran) with grant code (IR-UOZ-GR-0088). This research received no external funding.

**Institutional Review Board Statement:** Not applicable.

**Informed Consent Statement:** Not applicable.

**Data Availability Statement:** The data can be available upon request.

**Acknowledgments:** We are thankful to all individuals who kindly helped in indoor dust sampling. D.G.K. acknowledges the support of the project "PANhellenic infrastructure for Atmospheric Composition and climate change" PANACEA (MIS 5021516), funded by the Operational Program "Competitiveness, Entrepreneurship and Innovation" (NSRF 2014–2020) and co-financed by Greece and the European Union.

**Conflicts of Interest:** The authors declare no conflict of interest.

## References

1. Kim, K.-H.; Kabir, E.; Kabir, S. A review on the human health impact of airborne particulate matter. *Environ. Int.* **2015**, *74*, 136–143. [CrossRef] [PubMed]
2. Jain, S.; Sharma, S.K.; Vijayan, N.; Mandal, T.K. Investigating the seasonal variability in source contribution to $PM_{2.5}$ and $PM_{10}$ using different receptor models during 2013–2016 in Delhi, India. *Environ. Sci. Pollut. Res.* **2021**, *28*, 4660–4675. [CrossRef] [PubMed]
3. Men, C.; Wang, Y.; Liu, R.; Wang, Q.; Miao, Y.; Jiao, L.; Shoaib, M.; Shen, Z. Temporal variations of levels and sources of health risk associated with heavy metals in road dust in Beijing from May 2016 to April 2018. *Chemosphere* **2021**, *270*, 129434. [CrossRef]
4. Theodosi, C.; Tsagkaraki, M.; Zarmpas, P.; Grivas, G.; Liakakou, E.; Paraskevopoulou, D.; Lianou, M.; Gerasopoulos, E.; Mihalopoulos, N. Multi-year chemical composition of the fine-aerosol fraction in Athens, Greece, with emphasis on the contribution of residential heating in wintertime. *Atmos. Chem. Phys.* **2018**, *18*, 14371–14391. [CrossRef]
5. Stafoggia, M.; Zauli-Sajani, S.; Pey, J.; Samoli, E.; Alessandrini, E.; Basagaña, X.; Cernigliaro, A.; Chiusolo, M.; Demaria, M.; Díaz, J.; et al. Desert dust outbreaks in Southern Europe: Contribution to daily PM10 concentrations and short-term associations with mortality and hospital admissions. *Environ. Health Perspect.* **2016**, *124*, 413–419. [CrossRef]
6. Shahsavani, A.; Tobías, A.; Querol, X.; Stafoggia, M.; Abdolshahnejad, M.; Mayvaneh, F.; Guo, Y.; Hadei, M.; Hashemi, S.S.; Khosravi, A.; et al. Short-term effects of particulate matter during desert and non-desert dust days on mortality in Iran. *Environ. Int.* **2020**, *134*, 105299. [CrossRef]
7. Massányi, P.; Massányi, M.; Madeddu, R.; Stawarz, R.; Lukáč, N. Effects of cadmium, lead, and mercury on the structure and function of reproductive organs. *Toxics* **2020**, *8*, 94. [CrossRef]
8. Alengebawy, A.; Abdelkhalek, S.T.; Qureshi, S.R.; Wang, M.-Q. Heavy metals and pesticides toxicity in agricultural soil and plants: Ecological risks and human health implications. *Toxics* **2021**, *9*, 42. [CrossRef]
9. Fadel, M.; Ledoux, F.; Afif, C.; Courcot, D. Human health risk assessment for PAHs, phthalates, elements, PCDD/Fs, and DL-PCBs in PM2.5 and for NMVOCs in two East-Mediterranean urban sites under industrial influence. *Atmos. Pollut. Res.* **2022**, *13*, 101261. [CrossRef]
10. Amato, F.; Rivas, I.; Viana, M.; Moreno, T.; Bouso, L.; Reche, C.; Àlvarez-Pedrerol, M.; Alastuey, A.; Sunyer, J.; Querol, X. Sources of indoor and outdoor $PM_{2.5}$ concentrations in primary schools. *Sci. Total Environ.* **2014**, *490*, 757–765. [CrossRef]
11. Canha, N.; Mandin, C.; Ramalho, O.; Wyart, G.; Ribéron, J.; Dassonville, C.; Hänninen, O.; Almeida, S.M.; Derbez, M. Assessment of ventilation and indoor air pollutants in nursery and elementary schools in France. *Indoor Air* **2016**, *26*, 350–365. [CrossRef] [PubMed]
12. Kalimeri, K.K.; Saraga, D.E.; Lazaridis, V.D.; Legkas, N.A.; Missia, D.A.; Tolis, E.I.; Bartzis, J.G. Indoor air quality investigation of the school environment and estimated health risks: Two-season measurements in primary schools in Kozani, Greece. *Atmos. Pollut. Res.* **2016**, *7*, 1128–1142. [CrossRef]
13. Halios, C.; Santamouris, M.; Helmi, A.; Kapsalaki, M.; Saliari, M.; Spanou, A.; Tsakos, D. Exposure to fine particulate matter in ten night clubs in Athens Greece: Studying the effect of ventilation, cigarette smoking and resuspension. *Sci. Total Environ.* **2009**, *407*, 4894–4901. [CrossRef] [PubMed]
14. Peng, Z.; Deng, W.; Tenorio, R. Investigation of indoor air quality and the identification of influential factors at primary schools in the North of China. *Sustainability* **2017**, *9*, 1180. [CrossRef]
15. Rajagopalan, P.; Goodman, N. Improving the indoor air quality of residential buildings during bushfire smoke events. *Climate* **2021**, *9*, 32. [CrossRef]

16. Ali, N.; Van den Eede, N.; Dirtu, A.C.; Neels, H.; Covaci, A. Assessment of human exposure to indoor organic contaminants via dust ingestion in Pakistan. *Indoor Air* **2012**, *22*, 200–211. [CrossRef]
17. Kurt-Karakus, P.B. Determination of heavy metals in indoor dust from Istanbul, Turkey: Estimation of the health risk. *Environ. Int.* **2012**, *50*, 47–55. [CrossRef]
18. Latif, M.T.; Othman, M.R.; Kim, C.L.; Murayadi, S.A.; Sahaimi, K.N.A. Composition of household dust in semi-urban areas in Malaysia. *Indoor Built Environ.* **2009**, *18*, 155–161. [CrossRef]
19. Kliucininkas, L.; Krugly, E.; Stasiulaitiene, I.; Radziuniene, I.; Prasauskas, T.; Jonusas, A.; Kauneliene, V.; Martuzevicius, D. Indoor–outdoor levels of size segregated particulate matter and mono/polycyclic aromatic hydrocarbons among urban areas using solid fuels for heating. *Atmos. Environ.* **2014**, *97*, 83–93. [CrossRef]
20. Manisalidis, I.; Stavropoulou, E.; Stavropoulos, A.; Bezirtzoglou, E. Environmental and health impacts of air pollution: A review. *Front. Public Health* **2020**, *8*, 14. [CrossRef]
21. Patel, S.; Sankhyan, S.; Boedicker, E.K.; DeCarlo, P.F.; Farmer, D.K.; Goldstein, A.H.; Katz, E.F.; Nazaroff, W.W.; Tian, Y.; Vanhanen, J.; et al. Indoor particulate matter during HOMEChem: Concentrations, size distributions, and exposures. *Environ. Sci. Technol.* **2020**, *54*, 7107–7116. [CrossRef] [PubMed]
22. Shrestha, P.M.; Humphrey, J.L.; Carlton, E.J.; Adgate, J.L.; Barton, K.E.; Root, E.D.; Miller, S.L. Impact of outdoor air pollution on indoor air quality in low-income homes during wildfire seasons. *Int. J. Environ. Res. Public Health* **2019**, *16*, 3535. [CrossRef] [PubMed]
23. Duong, T.T.; Lee, B.-K. Determining contamination level of heavy metals in road dust from busy traffic areas with different characteristics. *J. Environ. Manag.* **2011**, *92*, 554–562. [CrossRef] [PubMed]
24. Titos, G.; Lyamani, H.; Pandolfi, M.; Alastuey, A.; Alados-Arboledas, L. Identification of fine (PM1) and coarse (PM10-1) sources of particulate matter in an urban environment. *Atmos. Environ.* **2014**, *89*, 593–602. [CrossRef]
25. Squizzato, S.; Masiol, M.; Agostini, C.; Visin, F.; Formenton, G.; Harrison, R.M.; Rampazzo, G. Factors, origin and sources affecting PM1 concentrations and composition at an urban background site. *Atmos. Res.* **2016**, *180*, 262–273. [CrossRef]
26. Hu, X.; Zhang, Y.; Ding, Z.; Wang, T.; Lian, H.; Sun, Y.; Wu, J. Bioaccessibility and health risk of arsenic and heavy metals (Cd, Co, Cr, Cu, Ni, Pb, Zn and Mn) in TSP and $PM_{2.5}$ in Nanjing, China. *Atmos. Environ.* **2012**, *57*, 146–152. [CrossRef]
27. Cao, S.; Duan, X.; Zhao, X.; Ma, J.; Dong, T.; Huang, N.; Sun, C.; He, B.; Wei, F. Health risks from the exposure of children to As, Se, Pb and other heavy metals near the largest coking plant in China. *Sci. Total Environ.* **2014**, *472*, 1001–1009. [CrossRef]
28. Tashakor, M.; Modabberi, S.; Argyraki, A. Assessing the contamination level, sources and risk of potentially toxic elements in urban soil and dust of Iranian cities using secondary data of published literature. *Environ. Geochem. Health* **2022**, *44*, 645–675. [CrossRef]
29. Saeedi, M.; Li, L.Y.; Salmanzadeh, M. Heavy metals and polycyclic aromatic hydrocarbons: Pollution and ecological risk assessment in street dust of Tehran. *J. Hazard. Mater.* **2012**, *227*, 9–17. [CrossRef]
30. Norouzi, S.; Khademi, H.; Ayoubi, S.; Cano, A.F.; Acosta, J.A. Seasonal and spatial variations in dust deposition rate and concentrations of dust-borne heavy metals, a case study from Isfahan, central Iran. *Atmos. Pollut. Res.* **2017**, *8*, 686–699. [CrossRef]
31. Najmeddin, A.; Moore, F.; Keshavarzi, B.; Sadegh, Z. Pollution, source apportionment and health risk of potentially toxic elements (PTEs) and polycyclic aromatic hydrocarbons (PAHs) in urban street dust of Mashhad, the second largest city of Iran. *J. Geochem. Explor.* **2018**, *190*, 154–169. [CrossRef]
32. Abbasi, S.; Keshavarzi, B.; Moore, F.; Mahmoudi, M.R. Fractionation, source identification and risk assessment of potentially toxic elements in street dust of the most important center for petrochemical products, Asaluyeh County, Iran. *Environ. Earth Sci.* **2018**, *77*, 673. [CrossRef]
33. Nadal, M.; Schuhmacher, M.; Domingo, J.L. Levels of metals, PCBs, PCNs and PAHs in soils of a highly industrialized chemical/petrochemical area: Temporal trend. *Chemosphere* **2007**, *66*, 267–276. [CrossRef]
34. Kafaei, R.; Tahmasbi, R.; Ravanipour, M.; Vakilabadi, D.R.; Ahmadi, M.; Omrani, A.; Ramavandi, B. Urinary arsenic, cadmium, manganese, nickel, and vanadium levels of schoolchildren in the vicinity of the industrialised area of Asaluyeh, Iran. *Environ. Sci. Pollut. Res.* **2017**, *24*, 23498–23507. [CrossRef] [PubMed]
35. Abdollahi, S.; Raoufi, Z.; Faghiri, I.; Savari, A.; Nikpour, Y.; Mansouri, A. Contamination levels and spatial distributions of heavy metals and PAHs in surface sediment of Imam Khomeini Port, Persian Gulf, Iran. *Mar. Pollut. Bull.* **2013**, *71*, 336–345. [CrossRef]
36. Naderizadeh, Z.; Khademi, H.; Ayoubi, S. Biomonitoring of atmospheric heavy metals pollution using dust deposited on date palm leaves in southwestern Iran. *Atmósfera* **2016**, *29*, 141–155. [CrossRef]
37. Dahmardeh Behrooz, R.; Tashakor, M.; Asvad, R.; Esmaili-Sari, A.; Kaskaoutis, D.G. Characteristics and Health Risk Assessment of Mercury Exposure via Indoor and Outdoor Household Dust in Three Iranian Cities. *Atmosphere* **2022**, *13*, 583. [CrossRef]
38. Sofia, D.; Gioiella, F.; Lotrecchiano, N.; Giuliano, A. Mitigation strategies for reducing air pollution. *Environ. Sci. Pollut. Res.* **2020**, *27*, 19226–19235. [CrossRef]
39. Keshavarzifard, M.; Moore, F.; Keshavarzi, B.; Sharifi, R. Distribution, source apportionment and health risk assessment of polycyclic aromatic hydrocarbons (PAHs) in intertidal sediment of Asaluyeh, Persian Gulf. *Environ. Geochem. Health* **2018**, *40*, 721–735. [CrossRef]
40. Alipour, H.; Darabi, H.; Dabbaghmanesh, T.; Bonyani, M. Entomological study of sand flies (Diptera: Psychodidae: Phlebotominae) in Asalouyeh, the heartland of an Iranian petrochemical industry. *Asian Pac. J. Trop. Biomed.* **2014**, *4*, S242–S245. [CrossRef]

41. Hashemi, S.; Taheri, H. The effect on quality of life of the people in the Pars Special Economic Energy Zone. *MAGNT Res. Rep.* **2013**, *2*, 570–578.
42. Davoudi, M.; Rahimpour, M.; Jokar, S.; Nikbakht, F.; Abbasfard, H. The major sources of gas flaring and air contamination in the natural gas processing plants: A case study. *J. Nat. Gas Sci. Eng.* **2013**, *13*, 7–19. [CrossRef]
43. Saidi, M.; Siavashi, F.; Rahimpour, M. Application of solid oxide fuel cell for flare gas recovery as a new approach; a case study for Asalouyeh gas processing plant, Iran. *J. Nat. Gas Sci. Eng.* **2014**, *17*, 13–25. [CrossRef]
44. Dehghani, M.; Nabipour, I.; Dobaradaran, S.; Godarzi, H. Cd and Pb concentrations in the surface sediments of the Asaluyeh Bay, Iran. *J. Community Health Res.* **2014**, *3*, 22–30.
45. Hamzeh, N.H.; Kaskaoutis, D.G.; Rashki, A.; Mohammadpour, K. Long-Term Variability of Dust Events in Southwestern Iran and Its Relationship with the Drought. *Atmosphere* **2021**, *12*, 1350. [CrossRef]
46. Darus, F.M.; Nasir, R.A.; Sumari, S.M.; Ismail, Z.S.; Omar, N.A. Heavy metals composition of indoor dust in nursery schools building. *Procedia-Soc. Behav. Sci.* **2012**, *38*, 169–175. [CrossRef]
47. Wan, D.; Han, Z.; Liu, D.; Yang, J. Risk assessments of heavy metals in house dust from a typical industrial area in Central China. *Hum. Ecol. Risk Assess. Int. J.* **2016**, *22*, 489–501. [CrossRef]
48. Zhang, Y.; Ji, X.; Ku, T.; Li, G.; Sang, N. Heavy metals bound to fine particulate matter from northern China induce season-dependent health risks: A study based on myocardial toxicity. *Environ. Pollut.* **2016**, *216*, 380–390. [CrossRef]
49. USEPA. *Risk Assessment Guidance for Superfund: Volume III—Part A, Process for Conducting Probabilistic Risk Assessment*; EPA 540-R-02-002; US Environmental Protection Agency: Washington, DC, USA, 2001.
50. USEPA. *Child Specific Exposure Factors Handbook*; EPA-600-P-00-002B; National Center for Environmental Assessment: Washington, DC, USA, 2002.
51. Pan, L.-b.; Ma, J.; Wang, X.-l.; Hou, H. Heavy metals in soils from a typical county in Shanxi Province, China: Levels, sources and spatial distribution. *Chemosphere* **2016**, *148*, 248–254. [CrossRef]
52. Chen, H.; Lu, X.; Li, L.Y.; Gao, T.; Chang, Y. Metal contamination in campus dust of Xi'an, China: A study based on multivariate statistics and spatial distribution. *Sci. Total Environ.* **2014**, *484*, 27–35. [CrossRef]
53. Grivas, G.; Cheristanidis, S.; Chaloulakou, A.; Koutrakis, P.; Mihalopoulos, N. Elemental composition and source apportionment of fine and coarse particles at traffic and urban background locations in Athens, Greece. *Aerosol Air Qual. Res.* **2018**, *18*, 1642–1659. [CrossRef]
54. Liu, R.; Wang, Q.; Lu, X.; Fang, F.; Wang, Y. Distribution and speciation of mercury in the peat bog of Xiaoxing'an Mountain, northeastern China. *Environ. Pollut.* **2003**, *124*, 39–46. [CrossRef]
55. Wang, F.; Meng, D.; Li, X.; Tan, J. Indoor-outdoor relationships of $PM_{2.5}$ in four residential dwellings in winter in the Yangtze River Delta, China. *Environ. Pollut.* **2016**, *215*, 280–289. [CrossRef] [PubMed]
56. Al Hejami, A.; Davis, M.; Prete, D.; Lu, J.; Wang, S. Heavy metals in indoor settled dusts in Toronto, Canada. *Sci. Total Environ.* **2020**, *703*, 134895. [CrossRef]
57. Kabata-Pendias, A. *Trace Elements in Soils and Plants*, 4th ed.; CRS Press: Roca Raton, FL, USA, 2011.
58. Aghadadashi, V.; Neyestani, M.R.; Mehdinia, A.; Bakhtiari, A.R.; Molaei, S.; Farhangi, M.; Esmaili, M.; Marnani, H.R.; Gerivani, H. Spatial distribution and vertical profile of heavy metals in marine sediments around Iran's special economic energy zone; Arsenic as an enriched contaminant. *Mar. Pollut. Bull.* **2019**, *138*, 437–450. [CrossRef]
59. Youssef, M.; El-Sorogy, A.; Al Kahtany, K.; Al Otiaby, N. Environmental assessment of coastal surface sediments at Tarut Island, Arabian Gulf (Saudi Arabia). *Mar. Pollut. Bull.* **2015**, *96*, 424–433. [CrossRef]
60. El-Sorogy, A.S.; Youssef, M.; Al-Kahtany, K.; Al-Otaiby, N. Assessment of arsenic in coastal sediments, seawaters and molluscs in the Tarut Island, Arabian Gulf, Saudi Arabia. *J. Afr. Earth Sci.* **2016**, *113*, 65–72. [CrossRef]
61. Neisi, A.; Goudarzi, G.; Akbar Babaei, A.; Vosoughi, M.; Hashemzadeh, H.; Naimabadi, A.; Mohammadi, M.J.; Hashemzadeh, B. Study of heavy metal levels in indoor dust and their health risk assessment in children of Ahvaz city, Iran. *Toxin Rev.* **2016**, *35*, 16–23. [CrossRef]
62. Naimabadi, A.; Gholami, A.; Ramezani, A.M. Determination of heavy metals and health risk assessment in indoor dust from different functional areas in Neyshabur, Iran. *Indoor Built Environ.* **2021**, *30*, 1781–1795. [CrossRef]
63. Hassan, S.K.M. Metal concentrations and distribution in the household, stairs and entryway dust of some Egyptian homes. *Atmos. Environ.* **2012**, *54*, 207–215. [CrossRef]
64. Abdulraheem, M.O.; Adeniran, J.A.; Ameen, H.A.; Odediran, E.T.; Yusuf, M.-N.O.; Abdulraheem, K.A. Source identification and health risk assessments of heavy metals in indoor dusts of Ilorin, North central Nigeria. *J. Environ. Health Sci. Eng.* **2022**, *20*, 315–330. [CrossRef] [PubMed]
65. Cheng, Z.; Chen, L.-J.; Li, H.-H.; Lin, J.-Q.; Yang, Z.-B.; Yang, Y.-X.; Xu, X.-X.; Xian, J.-R.; Shao, J.-R.; Zhu, X.-M. Characteristics and health risk assessment of heavy metals exposure via household dust from urban area in Chengdu, China. *Sci. Total Environ.* **2018**, *619*, 621–629. [CrossRef] [PubMed]
66. Tong, S.T.; Lam, K.C. Home sweet home? A case study of household dust contamination in Hong Kong. *Sci. Total Environ.* **2000**, *256*, 115–123. [CrossRef]
67. Lisiewicz, M.; Heimburger, R.; Golimowski, J. Granulometry and the content of toxic and potentially toxic elements in vacuum-cleaner collected, indoor dusts of the city of Warsaw. *Sci. Total Environ.* **2000**, *263*, 69–78. [CrossRef]

68. Turner, A.; Simmonds, L. Elemental concentrations and metal bioaccessibility in UK household dust. *Sci. Total Environ.* **2006**, *371*, 74–81. [CrossRef] [PubMed]
69. Rasmussen, P.; Subramanian, K.; Jessiman, B. A multi-element profile of house dust in relation to exterior dust and soils in the city of Ottawa, Canada. *Sci. Total Environ.* **2001**, *267*, 125–140. [CrossRef]
70. Yoshinaga, J.; Yamasaki, K.; Yonemura, A.; Ishibashi, Y.; Kaido, T.; Mizuno, K.; Takagi, M.; Tanaka, A. Lead and other elements in house dust of Japanese residences–Source of lead and health risks due to metal exposure. *Environ. Pollut.* **2014**, *189*, 223–228. [CrossRef]
71. Kim, N.; Fergusson, J. Concentrations and sources of cadmium, copper, lead and zinc in house dust in Christchurch, New Zealand. *Sci. Total Environ.* **1993**, *138*, 1–21. [CrossRef]
72. Chattopadhyay, G.; Lin, K.C.-P.; Feitz, A.J. Household dust metal levels in the Sydney metropolitan area. *Environ. Res.* **2003**, *93*, 301–307. [CrossRef]
73. Yao, Q.; Wang, X.; Jian, H.; Chen, H.; Yu, Z. Characterization of the particle size fraction associated with heavy metals in suspended sediments of the Yellow River. *Int. J. Environ. Res. Public Health* **2015**, *12*, 6725–6744. [CrossRef]
74. Baccarelli, A.; Martinelli, I.; Pegoraro, V.; Melly, S.; Grillo, P.; Zanobetti, A.; Hou, L.; Bertazzi, P.A.; Mannucci, P.M.; Schwartz, J. Living near major traffic roads and risk of deep vein thrombosis. *Circulation* **2009**, *119*, 3118–3124. [CrossRef] [PubMed]
75. Wang, Z.; Lu, Q.-C.; He, H.-D.; Wang, D.; Gao, Y.; Peng, Z.-R. Investigation of the spatiotemporal variation and influencing factors on fine particulate matter and carbon monoxide concentrations near a road intersection. *Front. Earth Sci.* **2017**, *11*, 63–75. [CrossRef]
76. Hussain, Z.; Khan, M.S.; Kundi, K.; Alaf, K.; Ullah, Y. Assessment of integrated indoor environmental air quality parameters in selected church buildings of Faisalabad city: A statistical based comparative study. *Sci. Rev. Eng. Environ. Sci.* **2021**, *30*, 134–147. [CrossRef]
77. Basaran, B.; Yılmaz Civan, M. Investigating of primary components and source apportionment of persistent organic pollutants of indoor dust. *Int. J. Environ. Sci. Technol.* **2021**, *18*, 2145–2160. [CrossRef]
78. Arif, M.; Kumar, R.; Kumar, R.; Zusman, E.; Singh, R.P.; Gupta, A. Assessment of indoor & outdoor black carbon emissions in rural areas of Indo-Gangetic Plain: Seasonal characteristics, source apportionment and radiative forcing. *Atmos. Environ.* **2018**, *191*, 227–240. [CrossRef]
79. Du, W.; Wang, G. Indoor air pollution was nonnegligible during COVID-19 lockdown. *Aerosol Air Qual. Res.* **2020**, *20*, 1851–1855. [CrossRef]
80. Ezani, E.; Brimblecombe, P.; Asha'ari, Z.H.; Fazil, A.A.; Ismail, S.N.S.; Ramly, Z.T.A.; Khan, M.F. Indoor and outdoor exposure to $PM_{2.5}$ during COVID-19 lockdown in suburban Malaysia. *Aerosol Air Qual. Res.* **2021**, *21*, 200476. [CrossRef]
81. Riley, W.J.; McKone, T.E.; Lai, A.C.; Nazaroff, W.W. Indoor particulate matter of outdoor origin: Importance of size-dependent removal mechanisms. *Environ. Sci. Technol.* **2002**, *36*, 200–207. [CrossRef]
82. Tian, L.; Zhang, G.; Lin, Y.; Yu, J.; Zhou, J.; Zhang, Q. Mathematical model of particle penetration through smooth/rough building envelop leakages. *Build. Environ.* **2009**, *44*, 1144–1149. [CrossRef]
83. Liu, E.; Yan, T.; Birch, G.; Zhu, Y. Pollution and health risk of potentially toxic metals in urban road dust in Nanjing, a mega-city of China. *Sci. Total Environ.* **2014**, *476*, 522–531. [CrossRef]
84. Behrooz, R.D.; Kaskaoutis, D.; Grivas, G.; Mihalopoulos, N. Human health risk assessment for toxic elements in the extreme ambient dust conditions observed in Sistan, Iran. *Chemosphere* **2021**, *262*, 127835. [CrossRef] [PubMed]
85. Tashakor, M.; Modabberi, S. Human Health Risks Associated with Potentially Harmful Elements from Urban Soils of Hamedan City, Iran. *Pollution* **2021**, *7*, 709–722. [CrossRef]
86. Shi, G.; Chen, Z.; Bi, C.; Wang, L.; Teng, J.; Li, Y.; Xu, S. A comparative study of health risk of potentially toxic metals in urban and suburban road dust in the most populated city of China. *Atmos. Environ.* **2011**, *45*, 764–771. [CrossRef]
87. Liu, L.; Zhang, X.; Zhong, T. Pollution and health risk assessment of heavy metals in urban soil in China. *Hum. Ecol. Risk Assess. Int. J.* **2016**, *22*, 424–434. [CrossRef]
88. Layton, D.W.; Beamer, P.I. Migration of contaminated soil and airborne particulates to indoor dust. *Environ. Sci. Technol.* **2009**, *43*, 8199–8205. [CrossRef]
89. Roberts, J.W.; Wallace, L.A.; Camann, D.E.; Dickey, P.; Gilbert, S.G.; Lewis, R.G.; Takaro, T.K. Monitoring and reducing exposure of infants to pollutants in house dust. In *Reviews of Environmental Contamination and Toxicology Vol 201*; Springer: Boston, MA, USA, 2009; pp. 1–39. [CrossRef]
90. de Burbure, C.; Buchet, J.-P.; Leroyer, A.; Nisse, C.; Haguenoer, J.-M.; Mutti, A.; Smerhovský, Z.; Cikrt, M.; Trzcinka-Ochocka, M.; Razniewska, G.; et al. Renal and neurologic effects of cadmium, lead, mercury, and arsenic in children: Evidence of early effects and multiple interactions at environmental exposure levels. *Environ. Health Perspect.* **2006**, *114*, 584–590. [CrossRef]
91. Rangel-Mendez, J.A.; Arcega-Cabrera, F.E.; Fargher, L.F.; Moo-Puc, R.E. Mercury levels assessment and its relationship with oxidative stress biomarkers in children from three localities in Yucatan, Mexico. *Sci. Total Environ.* **2016**, *543*, 187–196. [CrossRef]
92. Faustman, E.M.; Silbernagel, S.M.; Fenske, R.A.; Burbacher, T.M.; Ponce, R.A. Mechanisms underlying Children's susceptibility to environmental toxicants. *Environ. Health Perspect.* **2000**, *108*, 13–21. [CrossRef]
93. Gao, Y.; Ji, H. Microscopic morphology and seasonal variation of health effect arising from heavy metals in $PM_{2.5}$ and $PM_{10}$: One-year measurement in a densely populated area of urban Beijing. *Atmos. Res.* **2018**, *212*, 213–226. [CrossRef]

94. Huang, H.; Lin, C.; Yu, R.; Yan, Y.; Hu, G.; Li, H. Contamination assessment, source apportionment and health risk assessment of heavy metals in paddy soils of Jiulong River Basin, Southeast China. *RSC Adv.* **2019**, *9*, 14736–14744. [CrossRef]
95. Guan, Q.; Wang, F.; Xu, C.; Pan, N.; Lin, J.; Zhao, R.; Yang, Y.; Luo, H. Source apportionment of heavy metals in agricultural soil based on PMF: A case study in Hexi Corridor, northwest China. *Chemosphere* **2018**, *193*, 189–197. [CrossRef] [PubMed]
96. Zhao, R.; Guan, Q.; Luo, H.; Lin, J.; Yang, L.; Wang, F.; Pan, N.; Yang, Y. Fuzzy synthetic evaluation and health risk assessment quantification of heavy metals in Zhangye agricultural soil from the perspective of sources. *Sci. Total Environ.* **2019**, *697*, 134126. [CrossRef] [PubMed]
97. Sijko, M.; Kozłowska, L. Influence of Dietary Compounds on Arsenic Metabolism and Toxicity. Part I—Animal Model Studies. *Toxics* **2021**, *9*, 258. [CrossRef] [PubMed]
98. Ferreira-Baptista, L.; De Miguel, E. Geochemistry and risk assessment of street dust in Luanda, Angola: A tropical urban environment. *Atmos. Environ.* **2005**, *39*, 4501–4512. [CrossRef]
99. Satarug, S.; Gobe, C.G.; Vesey, A.D.; Phelps, K.R. Cadmium and lead exposure, nephrotoxicity, and mortality. *Toxics* **2020**, *8*, 86. [CrossRef]
100. Silvera, S.A.N.; Rohan, T.E. Trace elements and cancer risk: A review of the epidemiologic evidence. *Cancer Causes Control* **2007**, *18*, 7–27. [CrossRef]
101. Pauraitė, J.; Garbarienė, I.; Minderytė, A.; Dudoitis, V.; Mainelis, G.; Davulienė, L.; Uogintė, I.; Plauškaitė, K.; Byčenkienė, S. Effect of spring grass fires on indoor air quality in air-conditioned office building. *Lith. J. Phys.* **2021**, *61*, 191–204. [CrossRef]
102. De Gennaro, G.; Farella, G.; Marzocca, A.; Mazzone, A.; Tutino, M. Indoor and outdoor monitoring of volatile organic compounds in school buildings: Indicators based on health risk assessment to single out critical issues. *Int. J. Environ. Res. Public Health* **2013**, *10*, 6273–6291. [CrossRef]
103. Razali, N.Y.Y.; Latif, M.T.; Dominick, D.; Mohamad, N.; Sulaiman, F.R.; Srithawirat, T. Concentration of particulate matter, CO and $CO_2$ in selected schools in Malaysia. *Build. Environ.* **2015**, *87*, 108–116. [CrossRef]
104. Hamzeh, N.H.; Karami, S.; Kaskaoutis, D.G.; Tegen, I.; Moradi, M.; Opp, C. Atmospheric dynamics and numerical simulations of six frontal dust storms in the Middle East region. *Atmosphere* **2021**, *12*, 125. [CrossRef]
105. Gopalan, H.N. Environmental health in developing countries: An overview of the problems and capacities. *Environ. Health Perspect.* **2003**, *111*, A446–A447. [CrossRef] [PubMed]
106. Burstein, R.; Henry, N.J.; Collison, M.L.; Marczak, L.B.; Sligar, A.; Watson, S.; Marquez, N.; Abbasalizad-Farhangi, M.; Abbasi, M.; Abd-Allah, F.; et al. Mapping 123 million neonatal, infant and child deaths between 2000 and 2017. *Nature* **2019**, *574*, 353–358. [CrossRef] [PubMed]
107. Yu, B.; Hu, Z.; Liu, M.; Yang, H.; Kong, Q.; Liu, Y. Review of research on air-conditioning systems and indoor air quality control for human health. *Int. J. Refrig.* **2009**, *32*, 3–20. [CrossRef]
108. Sundell, J.; Levin, H.; Nazaroff, W.W.; Cain, W.S.; Fisk, W.J.; Grimsrud, D.T.; Gyntelberg, F.; Li, Y.; Persily, A.; Pickering, A.; et al. Ventilation rates and health: Multidisciplinary review of the scientific literature. *Indoor Air* **2011**, *21*, 191–204. [CrossRef]
109. Du, L.; Leivo, V.; Prasauskas, T.; Täubel, M.; Martuzevicius, D.; Haverinen-Shaughnessy, U. Effects of energy retrofits on Indoor Air Quality in multifamily buildings. *Indoor Air* **2019**, *29*, 686–697. [CrossRef]
110. Spinazzè, A.; Campagnolo, D.; Cattaneo, A.; Urso, P.; Sakellaris, I.A.; Saraga, D.E.; Mandin, C.; Canha, N.; Mabilia, R.; Perreca, E.; et al. Indoor gaseous air pollutants determinants in office buildings—The OFFICAIR project. *Indoor Air* **2020**, *30*, 76–87. [CrossRef]
111. Bluyssen, P.M.; Ortiz, M.; Zhang, D. The effect of a mobile HEPA filter system on 'infectious' aerosols, sound and air velocity in the SenseLab. *Build. Environ.* **2021**, *188*, 107475. [CrossRef]
112. Stavroulas, I.; Grivas, G.; Michalopoulos, P.; Liakakou, E.; Bougiatioti, A.; Kalkavouras, P.; Fameli, K.M.; Hatzianastassiou, N.; Mihalopoulos, N.; Gerasopoulos, E. Field evaluation of low-cost PM sensors (Purple Air PA-II) under variable urban air quality conditions, in Greece. *Atmosphere* **2020**, *11*, 926. [CrossRef]
113. Spinelle, L.; Gerboles, M.; Kok, G.; Persijn, S.; Sauerwald, T. Review of portable and low-cost sensors for the ambient air monitoring of benzene and other volatile organic compounds. *Sensors* **2017**, *17*, 1520. [CrossRef]

International Journal of
*Environmental Research and Public Health*

MDPI

*Review*

# Impact of Climate Change on Indoor Air Quality: A Review

**Aya Mansouri [1,2,*], Wenjuan Wei [1], Jean-Marie Alessandrini [1], Corinne Mandin [1] and Patrice Blondeau [2]**

[1] Scientific and Technical Centre for Building (CSTB), Health and Comfort Department, 84 Avenue Jean Jaurès, 77447 Marne-la-Vallée, France
[2] Laboratoire des Sciences de l'Ingénieur pour l'Environnement (LaSIE), UMR CNRS 7356, La Rochelle University, 17042 La Rochelle, France
* Correspondence: aya.mansouri@cstb.fr

**Abstract:** Climate change can affect the indoor environment due to heat and mass transfers between indoor and outdoor environments. To mitigate climate change impacts and adapt buildings to the changing environment, changes in building characteristics and occupants' behavior may occur. To characterize the effects of climate change on indoor air quality (IAQ), the present review focused on four aspects: (1) experimental and modeling studies that relate IAQ to future environmental conditions, (2) evolution of indoor and outdoor air concentrations in the coming years with regard to temperature rise, (3) climate change mitigation and adaptation actions in the building sector, and (4) evolution of human behavior in the context of climate change. In the indoor environment, experimental and modeling studies on indoor air pollutants highlighted a combined effect of temperature and relative humidity on pollutant emissions from indoor sources. Five IAQ models developed for future climate data were identified in the literature. In the outdoor environment, the increasing ambient temperature may lead directly or indirectly to changes in ozone, particle, nitrogen oxides, and volatile organic compound concentrations in some regions of the world depending on the assumptions made about temperature evolution, anthropogenic emissions, and regional regulation. Infiltration into buildings of outdoor air pollutants is governed by many factors, including temperature difference between indoors and outdoors, and might increase in the years to come during summer and decrease during other seasons. On the other hand, building codes in some countries require a higher airtightness for new and retrofitted buildings. The building adaptation actions include the reinforcement of insulation, implementation of new materials and smart building technologies, and a more systematic and possibly longer use of air conditioning systems in summer compared to nowadays. Moreover, warmer winters, springs, and autumns may induce an increasing duration of open windows in these seasons, while the use of air conditioning in summer may reduce the duration of open windows.

**Keywords:** global warming; temperature; humidity; air pollutants; IAQ

check for updates

**Citation:** Mansouri, A.; Wei, W.; Alessandrini, J.-M.; Mandin, C.; Blondeau, P. Impact of Climate Change on Indoor Air Quality: A Review. *Int. J. Environ. Res. Public Health* **2022**, *19*, 15616. https://doi.org/10.3390/ijerph192315616

Academic Editor: Paul B. Tchounwou

Received: 10 October 2022
Accepted: 20 November 2022
Published: 24 November 2022



## 1. Introduction

The mean global surface temperature in 2020 was 1.02 °C warmer than the baseline 1951–1980 mean [1]. This temperature rise has several direct health effects on humans, such as risks of hyperthermia due to exposure to high temperatures, and indirect health effects due to bad air quality [2]. Vardoulakis et al. [2] indicated that the increase in temperature may lead to higher indoor concentrations of airborne pollutants causing higher risks of allergy, cancer, and endocrine disruption.

Nazaroff [3] classified the factors influencing indoor air quality (IAQ) in response to climate change into three categories: (1) factors related to pollutants such as the transfer of outdoor pollutants into the indoor environment, the pollutant emission from indoor materials and products, and pollutant partitioning between the gas and the adsorbed/absorbed phases; (2) factors related to the building properties, such as insulation, materials, heating and cooling systems, and availability of indoor air cleaners; (3) factors related to occupants' activities and behaviors.

Regarding pollutant-related factors, outdoor hygrothermal conditions can influence temperature and relative humidity indoors due to heat transfer through the building envelope. Changes in indoor thermal conditions can affect the mass transfer parameters, pollutant emissions from building materials, chemical reactivity, and pollutant partitioning between the gas and the adsorbed/absorbed phases (in materials, settled dust, airborne particles) [4]. Concerning building-related factors, energy issues and adaptation to climate change have already driven changes in the way of construction and designing heating, ventilation, and air conditioning (HVAC) systems in buildings [5]. These include changes in insulation materials, solar protection, and wider use of air conditioning systems to avoid overheating and promote thermal comfort, among others [5]. Yang et al. [6] found that the cooling demand would increase under future climate scenarios, and consequently the use of cooling systems. Outdoor-originated pollutants, such as polycyclic aromatic hydrocarbons (PAHs), benzene, and ozone, enter the indoor environment via the ventilation system, windows, and/or air infiltrations through cracks, as highlighted by Cheng et al. [7]. The indoor/outdoor pollutants transport can vary because of changes in building characteristics. The rise in outdoor temperatures may also change human behaviors. Weschler [8] showed that to handle hot weather, occupants of air-conditioned homes tended to operate their air conditioning systems and close windows rather than open windows and operate fans.

To address the influencing factors of IAQ facing climate change, the present work aimed to conduct a literature review on the pollutant-, building-, and human-related factors, with a specific focus on modeling studies, which were never reviewed before. The review aimed to retrieve answers from the literature to the four following questions: (1) How did experimental and theoretical work characterize IAQ under future environmental conditions? (2) How would indoor and outdoor air concentrations evolve in the coming years with regard to climate change? (3) How may building characteristics evolve to mitigate the effects of climate change and adapt to the future climate, and how will these changes affect IAQ? (4) How may human behavior evolve in the context of climate change and influence IAQ?

The impact on IAQ of extreme events generated by climate change, such as floods (development of mold in flooded buildings), hurricanes (carbon monoxide poisoning due to the temporary use of portable generators in the absence of electricity), or wildfires (huge increase in outdoor air particle concentrations) is also important [9] but is not addressed in this review. In a first attempt to model IAQ under global warming conditions, extreme events are not prioritized, as their predictions remain a challenge to date.

## 2. Materials and Methods

The literature review was conducted with ScienceDirect and Google Scholar search engines using the following keywords: ("climate change") AND ("building") AND (("air temperature") OR ("relative humidity") OR ("volatile organic compound") OR ("aldehyde") OR ("particle") OR ("SVOC") OR ("indoor air quality")), without year limitation. A total of 615 results were obtained. In addition, 6 papers identified from grey literature, including international guidelines and regulations for indoor air and health, were considered. After duplicate removal, 618 articles were screened, among which 432 were excluded by title screening because they were not relevant to the topic of the review. Finally, 186 articles were investigated in detail. After reading the full text, 61 articles were included in this paper (Figure 1).

The 61 selected articles were published between 1998 and 2022. Figure 2 shows the distribution of these articles over the total period. A growing interest can clearly be noticed. The studies were conducted in Australia, Chile, China, France, Germany, Japan, Malaysia, South Korea, Spain, Taiwan, the UK, and the US. Pollutants that were studied experimentally or numerically included volatile organic compounds (VOCs), semi-volatile organic compounds (SVOCs), ozone, nitrogen dioxide ($NO_2$), radon, and airborne particles. Among the 61 articles, 15 addressed the experimental and theoretical work that characterized IAQ under future environmental conditions, 15 addressed the indoor/outdoor air concentration evolution with regard to climate change, 19 addressed the building characteristics evolution for climate change mitigation or adaptation, and 14 addressed the human behavior evolution in the context of climate change.

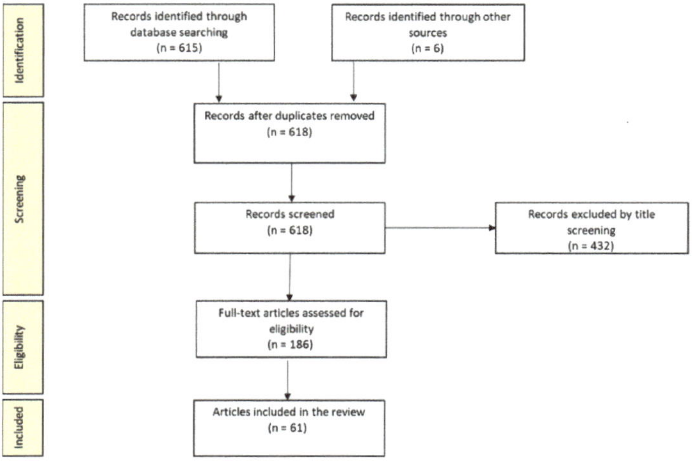


**Figure 1.** Review flow chart.

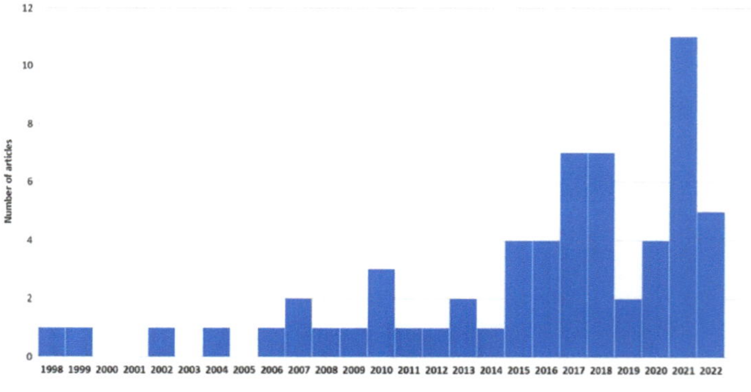


**Figure 2.** Evolution of the number of related articles with years.

## 3. Results and Discussion

### 3.1. Models to Predict Indoor Pollutant Concentrations in the Context of Climate Change

Five models that were developed or used to predict IAQ under future climate conditions were identified. The corresponding equations and scope of application of each model are summarized in Table 1. A mechanistic model was developed by Chang et al. [10] to assess climate change impacts on indoor air concentrations of chemical pollutants. The model only applies to VOCs. It was used to predict the VOC concentrations in Korean houses for the period of 2011–2100. This model calculates indoor VOC concentrations with a 1 min time step based on inputs including the meteorological data, the room and window sizes, the outdoor VOC concentrations, the indoor VOC concentrations in the adjoining room, the building crack sizes, the air handling system, the recirculation filter characteristics, and the chemical properties of the target pollutants. Temperature-dependent parameters such as the discharge coefficient for window opening, the pollutant diffusion coefficients in indoor source materials, the air density, and the pollutant vapor pressure, as well as the effect of the indoor and outdoor temperatures on the window opening patterns and the heat transfer through the building envelope (and thus on indoor emissions and air transfer), were considered as the main inputs. The seasonal duration of open windows and the seasonal average of formaldehyde concentrations were predicted in South Korea for three periods, 30 years each, from 2011 to 2100, under three VOC emission scenarios: (1) no emission from indoor sources, (2) low and continuous emissions from vinyl flooring, and (3) high and periodic emissions to designate cooking activities. The first scenario is not realistic, but it allows us to identify the trend in concentration variations of pollutants originated outdoors. Occupants were not considered as a source in any of the three scenarios. The predicted formaldehyde concentrations showed a slight increase in the annual average concentration by approximately 12% in the 2071–2100 period due to the increase in its outdoor concentrations, assuming no emission from indoor sources or low emissions from vinyl flooring (scenarios 1 and 2). In the case of strong indoor emissions, the annual formaldehyde concentration would decrease by 3% due to pollutant removal by natural ventilation. On a seasonal scale, the simulation under the high and periodic emission scenario (scenario 3) showed that the average formaldehyde concentration would increase in summer and decrease in the other seasons, because of a lower air change and a higher air change in the future compared to buildings of nowadays.

Ilacqua et al. [11] used a steady-state single-compartment model to predict the indoor air concentration of pollutants under future environmental conditions. The model only considers pollutant transports from outdoors to indoors by infiltrations through the building envelope. The infiltration in the paper of Ilacqua et al. [11] referred to the uncontrolled air flow through cracks and leaks in the building envelope and did not consider air change through mechanical ventilations or open windows. The infiltration rates were calculated based on the stack effect caused by the temperature difference between indoors and outdoors and the wind effect. The investigated pollutants were radon, $PM_{2.5}$, ultrafine particles (UFP), carbonyls, ozone, $NO_2$, and nitric acid ($HNO_3$). The simulations did not consider any change in the building envelope's airtightness in the future. The results suggested that the monthly mean infiltration rates in the 2040–2070 period would decrease in some American cities compared to the 1970–2000 period, except in summertime. Therefore, exposures to pollutants of outdoor origin would decrease while exposures to pollutants emitted by indoor sources would increase. The study showed that changes in buildings' air infiltration in the future affected occupants' pollutant exposure level. The decrease in infiltrations by 5% would increase the exposure to pollutants of indoor origin by 2 to 23% and decrease the exposure to pollutants of outdoor origin by 2 to 18%.

Salthammer et al. [12] developed a dynamic model for a single compartment to calculate indoor concentrations of ozone and particles in the 2040 in eight German cities located in different geographic zones. The model focuses on only outdoor sources and does not consider ozone and particle emissions from indoor sources. The indoor ozone concentration was calculated for a summer day with peak outdoor ozone concentrations. The diurnal indoor concentration did not exceed 100 $\mu g \cdot m^{-3}$, the WHO air quality guideline value [13], with an air exchange rate ranging from 0.5 to 3 $h^{-1}$ depending on the period of the day. The model also predicted a decrease in indoor $PM_{2.5}$ and $PM_{10}$ concentrations, primarily due to lower outdoor particle concentrations in future emission scenarios.

In 2022, Salthammer et al. [14] developed a comprehensive modeling framework considering two compartments (gas and particulate phases) to better estimate the effect of climate change on IAQ. The target chemical pollutants were 12 VOCs and SVOCs: limonene, isoprene, formaldehyde, n-butyl acetate, n-decane, acetic acid, acetaldehyde, toluene, benzophenone, triethyl phosphate (TEP), 2,2,4-trimethyl-1,3-pentanediol diisobutyrate (TXIB), and di-1-ethylhexyl adipate (DEHA). The model includes five submodels to estimate the heat and moisture transport in buildings, indoor emissions, physicochemical processes, mold growth, and human exposure, respectively. The inputs of the model are the thermal boundary conditions, building materials, and occupants' activities. The modeling framework has not been applied to future climates yet but the submodels were tested on existing data and past meteorological conditions. The thermal submodel was validated using indoor temperature and relative measured humidity in July and August 2020 in a house in Braunschweig, Germany. The outdoor temperature, relative humidity, and ozone concentration used in the simulations to validate the model were obtained from a monitoring station in the same city. The equations related to the chemical emissions and reactions were tested for limonene on 14 August 2020, and the emission rate of limonene was strongly and positively associated with indoor temperature.

Fazli and Stephens [15] developed a dynamic single-compartment model to estimate indoor air concentrations of $PM_{2.5}$, UFP, $NO_2$, ozone, VOCs, and aldehydes in 2050. The specificity of this model is that it considers changes in housing stocks, including the construction of millions of houses in the US over the 2010–2050 period, the fact that some existing houses will be renovated or demolished, and the population demography changes. The home construction and renovation are modeled as a change in their envelope, i.e., higher airtightness, and heating/cooling systems. The model takes into consideration infiltration, ventilation, deposition, reaction, and pollutant removal by HVAC filters. The results showed that the indoor annual mean concentrations of $PM_{2.5}$, UFP, and $NO_2$ would decrease due to the drop from both indoor (substitution of gas stoves by electric stoves) and outdoor sources (less infiltration and window opening). The indoor concentration of ozone originated outdoors would increase, thus promoting indoor chemical reactions. The indoor annual mean concentrations of formaldehyde, acetaldehyde, acrolein, 1,3-butadiene, benzene, and p-dichlorobenzene would increase due to the decrease in infiltration and duration of open windows.

**Table 1.** Models developed for the assessment of the influence of climate change on indoor air quality. (Note: The definitions of the variables are provided in the nomenclature.).

| References | Notes |
|---|---|
| Chang et al. [10] | Equation: $V\frac{d[C_V]}{dt} = (Q_{mv}C_{V_0} + Q_{ra}C_V)(1-e) + (Q_{cf} + Q_{window})C_{V_0} + Q_{ADJ_0}C_{V_{ADJ_0}} - [(Q_{mv} + Q_{ra} + Q_{cf} + Q_{ADJ_0} + Q_{window})C_V] - (\sum_i E_{s,i}A_i + Vk_VC_V + Q_{rcl}e_{rcl}C_V) + \sum_j E_{V,j}$; in the gas phase.<br>$V\frac{d[M(d)C_P(d)]}{dt} = [Q_{mv}M_0(d)C_{P_0}(d) + Q_{ra}M(d)C_P(d)][1-e(d)] + Q_{cf}M_0(d)C_{P_0}(d)P + Q_{window}M_0(d)C_{P_0}(d) + Q_{ADJ_0}M_{ADJ_0}(d)C_{P_{ADJ_0}}(d) - (Q_{mv} + Q_{ra} + Q_{cf} + Q_{ADJ_0} + Q_{window})M(d)C_P(d) - [\sum_i A_iv_{dep}M(d)C_P(d) + Q_{rcl}e_{rcl}(d)M(d)C_P(d)] + [\sum_k M_{pk}(d)C_{pk}(d) + R_{sus}M_f(d)C_f(d)A_{dtot}]$; in the PM phase.<br>Scope of application: VOCs and SVOCs.<br>Main features: Two-compartment dynamic model, suggesting that the effect of temperature on mass transfer parameters, pollutant emission from materials, and window opening behavior are the key factors to be considered in the context of climate change. |
| Ilacqua et al. [11] | Equation: $C_{in} = \frac{P\lambda}{\lambda+d_s}C_{out} + \frac{S}{V(\lambda+d_s)}$<br>Scope of application: radon, $PM_{2.5}$, carbonyls, ozone, $NO_2$, $HNO_3$, UFP<br>Main features: Single-compartment steady-state model considering infiltration as the only airflow path from outdoors to indoors. The model also considers the emission from indoor sources. However, it does not consider possible changes in the airtightness of the building envelopes and the interactions between vapor and solid phases (particles and surfaces). |
| Salthammer et al. [12] | Equation: $\frac{dC_{in}}{dt} = \lambda C_{out} - \lambda C_{in} - \lambda_d C_{in}$ for ozone<br>$\frac{dC_{in}}{dt} = \lambda PC_{out} - \lambda C_{in} - \lambda_d C_{in}$ for particles<br>Scope of application: ozone, particles<br>Main features: Single-compartment dynamic model. The model does not consider the evolution of the building envelope, window opening, or mechanical ventilation systems and the emissions from indoor sources. |
| Salthammer et al. [14] | Equation: $\frac{dC_{in}}{dt} = P\lambda C_{out} - C_{in} + \sum_{i=1}^{n}\frac{E_{s,i}A_i}{V} + \sum_{j=1}^{n}\frac{E_{V,j}}{V} \pm J_{coag} \pm \xi\Psi_{gas} \pm J_{SVOC}$<br>Scope of application: VOCs and SVOCs<br>Main features: Two-compartment dynamic model, combining 5 submodels to assess the building physics, emissions from indoor sources, reactions, particle/gas partitioning, mold growth, and human exposure. |
| Fazli and Stephens [15]<br>Fazli et al. [16] | Equation: $\frac{dC_{in}}{dt} = ((P\lambda_{inf} + \lambda_{nat})C_{out} + \frac{S}{V} - (\lambda_{inf} + \lambda_{nat}))C_{in} - \frac{Q_{exhaust}}{V}C_{in} - \beta C_{in} - f_{filt}\frac{\eta_{filt}Q_{filt}}{V}C_{in} - kC_{terp}C_{in}$<br>Scope of application: $PM_{2.5}$, UFP, $NO_2$, ozone, VOCs, aldehydes<br>Main features: Single-compartment dynamic model, that considers changes in building envelope airtightness, use of HVAC systems, and population demography changes. |

### 3.2. Pollutant-Related Factors

#### 3.2.1. Effect of Indoor Temperature and Humidity on Pollutant Emissions, Transport, and Partitioning Indoors

One of the effects of climate change is the increase in temperature and relative humidity outdoors and consequently indoors, with more frequent overheating periods. Hence, it is important to characterize the effect of temperature and relative humidity on the pollutant emission from indoor materials, as well as on other pollutant transport processes. Fang et al. [17] studied experimentally the impact of indoor temperature and humidity on the emission of total volatile organic compounds (TVOCs) from 5 types of building materials: PVC (polyvinyl chloride) flooring, loomed polyamide carpet, acrylic sealant, acrylic wall paint, and waterborne acrylic floor varnish. The emissions were measured for 9 combinations of temperature (18, 23, and 28 °C) and humidity (30, 50, and 70%). The results showed that for the wall paint and floor varnish, the one-week mean TVOC concentrations increased significantly by 1500 $\mu g \cdot m^{-3}$ when relative humidity increased from 30 to 70%; the one-week mean TVOC concentration emitted from the floor varnish also increased by 700 $\mu g \cdot m^{-3}$ when temperature increased from 18 to 28 °C. For the carpet, PVC flooring, and sealant, no clear impact of temperature and humidity on the emission of TVOCs was observed. Similar results on the same types of materials were reported by Wolkoff [18]: the effect of temperature and relative humidity is dependent on the building materials/products and the VOCs emitted. Based on 3675 air samples collected in two locations in China for 23 months, the combined effect of indoor temperature and humidity on indoor VOC concentrations was addressed in the model of Zhou et al. [19], where the estimated daily average concentration of a VOC ($mg \cdot m^{-3}$) in newly renovated residences is calculated as follows:

$$C = k_1 e^{n_1 T} \, k_2 e^{n_2 RH} + C_0 \tag{1}$$

where $T$ (°C) and $RH$ (%) are the room's daily average temperature and the relative humidity, respectively. $k_1$, $k_2$, $n_1$, and $n_2$ are constants related to the pollutant and other environmental conditions. $C_0$ ($mg \cdot m^{-3}$) is the concentration under initial temperature and relative humidity. As the equation indicates, the indoor VOC concentration is exponentially correlated with temperature and relative humidity. Zhou et al. [19] also showed that $k_1$, $k_2$, $n_1$, or $n_2$ were correlated with Henry's law constants (solubility in water) and molecule polarity for formaldehyde, butyl acetate, styrene, benzene, toluene, xylene, ethylbenzene, and n-undecane. Nevertheless, the equation has not been validated for long periods, e.g., for decades.

Xiong et al. [20] presented another model describing the combined effect of indoor temperature and relative humidity on emissions of formaldehyde and VOCs from building materials. The equation, theoretically developed from some physical and chemical equations, can be written as follows:

$$E = E_1 T^{0.75} e^{E_2 RH} - \frac{E_3}{T} \tag{2}$$

where $E$ is the steady-state emission rate ($\mu g \cdot m^{-2} \cdot h^{-1}$); $E_1$, $E_2$, and $E_3$ are positive constants related to the physical and chemical characteristics of the pollutant; $T$ (K) and $RH$ (%) are the indoor temperature and the relative humidity, respectively. This correlation was validated using experimental data from fiberboard and floor varnish emissions under different combinations of temperature and relative humidity. The constants were determined with an acceptable fitting degree.

The effect of temperature on VOC emissions is a widely discussed topic. Salthammer and Morrison [21] listed 36 articles, published between 1975 and 2021, on the temperature dependence of VOC emissions from different building materials. The review of the 36 articles covered various families of pollutants and compiled different methods to study the dependence of VOC emissions on indoor temperature.

The diffusion coefficient (diffusivity) of a pollutant from a surface to the air is an important factor primarily used to calculate the mass transfer coefficient. Guo [22] summarized 3 methods for the estimation of the diffusivity of VOCs in air, among them two are directly dependent on the temperature and are presented in the following equations:

$$D = \frac{10^{-3}T^{1.75}\sqrt{M_r}}{P_0\left(v_A^{\frac{1}{3}} + v_B^{\frac{1}{3}}\right)^2} \quad (3)$$

$$D = \frac{BT^{\frac{2}{3}}\sqrt{M_r}}{P_0\sigma_{AB}^2\gamma} \quad (4)$$

where $D$ is the diffusivity ($cm^2 \cdot s^{-1}$), $T$ is the temperature (K), $P_0$ is the atmospheric pressure (atm), $v_A$ and $v_B$ are the molar volumes for air and the studied compound ($cm^3 \cdot mol^{-1}$), $\sigma_{AB}$ is the characteristic length (Å), $\gamma$ is the collision integral for diffusion (dimensionless).

In Equations (3) and (4) $M_r$ and $B$ are defined as:

$M_r = \frac{M_a+M}{M_a M}$, $M_a$, and $M$ ($mol \cdot g^{-1}$) are the molecular weights of air and the studied compound.

$$B = 0.0217 - 0.0005M_r^{\frac{1}{2}}$$

The effect of temperature on the diffusion coefficient is quantified by applying the previous two equations. The results are shown in Table 2. A temperature increase of 10 °C can increase the diffusion coefficient by around 6%. This variation should not be neglected when calculating the mass transfer coefficient because it could have an important effect on the pollutant concentration in the gas phase.

**Table 2.** Effect of temperature on the diffusion coefficient.

| Change in Temperature (°C) | Increase in Diffusion Coefficient (%) | |
|---|---|---|
| | Equation (3) | Equation (4) |
| 15 to 25 | 6.2 | 2.3 |
| 25 to 35 | 6 | 2.2 |
| 35 to 45 | 5.8 | 2.2 |

Wei et al. [23] derived theoretically the equations of the particle/gas partition coefficient $K_p$ of some SVOCs as a function of temperature. Table 3 shows the values of $K_p$ of some phthalates at 20 °C and the percentage of variation of $K_p$ when the temperature rises from 20 °C to 30 °C. This percentage may reach 78%, which is an important variation for a realistic and possible temperature increase.

**Table 3.** Variation of $K_p$ of some phthalates with changing temperature.

| Phthalate | $K_p$ at 20 °C ($m^3 \cdot \mu g^{-1}$) | Variation (%) When $T$ Rises to 30 °C |
|---|---|---|
| DEHP | $4.3 \times 10^{-2}$ | −74 |
| DnBP | $3 \times 10^{-4}$ | −71 |
| DiBP | $2 \times 10^{-4}$ | −71 |
| DiNP | 34.082 | −78 |
| BBzP | $8 \times 10^{-4}$ | −69 |
| DEP | $7.6 \times 10^{-6}$ | −68 |
| DMP | $3.3 \times 10^{-6}$ | −62 |

Fadeyi [24] showed that ozone surface deposition velocity increased when relative humidity and temperature increased. For some materials such as concrete, the increase in ozone deposition velocity depended on the range of relative humidity; it appeared more clearly at humidity higher than 50% [24]. The ozone deposition velocity can increase by a factor of 17 when humidity increased from 50% to 90%, depending on the material type [25].

Salthammer and Morrison [21] highlighted the effect of temperature on the reaction rates of some pollutants with oxidants. They provided eight examples of indoor gas-phase reactions between ozone and hydroxyl as oxidants and nitrogen monoxide (NO), $NO_2$, limonene, α-pinene, and β-pinene, along with their first-order reaction rates at 25 °C and the percentages of variation of these rates when the temperature increased to 35 °C: the variation ranged from −5 to 30.6% for these eight reactions.

Beyond their own interest in the characterization of the impact of thermal conditions on pollutant emission, transport and partitioning, these relationships, when integrated into an IAQ model, will allow us to characterize the evolution of indoor pollutant concentrations following the evolution of indoor thermal conditions.

#### 3.2.2. Outdoor Concentration Evolution in the Future Climate Conditions

Outdoor air quality is an important determinant of IAQ since outdoor air is transferred into the building through ventilation and infiltration. Therefore, it is necessary to know how outdoor pollutant concentration will evolve under climate change conditions to anticipate the impact on IAQ.

Ozone is produced essentially outdoors [26] in the presence of ultraviolet light and precursors such as $NO_x$ [27]. High concentrations of ozone are recorded during heat waves [28] because high temperatures promote its formation [29]. Wang et al. [30] showed that, over East China, climate change alone, i.e., excluding the evolution of anthropogenic emissions, might be responsible for 8% of the total increase in annual-mean surface ozone until 2050. The same study [30] showed that in West China ozone concentration might decrease by 4% due to climate change. This ozone concentration difference between East and West China is because the West China region is less-industrialized and is therefore a low-$NO_x$ area [30]. The study by Hong et al. [31] in China showed that ozone concentration will increase in the urban regions in 2046–2050 by 16 $\mu g \cdot m^{-3}$, compared to the 2006–2010 years. In Europe, Meleux et al. [32] predicted the increase in ozone concentration under climate change conditions, i.e., with increased temperature and decreased cloudiness, considering current anthropogenic emissions, especially for the western and central regions. The number of days of 90 and 120 ppb ozone exceedance will double or triple in some European cities depending on the climate scenario [32]. Coelho et al. [33] predicted a decrease in ozone concentration (up to 30 $\mu g \cdot m^{-3}$) over Europe in 2031, compared to 2013, considering the effect of climate change and assuming constant anthropogenic emissions. Zhong et al. [34] studied the impact of climate change on ozone production from VOCs, methane, and carbon monoxide outdoors, and showed the effect of high temperatures on ozone formation. A higher temperature fosters photochemical reactions, VOC emission from vegetation, and evaporation, suggesting more humidity and OH radicals in the atmosphere, which leads to a higher outdoor ozone concentration [34].

By applying two climate models, Coelho et al. [33] predicted an increase in the annual mean $NO_2$ concentrations (up to 5 $\mu g \cdot m^{-3}$) in some European regions and a decrease in other European regions (up to 5 $\mu g \cdot m^{-3}$). Globally, atmospheric $NO_2$ concentrations are expected to increase in the 21st century despite the decreased cloudiness due to the increasing boundary layer height [33]. Giorgi and Meleux [35] showed that the $NO_x$ concentration should increase in most European regions in the 2071–2100 period by 0–1 ppb compared to the 1961–1990 period, under the high $CO_2$ emission scenario, due to lower mixing and dispersion over Europe and lower total deposition specifically in the capital cities, such as Paris, London, and Brussels.

Coelho et al. [33] predicted an increase in $PM_{2.5}$ and $PM_{10}$ concentrations (up to 30 $\mu g \cdot m^{-3}$) in Europe. In China, $PM_{2.5}$ concentration will increase in the urban regions in 2046–2050 by 8 $\mu g \cdot m^{-3}$ compared to the 2006–2010 years [31]. Particle concentration is expected to increase under the effect of the lower precipitation that reduces dispersion, dilution, and wet deposition of particles. Westervelt et al. [36] used a multiple linear regression (MLR) model to characterize the correlation between meteorological parameters and atmospheric $PM_{2.5}$ concentrations. The temperature and $PM_{2.5}$ concentrations were positively and significantly correlated. $PM_{2.5}$ concentration presented a negative correlation with wind speed due to the dilution effect. Other meteorological conditions such as precipitation, cloudiness, pressure, and relative humidity showed a negative poor correlation with $PM_{2.5}$ concentrations in most regions of the world. The decrease in heating demand due to warmer winters and the decrease in fossil fuel combustion could counterbalance the increase in $PM_{2.5}$ outdoor concentrations [9]. In addition, outdoor air pollution control regulation, reduction of sulfur dioxide emissions from coal-fired power plants due to climate change concerns, and the shift toward electric and hybrid vehicles could reduce particulate matter in outdoor air. On the other hand, outdoor particulate matter concentrations may also increase as a result of more frequent drought episodes leading to more windblown dust and wildfires [3].

Giorgi and Meleux [35] showed that global warming might lead to a higher concentration of isoprene (a biogenic VOC) in France, by 0–8 ppb from 1961–1990 to 2071–2100 period. The modeling results of Cao et al. [37] on isoprene showed an increasing global emission in the 21st century caused by a warming climate. The annual total isoprene emissions will rise by 100 to 250 million tons of carbon in 2100 compared to the 1850–1950 years.

The trends in the evolution of outdoor pollutant concentrations are summarized in Table 4. Overall, differences in the tendencies of variation of most of the outdoor concentrations can be noticed depending on the geographical zone, and the scenarios considered for their prediction.

**Table 4.** Predicted trends in outdoor pollutant concentrations under future climate conditions. ↑: increase, ↓: decrease, ↑ ↓: increase in some geographical zones and decrease in others.

| Pollutant | References | Tendency | Region |
|---|---|---|---|
| Ozone | Wang et al. [30] | ↑ ↓ | China |
| | Hong et al. [31] | ↑ | China |
| | Meleux et al. [32] | ↑ | Europe |
| | Coelho et al. [33] | ↓ | Europe |
| | Giorgi et Meleux [35] | ↑ | Europe |
| $NO_2$ | Coelho et al. [33] | ↑ ↓ | Europe |
| $NO_x$ | Giorgi et Meleux [35] | ↑ | Europe |
| Particles ($PM_{2.5}$ and $PM_{10}$) | Coelho et al. [33] | ↑ | Europe |
| $PM_{2.5}$ | Hong et al. [31] | ↑ | China |
| Isoprene | Giorgi et Meleux [35] | ↑ | Europe |
| | Cao et al. [37] | ↑ | World |

### 3.3. *Building-Related Factors*

#### 3.3.1. Climate Change Mitigation and Adaptation Actions

Some modifications in the insulation and construction materials can be made to mitigate the effect of climate change and adapt buildings. Substitution of concrete with wood in construction is a way to mitigate climate change by reducing carbon emissions [38,39]. However, the weakness of wood compared to concrete in terms of resistance to high humidity and fungi is still a major limitation of its use in construction [40]. Furthermore, wood products emit formaldehyde, terpenes, and BTEX (benzene, toluene, ethylbenzene, and xylene) [41,42]. The use of natural and bio-based insulating materials, e.g., cellulose, cork, straw bale, wood wool, and sheep wool, in insulation, has taken off in the last few years, besides the appearance of new materials [43]. Although straw bale houses are not VOC-free, they show lower VOC indoor concentrations than regular houses [44]. The integration of new insulation materials in buildings can impact indoor thermal conditions, and thus air infiltration and emissions of pollutants from indoor sources. Verichev et al. [45] presented the results of the energy simulation of an existing house in Chile and calculated the cost-effective optimum thickness of insulation materials for the period of 2020–2035. They found that the geographical area of usage of glass wool as insulation material can be doubled in 2020–2035 in comparison with 2006 in Chile. The different insulation materials do not emit the same quantity of pollutants, which can vary by two orders of magnitude [46,47].

The buildings will also evolve to counteract the high temperature indoors. Installation of shading devices could neutralize a proportion of the energy required for cooling in future years [48]. Techniques to mitigate the intensity of urban heat islands (UHI) and reduce both outdoor and indoor temperatures are also developed [49]. For instance, the use of green technologies, e.g., urban green spaces, green roofs, and green walls, can provide a mitigation potential of outdoor temperature in the range of 0.3–2.5 °C [49]. This range can be greater when these green technologies are mixed with other techniques such as the use of reflective materials and water technologies. A profound change in the built environment is foreseen, for which the influence on IAQ remains unknown.

#### 3.3.2. Effect of Outdoor Temperature on Air Infiltration

In many countries, building regulations require airtight envelopes to prevent heat loss by infiltration and improve the effectiveness of mechanical ventilation systems [2]. This increase in building airtightness might degrade IAQ if mechanical ventilation is not properly installed or maintained [50], but it may also lower pollutant transport from outdoors to indoors.

Lee et al. [51] studied the $PM_{2.5}$ infiltration, aiming to quantify the relationship between future changes in outdoor temperature and fine particle infiltration in the Greater Boston area. The study used the indoor-outdoor sulfur ratio as an indicator of $PM_{2.5}$ infiltration due to the absence of indoor sulfur sources. An increase in outdoor temperature of 2–3 °C in summer corresponds to up to 0.06 increase in the indoor-outdoor sulfur ratio. A similar result was obtained by Ilacqua et al. [11], who suggested that infiltration rates could increase by up to 25% in the summer months in the 2040–2070 period, compared to the reference period (1970–2000), assuming a constant indoor temperature of 24.9 °C. These infiltration rates were calculated using an equation from the Lawrence Berkeley National Laboratory model, taking into consideration the indoor-outdoor temperature difference and the wind effect [11]. However, based on their residential energy and indoor air quality model, Fazli et al. [16] found that infiltration factors ("the equilibrium proportion of particles remaining suspended on penetrating indoors" [52]) of $PM_{2.5}$, UFP, $NO_2$, and ozone in 2050 would be similar to those in 2010 because the effects of climate and building stoke changes neutralize each other. On the other hand, indoor-outdoor concentration ratios will be higher in 2050 for $PM_{2.5}$, ultrafine particles, and $NO_2$ in homes equipped with gas stoves compared to the same ratios in 2010 due to a lower natural ventilation rate.

### 3.4. *Human-Related Factors*

Human activities (e.g., cooking, smoking, etc.) and behavior (e.g., window opening) can significantly influence IAQ [53]. Changes in window opening can modify the air change rate and airflow velocities at indoor surfaces, which directly affect indoor pollutant concentrations. Under future climate conditions, window opening could increase from autumn to spring because of warmer weather and decrease in summer to protect from heat waves. Moreover, it is expected that the use of cooling systems will increase; such an increase has already been observed in the past years all over the world [54,55]. With the widespread use of cooling systems, the duration of open windows could drastically decrease. Predicting IAQ in the context of climate change requires being able to model window opening considering these different aspects.

According to Huang et al. [56], the air change rate is the most influencing factor on VOC emission rates from source materials: it is associated with the emission rates of 16 VOCs among 43 VOCs studied, while temperature and relative humidity are associated with 7 and 6 VOCs, respectively. The materials in the test residences were composite and solid wood for flooring, and paints and wood boards for walls. Window opening may become less effective in terms of thermal comfort under climate change conditions and other ventilation and air conditioning strategies may substitute it or become predominant [57].

Window opening is often modeled using stochastic/probabilistic models. For example, Rijal et al. [58] developed a stochastic model to predict occupant behavior in Japanese dwellings, considering cases of heated, cooled, and free-running dwellings. The model relates the probability of windows being open to indoor and outdoor temperatures based on adaptive thermal comfort. Andersen et al. suggested including other indoor parameters, such as indoor illuminance and wind speed, for the determination of window opening probability [59], and a multivariate regression method was proposed in the following form:

$$\log\frac{p}{1-p} = a + b_1x_1 + b_2x_2 + \ldots + b_nx_n + c_{12}x_1x_2 + c_{13}x_1x_3 + \ldots \quad (5)$$

where $p$ is the probability of window opening/closing, $a$ and $b_{1\text{-n}}$ are empirical coefficients, and $x_{1\text{-n}}$ stand for the driving factors for windows changing state. The investigated variables included indoor and outdoor temperature, indoor and outdoor relative humidity, indoor illuminance, wind speed, solar radiation, and sunshine hours. Equation (5) was further simplified by ignoring the interactions between variables. Andersen et al. [59] used the simplified regression equation, identified the most influential variables, and provided their coefficients based on measurements conducted in 15 houses in Denmark in 2008. These stochastic/probabilistic models were commonly developed under current climate conditions, although some have considered climate change.

Liu et al. [60] applied a statistical method to calculate the variation of the time fraction of open windows in China in 2050 compared to that in 2015. The results showed a decreasing time fraction in summer by up to −5% depending on the climate scenario due to the rising use of air conditioning systems, and an increasing time fraction by up to 2.8% in winter because of a warmer outdoor climate.

Chang et al. [10] considered in their model that the outdoor temperature controls the heating and cooling system to maintain a comfortable temperature indoors. The heating and cooling system was turned on if the outdoor temperature was below 10 °C or above 30 °C, and windows were closed. Between 10 °C and 30 °C outdoors, the indoor temperature was mainly controlled through window opening. The application of this algorithm to the 2071–2100 climate conditions in South Korea resulted in a increase in window opening duration by 200% in winter, 50% in spring, and 20% in autumn compared to the 1976–2005 years. However, window opening duration dropped by up to 55% in summer. A study conducted in China by Du et al. [61] on 10 houses with different characteristics (area, year of construction, occupancy, floor, etc.) showed that the window opening probability by human control was governed by the daily and yearly periods, the indoor-outdoor temperature difference and the occupant perception. Other determinants of window opening, with different degrees of dependence on climate change, such as wind speed, solar radiation, age and gender of occupants, smoking activities, orientation, and size of windows, were described in detail by Fabi et al. [62]. In summary, all the investigated articles suggest that the duration of window opening will decrease in summer and increase in other seasons. Consequently, the reduced ventilation rate will increase exposure to radon and pollutants emitted from indoor sources such as materials, occupants, and combustions [3].

## 4. Conclusions

The existing studies provide knowledge about most of the phenomena associated with climate change and the way they can influence indoor air quality. However, it is still difficult to get a quantitative evaluation of the combined and possibly antagonistic effects of climate change on indoor pollutant concentrations. Modeling is a suitable method to achieve this goal, and the literature review shows that IAQ modeling coupled with heat and airflow modeling is needed. Across the IAQ models that have been developed and used to predict indoor pollutant concentrations, those that are the most advanced in terms of modeling all physical and chemical processes on a wide range of pollutants, are not yet applied to future climate scenarios. Moreover, parameters have been disregarded in some models without prior analysis of their sensitivity to the output, which brings conclusions to caution.

Measurements of VOC emissions from materials suggest that the VOC emission rate can increase significantly under increased temperature, thus global warming and heat waves in summer may lead to higher indoor VOC source emissions in the years to come. Most studies described only qualitatively the impact of environmental, building, and human factors on indoor pollutant concentrations. Empirical laws and stochastic models of window opening were developed and could be used for further in-depth analysis of the influence of ventilation on IAQ in the context of climate change.

In summary, the influence of climate change on IAQ remains largely unknown and the evolution of many influencing factors is unpredictable, such as the technological development in the formulation and manufacturing of building materials or the evolution of building codes and building stocks in the medium term (2050) and the long term (2100). This influence is likely to vary across the countries because of the differences in building, socio-economic and cultural characteristics, public policies, and the local evolution of the climate.

**Author Contributions:** Conceptualization, A.M. and W.W.; methodology, A.M. and J.-M.A.; investigation, A.M.; writing—original draft preparation, A.M.; writing—review and editing, W.W., J.-M.A., C.M. and P.B.; supervision, W.W., P.B. and C.M.; project administration, W.W., C.M. and P.B.; funding acquisition, W.W. and C.M. All authors have read and agreed to the published version of the manuscript.

**Funding:** This work was funded by the CSTB research program on climate change.

**Conflicts of Interest:** The authors declare no conflict of interest.

## Nomenclature

|  |  |
|---|---|
| $A_{dtot}$ | The total area of the downward-facing interior surface ($m^2$) |
| $A_i$ | Area of interior surface *i* ($m^2$) |
| $C_f(d)$ | Concentration associated with $M_f(d)$ ($\mu g \cdot \mu g^{-1}$ Particulate matter) |
| $C_{in}$ | Indoor concentration ($\mu g \cdot m^{-3}$) |
| $C_{out}$ | Outdoor concentration ($\mu g \cdot m^{-3}$) |
| $C_p$ (d) | Mass concentration sorbed to particulate matter (PM) in the air compartment ($\mu g \cdot \mu g^{-1}$ PM) |
| $C_{p0}$ (d) | Mass concentration sorbed to PM in the outdoor air ($\mu g \cdot \mu g^{-1}$ PM) |
| $C_{pADJ0}$ (d) | Mass concentration sorbed to PM in the adjoining room ($\mu g \cdot \mu g^{-1}$ PM) |
| $C_{Pk}(d)$ | Concentration associated with $M_{Pk}(d)$ |
| $C_{terp}$ | Concentration of a reactant ($\mu g \cdot m^{-3}$) |
| $C_V$ | Concentration in the vapor phase of the air compartment ($\mu g \cdot m^{-3}$) |
| $C_{V0}$ | Concentration in the vapor phase of the outdoor air ($\mu g \cdot m^{-3}$) |
| $C_{VADJ0}$ | Concentration in the vapor phase of the adjoining room ($\mu g \cdot m^{-3}$) |
| d | Aerodynamic diameter of PM ($\mu m$) |
| $d_s$ | First-order decay rate constant combining all sink processes ($h^{-1}$) |
| e | Filter efficiency of the mechanical ventilation system (dimensionless) |
| e(d) | Filter efficiency of the mechanical ventilation system of PM (dimensionless) |
| $e_{rcl}$ | Efficiency of the indoor filter device (dimensionless) |
| $e_{rcl}$ (d) | Filter efficiency of the indoor recirculation device of PM (dimensionless) |
| $E_{si}$ | Emission rate from the interior surface i ($\mu g \cdot h^{-1} \cdot m^{-2}$) |
| $E_{Vj}$ | Emission rate from indoor source *j* other than interior surfaces ($\mu g \cdot h^{-1}$) |
| $f_{filt}$ | Fractional runtime of the HVAC system if applicable (dimensionless, ranging from 0 to 1) |
| $J_{coag}$ | Coagulation term for gain or loss of the target size particle ($\mu g \cdot m^{-3} \cdot h^{-1}$) |
| $J_{SVOC}$ | SVOC loss as gas phase and gain as particle phase ($\mu g \cdot m^{-3} \cdot h^{-1}$) |
| k | Bimolecular reaction rate constant between two gas-phase compounds($m^3 \cdot \mu g^{-1} \cdot h^{-1}$) |
| $K_V$ | First-order reaction rate constant in the gas phase ($h^{-1}$) |
| M (d) | PMs mass concentration in the air compartment ($\mu g$ PM$\cdot m^{-3}$) |
| $M_0$ (d) | PMs mass concentration in the outdoor air ($\mu g$ PM$\cdot m^{-3}$) |
| $M_{ADJ0}$ (d) | PMs mass concentration in the air of the adjoining room ($\mu g$ PM$\cdot m^{-3}$) |

| | |
|---|---|
| $M_f(d)$ | PMs concentration on the floor surface (μg PM·m$^{-2}$) |
| $M_{Pk}(d)$ | PMs emission rate from indoor source k (μg PM·h$^{-1}$) |
| P | Penetration factor through cracks (dimensionless, often close to 1) |
| $Q_{ADJ0}$ | Volumetric flow rate from the adjoining room (m$^3$·h$^{-1}$) |
| $Q_{cf}$ | Volumetric flow rate of air through cracks (m$^3$·h$^{-1}$) |
| $Q_{exhaust}$ | Airflow rate of any mechanical exhaust ventilation system (m$^3$·h$^{-1}$) |
| $Q_{filt}$ | Airflow rate through the central HVAC filter if applicable (m$^3$·h$^{-1}$) |
| $Q_{mv}$ | Volumetric flow rate of air drawn to the mechanical ventilation system (m$^3$·h$^{-1}$) |
| $Q_{ra}$ | Volumetric flow rate of return air to the mechanical ventilation system (m$^3$·h$^{-1}$) |
| $Q_{rcl}$ | Air recirculation rate of the indoor filter device (m$^3$·h$^{-1}$) |
| $Q_{window}$ | Natural airflow through the window opening (m$^3$·h$^{-1}$) |
| $R_{sus}$ | Resuspension rate coefficient from surface to indoor air (h$^{-1}$) |
| S | Indoor source strength (mass emission rate) (μg·h$^{-1}$) |
| t | Time (h) |
| V | Volume of the compartment (m$^3$) |
| $v_{dep}$ | Deposition velocity from indoor air to surface (m·h$^{-1}$) |
| $\beta$ | The first-order indoor loss rate by deposition to surfaces and/or surface reactions (h$^{-1}$) |
| $\eta_{filt}$ | Removal efficiency of a filter installed in the HVAC system if applicable (dimensionless, ranging from 0 to 1) |
| $\xi$ | Secondary organic aerosols yield |
| $\Psi_{gas}$ | The production or removal rate of a species due to gas-phase reaction (μg·m$^{-3}$·h$^{-1}$) |
| $\lambda$ | Air exchange rate (h$^{-1}$) |
| $\lambda_d$ | Deposition rate of particles or gas species on indoor surfaces (h$^{-1}$) |
| $\lambda_{inf}$ | Air exchange rate due to infiltration alone (h$^{-1}$) |
| $\lambda_{nat}$ | Air exchange rate due to the natural ventilation (h$^{-1}$) |

## References


1. NASA. 2020 Tied for Warmest Year on Record, NASA Analysis Shows. 2021. Available online: https://www.giss.nasa.gov/research/news/20210114/ (accessed on 22 November 2021).
2. Vardoulakis, S.; Dimitroulopoulou, C.; Thornes, J.; Lai, K.-M.; Taylor, J.; Myers, I.; Heaviside, C.; Mavrogianni, A.; Shrubsole, C.; Chalabi, Z.; et al. Impact of climate change on the domestic indoor environment and associated health risks in the UK. *Environ. Int.* **2015**, *85*, 299–313. [CrossRef]
3. Nazaroff, W.W. Exploring the consequences of climate change for indoor air quality. *Environ. Res. Lett.* **2013**, *8*, 015022. [CrossRef]
4. Wei, W.; Mandin, C.; Ramalho, O. Influence of indoor environmental factors on mass transfer parameters and concentrations of semi-volatile organic compounds. *Chemosphere* **2018**, *195*, 223–235. [CrossRef] [PubMed]
5. Aysha, S.; Mani, M. Adaptation of buildings to climate change. In *Encyclopedia of Sustainable Technologies*; Elsevier: Amsterdam, The Netherlands, 2017; pp. 331–349. [CrossRef]
6. Yang, Y.; Javanroodi, K.; Nik, V.M. Climate change and energy performance of European residential building stocks—A comprehensive impact assessment using climate big data from the coordinated regional climate downscaling experiment. *Appl. Energy* **2021**, *298*, 117246. [CrossRef]
7. Cheng, X.; Zhang, H.; Pan, W.; Liu, S.; Zhang, M.; Long, Z.; Zhang, T.; Chen, Q. Field study of infiltration rate and its influence on indoor air quality in an apartment. *Procedia Eng.* **2017**, *205*, 3954–3961. [CrossRef]
8. Weschler, C.J. Changes in indoor pollutants since the 1950s. *Atmos. Environ.* **2009**, *43*, 153–169. [CrossRef]
9. Kinney, P.L. Indoor Air Quality in the Context of Climate Change. In *Handbook of Indoor Air Quality*; Zhang, Y., Hopke, P.K., Mandin, C., Eds.; Springer: Singapore, 2022. [CrossRef]
10. Chang, L.; Lee, Y.; Kim, C.-K.; Lee, D.S. Development of a multimedia model (IIAQ-CC) to assess climate change influences on volatile and semi-volatile organic compounds in indoor environments. *Build. Environ.* **2018**, *143*, 217–226. [CrossRef]
11. Ilacqua, V.; Dawson, J.; Breen, M.; Singer, S.; Berg, A. Effects of climate change on residential infiltration and air pollution exposure. *J. Expo. Sci. Environ. Epidemiol.* **2017**, *27*, 16–23. [CrossRef]
12. Salthammer, T.; Schieweck, A.; Gu, J.; Ameri, S.; Uhde, E. Future trends in ambient air pollution and climate in Germany—Implications for the indoor environment. *Build. Environ.* **2018**, *143*, 661–670. [CrossRef]
13. WHO Global Air Quality Guidelines: Particulate Matter ($PM_{2.5}$ and $PM_{10}$), Ozone, Nitrogen Dioxide, Sulfur Dioxide and Carbon Monoxide: Executive Summary. 2021. Available online: https://apps.who.int/iris/handle/10665/345334 (accessed on 22 November 2021).
14. Salthammer, T.; Zhao, J.; Schieweck, A.; Uhde, E.; Hussein, T.; Antretter, F.; Künzel, H.; Pazold, M.; Radon, J.; Birmili, W. A holistic modeling framework for estimating the influence of climate change on indoor air quality. *Indoor Air* **2022**, *32*, e13039. [CrossRef] [PubMed]

15. Fazli, T.; Stephens, B. Development of a nationally representative set of combined building energy and indoor air quality models for U.S. residences. *Build. Environ.* **2018**, *136*, 198–212. [CrossRef]
16. Fazli, T.; Dong, X.; Fu, J.S.; Stephens, B. Predicting U.S. residential building energy use and indoor pollutant exposures in the mid-21st century. *Environ. Sci. Technol.* **2021**, *55*, 3219–3228. [CrossRef] [PubMed]
17. Fang, L.; Clausen, G.; Fanger, P.O. Impact of temperature and humidity on chemical and sensory emissions from building materials. *Indoor Air* **1999**, *9*, 193–201. [CrossRef] [PubMed]
18. Wolkoff, P. Impact of air velocity, temperature, humidity, and air on long-term VOC emissions from building products. *Atmos. Environ.* **1998**, *32*, 2659–2668. [CrossRef]
19. Zhou, C.; Zhan, Y.; Chen, S.; Xia, M.; Ronda, C.; Sun, M.; Chen, H.; Shen, X. Combined effects of temperature and humidity on indoor VOCs pollution: Intercity comparison. *Build. Environ.* **2017**, *121*, 26–34. [CrossRef]
20. Xiong, J.; Zhang, P.; Huang, S.; Zhang, Y. Comprehensive influence of environmental factors on the emission rate of formaldehyde and VOCs in building materials: Correlation development and exposure assessment. *Environ. Res.* **2016**, *151*, 734–741. [CrossRef] [PubMed]
21. Salthammer, T.; Morrison, G.C. Temperature and indoor environments. *Indoor Air* **2022**, *32*, e13022. [CrossRef] [PubMed]
22. Guo, Z. Review of indoor emission source models. Part 2. Parameter estimation. *Environ. Pollut.* **2002**, *120*, 551–564. [CrossRef]
23. Wei, W.; Mandin, C.; Blanchard, O.; Mercier, F.; Pelletier, M.; le Bot, B.; Glorennec, P.; Ramalho, O. Temperature dependence of the particle/gas partition coefficient: An application to predict indoor gas-phase concentrations of semi-volatile organic compounds. *Sci. Total Environ.* **2016**, *563–564*, 506–512. [CrossRef] [PubMed]
24. Fadeyi, M.O. Ozone in indoor environments: Research progress in the past 15 years. *Sustain. Cities Soc.* **2015**, *18*, 78–94. [CrossRef]
25. Grøntoft, T.; Henriksen, J.F.; Seip, H.M. The humidity dependence of ozone deposition onto a variety of building surfaces. *Atmos. Environ.* **2004**, *38*, 59–68. [CrossRef]
26. Weschler, C.J. Ozone's impact on public health: Contributions from indoor exposures to ozone and products of ozone-initiated chemistry. *Environ. Health Perspect.* **2006**, *114*, 1489–1496. [CrossRef]
27. Fisk, W.J. Review of some effects of climate change on indoor environmental quality and health and associated no-regrets mitigation measures. *Build. Environ.* **2015**, *86*, 70–80. [CrossRef]
28. Solberg, S.; Hov, Ø.; Søvde, A.; Isaksen, I.S.A.; Coddeville, P.; de Backer, H.; Forster, C.; Orsolini, Y.; Uhse, K. European surface ozone in the extreme summer 2003. *J. Geophys. Res.* **2008**, *113*, D07307. [CrossRef]
29. Poole, J.A.; Barnes, C.S.; Demain, J.G.; Bernstein, J.A.; Padukudru, M.A.; Sheehan, W.J.; Fogelbach, G.G.; Wedner, J.; Codina, R.; Levetin, E.; et al. Impact of weather and climate change with indoor and outdoor air quality in asthma: A Work Group Report of the AAAAI Environmental Exposure and Respiratory Health Committee. *J. Allergy Clin. Immunol.* **2019**, *143*, 1702–1710. [CrossRef] [PubMed]
30. Wang, Y.; Shen, L.; Wu, S.; Mickley, L.; He, J.; Hao, J. Sensitivity of surface ozone over China to 2000–2050 global changes of climate and emissions. *Atmos. Environ.* **2018**, *75*, 374–382. [CrossRef]
31. Hong, C.; Zhang, Q.; Zhang, Y.; Davis, S.J.; Tong, D.; Zheng, Y.; Liu, Z.; Guan, D.; He, K.; Schellnhuber, H.J. Impacts of climate change on future air quality and human health in China. *Proc. Natl. Acad. Sci. USA* **2019**, *116*, 17193–17200. [CrossRef] [PubMed]
32. Meleux, F.; Solmon, F.; Giorgi, F. Increase in summer European ozone amounts due to climate change. *Atmos. Environ.* **2007**, *41*, 7577–7587. [CrossRef]
33. Coelho, S.; Rafael, S.; Lopes, D.; Miranda, A.I.; Ferreira, J. How changing climate may influence air pollution control strategies for 2030? *Sci. Total Environ.* **2021**, *758*, 143911. [CrossRef] [PubMed]
34. Zhong, L.; Lee, C.-S.; Haghighat, F. Indoor ozone and climate change. *Sustain. Cities Soc.* **2017**, *28*, 466–472. [CrossRef]
35. Giorgi, F.; Meleux, F. Modelling the regional effects of climate change on air quality. *C. R. Geosci.* **2007**, *339*, 721–733. [CrossRef]
36. Westervelt, D.M.; Horowitz, L.W.; Naik, V.; Tai, A.P.K.; Fiore, A.M.; Mauzerall, D.L. Quantifying $PM_{2.5}$-meteorology sensitivities in a global climate model. *Atmos. Environ.* **2016**, *142*, 43–56. [CrossRef]
37. Cao, Y.; Yue, X.; Liao, H.; Yang, Y.; Zhu, J.; Chen, L.; Tian, C.; Lei, Y.; Zhou, H.; Ma, Y. Ensemble projection of global isoprene emissions by the end of 21st century using CMIP6 models. *Atmos. Environ.* **2021**, *267*, 118766. [CrossRef]
38. Himes, A.; Busby, G. Wood buildings as a climate solution. *Dev. Built Environ.* **2020**, *4*, 100030. [CrossRef]
39. Panaloza, D.; Erlandsson, M.; Berlin, J.; Wanlinder, M.; Falk, A. Future scenarios for climate mitigation of new construction in Sweden: Effects of different technological pathways. *J. Clean. Prod.* **2018**, *187*, 1025–1035. [CrossRef]
40. Marsono, A.K.B.; Balasbaneh, A.T. Combinations of building construction material for residential building for the global warming mitigation for Malaysia. *Constr. Build. Mater.* **2015**, *85*, 100–108. [CrossRef]
41. Déoux, S. Are environmentally-friendly building and decorating materials allergenic? *Rev. Française D'allergologie* **2010**, *50*, 481–484. [CrossRef]
42. Ruiz-Jimenez, J.; Heiskanen, I.; Tanskanen, V.; Hartonen, K.; Riekkola, M. Analysis of indoor air emissions: From building materials to biogenic and anthropogenic activities. *J. Chromatogr. Open* **2022**, *2*, 100041. [CrossRef]
43. Bozsaky, D. The historical development of thermal insulation materials. *Architecture* **2010**, *41*, 49–56. Available online: http://www.pp.bme.hu/ar (accessed on 22 November 2021). [CrossRef]
44. Dudzinska, M.R.; Staszowska, A. Assessment of chemical pollutant levels in the indoor air of straw bale homes. A case study. *Int. J. Conserv. Sci.* **2021**, *12*, 817–826. Available online: https://ijcs.ro/public/IJCS-21-61_Dudzinska.pdf (accessed on 22 November 2021).

45. Verichev, K.; Zamorano, M.; Fuentes-Sepúlveda, A.; Cárdenas, N.; Carpio, M. Adaptation and mitigation to climate change of envelope wall thermal insulation of residential buildings in a temperate oceanic climate. *Energy Build.* **2021**, *235*, 110719. [CrossRef]
46. Wi, S.; Kang, Y.; Yang, S.; Kim, Y.U.; Kim, S. Hazard evaluation of indoor environment based on long-term pollutant emission characteristics of building insulation materials: An empirical study. *Environ. Pollut.* **2021**, *285*, 117223. [CrossRef]
47. Wi, S.; Park, J.H.; Kim, Y.U.; Kim, S. Evaluation of environmental impact on the formaldehyde emission and flame-retardant performance of thermal insulation materials. *J. Hazard. Mater.* **2021**, *402*, 123463. [CrossRef] [PubMed]
48. Huang, K.T.; Hwang, R.L. Future trends of residential building cooling energy and passive adaptation measures to counteract climate change: The case of Taiwan. *Appl. Energy* **2016**, *184*, 1230–1240. [CrossRef]
49. Yenneti, K.; Ding, L.; Prasad, D.; Ulpiani, G.; Paolini, R.; Haddad, S.; Santamouris, M. Urban Overheating and Cooling Potential in Australia: An Evidence-Based Review. *Climate* **2020**, *8*, 126. [CrossRef]
50. Ortiz, M.; Itard, L.; Bluyssen, P.M. Indoor environmental quality related risk factors with energy-efficient retrofitting of housing: A literature review. *Energy Build.* **2020**, *221*, 110102. [CrossRef]
51. Lee, W.C.; Shen, L.; Catalano, P.J.; Mickley, L.J.; Koutrakis, P. Effects of future temperature change on $PM_{2.5}$ infiltration in the Greater Boston area. *Atmos. Environ.* **2017**, *150*, 98–105. [CrossRef]
52. Chen, C.; Zhao, B. Review of relationship between indoor and outdoor particles: I/O ratio, infiltration factor and penetration factor. *Atmos. Environ.* **2011**, *45*, 275–288. [CrossRef]
53. Lin, B.; Huangfu, Y.; Lima, N.; Jobson, B.; Kirk, M.; O'Keeffe, P.; Pressley, S.N.; Walden, V.; Lamb, B.; Cook, D.J. Analyzing the relationship between human behavior and indoor air quality. *J. Sens. Actuator Netw.* **2017**, *6*, 13. [CrossRef]
54. Field, R.W. Climate Change and Indoor Air Quality, Contractor Report. U.S. Environmental Protection Agency, Office of Radiation and Indoor Air, 2010. Available online: https://www.epa.gov/sites/default/files/2014-08/documents/field_climate_change_iaq.pdf (accessed on 22 November 2021).
55. Zhang, X.B.; Sun, J.; Fei, Y.; Wei, C. Cooler rooms on a hotter planet? Household coping strategies, climate change, and air conditioning usage in rural China. *Energy Res. Soc. Sci.* **2020**, *68*, 101605. [CrossRef]
56. Huang, L.; Wei, Y.; Zhang, L.; Ma, Z.; Zhao, W. Estimates of emission strengths of 43 VOCs in wintertime residential indoor environments, Beijing. *Sci. Total Environ.* **2021**, *793*, 148623. [CrossRef] [PubMed]
57. Wang, H.; Chen, Q. Impact of climate change heating and cooling energy use in buildings in the United States. *Energy Build.* **2014**, *82*, 428–436. [CrossRef]
58. Rijal, H.B.; Humphreys, M.A.; Nicol, J.F. Development of a window opening algorithm based on adaptive thermal comfort to predict occupant behavior in Japanese dwellings. *Jpn. Archit. Rev.* **2018**, *1*, 310–321. [CrossRef]
59. Andersen, R.; Fabi, V.; Toftum, J.; Corgnati, S.P.; Olesen, B.W. Window opening behaviour modelled from measurements in Danish dwellings. *Build. Environ.* **2013**, *69*, 101–113. [CrossRef]
60. Liu, Y.; Liu, S.; Wang, S.; Zhao, B. How will window opening change under global warming: A study for China residence. *Build. Environ.* **2022**, *209*, 108672. [CrossRef]
61. Du, C.; Yu, W.; Ma, Y.; Cai, Q.; Li, B.; Li, N.; Wang, W.; Yao, R. A holistic investigation into the seasonal and temporal variations of window opening behavior in residential buildings in Chongqing. *China Energy Build.* **2021**, *231*, 110522. [CrossRef]
62. Fabi, V.; Andersen, R.V.; Corgnati, S.; Olesen, B.W. Occupants' window opening behavior: A literature review of factors influencing occupant behavior and models. *Build. Environ.* **2012**, *58*, 188–198. [CrossRef]

International Journal of
*Environmental Research and Public Health*

MDPI

*Article*

# Distribution of Minor and Major Metallic Elements in Residential Indoor Dust: A Case Study in Latvia

**Agnese Araja *, Maris Bertins, Gunita Celma, Lauma Busa and Arturs Viksna**

Faculty of Chemistry, University of Latvia, Jelgavas Str.1, LV-1004 Riga, Latvia
* Correspondence: agnese.araja@lu.lv

**Abstract:** The coronavirus disease 2019 (COVID-19) pandemic has not only brought considerable and permanent changes to economies and healthcare systems, but it has also greatly changed the habits of almost the entire society. During the lockdowns, people were forced to stay in their dwellings, which served as a catalyst for the initiation of a survey on the estimation of the metallic element content in residential indoor dust in different parts of Latvia. This article presents the study results obtained through the analysis of collected dust samples from 46 dwellings, both in the capital of Latvia, Riga, and in smaller cities. Two methods were employed for indoor dust collection: vacuum sampling and manual sampling with a brush and plastic spatula. After microwave-assisted acid extraction, the samples were analyzed using inductively coupled plasma mass spectrometry (ICP-MS) in terms of the major (Na, K, Ca, Mg, Al and Fe) and minor (Mn, Ni, Co, Pb, Cr, As, Ba, Li, Be, B, V, Cu, Zn, Se, Rb, Sr, Cd, La, Ce and Bi) elements. For the data analysis, principal component analysis was performed. Among the measured metals, the highest values were determined for the macro and most abundant elements (Na > K > Ca > Fe > Mg > Al). The concentration ranges of the persistently detected elements were as follows: Pb, 0.27–1200 mg kg$^{-1}$; Cd, 0.01–6.37 mg kg$^{-1}$; Ni, 0.07–513 mg kg$^{-1}$; As, 0.01–69.2 mg kg$^{-1}$; Cu, 5.71–1900 mg kg$^{-1}$; Zn, 53.6–21,100 mg kg$^{-1}$; and Cr, 4.93–412 mg kg$^{-1}$. The critical limit values of metallic elements in soil defined by the legislation of the Republic of Latvia (indicating the level at or above which the functional characteristics of soil are disrupted, or pollution poses a direct threat to human health or the environment) were exceeded in the following numbers of dwellings: Pb = 4, Ni = 2, As = 1, Cu = 16, Cr = 1 and Zn = 28.

**Keywords:** indoor dust; metallic elements; dust sampling; ICP-MS; PCA

check for updates

**Citation:** Araja, A.; Bertins, M.; Celma, G.; Busa, L.; Viksna, A. Distribution of Minor and Major Metallic Elements in Residential Indoor Dust: A Case Study in Latvia. *Int. J. Environ. Res. Public Health* **2023**, *20*, 6207. https://doi.org/10.3390/ijerph20136207

Academic Editors: Nuno Canha, Marta Almeida, Evangelia Diapouli, Joseph Pinto and Atin Adhikari

Received: 11 March 2023
Revised: 14 June 2023
Accepted: 19 June 2023
Published: 22 June 2023



## 1. Introduction

Exposure to indoor air pollutants has a significant impact, firstly, since it has been reported that people in industrialized countries spend an average of 90% of their time in enclosed microenvironments [1,2]. Secondly, in accordance with the World Health Organization (WHO) [3], indoor microenvironment air contamination is one of the main current environmental health risks, as indoor environments frequently have even higher levels of air pollutants than the outdoor environment [4,5]. Recent studies [6–9] have demonstrated the extent of worldwide interest in human exposure to indoor pollutants; in particular, residential dust may be considered one of the important pathways of human exposure to toxic trace elements. Turner defined household dust as fine ($\leq$100 µm) settled or airborne particulate material encountered in the indoor domestic setting [10]. However, it must be pointed out that household dust is a heterogeneous mixture of materials of different sizes and shapes, organic and inorganic in origin, including different ultrafine fibers, mold, allergens, soot, animal fur, skin particles, and heating and building residues [11]. Indoor dust is considered a complex matrix in terms of its numerically remarkable possible origin sources, including the infiltration of outdoor contaminants, heating, cooking and household electronic devices, building and reconstruction materials, consumer products, smoking and incense burning, and other sources, which might be closely linked to inhabitants' activities [12–14]. The ability of indoor dust to act as both a sink and a transport medium for various persistent chemical contaminants, such as heavy metals, has been demonstrated in many studies [15–17]. Foremost among the various other contaminants detected in residential dust, heavy metals are of well-grounded concern due to long-term exposure, their adverse health effects and such specific features as their high toxicity, non-biodegradability, persistence, bioaccumulation potential and long biological half-life [18,19]. It has been suggested that household dust-bound heavy metals accumulate in humans through inhalation, non-dietary ingestion or dermal contact. The absorption and penetration of metals into human fatty tissues and circulatory system tissues, as well as other parts of the body, will probably result in metal-associated diseases later in life [20–22]. Children at a young age are more frequently exposed to interior contaminants than people in other age groups due to their particular behaviors, such as crawling and hand-to-mouth practices. The intensive growth and rapid development of children's organs in the first years of life and their immature immune system, higher inhalation rates per body mass and lower tolerance to pollutants make children much more susceptible to heavy metal exposure than adults [23,24].

This is the first study to describe the analysis results of 120 indoor dust samples gathered from 46 residential dwellings in Latvia in terms of their metallic element concentrations. In this paper, we report the concentrations of 25 metallic elements in milligrams per kilogram (mg $kg^{-1}$) of total dust in the particle size range < 0.2 mm, emphasizing the concentrations of non-carcinogenic elements and carcinogenic elements such as copper (Cu), nickel (Ni), zinc (Zn), manganese (Mn), arsenic (As), cadmium (Cd), chromium (Cr) and lead (Pb), as these elements present greater health risks for both adults and children [23]. The objective of this study was to determine the content of 25 major, or most abundant, and minor, or trace, metallic elements in residential indoor dust in Latvia. Overall, this study targets to answer the following questions:

- Which are the minor and major metallic elements in residential indoor dust in Latvia?
- Which are the concentration ranges of toxic elements adhered to the investigated household dust from dwellings located in urban, suburban and rural areas in Latvia?
- Can the results from two non-standardized dust-sampling approaches—dust vacuuming and manual dust sampling with a brush and plastic spatula—be comparable?
- Which trace elements detected in residential indoor dust could be of greater concern for human health?

## 2. Materials and Methods

### 2.1. Study Area

The current study was conducted in Latvia (Figure 1), which is a country in the Baltic region in the northern part of Europe, covering an area of 64,589 $km^2$ with a population of 1.9 million. Part of the studied residential indoor dust was collected in the capital of Latvia, Riga, which is a significantly larger city than the others with a population of 671,000 inhabitants. Riga's territory covers 307.17 $km^2$ and lies 1–10 m above sea level, on a flat and sandy plain. The average values of the air quality index (AQI) during the sampling period in the Riga central area ranged from 76 to 88 (poor air quality in terms of $PM_{10}$, but good to very good air quality in terms of $SO_2$, $NO_2$ and $O_3$) with a maximum of 198 [25,26]. The remainder of the dust samples were collected in four smaller Latvian cities (Liepaja, Jurmala, Ogre and Marupe) and ten rural areas (Dobele, Baldone, Kandava, Krimulda, Lielvarde, Ropazi, Saulkrasti, Zvejniekciems, Viesite and Velmeri) with a population more than 10 times less than that of the capital city.

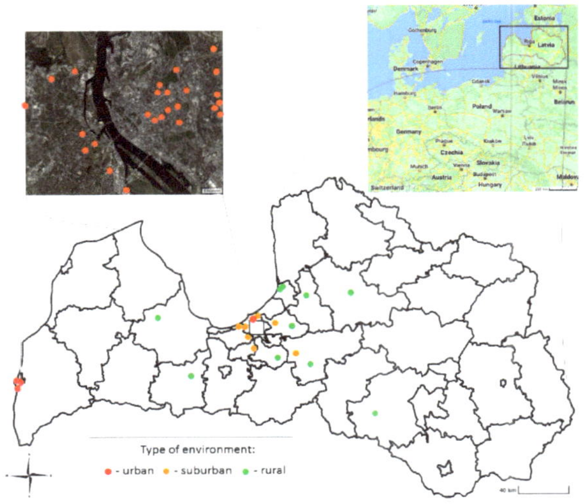


**Figure 1.** Site map of Latvia and Riga with the different types of environments represented by colored dots: red dots—urban, orange dots—suburban, and green dots—rural sampling sites [27].

### 2.2. Sampling Strategy

Sampling was performed between March and May 2022 at 46 sites to cover various areas in Latvia. One hundred twenty residential indoor dust samples were collected from 28 urban dwellings in a commercial area with heavy city traffic nearby, 8 suburban homes and 10 rural residences surrounded by less industrial infrastructure and traffic, but a lot more greenery. The choice of sampling sites, the number of samples per dwelling and the choice of the room where sampling was performed were based on the inhabitants' willingness to take part in the current study and their possibilities. The participants were invited to deliver collected indoor dust samples along with the completed short questionnaire about their dwelling environment. The aggregated data of the questionnaires are summarized in Table 1, and they include parameters characterizing housing, the number of occupants, their hobbies and habits, age of the property and type of construction materials used, type of fuel used for heating and presence of pets.

Two sampling methods were applied for indoor dust collecting: household vacuum cleaners with an unused dust container for each sample, and a previously cleaned brush with a plastic spatula. Both dust sampling methods in parallel were used in 16 dwellings. In 23 dwellings, dust samples were obtained by manual sweeping, while in the remaining 7 residences, vacuuming was used for dust collection. The numbers of samples collected in each type of room from the different sampling sites are presented in Table 1, as well as in the Supplementary Materials Table S1. Residential indoor dust samples were collected from dwelling floors, carpets, windowsills, tables and shelves, and included lamp covers and other surfaces that had not been cleaned for some time. Immediately after collection, each dust sample was transferred to an unused, labeled sealed plastic bag for safe transportation and storage, avoiding cross-contamination.

**Table 1.** The characteristics of the dwellings where the indoor dust sampling was performed.

<table>
<tr><th>Parameter</th><th colspan="6">Data from Questionnaires (Number of Dwellings * or Samples in Urban (U), Suburban (S) and Rural (R) Sites)</th></tr>
<tr><td>Sampling site</td><td colspan="2">Urban<br>28 *</td><td colspan="2">Suburban<br>8 *</td><td colspan="2">Rural<br>10 *</td></tr>
<tr><td>Structure of the building</td><td colspan="2">Bricks<br>19 *</td><td colspan="2">Concrete<br>23 *</td><td colspan="2">Wood<br>4 *</td></tr>
<tr><td>Age of building</td><td colspan="2"><5 years<br>3 *</td><td colspan="2">6–10 years<br>5 *</td><td colspan="2">>10 years<br>38 *</td></tr>
<tr><td>Type of heating</td><td colspan="2">Central heating<br>33 *</td><td colspan="2">Wood stove heating<br>8 *</td><td colspan="2">Briquette or pellet stove heating<br>5 *</td></tr>
<tr><td>Type of room</td><td>Corridor<br>6<br>(U2; S2; R2)</td><td>Living room<br>33<br>(U16; R17)</td><td>Living room and kitchen<br>21<br>(S21)</td><td>Kitchen<br>10<br>(U6; R4)</td><td>Bedroom<br>45<br>(U24; S10; R11)</td><td>All rooms<br>5<br>(U5)</td></tr>
<tr><td>Floor cover</td><td colspan="2">Hardwood<br>8</td><td>Laminate<br>42</td><td>Parquet<br>6</td><td>Tile<br>16</td><td>Other<br>48</td></tr>
<tr><td>Smoking</td><td colspan="2">No<br>27 *</td><td colspan="2">Yes<br>15 *</td><td colspan="2">No information<br>4 *</td></tr>
<tr><td>Pets in home</td><td colspan="2">No<br>29 *</td><td colspan="4">Yes<br>17 *</td></tr>
<tr><td>Cleaning frequency</td><td>Several times per week<br>5 *</td><td>Twice a week<br>7 *</td><td>Once a week<br>21 *</td><td>Once per two weeks<br>8 *</td><td>Once per month<br>3 *</td><td>Less often<br>2 *</td></tr>
</table>

* Number of Dwellings.

### 2.3. Chemical Analysis of Indoor Dust

Prior to chemical analysis, the manual removal of pet and human hair, as well as larger refuse, was performed. Afterward, the dust samples were sieved through a 0.2 mm sieve (Rotilabo). An amount of ~0.20 g of each air-dried dust sample was subjected to conventional microwave-assisted acid digestion (2.00 mL conc. $H_2O_2$, for trace analysis, Fischer Chemical, 30% and 6.00 mL conc. $HNO_3$, trace metal grade, Fisher Chemical, 69%) performed with a microwave oven (Milestone Start E) under pressure conditions. The maximum pressure at the peak point was approximately 200 psi (13.8 bar). The heating program was set as follows: heating for 15 min to 160 °C and holding at 160 °C for 30 min. Acidic extract solutions of dust samples were diluted to a volume of 15.0 mL with deionized water (0.055 μS $cm^{-1}$, Adrona) and measured by inductively coupled plasma mass spectrometry (ICP-MS, Agilent 8900 Triple Quadrupole). The instrumental parameters of ICP-MS were set as summarized in Table 2.

For the purpose of elemental quantification, the calibration graph was constructed using six standard solutions in a concentration range from 0.1 μg $L^{-1}$ to 100.0 μg $L^{-1}$, which contained all the elements of interest (Al, As, Ba, Bi, B, Cd, Ca, Ce, Cr, Co, Cu, Fe, La, Pb, Rb, Li, Mg, Mn, Ni, K, Se, Na, Sr, V and Zn) dissolved in 2% $HNO_3$. Analytical standard stock solutions were prepared from a certified reference material (HPS, ICP-MS-68A, 10 mg $L^{-1}$, traceable to NIST SRM 3100). The element concentrations in the samples were calculated using the external calibration graph method, and a blank correction for each sample was applied. When dust samples were prepared using microwave-assisted acid digestion performed with a microwave oven, each set consisted of 10 samples, including one designated as a blank sample. The blank sample solely contained added reagents ($H_2O_2$ and $HNO_3$), and the blank correction was applied to all samples, irrespective of the sampling method employed. The blank levels in our study for most elements (Li, V, Cr, Mn, Co, Ni, Cu, Zn, As, Rb, Sr, Cd, Ba, La, Ce, Pb and Bi) were determined to be below the limit of detection (<0.1 mg/L). However, in the case of B, Na, K, Ca and Fe, some background levels could reach up to 100 μg/L. An internal standard solution (10.0 μg $L^{-1}$, Agilent) was used for system stability control during the measurements. Two standard solutions (10.0 μg $L^{-1}$) were used between every ten samples to verify system stability. Two standard reference materials were used to verify the accuracy of the sample preparation and the elemental quantification methods: the standard reference material A2 (LGC, Elements on Filter Media/Work-room Air, Filter No. A2-1365) and the standard reference material B4 (LGC, Elements on Filter Media/Work-room Air, Filter No. B148).

**Table 2.** The instrumental parameters of inductively coupled plasma mass spectrometry (ICP-MS).

| Parameter | Setting | Parameter | Setting |
|---|---|---|---|
| RF power (W) | 1550 | Extraction 1 lens (V) | −5.0 |
| Sampling depth (mm) | 8.0 | Extraction 2 lens (V) | 7.0 |
| Plasma gas flow rate (L $min^{-1}$) | 15.0 | Omega lens (V) 7.0 | −200 |
| Nebulizer gas flow rate (mL $min^{-1}$) | 0.90 | Omega bias lens (V) | −110 |
| Makeup gas flow rate (mL $min^{-1}$) | 0.0 | Octopole bias (V) | −3.0 |
| He cell gas flow (mL $min^{-1}$) | 5.0 | Cell gas flow rate (% of full scale) | 20 |

### 2.4. Statistical Analysis

Statistical analysis was performed using Microsoft Excel version 2304 and SPSS software version 29.0. Metallic element concentration data were initially explored by descriptive statistics (arithmetic mean, median, standard deviation and coefficient of variation). Further, the Kruskal–Wallis test, which is a non-parametric statistical test, was used to determine the statistically significant differences between the medians of metallic element concentrations of the different dust sample groups. These dust sample groups were divided based on the sampling sites (urban, suburban and rural) and different premises within the dwellings. Statistically significant differences between dust samples divided based on indoor smoking habit were assessed using the Mann–Whitney U test, which is a non-parametric test used to compare two independent samples. To evaluate the statistically significant differences between concentrations of metallic elements of the dust samples collected by two sampling techniques, the Mann–Whitney U test was performed. Besides that, for the mutual comparison of the results of the samples collected using in parallel both sampling techniques, the relative differences in the determined concentrations of metallic elements were calculated. The relative differences for each of the selected 16 dwellings were obtained using the mathematical equation 1, where $C_M$ is the concentration of a metallic element in a manually collected dust sample and $C_V$ is the concentration of a metallic element in a sample obtained by vacuuming.

$$\Delta D = \frac{C_M - C_V}{C_{mean(M,V)}} \quad (1)$$

In the current study, principal component analysis (PCA), which is a well-established tool for the statistical treatment of multivariate datasets, was also performed. For the evaluation of the obtained indoor dust data, the chemometric (Chemometric Agile Tool (PCA (R-based Chemometric Agile Tool (CAT software), R version 3.1.2 (31 October 2014), University of Genoa, Italia)) approach was used with the aim to determine a collection of linear combinations of the original variables, also known as principal components, which can capture the most variation in the data. During ICP-MS analysis, data processing and collection and calculation of results were carried out using the MassHunter Workstation program, including its subprograms—Instrument control and Offline data analysis.

## 3. Results

### 3.1. Concentrations of Major and Minor Elements in Residential Dust

The metal contents of 120 residential dust samples from 46 dwellings in Latvia are summarized in Table 3. In general, the concentrations of metals in the residential dust varied significantly across the indoor locations, as determined by the coefficients of variation (CVs) and the arithmetic mean and median concentrations. Ca had the highest mean concentration, followed by Na > Mg > K > Fe > Al > Zn. From those, the most studied elements in household dust are Fe, Al and Zn, as they are more closely associated with anthropogenic sources, in contrast to Na, K, Mg and Ca, which are mostly of natural origin. For example, Ca is the major element of the mineral calcite, which is often detected in indoor dust [28]. Regarding the concentration ranges of Al and Fe, only results from Malaysia [29] reported a lower mean concentration of Al at 1230 (mg kg$^{-1}$), compared with the current study, but data obtained from other studies were even one order higher [7,30,31]. The mean values for Zn from this study compare well with published data from Turkey [23] and Poland [32], but are several times higher than the concentrations found in Saudi Arabia [20] and Malaysia [29]. Higher concentrations of Al and Zn in Latvia could be attributed to the fact that the sampling took place in the spring when the resuspension of street dust is more pronounced. This could also be a significant reason for Al contributing to the composition of indoor dust, together with Zn from automobile emissions, i.e., the wear and tear of rubber tires and galvanized vehicular parts [8,12].

**Table 3.** Arithmetic mean concentrations (mg kg$^{-1}$) and other descriptive parameters of metallic elements from indoor dust ($n$ = 120 samples).

| | Arithmetic Mean | Min | Max | Median | Standard Deviation | CV, % |
|---|---|---|---|---|---|---|
| Li | 3.43 | 0.01 | 15.7 | 2.60 | 3.00 | 87.1 |
| B | 68.7 | 2.94 | 406 | 41.7 | 71.7 | 104 |
| Na | 5580 | 692 | 52,000 | 4270 | 6090 | 109 |
| Mg | 4250 | 500 | 15,900 | 3590 | 2690 | 63.2 |
| Al | 2160 | 8.20 | 10,620 | 1660 | 1930 | 89.0 |
| K | 4210 | 122 | 12,800 | 3980 | 2030 | 48.3 |
| Ca | 17,970 | 851 | 57,000 | 14,740 | 12,300 | 68.4 |
| V | 4.95 | 27.0 | 0.01 | 3.93 | 4.60 | 93.1 |
| Cr | 53.1 | 4.93 | 412 | 41.6 | 48.4 | 91.1 |
| Mn | 117 | 6.82 | 865 | 74.5 | 133 | 114 |
| Fe | 3890 | 187 | 19,380 | 2690 | 3500 | 90.0 |
| Co | 3.35 | 0.26 | 26.6 | 2.53 | 3.65 | 110 |
| Ni | 21.5 | 0.07 | 513 | 12.4 | 55.2 | 256 |
| Cu | 117 | 5.71 | 1880 | 69.3 | 214 | 182 |
| Zn | 1010 | 53.6 | 21,100 | 372 | 2430 | 240 |
| As | 2.19 | 0.01 | 69.2 | 0.78 | 7.33 | 334 |
| Se | 0.44 | 0.01 | 1.61 | 0.29 | 0.41 | 92.7 |
| Rb | 7.07 | 0.20 | 34.9 | 5.92 | 4.80 | 67.9 |
| Sr | 47.6 | 11.1 | 309 | 35.1 | 43.2 | 90.8 |
| Cd | 0.99 | 0.01 | 6.37 | 0.53 | 1.26 | 127 |

**Table 3.** *Cont.*

| | Arithmetic Mean | Min | Max | Median | Standard Deviation | CV, % |
|---|---|---|---|---|---|---|
| Ba | 209 | 7.01 | 1800 | 109 | 283 | 136 |
| La | 4.95 | 0.07 | 23.4 | 3.36 | 4.97 | 101 |
| Ce | 9.25 | 0.33 | 44.6 | 6.18 | 8.99 | 97.1 |
| Pb | 61.2 | 0.27 | 1180 | 21.4 | 138 | 225 |
| Bi | 12.2 | 0.04 | 421 | 1.11 | 56.6 | 463 |

Zn is one of those elements to which attention is paid in studies on the toxic components in indoor dust, and in this study, Zn has one of the highest coefficients of variation (CVs), while the concentrations of other previously described elements were more homogeneous. In similar studies investigating exposure to toxic contaminants in indoor dust, besides Zn, which was already highlighted above, attention more often focuses on Pb, Cu, Cd, Cr, Ni, Mn and As. Among these elements, Mn and Cu had the highest mean concentration (117 mg $kg^{-1}$), followed by Pb (61.2 mg $kg^{-1}$), Cr (53.1 mg $kg^{-1}$), Ni (21.5 mg $kg^{-1}$), As (2.19 mg $kg^{-1}$) and Cd (0.99 mg $kg^{-1}$). Pb, Cr and Ni were detected in all collected 120 dust samples, whereas As and Cd were present in 81% and 49% of the residential dust samples, respectively. In general, compared with data obtained from around the world, these results were consistent only with reported results from Turkey (except for Ni (263 mg $kg^{-1}$) [23]) and results from Australia (except for Pb (386 mg $kg^{-1}$) and Cd (4.40 mg $kg^{-1}$) [33]). It is noteworthy that concentrations of the metallic elements listed above were lower in Latvia than those reported for other regions of the world [6,13,34,35]. The arithmetic mean concentrations of other less abundant trace elements are presented in the following descending order: Ba > B > Sr > Bi > Ce > Rb > La > V > Li > Co > Se.

### 3.2. Different Sampling Approaches

It is well known that the sampling methodology is fundamental in obtaining representative samples. Variations in indoor dust surface sampling practices should be of special concern since the sampling process might reflect greatly on measurement uncertainty. If the chosen sampling method is not standardized, analytical results from different studies might not be comparable. A standardized indoor dust collection method [36,37] is not always available. In order to cover a sufficient number of indoor areas, studies have used the following indoor dust collection techniques most frequently:

- Filters from heating, ventilation and air conditioning systems [20,38];
- Dust collection from the dust bags of domestic vacuum cleaners and electrical brooms [6,8,23,31–34,39,40];
- Usage of high-volume small-surface vacuum samplers, mini-volume samplers and fine particulate dust samplers [7,12,21,41];
- Settled dust collection with wet wipes [42];
- Dust collection by gently sweeping with a brush and plastic spatula [9], soft paintbrush [5] and fingers [43].

As one of the objectives of this study was to evaluate whether the concentrations of metallic elements in indoor dust collected using two different approaches are comparable, we collected dust samples in 16 dwellings (9 urban, 2 suburban and 5 rural areas) using both methods—vacuuming and manual collection with a brush and plastic spatula—simultaneously under the same conditions. The Mann–Whitney U test was performed to compare determined concentrations of metallic elements of the samples collected by both sampling techniques. The statistical analysis did not show a statistically significant difference except for Cd ($p < 0.05$) (see Table 4). The relative differences in the determined concentrations of metallic elements in the samples collected using both sampling techniques in parallel were used for the mutual comparison of the results from each dwelling. The relative differences (ΔDs) for each of the selected 16 dwellings are summarized in Table 4.

**Table 4.** The relative concentration differences (ΔDs) of some metallic elements determined in indoor dust collected using vacuuming and manual collection and the corresponding *p*-values.

| *p*-Values | Zn | Mn | Cu | Pb | Cr | Ni | As | Cd |
|---|---|---|---|---|---|---|---|---|
| | **0.66** | **0.10** | **0.41** | **0.31** | **0.58** | **0.22** | **0.12** | **0.002** |
| **Sampling Site** | | | | ΔD | | | | |
| 1-Urb | 1.00 | −0.37 | 0.31 | 0.39 | 0.67 | 0.06 | −0.48 | nd (M) * |
| 2-Urb | 1.08 | 0.28 | 0.39 | 0.17 | −0.57 | 0.36 | −0.62 | nd (M) |
| 3-Urb | 0.32 | 0.22 | 0.28 | 0.57 | 0.63 | 0.08 | 0.31 | nd (M) |
| 4-Urb | −0.39 | −0.55 | −0.23 | −0.97 | 0.14 | −0.59 | nd (M,V) | nd (M,V) * |
| 5-Urb | −0.54 | −1.15 | −0.88 | −1.62 | −0.11 | −0.86 | nd (M) | nd (M) |
| 6-Urb | −0.20 | 0.43 | 0.20 | −0.52 | 0.36 | 0.76 | −0.24 | nd (M,V) |
| 7-Urb | −0.44 | −1.27 | −0.06 | −0.51 | −0.64 | −1.98 | nd (M) | nd (M) |
| 8-Urb | 1.44 | 0.23 | 0.62 | 0.30 | 0.62 | 0.15 | 0.97 | nd (M) |
| 9-Urb | 0.24 | −0.18 | 0.24 | −0.97 | −0.27 | −0.37 | 0.08 | nd (M) |
| 10-Sub | −1.12 | −0.50 | −1.54 | 0.56 | −1.75 | −1.66 | nd (V) * | nd (M,V) |
| 11-Sub | 0.13 | 0.19 | −0.15 | 0.38 | 0.27 | 1.10 | nd (V) | nd (M,V) |
| 12-Rur | −0.13 | −0.51 | −0.15 | −0.85 | 0.09 | −1.47 | nd (M) | nd (M) |
| 13-Rur | −0.44 | −0.42 | −0.32 | −1.12 | 0.05 | −0.81 | nd (M) | nd (M) |
| 14-Rur | 0.03 | −0.54 | −0.22 | −0.75 | 0.13 | −0.53 | nd (M) | nd (M) |
| 15-Rur | −0.26 | −0.81 | −0.76 | −1.00 | 0.44 | −0.81 | −1.10 | nd (M) |
| 16-Rur | 1.72 | −0.66 | 1.36 | 0.32 | −0.27 | 1.23 | −0.24 | nd (M) |

* nd (M)—not detected in manual dust collection. * nd (V)—not detected in dust samples using vacuuming. * nd (M,V)—not detected in both dust samples.

A distinct trend was observed in the case of Na, K Ca, and Mg, since in all 16 sampling zones, higher concentrations of these elements were determined in the dust samples collected with commercial vacuum cleaners (see Supplementary Materials Table S2). Cr and Cu had the lowest relative concentration differences between the two samples, obtained in parallel using both sampling approaches. The Mn, Fe, Co, Ni and Pb concentrations were, on average, 30% higher in the samples collected by vacuuming (see Table S2). In contrast, higher concentrations of Zn were observed in the swept dust samples. In these parallel samples, Cd was the only element detected only in the dust samples collected by the vacuum cleaners. Cd is associated with fine particles [44–46], while Na, K and Ca are associated with larger particles in both outdoor and indoor dust [47–49]. Therefore, using a commercial vacuum cleaner might ensure more efficient dust sampling than collection by dust sweeping.

### *3.3. Distribution of Elements Depending on the Sampling Site and Zones within the Dwellings*

The metallic element concentrations in residential dust vary between the zones of a given dwelling and among geographical locations. Variations in the concentrations of metallic elements in different geographical locations are more representative of the impact of the outdoor sources associated with surrounding industrial sites or with increased vehicle emissions in densely populated areas. In Figure 2, arithmetic mean concentrations with the relevant standard deviations of selected metallic elements are summarized, depending on the splitting of indoor dust sampling sites into urban ($n$ = 50 samples), suburban ($n$ = 23 samples) and rural ($n$ = 31 sample). In the calculation of arithmetic mean concentrations, the results of 16 dust samples obtained from 15 apartments whose residents smoked were not included. In addition, data were subjected to the Kruskal–Wallis test, which showed that statistically significant differences ($p < 0.05$) were observed only between the suburban–urban and suburban–rural dust sample groups for Al ($p = 0.01$ and $p = 0.03$) and Sr ($p = 0.02$ and $p = 0.001$) (see Supplementary Materials Table S3).

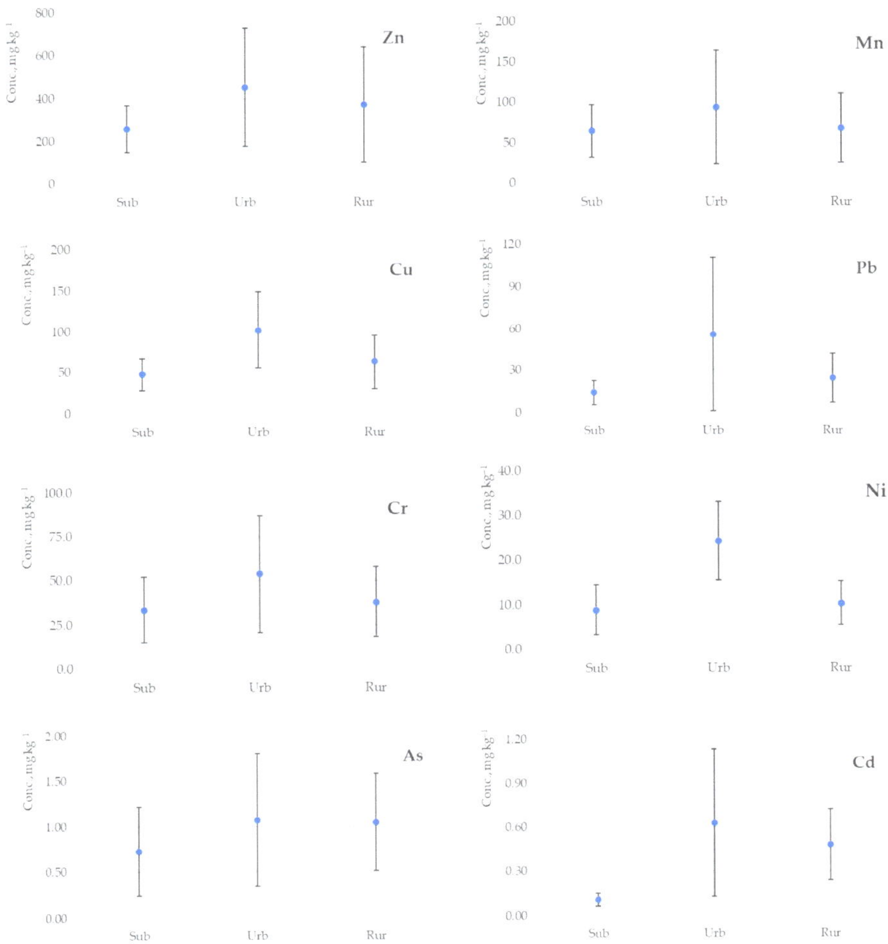


**Figure 2.** Arithmetic mean concentrations (mg kg$^{-1}$) with the relevant standard deviations of selected metallic elements in different sampling locations: urban (Urb) ($n = 50$), suburban (Sub) ($n = 23$) and rural (Rur) ($n = 31$).

The concentrations of all the above-listed metallic elements, depending on the dust sampling location, have the following increasing trend from suburban sampling sites to rural and urban: Sub < Rur < Urb. Meanwhile, the content of As is quite similar in urban and rural locations. Comparatively high concentrations of As can manifest in rural sites due to metal-containing pesticides or soil fertilization. Such agricultural activities are common in the rural regions of Latvia, and this is evidenced by the fact that the target values of not only As, but also the Zn, Ni and Cd content in the soil, exceeding which cannot ensure sustainable soil quality, have been exceeded both in the topsoil and in the deeper soil layers [50]. Elevated levels of Zn and Pb in urban sites can be associated with more intense vehicular traffic and industrial activities in the cities [51]. Latif et al. found that the concentrations of Pb and Zn decreased as the distance of the houses from major roads increased in urban areas [39]. In the current study, a similar trend was observed—in dust samples from dwellings located in the center of Riga, average values of Zn (515 mg $kg^{-1}$) and Pb (23.5 mg $kg^{-1}$) were higher than values of Zn (390 mg $kg^{-1}$) and Pb (5.36 mg $kg^{-1}$) in dwellings located apart from intensive traffic streets.

Dissimilarities in element distribution among different sampling rooms can indicate various internal pollution sources, which are largely affected by the household customs of the inhabitants, as well as the age of the particular building and the materials used in construction or renovation. In Figure 3, the percentages of selected metallic elements in the indoor dust of samples gathered from different rooms (corridor $n$ = 6 samples; kitchen $n$ = 10 samples; living room $n$ = 33 samples; bedroom $n$ = 45 samples) within dwellings are compared, except 5 dust samples collected from whole dwellings and 21 samples collected in the kitchen and living room together. A more detailed allocation of the metallic element concentrations obtained from different dwelling zones, in each urban, suburb and rural location, separating dwellings with smoking inside is summarized in Supplementary Materials Table S1. For all zones of dwellings, Zn accounted for the highest percentage of selected metallic elements in the bedroom area, and consequently, the percentages of the rest of the selected elements in this zone were lower. The use of various zinc-containing products, such as bed linen detergents or skin care products, which are used before bedtime, could contribute to increased concentrations of zinc. Looking closer, the mean concentrations of Zn in bedrooms located in urban, suburban and rural areas were 826 mg $kg^{-1}$, 754 mg $kg^{-1}$ and 701 mg $kg^{-1}$, respectively, while in kitchens and living rooms, the concentrations of Zn were 2–3 times lower, except one urban corridor area, which had an increased value (see Table S1). The entry corridor was the area with the highest average percentages of Pb, Cr and As.

The percentages of Pb and As in the entry areas were around 3 times higher than those in the bedroom or kitchen areas, although these elements are not typically associated with predominantly outdoor origin dust, which is tracked indoors by adhering to footwear, containing elements such as Mg, Al, V, Mn and Fe [30]. The results of the Kruskal–Wallis test demonstrated statistically significant differences between dust samples from bedrooms and kitchens only for B ($p = 0.01$), Mg (0.004) and Mn (0.003). Statistically significant differences in B between samples from living rooms and kitchens ($p = 0.01$) as well as kitchens and corridors ($p = 0.01$) were also observed, probably due to the application of B-containing bleaches and detergents (see Table S4). Relatively higher concentrations of Cd were observed either in entrance corridors or in living room areas, double those concentrations found in kitchen and bedroom zones. The kitchen areas had the highest concentrations of Mn, Cu and Ni (see Table S1). The influence of cooking and smoking activities has been indicated in indoor dust with regard to the amount of particulate matter and higher amounts of Cd and Ni [39,52]. In the current study, the smoking effect was observed by comparing the average concentrations of the metallic elements determined in the dust of non-smokers' (104 samples) and smokers' apartments (16 samples). Elevated mean concentrations of V (7.19 mg $kg^{-1}$), Cu (290 mg $kg^{-1}$), Zn (1090 mg $kg^{-1}$), As (1.39 mg $kg^{-1}$), Cd (1.52 mg $kg^{-1}$) and Pb (187 mg $kg^{-1}$) in indoor dust obtained from dwellings with smoking inside were observed (see Table S1). The performed Mann–Whitney U test rejected the null hypothesis only in the case of Cr ($p = 0.004$) and Pb (0.003) (see Table S5). The literature review revealed the relationship between house age and dust heavy metal concentrations, which was significant for Cd, Pb and Zn ($p < 0.001$), but not for As, Cr, Cu or Ni [34]. Another investigation linked increased concentrations of As, Cu, Pb and Zn in indoor dust from homes > 50 years old compared with homes < 50 years old [17]. However, in the current research, the dust results from four old, segregated buildings (without smoking inside) indicated increased values. Elevated concentrations of Cr were observed in two of the buildings, with a mean concentration of 111 mg $kg^{-1}$. High concentrations of Cu, Cd and Pb were found in three of the buildings at 116 mg kg$^{-1}$, 2.70 mg kg$^{-1}$ and 160 mg kg$^{-1}$, respectively, whereas elevated concentrations of Zn (2610 mg kg$^{-1}$) and As (5.63 mg kg$^{-1}$) were detected in all four houses. Cd-, Pb- and Cr-containing paints or construction materials with coatings containing trace metals used in older buildings can contribute to higher levels of these elements [33,53]. Along with this, the combined effect of other mixed indoor sources (fumes from heaters, cooking, carpets or other textiles) has a significant impact [54], most notably since the presence of finer particles indoors creates an additional opportunity for contaminants to chain to dust, resulting in increased concentrations of trace elements [17].

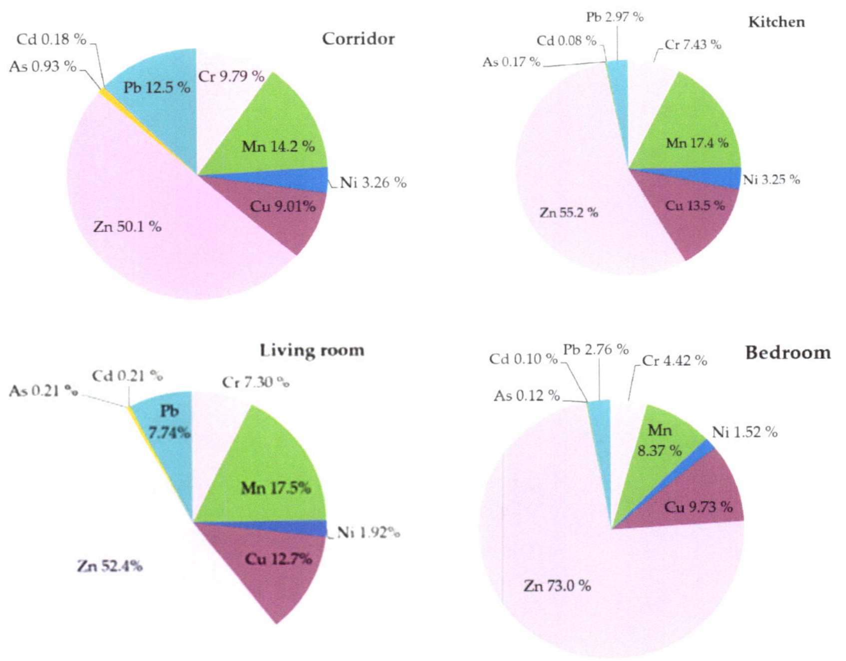


**Figure 3.** Percentages of selected metallic elements in indoor dust gathered from different rooms within dwellings (corridor $n$ = 6 samples; kitchen $n$ = 10 samples; living room $n$ = 33 samples; bedroom $n$ = 45 samples).

### *3.4. Principal Component Analysis*

In this study, experimental data from 120 samples were analyzed for 25 elements (Li, B, Na, Mg, Al, K, Ca, V, Cr, Mn, Fe, Co, Ni, Cu, Zn, As, Se, Rb, Sr, Cd, Ba, La, Ce, Pb and Bi) across three sampling locations with central heating systems, central heating systems with a gas stove and wood stove heating, using PCA (CAT software, manufacturer, city and country). PCA can help to decrease the dimensionality of a dataset while retaining critical information by detecting and retaining the principal components that capture the most variation in the data [55]. Concentration of chemical elements is the loading parameter for this PCA, which shows the influence extent of each element of the PCA score plot (samples). The further away from the center the element is, the more it affects the resulting score. Current results of PCA showed that the overall deviances between the samples can be characterized by component 1 (36.2%) and component 2 (12.3%). The obtained PCA plot, shown in Figure 4, shows that most differences between the groups can be observed in component 2, where all studied elements are separated into three groups: Cd, Al, Mn, V and Fe in the first group; Zn, As and Cu in the second; and Cr, Ni and Co in the third group, which had a negative correlation with the first group. In addition, PCA showed that samples from dwellings with central heating systems operated with gas tended to differ from the samples from dwellings with wood stove heating. The main differences between these two groups were in the concentrations of Zn, Cu, Cr, Ni and Co, which were lower in the dwellings with wood stove heating, where the concentrations of Al, Mn, Fe and As were higher, as explained by the higher amount of these elements in wood.

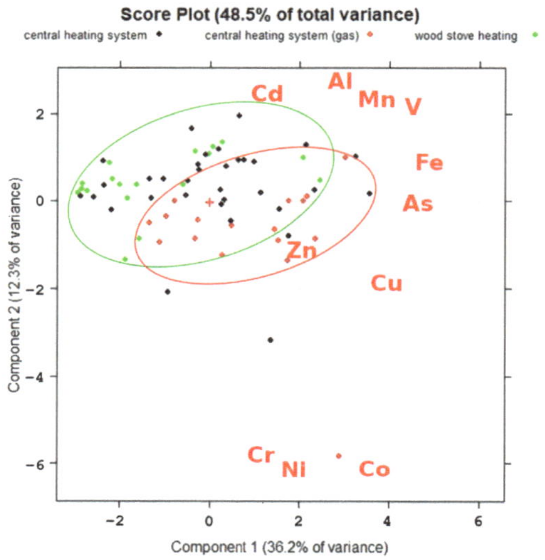


**Figure 4.** Principal component analysis of selected metallic elements in indoor dust gathered from dwellings with different heating systems.

## 4. Discussion

Even though soil is considered a major contributor to household dust, its characterization is a challenge due to the complexity of the many variations and combinations of its natural and anthropogenic origin, which differ from one residential space to another [56]. In this study, higher than average concentrations were detected for the elements Ca > Na > Mg > K > Fe > Al in residential dust. These elements are major components of the Earth's crust and natural soils, but other additional contributing sources have been observed. Based on an examination of major element correlation, Zararsiz and Öztürk reported that Al has a different origin from other crust metals [8]. For the most part, in the analyzed samples in the current study, the average range of Al concentrations varied from 1700 to 2200 mg kg$^{-1}$, but in six dwellings, 3 to 5 times higher concentrations of Al were observed in the living rooms, bathrooms and kitchens. Additional contributors to Al sources indoors can be pharmaceuticals, cosmetics, catalysts, pigments, paints, packaging and building materials [8]. The wide use of Ca in various cosmetic products and food additives has been reported, and accordingly, Ca, along with Al, could originate from the use of cosmetic products indoors [57]. Furthermore, the average concentrations of all the previously mentioned major elements along with Mn tend to be 1.2 to 1.7 times higher in homes with pets. This could be attributed to outdoor dust tracked indoors by pets, since traditionally in Latvia it is customary to take off outdoor shoes immediately after entering the premises of dwellings.

Indoor dust can act as a medium for the deposition of metallic elements. Among other contaminants, heavy metallic elements are of concern in the context of human health issues, considering their toxicity, low degradation potential, and ability to enter and accumulate in the body [18,19]. Considering the above, in combination with the health risk evaluation using the hazard index [14], we focused on Pb followed by Zn, Cu, Cd, Cr, Ni, Mn and As as the elements for exposure research on indoor dust. The highest concentrations of these metals were observed in urban areas. Concentrations of Mn, Ni, Cu, Pb and Cr in urban zones were roughly twice as high as their concentrations in suburban and rural areas. The exceptions were As, Zn and Cd, with quite similar concentrations in urban and rural areas. Mn and As indicated the lowest concentration differences in terms of different geographical locations: 94.0 mg $kg^{-1}$ (Urb) > 67.7 mg $kg^{-1}$ (Rur) > 64.6 mg $kg^{-1}$ (Sub) and 1.08 mg $kg^{-1}$ (Urb) > 1.05 mg $kg^{-1}$ (Rur) > 0.73 mg $kg^{-1}$ (Sub), respectively. Zn, Cu and Mn were the elements with the highest contribution from among the above set of metallic elements, i.e., Pb, Zn, Cu, Cd, Cr, Ni, Mn and As. Contrary to studies postulating that Cu and Zn are linked to industrial and traffic sources, such as the wear and tear of tires, lubricating oils, and the corrosion of alloys and galvanized vehicle parts [8,12,58], Dingle et al. [30] reported a higher indoor contribution. This could be related to the extensive use of both Cu and Zn in household items and in construction materials [30]. In other research, Mn and Zn, along with Pb and As, were significantly associated with the concentration levels of residential soil, in contrast with Cu, Cr and Ni, indicating that indoor sources may be more dominant in the relative contributions [6]. On the other hand, comparing concentrations in different dwelling zones, located in different areas, the lowest mean concentrations were mostly observed in suburban bedrooms: Cr 33.8 ± 17.0 mg $kg^{-1}$, Mn 57.7 ± 19.1 mg $kg^{-1}$, Ni 8.91 ± 5.54 mg $kg^{-1}$ and Cu 59.3 ± 32.0 mg $kg^{-1}$. Meanwhile, the highest concentrations of Cr and Mn were detected in urban kitchens at 69.4 ± 60.7 mg $kg^{-1}$ and 156 ± 135 mg $kg^{-1}$, respectively. For Cu, higher concentrations were observed in rural living rooms at 159 ± 354 mg $kg^{-1}$, (see Table S1). The effect of smoking amounted to an increase in the mean concentrations of some elements several times. Specifically, V, Cr, As, Zn and Cd increased around 2 times, but Cu and Pb increased even more, at 4 and 8 times, respectively (see Table S1).

The most challenging procedure with regard to indoor dust study is sampling to obtain representative and homogeneous data. Indoor dust sampling can be achieved in several ways, thus resulting in different units of metallic element concentrations. Besides the more often used concentration units (mg metal $kg^{-1}$ dust), metal loadings are expressed in μg metal $m^{-2}$ floor [42]. To avoid additional errors in the sampling procedure, measurements of the surface area to be cleaned or the duration of vacuuming were not required from the volunteers who provided us with the collected indoor dust. To compare the results of the metal loadings with the metal concentrations, the metal loading expressed in μg $m^{-2}$ needed to be combined with the dust deposition value and the dust deposition rate [34]. During the current study, using two approaches—commercial vacuum cleaner and manual collection with a brush and plastic spatula—parallel dust samples were collected. In general, the metallic element concentrations in parallel dust samples were mutually comparable, although the concentrations (mg $kg^{-1}$) of Na, K, Ca, Mg, Mn, Fe, Co, Ni and Pb were higher in the dust samples obtained by vacuuming. Cr and Cu showed the lowest concentration differences between parallel samples, while a higher Zn content was detected in the swept dust samples. Referring to the research done by Colt et al., household vacuum cleaners could be considered a reasonable alternative to the high-volume sampler [59]. However, after testing four types of vacuum cleaners (washable filter bagless, wet, bagged and HEPA filter-equipped robot), Vicente et al. reported that vacuuming itself is recognized as a source of indoor particle generation, and since numerous models of vacuuming devices are available on the market, further investigations are necessary [60].

Pb, Cd, As and Cr are systemic toxicants that are known to induce multiple organ damage, even at lower levels of exposure [18]. Ni has an extensive range of carcinogenic mechanisms, among them the controlled expressions of certain genes, while Mn can increase oxidative stress, thereby contributing to Mn cellular toxicity. The long-term exposure of a human body to the previously listed trace elements can progressively lead to diseases such as Parkinson's disease, multiple sclerosis, muscular dystrophy and Alzheimer's disease [19].

As there are no standard values for heavy metals in residential dust, the Regulations Regarding Quality Standards for Soil and Ground [61] were used as a reference to assess the contamination levels of heavy metals in indoor dust in Latvia. In this study, only the obtained mean concentration of Zn exceeded the critical limit value (700 mg $kg^{-1}$), while the arithmetic mean concentrations of other heavy metallic elements were below the defined critical guidelines (As, 40 mg $kg^{-1}$; Pb, 500 mg $kg^{-1}$; Cr, 350 mg $kg^{-1}$; Ni, 200 mg $kg^{-1}$; Cu, 150 mg $kg^{-1}$; and Cd, 8 mg $kg^{-1}$). Taking a closer look at the results of each sampling site, only the Cd concentrations in all sites were below the critical limit value, although at two locations, the concentrations were close to the critical limit value at 5.23 mg $kg^{-1}$ and 6.37 mg $kg^{-1}$, respectively. Corresponding critical limit values were surpassed in the following numbers of locations: As (Sub 1); Cu (Sub 3, Rur 2, Urb 11); Pb (Rur 2, Urb 2); Zn (Sub 5, Rur 8, Urb 15); Ni (Rur 1, Urb 1); and Cr (Sub 1). This means that the concentrations of As, Cu, Pb, Zn, Ni and Cr surpassed the critical limit values in 1%, 13%, 3%, 23%, 2% and 1% of all studied dwellings, respectively. Since the comparison is between obtained concentrations and the critical limit values, when remedial measures should be taken (in the case of soil), perhaps the comparison should be made more correctly with the precautionary limit values [61], which indicate the maximum level of pollution above which it is likely to have a negative impact on human health or the environment. If the WHO environmental quality standards for soils are used for the evaluation of heavy metal contamination in residential dust, the number of excesses is more critical. The mean concentrations of As, Cu, Pb, Zn, Ni and Cr exceed the corresponding permissible limits (20 mg $kg^{-1}$, 3 mg $kg^{-1}$, 100 mg $kg^{-1}$, 100 mg $kg^{-1}$, 50 mg $kg^{-1}$, 100 mg $kg^{-1}$ and 300 mg $kg^{-1}$ [56]) in 1%, 30%, 13%, 60%, 3% and 8% of dwellings in the investigated areas, respectively. Consequently, the highest risks of exposure from among the selected metallic elements could be applicable to Zn and Cu, followed by Pb and Ni.

## 5. Conclusions

The present study provided the first results of metallic element contaminations in residential indoor dust in Latvia. Indoor dust samples were collected at 46 sites to cover various areas in Latvia. A total of 120 dust samples from 28 urban dwellings, 8 suburban homes and 10 rural residences were gathered and subjected to inductively coupled plasma mass spectrometry (ICP-MS) analysis. The results of the analyzed indoor dust samples showed the following trends in terms of the major and minor metallic elements:

- High amounts ($10^4$–$10^3$ mg $kg^{-1}$) for Ca > Na > Mg > K > Fe > Al > Zn;
- Moderate amounts ($10^2$–$10^1$ mg $kg^{-1}$) for Ba > Cu > Mn > B > Pb > Cr > Sr > Ni > Bi;
- Lower concentrations ($10^0$–$10^{-1}$ mg $kg^{-1}$) for Ce > Rb > La > V > Li > Co > As > Cd > Se.

For the concentrations of the selected metallic elements with regard to dust sampling location, the trend Sub < Rur < Urb was representative, except for As, which had almost similar average concentrations in the urban and rural areas.

The two applied non-standardized indoor dust sampling approaches were appropriate, easy-to-implement dust sampling techniques. However, after the comparison of metallic concentrations from parallel sampling, a well-grounded conclusion about the most effective method cannot be made, emphasizing that there is a wide variety of vacuum cleaners available on the market.

The number of heavy metal limit value exceedances indicated a higher risk for the inhabitants of urban areas, given that the combined effects of outdoor and interior sources affected the majority of the metallic elements found in household dust. The higher risks of exposure to Zn and Cu were observed not only in urban areas, but also in suburban and rural regions, with the total exceeding the critical limit values in 44 samples. The risks of exposure to Ni and Pb were observed to a lesser extent in the urban and rural sampling locations, with the critical limit values exceeded in six samples.

**Supplementary Materials:** The following supporting information can be downloaded at: https://www.mdpi.com/article/10.3390/ijerph20136207/s1, Table S1: Arithmetic mean concentrations of metallic elements (mg $kg^{-1}$) with standard deviations and minimal and maximal values detected in indoor dust samples obtained from different parts of dwellings in urban (Urb), suburban (Sub) and rural (Rur) regions of Latvia. Table S2: The determined concentrations of metallic elements in the dust samples collected using both sampling techniques—vacuuming and manual collection with a brush and plastic spatula— in parallel. Table S3: Results of the Kruskal–Wallis test and the Mann–Whitney test for different dust sample groups depending on sampling site. Table S4: Results of the Kruskal–Wallis test and the Mann–Whitney test for different dust sample groups depending on sampling location within the dwelling. Table S5: Results of the Mann–Whitney test for two dust sample groups depending on smoking inside the dwelling.

**Author Contributions:** Conceptualization, A.A. and A.V.; methodology, A.A.; software, G.C. and M.B.; validation, M.B. and L.B.; formal analysis, M.B.; investigation, A.A.; resources, A.A.; data curation, A.A.; writing—original draft preparation, A.A.; writing—review and editing, L.B. and A.V.; visualization, G.C. and M.B.; supervision, A.V. All authors have read and agreed to the published version of the manuscript.

**Funding:** This research received no external funding.

**Institutional Review Board Statement:** Not applicable.

**Informed Consent Statement:** Not applicable.

**Data Availability Statement:** Not applicable.

**Acknowledgments:** The authors would like to thank all the volunteers, who provided indoor dust samples and related information.

**Conflicts of Interest:** The authors declare no conflict of interest.

## References

1. Klepeis, N.E.; Nelson, W.C.; Ott, W.R.; Robinson, J.P.; Tsang, A.M.; Switzer, P.; Behar, J.V.; Hern, S.C.; Engelmann, W.H. The National Human Activity Pattern Survey (NHAPS): A resource for assessing exposure to environmental pollutants. *J. Expo. Sci. Environ. Epidemiol.* **2001**, *11*, 231–252. [CrossRef] [PubMed]
2. Schweizer, C.; Edwards, R.D.; Bayer-Oglesby, L.; Gauderman, W.J.; Ilacqua, V.; Jantunen, M.J.; Lai, H.K.; Nieuwenhuijsen, M.; Künzli, N. Indoor time–microenvironment–activity patterns in seven regions of Europe. *J. Expo. Sci. Environ. Epidemiol.* **2007**, *17*, 170–181. [CrossRef] [PubMed]
3. World Health Organization (WHO). Burden of Disease from Household Air Pollution for 2016. 2018. Available online: https://cdn.who.int/media/docs/default-source/air-quality-database/aqd-2018/hap_bod_methods_may2018.pdf?sfvrsn=d277d739_32018/hap_bod_methods_may2018.pdf?sfvrsn=d277d739_3 (accessed on 28 July 2022).
4. Lucialli, P.; Marinello, S.; Pollini, E.; Scaringi, M.; Sajani, S.Z.; Marchesi, S.; Cori, L. Indoor and outdoor concentrations of benzene, toluene, ethylbenzene and xylene in some Italian schools evaluation of areas with different air pollution. *Atmos. Pollut. Res.* **2020**, *11*, 1998–2010. [CrossRef]
5. Latif, M.T.; Yong, S.M.; Saad, A.; Mohamad, N.; Baharudin, N.H.; Mokhtar, M.B.; Tahir, N.M. Composition of heavy metals in indoor dust and their possible exposure: A case study of preschool children in Malaysia. *Air Qual. Atmos. Health* **2014**, *7*, 181–193. [CrossRef]
6. Doyi, I.N.Y.; Isley, C.F.; Soltani, N.S.; Taylor, M.P. Human exposure and risk associated with trace element concentrations in indoor dust from Australian homes. *Environ. Int.* **2019**, *133*, 105125. [CrossRef] [PubMed]
7. Jeleńska, M.; Górka-Kostrubiec, B.; Werner, T.; Kądziałko-Hofmokl, M.; Szczepaniak-Wnuk, I.; Gonet, T.; Szwarczewski, P. Evaluation of indoor/outdoor urban air pollution by magnetic, chemical and microscopic studies. *Atmos. Pollut. Res.* **2017**, *8*, 754–766. [CrossRef]
8. Zararsız, A.; Öztürk, F. Estimation of Health Risks Associated with Household Dust Contamination in Bolu (Turkey). *Düzce Univ. J. Sci. Technol.* **2020**, *8*, 2245–2265. [CrossRef]
9. Sabzevari, E.; Sobhanardakani, S. Analysis of Selected Heavy Metals in Indoor Dust in Khorramabad City, Iran: A Case Study. *Jundishapur J. Health Sci.* **2018**, *10*, e67382. [CrossRef]
10. Turner, A. Oral bioaccessibility of trace metals in household dust: A review. *Environ. Geochem. Health* **2011**, *33*, 331–341. [CrossRef]
11. Paustenbach, D.J.; Finley, B.L.; Long, T.F. The critical role of house dust in understanding the hazards posed by contaminated soils. *Int. J. Toxicol.* **1997**, *16*, 339–362. [CrossRef]
12. Kulshrestha, A.; Massey, D.D.; Masih, J.; Taneja, A. Source Characterization of Trace Elements in Indoor Environments at Urban, Rural and Roadside Sites in a Semi Arid Region of India. *Aerosol Air Qual. Res.* **2014**, *14*, 1738–1751. [CrossRef]

13. Khoder, M.I.; Hassan, S.K.; El-Abssawy, A.A. An evaluation of loading rate of dust, Pb, Cd, and Ni and metals mass concentration in the settled surface dust in domestic houses and factors affecting them. *Indoor Built Environ.* **2010**, *19*, 391–399. [CrossRef]
14. Tan, S.Y.; Praveena, S.M.; Abidin, E.Z.; Cheema, M.S. A review of heavy metals in indoor dust and its human health-risk implications. *Rev. Environ. Health* **2016**, *31*, 447–456. [CrossRef]
15. Salwa, K.M.H. Metal concentrations and distribution in the household, stairs and entryway dust of some Egyptian homes. *Atmos. Environ.* **2012**, *54*, 207–215. [CrossRef]
16. Isley, C.F.; Fry, K.F.; Liu, X.; Filippelli, G.M.; Entwistle, J.A.; Martin, A.P.; Kah, M.; Meza-Figueroa, D.; Shukle, J.T.; Jabeen, K.; et al. International Analysis of Sources and Human Health Risk Associated with Trace Metal Contaminants in Residential Indoor Dust. *Environ. Sci. Technol.* **2022**, *56*, 1053–1068. [CrossRef]
17. Ibañez-Del Rivero, C.; Fry, K.L.; Gillings, M.M.; Barlow, C.F.; Aelion, C.M.; Taylor, M.P. Sources, pathways and concentrations of potentially toxic trace metals in home environments. *Environ. Res.* **2023**, *220*, 115173. [CrossRef]
18. Tchounwou, P.B.; Yedjou, C.G.; Patlolla, A.K.; Sutton, D.J. Heavy metal toxicity and the environment. *EXS* **2012**, *101*, 133–164. [CrossRef]
19. Engwa, G.; Ferdinand, P.; Nwalo, N.; Unachkwu, M. Mechanism and health effects of heavy metal toxicity in humans. In *Poisoning in the Modern World—New Tricks for an Old Dog?* Ozgur, K., Banu, A., Eds.; Intech Open: London, UK, 2019; pp. 77–99. [CrossRef]
20. Alghamdi, M.A.; Hassan, S.K.; Alzahrani, N.A.; Almehmadi, F.M.; Khoder, M.I. Risk assessment and implications of schoolchildren exposure to classroom heavy metals particles in Jeddah, Saudi Arabia. *Int. J. Environ. Res. Public Health* **2019**, *16*, 5017. [CrossRef]
21. Zhang, H.; Mao, Z.; Huang, K.; Wang, X.; Cheng, L.; Zeng, L.; Zhou, Y.; Jing, T. Multiple exposure pathways and health risk assessment of heavy metal(loid)s for children living in fourth-tier cities in Hubei Province. *Environ. Int.* **2019**, *129*, 517–524. [CrossRef]
22. Dockery, D.; Pope, A. Epidemiology of acute health effects: Summary of time series studies. In *Particles in Our Air—Concentration and Health Effects*; Wilson, R., Spengler, J.D., Eds.; Harvard University Press: Cambridge, MA, USA, 1996; pp. 123–147.
23. Kurt-Karakus, P.B. Determination of heavy metals in indoor dust from Istanbul, Turkey: Estimation of the health risk. *Environ. Int.* **2012**, *50*, 47–55. [CrossRef]
24. Tong, S.T.Y.; Lam, K.C. Home sweet home? A case study of household dust contamination in Hong Kong. *Sci. Total Environ.* **2000**, *256*, 115–123. [CrossRef] [PubMed]
25. Riga Municipality Portal. Available online: https://www.riga.lv/en/article/welcome-new-portal-riga-municipality (accessed on 28 September 2022).
26. Riga Air Quality Index. Available online: http://gmsd.riga.lv/main.php?i_lang=2 (accessed on 28 September 2022).
27. Google Maps. Latvia. Riga. Available online: https://www.google.com/maps/ (accessed on 29 September 2022).
28. Doyi, I.N.Y.; Strezov, V.; Isley, C.F.; Yazdanparast, T.; Taylor, M.P. The relevance of particle size distribution and bioaccessibility on human health risk assessment for trace elements measured in indoor dust. *Sci. Total Environ.* **2020**, *733*, 137931. [CrossRef] [PubMed]
29. Darus, F.M.; Nasir, R.A.; Sumari, S.M.; Ismail, Z.S.; Omar, N.A. Heavy Metals Composition of Indoor Dust in Nursery Schools Building. *Procedia Soc. Behav. Sci.* **2012**, *38*, 169–175. [CrossRef]
30. Dingle, J.H.; Kohl, L.; Khan, N.; Meng, M.; Shi, Y.A.; Pedroza-Brambila, M.; Chow, C.; Chan, A.W.H. Sources and composition of metals in indoor house dust in a mid-size Canadian city. *Environ. Pollut.* **2021**, *289*, 117867. [CrossRef]
31. Rashed, M.N. Total and extractable heavy metals in indoor, outdoor and street dust from Aswan city, Egypt. *Clean* **2008**, *36*, 850–857. [CrossRef]
32. Lisiewicz, M.; Heimburger, R.; Golimowski, J. Granulometry and the content of toxic and potentially toxic elements in vacuum-cleaner collected, indoor dusts of the city of Warsaw. *Sci. Total Environ.* **2000**, *263*, 69–78. [CrossRef]
33. Chattopadhyay, G.; Lin, K.C.P.; Feitz, A.J. Household dust metal levels in the Sydney metropolitan area. *Environ. Res.* **2003**, *93*, 301–307. [CrossRef]
34. Rasmussen, P.E.; Levesque, C.; Chénier, M.; Gardner, H.D.; Jones-Otazo, H.; Petrovic, S. Canadian House Dust Study: Population-based concentrations, loads and loading rates of arsenic, cadmium, chromium, copper, nickel, lead, and zinc inside urban homes. *Sci. Total Environ.* **2013**, *443*, 520–529. [CrossRef]
35. Turner, A.; Simmonds, L. Elemental concentrations and metal bioaccessibility in UK household dust. *Sci. Total Environ.* **2006**, *371*, 74–81. [CrossRef]
36. Glorennec, P.; Lucas, J.P.; Mandin, C.; le Bot, B. French children's exposure to metals via ingestion of indoor dust, outdoor playground dust and soil: Contamination data. *Environ. Int.* **2012**, *45*, 129–134. [CrossRef]
37. Ashley, K.; Brisson, M.J.; White, K.T. Review of standards for surface and dermal sampling. *J. ASTM Int.* **2011**, *1533*, 3–16. [CrossRef]
38. Al-Harbi, M.; Alhajri, I.; Whalen, J.K. Characteristics and health risk assessment of heavy metal contamination from dust collected on household HVAC air filters. *Chemosphere* **2021**, *277*, 130276. [CrossRef] [PubMed]
39. Latif, M.T.; Othman, M.R.; Kim, C.L.; Murayadi, S.A.; Sahaimi, K.N.A. Composition of household dust in semi-urban areas in Malaysia. *Indoor Built Environ.* **2009**, *18*, 155–161. [CrossRef]
40. Tashakor, M.; Behrooz, R.D.; Asvad, S.R.; Kaskaoutis, D.G. Tracing of Heavy Metals Embedded in Indoor Dust Particles from the Industrial City of Asaluyeh, South of Iran. *Int. J. Environ. Res. Public Health* **2022**, *19*, 7905. [CrossRef] [PubMed]

41. Plumejeaud, S.; Reis, A.P.; Tassistro, V.; Patinha, C.; Noack, Y.; Orsière, T. Potentially harmful elements in house dust from Estarreja, Portugal: Characterization and genotoxicity of the bioaccessible fraction. *Environ. Geochem. Health* **2018**, *40*, 127–144. [CrossRef] [PubMed]
42. Barrio-Parra, F.; de Miguel, E.; Lázaro-Navas, S.; Gómez, A.; Izquierdo, M. Indoor Dust Metal Loadings: A Human Health Risk Assessment. *Expo. Health* **2018**, *10*, 41–50. [CrossRef]
43. Al Hejami, A.; Davis, M.; Prete, D.; Lu, J.; Wang, S. Heavy metals in indoor settled dusts in Toronto, Canada. *Sci. Total Environ.* **2020**, *703*, 134895. [CrossRef]
44. Gemenetzis, P.; Moussas, P.; Arditsoglou, A.; Samara, C. Mass concentration and elemental composition of indoor $PM_{2.5}$ and $PM_{10}$ in University rooms in Thessaloniki, Northern Greece. *Atmos. Environ.* **2006**, *40*, 3195–3206. [CrossRef]
45. Mohammed, G.; Karani, G.; Mitchell, D. Trace Elemental Composition in $PM_{10}$ and $PM_{2.5}$ Collected in Cardiff, Wales. *Energy Procedia* **2017**, *111*, 540–547. [CrossRef]
46. Yu, P.; Han, Y.; Wang, M.; Zhu, Z.; Tong, Z.; Shao, X.; Peng, J.; Hamid, Y.; Yang, X.; Deng, Y.; et al. Heavy metal content and health risk assessment of atmospheric particles in China: A meta-analysis. *Sci. Total Environ.* **2023**, *867*, 161556. [CrossRef]
47. Oliveira, M.; Slezakova, K.; Delerue-Matos, C.; Pereira, M.C.; Morais, S. Assessment of air quality in preschool environments (3–5 years old children) with emphasis on elemental composition of $PM_{10}$ and $PM_{2.5}$. *Environ. Pollut.* **2016**, *214*, 430–439. [CrossRef] [PubMed]
48. Zwoździak, A.; Sówka, I.; Krupińska, B.; Zwoździak, J.; Nych, A. Infiltration or indoor sources as determinants of the elemental composition of particulate matter inside a school in Wrocław, Poland? *Build. Environ.* **2013**, *66*, 173–180. [CrossRef]
49. Evagelopoulos, V.; Charisiou, N.D.; Zoras, S. Dataset of Polycyclic aromatic hydrocarbons and trace elements in $PM_{2.5}$ and $PM_{10}$ atmospheric particles from two locations in North-Western Greece. *Data Brief.* **2022**, *42*, 108266. [CrossRef]
50. Bārdule, A. Micro-and Macro Element Flows in Short Rotation Hybrid Aspen (*Populus tremuloides* Michx. * *Populus tremula* L.) Plantation in Agricultural Land. Ph.D. Thesis, Latvian State Forest Research Institute "Silava", University of Latvia, Riga, Latvia, 2019; p. 124. (In Latvian).
51. Reis, A.; Cave, M.R.; Sousa, A.J.; Wragg, J.; Rangel, M.J.; Oliveira, A.R.; Patinha, C.; Rocha, F.; Orsiére, T.; Noack, Y. Lead and zinc concentrations in household dust and toenails of the residents (Estarreja, Portugal): A source-pathway-fate model. *Environ. Sci. Processes Impacts* **2018**, *20*, 1210–1224. [CrossRef]
52. Chao, C.Y.H.; Tung, T.C.W.; Burnett, J. Influence of Different Indoor Activities on the Indoor Particulate Levels in Residential Buildings. *Indoor Built Environ.* **1998**, *7*, 110–121. [CrossRef]
53. Ogilo, J.K.; Onditi, A.O.; Salim, A.M.; Yusuf, A.O. Assessment of Levels of Heavy Metals in Paints from Interior Walls and Indoor Dust from Residential Houses in Nairobi City County, Kenya. *Chem. Sci. Int. J.* **2017**, *21*, 37392. [CrossRef]
54. Chu, H.; Liu, Y.; Xu, N.; Xu, J. Concentration, sources, influencing factors and hazards of heavy metals in indoor and outdoor dust: A review. *Environ. Chem. Lett.* **2022**, *21*, 1203–1230. [CrossRef]
55. Gruppo di Chemiometria della Divisione di Chimica Analitica della Società Chimica Italiana (SCI). Chemometric Agile Tool (CAT). Available online: http://gruppochemiometria.it/index.php/software (accessed on 1 June 2022).
56. Shi, T.; Wang, Y. Heavy metals in indoor dust: Spatial distribution, influencing factors, and potential health risks. *Sci. Total Environ.* **2021**, *755*, 142367. [CrossRef]
57. Larsen, P.B.; Christensen, F.; Jensen, K.A.; Brinch, A.; Mikkelsen, S.H.; Exposure Assessment of Nanomaterials in Consumer Products. Environmental Project No. 1636. 2015. Available online: https://www2.mst.dk/Udgiv/publications/2015/01/978-87-93283-57-2.pdf (accessed on 9 September 2022).
58. Olowoyo, J.O.; Mugivhisa, L.L.; Magoloi, Z.G. Composition of Trace Metals in Dust Samples Collected from Selected High Schools in Pretoria, South Africa. *Appl. Environ. Soil Sci.* **2016**, *2016*, 5829657. [CrossRef]
59. Colt, J.S.; Gunier, R.B.; Metayer, C.; Nishioka, M.G.; Bell, E.M.; Reynolds, P.; Buffler, P.A.; Ward, M.H. Household vacuum cleaners vs. the high-volume surface sampler for collection of carpet dust samples in epidemiologic studies of children. *Environ. Health* **2008**, *7*, 6. [CrossRef]
60. Vicente, E.D.; Vicente, A.M.; Evtyugina, M.; Calvo, A.I.; Oduber, F.; Blanco Alegre, C.; Castro, A.; Fraile, R.; Nunes, T.; Lucarelli, F.; et al. Impact of vacuum cleaning on indoor air quality. *Build. Environ.* **2020**, *180*, 107059. [CrossRef]
61. Government of Latvia. *Regulations Regarding Quality Standards for Soil and Ground. Republic of Latvia Cabinet Regulation No. 804 Adopted 25 October 2005*; Government of Latvia: Riga, Latvia, 2005. Available online: https://likumi.lv/ta/en/en/id/120072 (accessed on 15 November 2022).

International Journal of *Environmental Research and Public Health*

MDPI

*Article*

# Health Risk Assessment of Exposure to Air Pollutants Exceeding the New WHO Air Quality Guidelines (AQGs) in São Paulo, Brazil

**Caroline Fernanda Hei Wikuats [1,*], Thiago Nogueira [2], Rafaela Squizzato [1], Edmilson Dias de Freitas [1] and Maria de Fatima Andrade [1]**

[1] Departamento de Ciências Atmosféricas, Instituto de Astronomia, Geofísica de Ciências Atmosféricas, Universidade de São Paulo, São Paulo 05508-090, Brazil

[2] Departamento de Saúde Ambiental, Faculdade de Saúde Pública, Universidade de São Paulo, São Paulo 01246-904, Brazil

* Correspondence: caroline.wikuats@usp.br

**Abstract:** We applied the AirQ+ model to analyze the 2021 data within our study period (15 December 2020 to 17 June 2022) to quantitatively estimate the number of specific health outcomes from long- and short-term exposure to atmospheric pollutants that could be avoided by adopting the new World Health Organization Air Quality Guidelines (WHO AQGs) in São Paulo, Southeastern Brazil. Based on temporal variations, $PM_{2.5}$, $PM_{10}$, $NO_2$, and $O_3$ exceeded the 2021 WHO AQGs on up to 54.4% of the days during sampling, mainly in wintertime (June to September 2021). Reducing $PM_{2.5}$ values in São Paulo, as recommended by the WHO, could prevent 113 and 24 deaths from lung cancer (LC) and chronic obstructive pulmonary disease (COPD) annually, respectively. Moreover, it could avoid 258 and 163 hospitalizations caused by respiratory (RD) and cardiovascular diseases (CVD) due to $PM_{2.5}$ exposure. The results for excess deaths by RD and CVD due to $O_3$ were 443 and 228, respectively, and 90 RD hospitalizations due to $NO_2$. Therefore, AirQ+ is a useful tool that enables further elaboration and implementation of air pollution control strategies to reduce and prevent hospital admissions, mortality, and economic costs due to exposure to $PM_{2.5}$, $O_3$, and $NO_2$ in São Paulo.

**Keywords:** air pollutants; ambient air quality standards; health outcomes; AirQ+ software



**Citation:** Wikuats, C.F.H.; Nogueira, T.; Squizzato, R.; de Freitas, E.D.; Andrade, M.d.F. Health Risk Assessment of Exposure to Air Pollutants Exceeding the New WHO Air Quality Guidelines (AQGs) in São Paulo, Brazil. *Int. J. Environ. Res. Public Health* **2023**, *20*, 5707. https://doi.org/10.3390/ijerph20095707

Academic Editors: José Ángel Fernández, Nuno Canha, Marta Almeida and Evangelia Diapouli

Received: 28 February 2023
Revised: 25 April 2023
Accepted: 27 April 2023
Published: 2 May 2023



## 1. Introduction

Air pollution is one of the greatest environmental risks to human health increasing morbidity and mortality and reducing life expectancy [1,2]. In 2019, air pollution ranked fourth as the world's leading risk factor for early death causing 6.7 million deaths worldwide, of which 4.1 million were due to ambient air pollution [3–6].

According to the World Health Organization (WHO), in 2019, 37% of premature deaths related to outdoor air pollution occurred due to ischemic heart disease (IHD) and stroke, while 23 and 18% were caused by acute lower respiratory infections and chronic obstructive pulmonary disease (COPD), respectively, and 11% due to lung cancer (LC) [2]. Furthermore, global deaths (in millions of people) resulting from exposure to ambient air pollution have increased by 51% since 1990 and are estimated to double by 2050 if more relevant interventions do not occur [5,7].

Air pollution consists of both gaseous and particulate primary pollutants released directly into the atmosphere, such as nitrogen oxides ($NO_x$), carbon monoxide (CO), and sulfur dioxide ($SO_2$), as well as secondary pollutants formed in the atmosphere, including fine particulate matter ($PM_{2.5}$) and ozone ($O_3$). Particulate matter is the most studied pollutant presenting consistent evidence of adverse health effects [8,9]. Exposure to both coarse and fine particles ($PM_{10}$ and $PM_{2.5}$, respectively) is harmful to health, but the latter represents the most robust predictor of mortality from lung cancer, respiratory, cardiovascular, and other diseases [1,3]. Numerous studies also report a relationship between exposure to gaseous pollutants, such as $O_3$, and increased morbidity and mortality [5].

In 2019, more than 90% of the world's population was living in areas, mainly in low- and middle-income countries (e.g., Brazil), where ambient air pollution concentrations surpassed the 2005 WHO guideline for $PM_{2.5}$ (annual mean: 10 μg $m^{-3}$) [3]. The WHO Air Quality Guidelines (AQGs) provide evidence-based guidance for air pollutant levels to reduce exposure and protect public health. In 2021, the AQGs were updated due to the ongoing risk of air pollution to human health, with more restrictive concentrations for pollutants such as $PM_{2.5}$ (annual mean: from 10 to 5 μg $m^{-3}$), $PM_{10}$, CO, and nitrogen dioxide ($NO_2$) [10,11].

In Brazil, despite the reductions in primary pollutant concentrations achieved through the implementation of air pollution control programs in recent years, several levels remain above the WHO AQGs. Additionally, although public policies such as the Program for the Control of Air Pollution Emissions by Motor Vehicles (PROCONVE, established in 1986) have led to a decrease in vehicular emissions, the growth of the vehicle fleet (as reported by Nogueira et al. [12]) has offset these gains. The concentrations of secondary pollutants such as $PM_{2.5}$ and $O_3$ remain a concern and are not yet effectively controlled, as noted by Andrade et al. [13]. Further studies on outdoor exposure to mainly secondary pollutants are necessary to develop or adjust policies, regulate their concentrations, improve air quality, and potentially reduce adverse health effects.

Moreover, there is a scarcity of studies conducted in Brazil focusing on the quantitative assessment of air pollution effects on hospitalizations and mortality. Considering existing research based on data from the 1990s and 2000s, a positive association was observed between Disability-Adjusted Life Years (DALYs) and air pollution in São Paulo adding up to 28,212 years annually and more than 5000 deaths that could be prevented if $PM_{2.5}$ levels were reduced to the 2005 WHO AQG of 10 μg $m^{-3}$ [14,15]. These studies present relevant results related to air pollution and health effects in the Metropolitan Area of São Paulo (MASP), but it is also important to evaluate the outcomes with the new WHO AQGs, taking into account the reduction in the levels of air pollutants observed recently. To the best of our knowledge, this is the first study assessing the health risks of air pollutants in the region using AirQ+, which is a tool for quantifying the health burden and impact of air pollution developed by the WHO Regional Office for Europe [16].

The progress toward achieving the Sustainable Development Goals (SDG) outlined in the 2030 Agenda has been insufficient and unequal globally [17]. In Brazil, these inadequacies are apparent in the statistical monitoring of the 2030 Agenda, which fails to report data on mortality rates caused by household and outdoor air pollution. To address this issue, we employed the AirQ+ model to quantify the number of health outcomes that could be prevented by adopting the new WHO AQGs for long- and short-term exposure to $PM_{2.5}$, $NO_2$, and $O_3$ in São Paulo, Brazil, based on their temporal variations.

## 2. Materials and Methods

### 2.1. Study Site

São Paulo is South America's largest and most developed and industrialized region. It has an estimated population of approximately 12.4 million individuals, a territorial unit area of 1521 $km^2$, a population density of around 7398 inhabitants.$km^{-2}$, and a Human Development Index (HDI) of 0.805 [18]. In addition, it has a Gross Domestic Product (GDP) of around USD 150 billion, corresponding to 33% from the state of São Paulo and 58% from the MASP [19,20].

According to Köppen's classification, the region presents a humid subtropical climate (Cwa) with dry and cool winters (temperatures below 18 °C) and wet and warm summers (temperatures above 22 °C) [21]. São Paulo is surrounded by hills of 1200 m and is located on a plateau at 860 m above sea level. The frequent occurrence of subsidence layers and thermal inversion makes the dispersion of pollutants difficult, especially during winter [22].

Furthermore, São Paulo has a fleet of around 8.9 million vehicles responsible for the emissions of 96% of CO, 73% of hydrocarbons (HC), 65% of $NO_x$, 40% of PM, and 11% of $SO_x$ [23,24]. We carried out measurements with a mobile laboratory at the School of Medicine of the University of São Paulo (FMUSP) (23°33′16.2″ S, 46°40′19.7″ W) (Figure 1), where the influence of emissions from mobile sources, such as light- and heavy-duty vehicles (LDVs and HDVs, respectively), is significant.

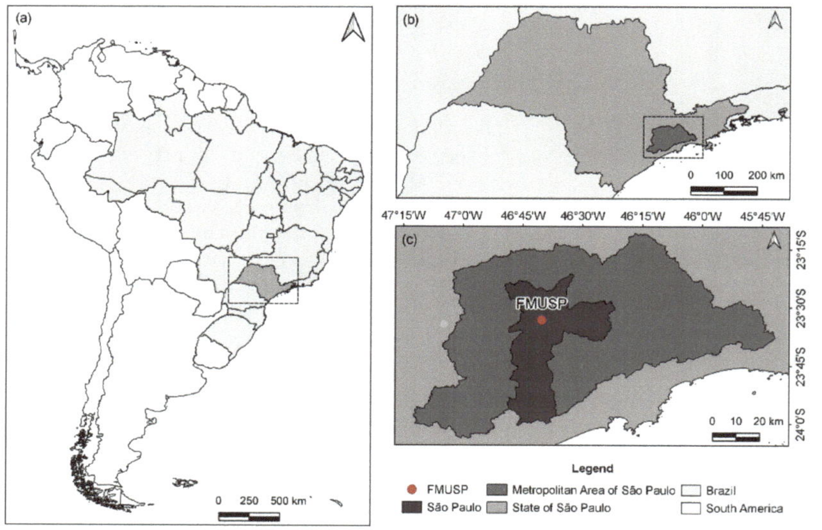


**Figure 1.** Location of Brazil and the state of São Paulo (dashed line rectangle) (**a**), the Metropolitan Area of São Paulo (MASP) (dashed line rectangle) within the state (**b**), the city of São Paulo and the School of Medicine of the University of São Paulo (FMUSP) (**c**).

### 2.2. Data Sampling and Analysis

We performed the sampling campaign of air pollutants with the Mobile Laboratory for Research and Monitoring of Air Quality (LuMIAR) at the FMUSP continuously between 15 December 2020 and 17 June 2022 (a total of 550 days). LuMIAR is equipped with monitors for measuring $PM_{2.5}$, $PM_{10}$, $SO_2$, CO, $NO_2$, and $O_3$, and the data are reported on a 1-h average. In Table 1, we present information from the instruments considered for this study [25]. Concentration units for $PM_{2.5}$, $PM_{10}$, $SO_2$, $NO_2$, and $O_3$ are presented in $\mu g\ m^{-3}$, while CO is presented in $mg\ m^{-3}$ to compare with the AQGs. The Weather Station of the Institute of Astronomy, Geophysics, and Atmospheric Sciences of the University of São Paulo provided the meteorological variables (air temperature, precipitation, wind speed, and wind direction).

We performed all the data organization and statistical analyses in the R Software (version 4.0.3) and used the openair package (version 2.10-0) to plot time series, temporal variations, and wind rose graphs for the pollutants and meteorological variables [26,27]. We used the AirQ+ software tool (version 2.1.1) to conduct health risk assessments for long- and short-term effects [16] based on temporal trends for $PM_{2.5}$, $NO_2$, and $O_3$. It can be applied for any region to evaluate how much a specific health outcome is attributable to certain air pollutant concentrations and the health effects if levels change in the future compared to the present. AirQ+ calculations were based on methodologies and concentration-response functions determined by epidemiological studies available up to 2013 and their meta-analysis [28]. Some limitations of the software include the following: calculations do not consider multiple exposure cases or multi-pollutant scenarios; it uses ambient data as a proxy indicator of population exposure; and "morbidity estimates present low reliability due to difficult conformity in the assessment of health outcomes related to hospital admissions" [28].

**Table 1.** Information about the Mobile Laboratory for Research and Monitoring of Air Quality (LuMIAR) equipment.

| Equipment. | Parameter | Model | Manufacturer | Unit |
|---|---|---|---|---|
| TEOM Continuous Ambient Particulate Monitor | $PM_{2.5}$, $PM_{10}$ | 1405 | Thermo Fisher Scientific | $\mu g\ m^{-3}$ |
| $SO_2$ Analyzer | $SO_2$ | 43i | Thermo Fisher Scientific | $\mu g\ m^{-3}$ |
| Photoacoustic Gas Monitor | CO | Innova 1512 | LumaSense Technologies | $mg\ m^{-3}$ |
| $NO_x$ Analyzer | $NO_2$ | 42i | Thermo Fisher Scientific | $\mu g\ m^{-3}$ |
| $O_3$ Analyzer | $O_3$ | 49i | Thermo Fisher Scientific | $\mu g\ m^{-3}$ |

We obtained population data for the city of São Paulo from the Brazilian Institute of Geography and Statistics (IBGE) [18], and mortality and hospital admission data from the Department of Informatics of the Brazilian Unified Health System (DATASUS) [29]. We considered the period from 1 January to 31 December 2021, for the incidence per 100,000 inhabitants at risk per year based on the mortality and hospitalization statistics, as it was the only complete year available in the dataset. Mortality data included LC, COPD, IHD, and stroke information for different age groups. On the other hand, hospitalization and mortality data considered respiratory (RD) and cardiovascular diseases (CVD) for the total population.

Furthermore, we used 24-h mean values for the entire year, location area size, total population, the population at risk, incidence per 100,000 inhabitants, and cut-off concentration as the input data in the software. For long-term exposure, we applied the annual WHO AQGs that denoted "the lower end of the range in which significant effects on survival have been observed" [30]. On the other hand, the 24-h mean WHO AQGs were considered for short-term effects. The AirQ+ software already included default relative risk (RR) values. We utilized the log-linear and the Global Burden of Disease (GBD) 2015/2016 (integrated function 2016 vs. 2005 WHO AQG value) methods to estimate the results for the pollutants. The optional input data included latitude, longitude, and location identification. We assessed different health-related results using this software tool, including the attributable proportion of cases, the number of attributable cases, the number of attributable cases per 100,000 population at risk, the proportion of cases in the pollutant concentration range, and the cumulative distribution by air pollutant concentration [16,31,32]. We also used Origin (version 2020) to plot bar charts for the attributable cases of mortality and hospital admissions.

## 3. Results and Discussion

### *3.1. Pollutant Concentrations and Meteorological Variables Overview*

Descriptive statistics and time series for the pollutants measured at the FMUSP are shown in Table 2 and Figure 2, respectively. Mean values for all pollutants were similar in 2021, 2022, and throughout the entire sampling period (15 December 2020 to 17 June 2022) (Table 2). Abe and Miraglia [15] reported concentrations of $21 \pm 10\ \mu g\ m^{-3}$ for $PM_{2.5}$, $36 \pm 17\ \mu g\ m^{-3}$ for $PM_{10}$, and $83 \pm 36\ \mu g\ m^{-3}$ for $O_3$ between 2009 and 2011 in São Paulo. Compared with the overall mean values obtained at the FMUSP, these three pollutants observed significant decreases of 47.1, 47.8, and 55.9% over ten years.

**Table 2.** Descriptive statistics of the pollutants sampled at the FMUSP between 15 December 2020 and 17 June 2022. Data uncertainty is expressed as the standard deviation (±values in parenthesis).

| Pollutant | Overall Mean | 2021 Mean | 2022 Mean |
|---|---|---|---|
| $PM_{2.5}$ (μg m$^{-3}$) | 11.1 (±6.8) | 11.6 (±7.4) | 10.5 (±5.3) |
| $PM_{10}$ (μg m$^{-3}$) | 18.8 (±12.1) | 19.5 (±13.3) | 18.0 (±9.1) |
| $SO_2$ (μg m$^{-3}$) | 2.8 (±1.0) | 2.8 (±1.0) | 2.9 (±1.0) |
| CO (mg m$^{-3}$) | 0.7 (±0.5) | 1.2 (±0.4) | 1.5 (± 0.4) |
| $NO_2$ (μg m$^{-3}$) | 32.2 (±15.0) | 32.2 (±15.4) | 33.8 (±13.9) |
| $O_3$ (μg m$^{-3}$) | 36.6 (±22.2) | 37.9 (±21.8) | 33.2 (±22.9) |

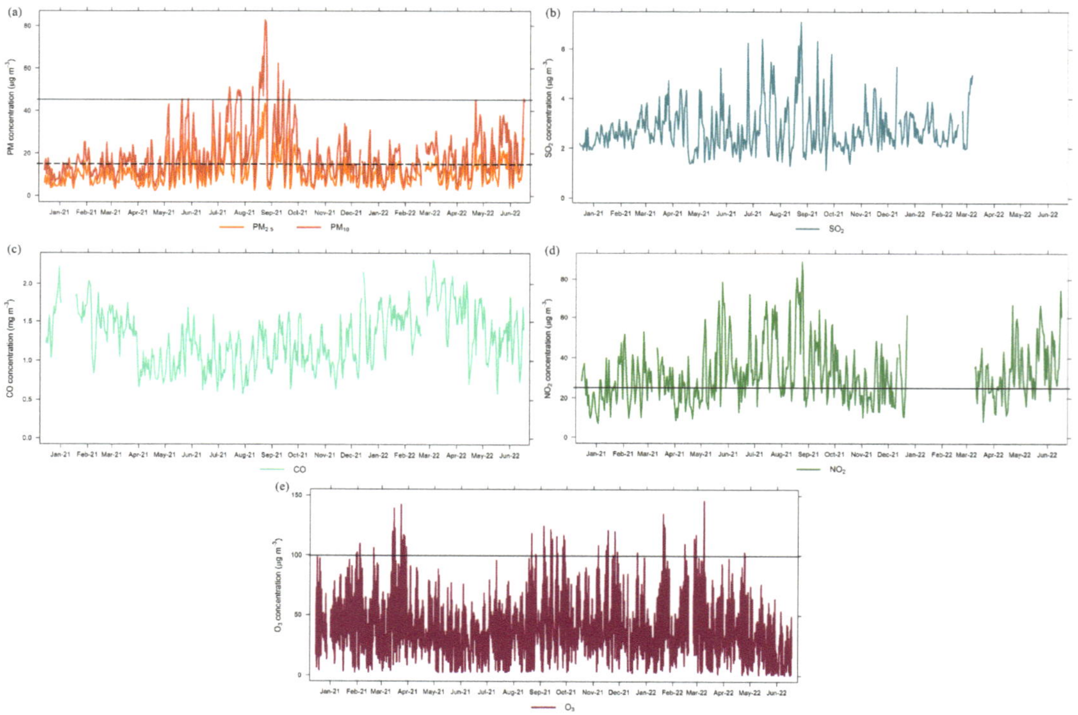


**Figure 2.** Twenty-four-hour time series for $PM_{2.5}$ and $PM_{10}$ (**a**), $SO_2$ (**b**), CO (**c**), and $NO_2$ (**d**), and 8-h time series for $O_3$ (**e**) during the sampling period. The black horizontal lines indicate the WHO Air Quality Guidelines. 24-h mean for $PM_{2.5}$: 15 μg m$^{-3}$, $PM_{10}$: 45 μg m$^{-3}$, $NO_2$: 25 μg m$^{-3}$; and 8-h mean for $O_3$: 100 μg m$^{-3}$.

Nevertheless, $PM_{2.5}$, $PM_{10}$, $NO_2$, and $O_3$ still exceeded the 2021 WHO AQGs on 117 (21.3%), 26 (4.7%), 299 (54.4%), and 171 (31.1%) days, mainly during wintertime (June to September 2021), considering the 550 days of the sampling campaign (Figure 2). For $NO_2$, 54.4% of the data exceeded the AQG for the 24-h mean, while it was not surpassed when considering the 1-h mean (not shown). The graphs for $SO_2$ and CO do not include black horizontal lines indicating the WHO AQGs, as the values registered at the FMUSP were much lower. Over the years, an expressive reduction in $SO_2$ concentrations in São Paulo has occurred due to the diminished sulfur content in vehicular and industrial fuels [33].

The WHO AQGs are much more restrictive than the Brazilian National Air Quality Standards (NAQS) recommended by the Brazilian National Environmental Council (CONAMA) Resolution 491/2018 [34] (Table 3), which was established considering the 2005 WHO AQGs as a reference. Nonetheless, the implementation criteria consider intermediate NAQS, i.e., standards determined as temporary values to be completed in phases, although the period of use of each one before reaching the final AQGs (from 2005) has not been specified. Therefore, the values for intermediate phase 1 are currently in effect. Comparing our results with the NAQS, none of the pollutants surpassed the limits (not shown), which highlights the differences between the Brazilian legislation and the WHO recommendations. Hence, there is a need for monitoring the concentrations of atmospheric pollutants in the region and for more effective policies to implement restrictive limits following the WHO guidelines, reducing the delay that is currently observed.

**Table 3.** Air quality guidelines recommended by the WHO and CONAMA.

| Pollutant | | Averaging Time | WHO | CONAMA |
|---|---|---|---|---|
| $PM_{2.5}$ | ($\mu g\ m^{-3}$) | Annual | 5 | 20 |
| | | 24-h | 15 | 60 |
| $PM_{10}$ | | Annual | 15 | 40 |
| | | 24-h | 45 | 120 |
| $SO_2$ | | 24-h | 40 | 125 |
| CO | ($mg\ m^{-3}$) | | 4 | - |
| | (ppm) | 8-h | 9 | 9 |
| $NO_2$ | ($\mu g\ m^{-3}$) | Annual | 10 | 60 |
| | | 24-h | 25 | - |
| | | 1-h | 200 | 260 |
| $O_3$ | | 8-h | 100 | 140 |

WHO—World Health Organization; CONAMA—Brazilian National Environmental Council (Conselho Nacional do Meio Ambiente).

Historically, cities in Brazil (including São Paulo) have suffered from rampant violations of AQGs, particularly in the 1970s and 1980s, due to a lack of control and regulation of air pollution sources. However, following international initiatives, governmental organizations have implemented various measures to address this issue, focusing on the reduction in emissions from mobile sources in the transport sector [13]. The most successful initiative was implementing the PROCONVE program, based on CONAMA Resolutions 18/1986 and 297/2002, effectively reducing concentrations of primary pollutants, such as CO, $PM_{10}$, and $SO_2$ [35,36]. The program was responsible for decreasing 90% of emissions from LDVs and 80% from HDVs based on enforceable legislation to promote national technological development in automobile engineering and methods and equipment for testing and measuring pollutant emissions, to encourage the large-scale use of biofuels, and to reduce the sulfur content in fuels [13,33,35,36]. For example, it resulted in the introduction of flex-fuel LDVs (powered by gasohol, ethanol, or any mixture of both) considering new exhaust systems and catalytic converters, and the addition of biodiesel to diesel for HDVs. Since 2012, post-treatment with selective catalytic reduction (SCR) systems for the $NO_x$ emissions of HDVs has also been mandatory [12].

In 1990, CONAMA Resolution 003/1990 established NAQS considering the need to monitor and control air pollutants in Brazil, which was then repealed by CONAMA Resolution 491/2018 to follow the 2005 WHO AQGs as mentioned above. Furthermore, the city of São Paulo implemented driving restrictions in the 1990s based on the last digits of the license plate of cars on pre-established days during peak hours (from 7 to 10 h and from 17 to 20 h). This initiative was introduced to decrease air pollution emissions and alleviate traffic congestion [13,33].

Regarding industrial emissions in the MASP, currently, there is a small contribution to the concentrations of air pollutants due to a moving process of point sources to other cities, which had fewer restrictions to install potentially polluting industries, that occurred in the 1980s and 1990s. On the other hand, policies have been implemented to reduce industrial emissions since the 1980s in the state of São Paulo. For example, the change from oil-fired to electric boilers to generate energy for industries was remarkable in reducing $SO_2$ ambient concentrations. In 2007, another important action was to institute rules for new facilities considering the local level of regular pollutants (CO, $NO_2$, $O_3$, $PM_{10}$, total particulate matter, $SO_2$, and smoke) [13,33].

Nogueira et al. [12] reported reductions of 4.9 and 5.1% per year for CO and of 5.5 and 4.2% per year for $NO_x$ from 2001 to 2018 in São Paulo considering emission factors for LDVs and HDVs, respectively. The authors discussed that, although overdue, the vehicular emission regulations adopted in Brazil have significantly improved air quality in the MASP. From 1996 to 2009, a downward trend was also observed for all pollutants (CO, $NO_x$, $SO_2$, and $PM_{10}$) monitored by the Environmental Company of the State of São Paulo (CETESB), except for $O_3$ [33]. Therefore, despite the reductions in primary pollutants levels achieved with air pollution regulation programs, controlling the concentrations and emissions of secondary pollutants such as $PM_{2.5}$ and $O_3$ is the greatest challenge currently faced by governmental organizations in São Paulo and Brazil.

The hourly time variations of pollutants are shown in Figure 3 for the entire sampling period (hours are shown as local time). Except for $O_3$, the pollutants presented similar hourly profiles, with two peaks (from 8 to 11 h and 19 to 23 h) due to emissions from mobile sources with increased traffic on the roads close to the sampling site. The peak associated with the nighttime rush hour indicated larger time variability among the pollutants, mainly for $PM_{10}$ and CO, which presented peaks earlier (from 16 to 18 h) and later (between 23 and 1 h) in the day, respectively. In addition, the Planetary Boundary Layer (PBL) stability in the evening hampers pollutants' dispersion, maintaining high concentrations during this period. Moreover, the hourly profiles follow the characteristics already observed in São Paulo [22,33].

For $O_3$, the highest concentrations occurred between 14 and 16 h, a different behavior from the other pollutants due to photochemical reactions between $O_3$ precursors ($NO_x$ and volatile organic compounds—VOCs) and solar radiation. The chemistry of $NO_x$ at night differs from that observed during the day, as $NO_2$ photolysis does not occur, i.e., NO present in the atmosphere at night can rapidly react with $O_3$, and nearly all $NO_x$ is converted into $NO_2$ [37]. As also observed by Carvalho et al. and Massambani and Andrade [33,38], between 3 and 4 h (Figure 3), a secondary peak appeared due to horizontal and vertical transport from other regions as there is no $O_3$ formation during the night [33,39].

In Figure S1 of the Supplementary Material, July, August, and September presented the highest monthly averages for $PM_{2.5}$, $PM_{10}$, $SO_2$, and $NO_2$, corresponding to winter in Brazil, a characteristic period of higher levels of air pollution due to stable atmospheric conditions and lower precipitation rates. It is possible to observe great variability in the CO data (Figure S1), with the highest concentrations in January and February at the FMUSP, which was not expected but probably occurred due to a specific local source still under investigation.

These results agree with the local meteorological conditions (Figures 4 and S2). The mean temperature values over the entire sampling period were 19.7 $\pm$ 3.4 °C, while precipitation was recorded on 266 days with a maximum of 71.2 mm/day. As expected, April to September (fall and winter) presented the lowest precipitation rates, since São Paulo has dry and cool winters [18]. The relative humidity was 79.9 $\pm$ 8.1%, and the wind speed was 1.5 $\pm$ 0.5 m $s^{-1}$ with a maximum of 4.9 m $s^{-1}$. Wind direction showed a consistent pattern coming from the southeast and northeast more than 50 and 30% of the time. These meteorological data follow the climate normal (1991–2020) for São Paulo according to information from the Brazilian National Institute of Meteorology (INMET) (not shown).

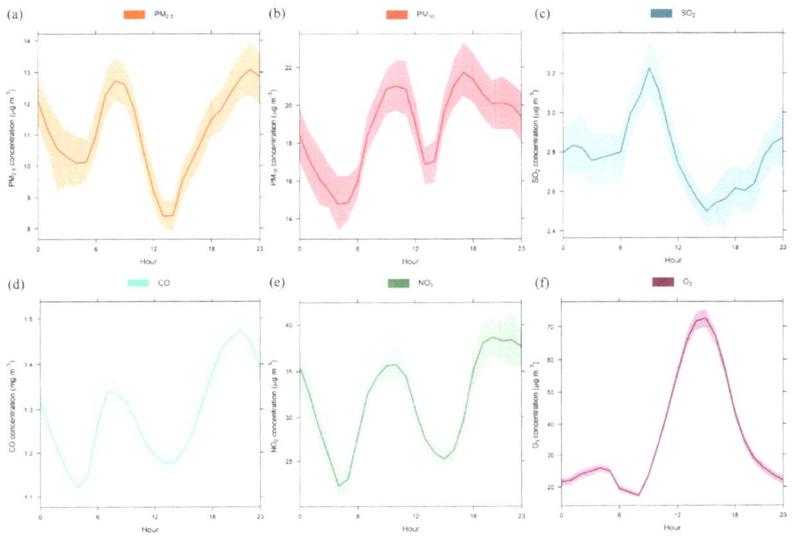


**Figure 3.** Hourly variations for $PM_{2.5}$ (**a**), $PM_{10}$ (**b**), $SO_2$ (**c**), CO (**d**), $NO_2$ (**e**), and $O_3$ (**f**) during the sampling period. The shaded areas represent the 95% confidence intervals of the mean.

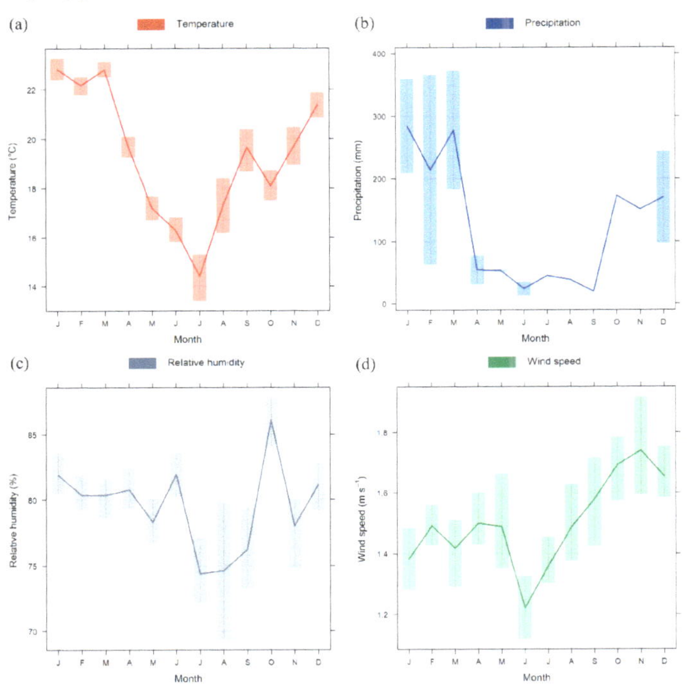


**Figure 4.** Monthly variations for temperature (**a**), precipitation (**b**), relative humidity (**c**), and wind speed (**d**) during the sampling period. The shaded areas represent the 95% confidence intervals of the mean.

Our analysis revealed a decrease in the concentrations of primary pollutants during the weekend, mainly on Sunday (Figure S1), as the number of vehicles in circulation can decrease up to 70% compared to weekdays, according to data from the Traffic Engineering Company of São Paulo [40]. On the other hand, $O_3$ levels were relatively higher during the weekends, a trend already documented in the region and associated with a VOC-limited atmosphere in urban areas [41,42]. This is because the substantial reduction in $NO_x$ emissions from HDVs on weekends leads to an increase in $O_3$ levels [25,37].

### 3.2. Health Risk Assessment

The main impact assessment results from AirQ+ are relative risk (RR) (and 95% confidence intervals—CI), attributable cases per 100,000 population at risk (B + c), and attributable cases (N + c).

#### 3.2.1. Outcomes from $PM_{2.5}$ Exposure

Regarding long-term health effects, we evaluated the following outcomes: mortality from LC and COPD. The RR of mortality from LC was 1.4 (95% CI: 0.6–2.1%) and 5.8% (95% CI: 2.6–9.0%) higher for people exposed to the annual 11.6 $\mu g\ m^{-3}$ registered in São Paulo compared to the annual WHO AQGs of 10 and 5 $\mu g\ m^{-3}$, respectively (Table 4). In addition, mortality due to LC attributable to long-term exposure to $PM_{2.5}$ concentrations above the 2005 WHO AQG scenario was 28 cases. For the 2021 WHO AQG, it was 113 cases (Figure 5), representing a 300% difference. Therefore, reducing $PM_{2.5}$ values in São Paulo from 11.6 to 5 $\mu g\ m^{-3}$, as recommended by the WHO, could prevent 113 deaths from lung cancer in the region annually. The RR of mortality from COPD in São Paulo was 1.7% (95% CI: 0.9–2.6%) higher for people exposed to the annual 11.6 $\mu g\ m^{-3}$ than 10 $\mu g\ m^{-3}$. Additionally, 24 excess mortality cases for this health outcome were observed considering long-term exposure to $PM_{2.5}$ above the 2005 annual WHO AQG (Figure 5).

We estimated the following outcomes concerning short-term health effects: hospital admissions for RD and CVD. For the former, excess hospitalizations increased from 57 to 258 when comparing the 2005 and 2021 WHO AQGs results (Figure 5 and Table 4). For the latter, the values were 36 and 163 excess hospital admission cases caused by $PM_{2.5}$ concentrations over the 2005 and 2021 24-h mean WHO AQGs, respectively. This indicates an increase of 350% in both health outcomes for the number of excess hospitalizations when comparing the WHO AQGs (from 2005 and 2021). On the other hand, the results were diluted in age groups for IHD and stroke; hence, the excess mortality varied from 0.36 (35–39 years old) to 3.65 cases (60–64 years old) and from 0.05 (25–29 years old) to 3.86 cases (80–84 years old), respectively, for long-term effects (Table 5).

**Table 4.** Impact assessment of health outcomes from long- and short-term exposure to $PM_{2.5}$ considering the 2005 and 2021 WHO Air Quality Guidelines.

| Health Endpoint | Timeframe | Age | 2005 WHO AQG | | | 2021 WHO AQG | | |
|---|---|---|---|---|---|---|---|---|
| | | | RR | B + c | N + c | RR | B + c | N + c |
| Mortality from LC | Long-term | 30+ | 1.0138 (1.0062–1.0210) | 0.46 (0.21–0.70) | 27.93 (12.76–42.32) | 1.0584 (1.0262–1.0902) | 1.89 (0.87–2.83) | 113.41 (52.42–169.94) |
| Mortality from COPD | | 25+ | 1.0166 (1.0087–1.0258) | 0.34 (0.18–0.53) | 24.37 (12.93–37.53) | - | - | - |
| Hospital admissions for RD | Short-term | Total population | 1.0008 (1.0000–1.0016) | 0.51 (0.00–1.07) | 57.07 (0.00–120.61) | 1.0034 (1.0000–1.0073) | 2.29 (0.00–4.84) | 257.65 (0.00–544.74) |
| Hospital admissions for CVD | | | 1.0004 (1.0001–1.0007) | 0.32 (0.06–0.59) | 36.20 (6.77–66.03) | 1.0017 (1.0003–1.0030) | 1.45 (0.27–2.65) | 163.42 (30.54–298.01) |

LC—lung cancer; COPD—chronic obstructive pulmonary disease; RD—respiratory disease; CVD—cardiovascular disease; RR—relative risk; B + c—estimated change of incidence (per 100,000) at a certain category of exposure; N + c—estimated number of cases attributable to a certain level of exposure.

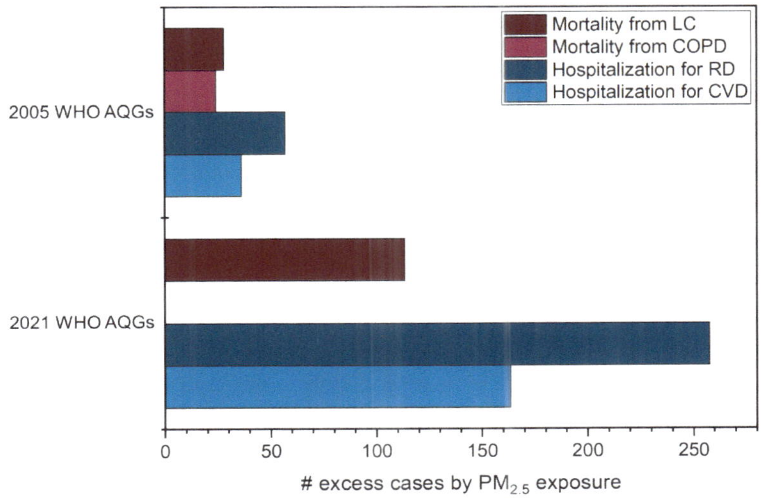


**Figure 5.** Attributable cases of mortality and hospital admissions caused by $PM_{2.5}$ exposure considering the 2005 and 2021 WHO Air Quality Guidelines. LC—lung cancer; COPD—chronic obstructive pulmonary disease; RD—respiratory disease; CVD—cardiovascular disease.

**Table 5.** Impact assessment of health outcomes from long-term exposure to $PM_{2.5}$ considering the 2005 WHO Air Quality Guideline.

| Timeframe | Age | IHD | | | Stroke | | |
|---|---|---|---|---|---|---|---|
| | | RR | B + c | N + c | RR | B + c | N + c |
| Long-term | 25–29 | 1.0220 (1.0122–1.0453) | 0.04 (0.02–0.08) | 0.43 (0.24–0.87) | 1.0174 (1.0075–1.0274) | 0.00 (0.00–0.01) | 0.05 (0.02–0.08) |
| | 30–34 | 1.0200 (1.0112–1.0418) | 0.04 (0.02–0.08) | 0.39 (0.22–0.80) | 1.0162 (1.0076–1.0262) | 0.02 (0.01–0.04) | 0.24 (0.11–0.38) |
| | 35–39 | 1.0185 (1.0100–1.0365) | 0.04 (0.02–0.08) | 0.36 (0.20–0.70) | 1.0153 (1.0074–1.0249) | 0.03 (0.01–0.04) | 0.23 (0.11–0.37) |
| | 40–44 | 1.0173 (1.0093–1.0377) | 0.16 (0.09–0.35) | 1.33 (0.72–2.83) | 1.0140 (1.0064–1.0231) | 0.09 (0.04–0.15) | 0.75 (0.34–1.22) |
| | 45–49 | 1.0159 (1.0088–1.0330) | 0.16 (0.09–0.34) | 1.22 (0.68–2.49) | 1.0131 (1.0064–1.0211) | 0.09 (0.05–0.15) | 0.70 (0.34–1.12) |
| | 50–54 | 1.0142 (1.0075–1.0298) | 0.37 (0.20–0.76) | 2.45 (1.31–5.07) | 1.0119 (1.0055–1.0190) | 0.25 (0.12–0.40) | 1.68 (0.79–2.67) |
| | 55–59 | 1.0128 (1.0073–1.0262) | 0.40 (0.23–0.81) | 2.21 (1.27–4.46) | 1.0108 (1.0053–1.0178) | 0.28 (0.14–0.46) | 1.53 (0.76–2.51) |
| | 60–64 | 1.0115 (1.0064–1.0233) | 0.86 (0.48–1.73) | 3.65 (2.03–7.32) | 1.0097 (1.0048–1.0160) | 0.62 (0.31–1.02) | 2.61 (1.30–4.30) |
| | 65–69 | 1.0102 (1.0058–1.0205) | 1.07 (0.61–2.14) | 3.25 (1.84–6.46) | 1.0088 (1.0040–1.0145) | 0.79 (0.36–1.29) | 2.39 (1.08–3.90) |
| | 70–74 | 1.0090 (1.0051–1.0191) | 1.31 (0.74–2.75) | 3.11 (1.76–6.53) | 1.0078 (1.0038–1.0127) | 1.12 (0.55–1.83) | 2.67 (1.30–4.35) |
| | 75–79 | 1.0077 (1.0048–1.0158) | 1.56 (0.97–3.18) | 2.67 (1.65–5.44) | 1.0066 (1.0034–1.0111) | 1.33 (0.69–2.23) | 2.28 (1.18–3.81) |
| | 80–84 | 1.0065 (1.0038–1.0118) | 2.77 (1.65–5.01) | 3.31 (1.97–5.99) | 1.0057 (1.0027–1.0098) | 3.23 (1.53–5.53) | 3.86 (1.83–6.60) |
| | 85–89 | 1.0054 (1.0031–1.0103) | 4.83 (2.80–9.21) | 2.76 (1.60–5.27) | 1.0047 (1.0023–1.0078) | 5.58 (2.70–9.24) | 3.19 (1.54–5.29) |
| | 90–94 | 1.0043 (1.0026–1.0081) | 10.30 (6.33–19.52) | 2.19 (1.35–4.14) | 1.0038 (1.0018–1.0062) | 12.17 (5.81–19.58) | 2.58 (1.23–4.16) |
| | 95+ | 1.0031 (1.0019–1.0056) | 24.65 (14.88–44.00) | 1.61 (0.97–2.87) | 1.0028 (1.0013–1.0046) | 29.05 (13.59–47.15) | 1.90 (0.89–3.08) |

IHD—ischemic heart disease; RR—relative risk; B + c—estimated change of incidence (per 100,000) at a certain category of exposure; N + c—estimated number of cases attributable to a certain level of exposure.

### 3.2.2. Outcomes from $O_3$ and $NO_2$ Exposure

For $O_3$, we calculated the sum of ozone means over 35 ppb (SOMO35) (or 70 μg $m^{-3}$), and added it as the input concentration, as indicated by the WHO [16]. A decrease in peak $O_3$ values in several regions in Europe occurred during the 1990s, but no trend was observed in the sum of maximum 8-h $O_3$ levels over 35 ppb (70 μg $m^{-3}$). Therefore, SOMO35 is a metric for health impact assessment with a recommended cut-off value of 70 μg $m^{-3}$ due to a statistically significant increase observed in mortality risk estimates considering $O_3$ concentrations above 50–70 μg $m^{-3}$. In addition, more consistent atmospheric model estimates were available for results above 70 μg $m^{-3}$ [43].

RD caused 443 and 93 excess annual mortality cases for long- and short-term effects, respectively, while for hospital admissions, the result was 404 incidents. $O_3$ was also attributable to 228 and 995 excess mortality and hospitalization results for CVD, respectively (Figure 6 and Table 6).

For $NO_2$, there were 90 excess hospital admission incidents due to RD (Figure 6 and Table 7). Thus, compliance with the 2021 WHO AQG for $NO_2$ could prevent 90 hospitalizations derived from respiratory diseases annually in São Paulo. Notably, this number is probably underestimated since there may be a high rate of non-hospitalization care due to respiratory symptoms. It was not possible to compare with values from 2005, as this metric was introduced in 2021. All these results indicate that better public policymaking for air quality in São Paulo and other large cities in Brazil is necessary to reduce the number of deaths and hospital admissions due to exposure to air pollutants.

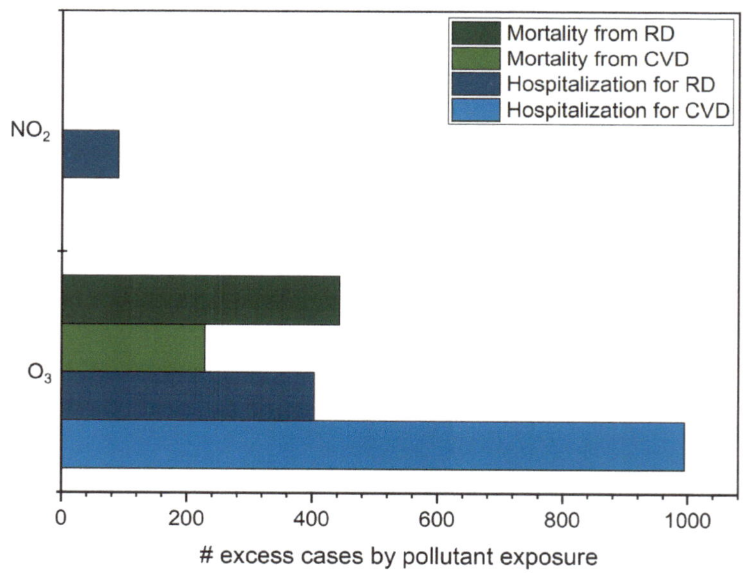


**Figure 6.** Attributable cases of mortality and hospital admissions caused by $O_3$ and $NO_2$ exposure considering the sum of ozone means over 35 ppb (SOMO35) and the 2021 WHO Air Quality Guideline, respectively. RD—respiratory disease; CVD—cardiovascular disease.

**Table 6.** Impact assessment of health outcomes from long- and short-term exposure to $O_3$ considering the sum of ozone means over 35 ppb (SOMO35).

| Health Endpoint | Timeframe | Age | RR | B + c | N + c |
|---|---|---|---|---|---|
| Mortality from RD | Long-term | Total population | 1.0290 (1.0103–1.0499) | 3.94 (1.43–6.65) | 443.23 (160.46–748.56) |
| | Short-term | | 1.0060 (1.0000–1.0144) | 0.83 (0.00–1.99) | 93.37 (0.00–223.97) |
| Hospital admissions for RD | | | 1.0090 (1.0014–1.0171) | 3.59 (0.57–6.73) | 403.58 (64.57–756.81) |
| Mortality from CVD | | | 1.0101 (1.0027–1.0175) | 2.03 (0.54–3.50) | 228.18 (60.87–393.67) |
| Hospital admissions for CVD | | | 1.0184 (1.0103–1.0263) | 8.84 (5.00–12.54) | 994.69 (562.13–1411.28) |

RD—respiratory disease; CVD—cardiovascular disease; RR—relative risk; B + c—estimated change of incidence (per 100,000) at a certain category of exposure; N + c—estimated number of cases attributable to a certain level of exposure.

**Table 7.** Impact assessment of health outcomes from short-term exposure to $NO_2$ considering the 2021 WHO Air Quality Guideline.

| Health Endpoint | Timeframe | Age | RR | B + c | N + c |
|---|---|---|---|---|---|
| Hospital admissions for RD | Short-term | Total population | 1.0020 (1.0013–1.0027) | 0.80 (0.51–1.08) | 89.61 (57.25–121.93) |

RD—respiratory disease; RR—relative risk; B + c—estimated change of incidence (per 100,000) at a certain category of exposure; N + c—estimated number of cases attributable to a certain level of exposure.

#### 3.2.3. Discussion of Health Outcomes Results

The well-known adverse health effects caused by $PM_{2.5}$ may be aggravated by exposure to $O_3$, which may cause lung epithelial damage and inflammatory response resulting in susceptibility to various infections [15,44]. Other epidemiologic studies also have shown significant health outcomes due to exposure to $O_3$, such as toxic effects on pulmonary tissues and mortality from respiratory and cardiovascular diseases [15,45]. For $NO_2$, several studies suggest that it may increase the risk for all-cause, respiratory, and cardiovascular mortality, besides asthma [46–49].

Other studies also evaluated the health effects caused by exposure to pollutants in different parts of the world using AirQ+ [31,50–54]. Kliengchuay et al. [50] reported 125, 27, and 26 deaths caused by COPD, IHD, and stroke due to long-term exposure to $PM_{2.5}$ in Ratchaburi, Thailand, where the annual mean was $26.9 \pm 18.7$ μg $m^{-3}$. The estimated number of premature deaths in Tehran, Iran, varied from 397 to 419 for IHD and from 86 to 102 for LC between 2016 and 2018, with mean $PM_{2.5}$ concentrations from $29.4 \pm 16.1$ to $31.6 \pm 16.2$ μg $m^{-3}$ [51]. Therefore, the results differ between studies depending on the concentrations registered in each region.

To the best of our knowledge, this is the first study assessing the health risks of air pollutants in São Paulo with AirQ+. However, several epidemiological studies have been conducted in the area to determine the adverse health effects of air pollution on the population based on acute effects in terms of hospital admissions, emergency room visits, and mortality in children and the elderly. According to Miraglia et al. [14], the key results of these studies showed a positive association even with concentrations below the AQGs. The life expectancy reduction due to air pollution was four years for RD, ten years for CVD, and 19 years for children with RD [14]. Health problems also generate high costs for the public health system. In the same study, Miraglia et al. [14] reported that the total health cost due to air pollution in São Paulo was more than USD 3.2 million. Nevertheless, these costs were not estimated in this investigation, which would be relevant to include in future analyses.

Abe and Miraglia [15] used the APHEKOM model and air pollutant levels between 2009 and 2011 to estimate that São Paulo could prevent more than 1500 cardiovascular and respiratory hospitalizations annually if a $PM_{10}$ concentration of 20 $\mu g\ m^{-3}$ (2005 WHO AQG) was reached. In addition, more than 5000 deaths could be postponed if $PM_{2.5}$ levels were reduced to the 2005 WHO AQG of 10 $\mu g\ m^{-3}$. The authors reported that life expectancy could increase by 15.8 months, corresponding to 266,486 life years gain and saving USD 15.1 billion annually. For $O_3$, more than 50 respiratory hospital admissions could be avoided (population $\geq$ 15 years old) following the 2005 WHO AQG (average daily maximum 8-h mean: 100 $\mu g\ m^{-3}$) [15]. In general, we presented lower results compared to the ones reported by Abe and Miraglia [15], although both studies were conducted in São Paulo; however, some significant aspects between the analyses caused this divergence. We used the WHO's AirQ+ model (vs. APHEKOM by Abe and Miraglia [15]), which resulted in variations due to slightly different methodologies. Moreover, average concentrations between 2009 and 2011 [15] were higher than in 2021 (21 $\pm$ 10 vs. 11.6 $\pm$ 7.4 $\mu g\ m^{-3}$ for $PM_{2.5}$, 36 $\pm$ 17 vs. 19.5 $\pm$ 13.3 $\mu g\ m^{-3}$ for $PM_{10}$, and 83 $\pm$ 36 vs. 37.9 $\pm$ 21.8 $\mu g\ m^{-3}$ for $O_3$) as downward trends in atmospheric pollutants have been observed in the MASP due to vehicular emission regulations adopted in Brazil.

In this study, we only investigated $PM_{2.5}$, $O_3$, and $NO_2$ due to the unavailability of robust data on the incidence of chronic bronchitis in adults, asthma symptoms in asthmatic children, and the prevalence of bronchitis in children required for the analysis of $PM_{10}$ in the software. Additionally, AirQ+ does not have the option to evaluate CO and $SO_2$. Hence, we used mortality and hospital admission data, which are less robust and more heterogeneous than the former. Mortality data are the best option for health risk assessments because they are accessible from high-quality records in São Paulo, reliable, and not subject to classification errors [15].

It is noteworthy that we obtained mortality and hospitalization data for São Paulo from DATASUS, which is a website for collecting, processing, and disseminating health information [29]. The Brazilian Ministry of Health classified COVID-19 as a severe acute respiratory syndrome (SARS), with another website for monitoring related cases and deaths that unfortunately presented remarkable numbers in the country [55]. In addition, the data for all health outcomes considered in our study did not show significant differences between 2017, 2018, 2019, 2020, and 2021 [29]. Thus, we believe there is a low probability of having COVID-19-related data among the information we used in the analyses.

Spatial averaging of the concentrations over an entire city may dampen the values and lower the mortality rates [51]. Therefore, using data from other monitoring stations in the city is important to have more robust results and identify trends related to the sources of air pollution emissions, which we suggest for future work. Nevertheless, monitoring stations do not correspond to the total exposure to air pollutants for all inhabitants in the region as it depends on different circumstances, such as indoor and outdoor activities and occupational exposure, among other factors [52].

This study also does not include other potential contributing factors to the relationship between air pollutants and hospitalization and mortality, such as body mass index, personal habits (e.g., smoking and drinking), physical activities, education, income, and medical history. Nonetheless, this study shows how many deaths and hospital admissions can be avoided in the area if more restrictive values for air pollutants are adopted, considering the 2021 WHO AQGs compared to those from 2005. It also should be noted that it accurately assesses health outcomes for different pollutants in the city of São Paulo, which exceed the 2021 WHO AQGs, and provides evidence to develop effective policies to enhance air quality and prevent health effects regarding hospitalizations and deaths.

## 4. Conclusions

In São Paulo, $PM_{2.5}$, $PM_{10}$, $NO_2$, and $O_3$ exceeded the new WHO AQGs, mainly during wintertime as July, August, and September presented the highest monthly averages for all pollutants, except for $O_3$, which presents a different behavior due to photochemical reactions. Winter is a characteristic period of higher levels of air pollution in the region due to stable atmospheric conditions and lower precipitation rates. Wind speed presented low values, which also hindered the dispersion of pollutants.

The air pollutants considered in this study present consistent evidence of adverse health effects, i.e., mortality from lung cancer, stroke, and other respiratory and cardiovascular diseases. Considering pollutant concentrations and health data for the city of São Paulo, we estimated that reducing $PM_{2.5}$ values could prevent 113 and 24 deaths from lung cancer and chronic obstructive pulmonary disease annually, respectively. In addition, it could avoid 258 and 163 hospitalizations caused by respiratory and cardiovascular diseases due to $PM_{2.5}$ exposure. The results for excess respiratory and cardiovascular deaths due to $O_3$ were 443 and 228, respectively, and 90 hospitalizations from respiratory diseases due to $NO_2$. These data provide valuable information for local authorities to design and implement effective policies aimed at promoting a healthy environment.

In this study, we only considered temporal (and not spatial) trends for the concentrations of pollutants. Thus, future studies are needed to assess the differences in time variation, spatial distribution, and attributable proportion of hospitalizations and deaths in different regions of the city of São Paulo. Calculating the health costs due to air pollution would also add more important information for policymakers to analyze the cost-effectiveness of interventions.

AirQ+ is a useful tool that enables further elaboration and implementation of air pollution control strategies to reduce and prevent hospital admissions, mortality, and economic costs due to exposure to $PM_{2.5}$, $O_3$, and $NO_2$ in São Paulo. The software is also helpful for national progress in implementing the 2030 Agenda and efficient public health regulations.

**Supplementary Materials:** The following supporting information can be downloaded at: https://www.mdpi.com/article/10.3390/ijerph20095707/s1, Figure S1: Hourly, monthly, and day-of-the-week variations for $PM_{2.5}$ (a), $PM_{10}$ (b), $SO_2$ (c), CO (d), $NO_2$ (e), and $O_3$ (f) during the sampling period. The shaded areas represent the 95% confidence intervals of the mean; Figure S2: Twenty-four-hour time series for temperature (a), precipitation (b), relative humidity (c), and wind speed (d) during the sampling period; Figure S3: Overall wind rose (a) and monthly wind roses (b) during the sampling period.

**Author Contributions:** Conceptualization, C.F.H.W. and T.N.; Methodology, C.F.H.W. and T.N.; Formal Analysis, C.F.H.W. and T.N.; Investigation, C.F.H.W. and R.S.; Resources, E.D.d.F. and M.d.F.A.; Writing—Original Draft Preparation, C.F.H.W. and T.N.; Writing—Review and Editing, C.F.H.W., T.N., R.S., E.D.d.F. and M.d.F.A.; Visualization, C.F.H.W.; Supervision, T.N. and M.d.F.A.; Project Administration, E.D.d.F. and M.d.F.A.; Funding Acquisition, E.D.d.F. and M.d.F.A. All authors have read and agreed to the published version of the manuscript.

**Funding:** This research was funded by Fundação de Amparo à Pesquisa do Estado de São Paulo (FAPESP, São Paulo Research Foundation) (grants #2020/07674-0, #2016/18438-0, #2015/03804-9, #2021/03069-8, and #2019/19433-0). This study was also financed in part by the Coordenação de Aperfeiçoamento de Pessoal de Nível Superior—Brasil (CAPES, Coordination for the Improvement of Higher Education Personnel)—Finance Code 001. T.N. and E.D.d.F. thank the Conselho Nacional de Desenvolvimento Científico e Tecnológico (CNPq, Brazilian National Council for Scientific and Technological Development) (grants #165393/2020-3, #309514/2019-3, and #313210/2022-5).

**Institutional Review Board Statement:** Not applicable.

**Informed Consent Statement:** Not applicable.

**Data Availability Statement:** Data presented in this study are available from the corresponding author upon request.

**Acknowledgments:** The authors acknowledge the School of Medicine of the University of São Paulo (FMUSP), which provided all the necessary support during the sampling campaign. The authors also would like to thank the Weather Station of the Institute of Astronomy, Geophysics, and Atmospheric Sciences of the University of São Paulo (IAG-USP) for making available the meteorological observations.

**Conflicts of Interest:** The authors declare no conflict of interest.

## References

1. Cohen, A.J.; Brauer, M.; Burnett, R.; Anderson, H.R.; Frostad, J.; Estep, K.; Balakrishnan, K.; Brunekreef, B.; Dandona, L.; Dandona, R.; et al. Estimates and 25-Year Trends of the Global Burden of Disease Attributable to Ambient Air Pollution: An Analysis of Data from the Global Burden of Diseases Study 2015. *Lancet* **2017**, *389*, 1907–1918. [CrossRef]
2. World Health Organization (WHO). Ambient (Outdoor) Air Pollution. Available online: https://www.who.int/news-room/fact-sheets/detail/ambient-(outdoor)-air-quality-and-health (accessed on 15 August 2022).
3. Health Effects Institute (HEI). *State of Global Air 2020*; HEI: Boston, MA, USA, 2020.
4. Murray, C.J.L.; Abbafati, C.; Abbas, K.M.; Abbasi-Kangevari, M.; Abd-Allah, F.; Abdelalim, A.; Abdollahi, M.; Abdollahpour, I.; Abegaz, K.H.; Abolhassani, H.; et al. Global Burden of 87 Risk Factors in 204 Countries and Territories, 1990–2019: A Systematic Analysis for the Global Burden of Disease Study 2019. *Lancet* **2020**, *396*, 1223–1249. [CrossRef] [PubMed]
5. Rajagopalan, S.; Landrigan, P.J. Pollution and the Heart. *N. Engl. J. Med.* **2021**, *385*, 1881–1892. [CrossRef] [PubMed]
6. Roth, G.A.; Mensah, G.A.; Johnson, C.O.; Addolorato, G.; Ammirati, E.; Baddour, L.M.; Barengo, N.C.; Beaton, A.; Benjamin, E.J.; Benziger, C.P.; et al. Global Burden of Cardiovascular Diseases and Risk Factors, 1990–2019: Update from the GBD 2019 Study. *J. Am. Coll. Cardiol.* **2020**, *76*, 2982–3021. [CrossRef] [PubMed]
7. Vohra, K.; Vodonos, A.; Schwartz, J.; Marais, E.A.; Sulprizio, M.P.; Mickley, L.J. Global Mortality from Outdoor Fine Particle Pollution Generated by Fossil Fuel Combustion: Results from GEOS-Chem. *Environ. Res.* **2021**, *195*, 110754. [CrossRef]
8. Pope, C.A., III; Coleman, N.; Pond, Z.A.; Burnett, R.T. Fine Particulate Air Pollution and Human Mortality: 25+ Years of Cohort Studies. *Environ. Res.* **2020**, *183*, 108924. [CrossRef]
9. Sokhi, R.S.; Moussiopoulos, N.; Baklanov, A.; Bartzis, J.; Coll, I.; Finardi, S.; Friedrich, R.; Geels, C.; Grönholm, T.; Halenka, T.; et al. Advances in Air Quality Research—Current and Emerging Challenges. *Atmos. Chem. Phys.* **2022**, *22*, 4615–4703. [CrossRef]
10. World Health Organization (WHO). *WHO Global Air Quality Guidelines: Particulate Matter (PM2.5 and PM10), Ozone, Nitrogen Dioxide, Sulfur Dioxide and Carbon Monoxide*; World Health Organization: Geneva, Switzerland, 2021.
11. World Health Organization (WHO). Air Pollution Data Portal. Available online: https://www.who.int/data/gho/data/themes/air-pollution (accessed on 15 August 2022).
12. Nogueira, T.; Kamigauti, L.Y.; Pereira, G.M.; Gavidia-Calderón, M.E.; Ibarra-Espinosa, S.; De Oliveira, G.L.; De Miranda, R.M.; Vasconcellos, P.D.C.; Freitas, E.D.; Andrade, M.F. Evolution of Vehicle Emission Factors in a Megacity Affected by Extensive Biofuel Use: Results of Tunnel Measurements in São Paulo, Brazil. *Environ. Sci. Technol.* **2021**, *55*, 6677–6687. [CrossRef]
13. Andrade, M.F.; Kumar, P.; Freitas, E.D.; Ynoue, R.Y.; Martins, J.; Martins, L.D.; Nogueira, T.; Perez-Martinez, P.; De Miranda, R.M.; Albuquerque, T.; et al. Air Quality in the Megacity of São Paulo: Evolution over the Last 30 Years and Future Perspectives. *Atmos. Environ.* **2017**, *159*, 66–82. [CrossRef]
14. Miraglia, S.G.E.K.; Saldiva, P.H.N.; Böhm, G.M. An Evaluation of Air Pollution Health Impacts and Costs in São Paulo, Brazil. *Environ. Manag.* **2005**, *35*, 667–676. [CrossRef]
15. Abe, K.C.; Miraglia, S.G.E.K. Health Impact Assessment of Air Pollution in São Paulo, Brazil. *Int. J. Environ. Res. Public Health* **2016**, *13*, 694. [CrossRef]
16. World Health Organization (WHO). *Health Impact Assessment of Air Pollution: Introductory Manual to AirQ+*; WHO Regional Office for Europe: Copenhagen, Denmark, 2020.
17. United Nations. *The Sustainable Development Goals Report 2022*; United Nations: New York, NY, USA, 2022.
18. Brazilian Institute of Geography and Statistics (IBGE). São Paulo. Available online: https://cidades.ibge.gov.br/brasil/sp/sao-paulo/panorama (accessed on 1 September 2022).
19. Brazilian Institute of Geography and Statistics (IBGE). Produto Interno Bruto Dos Municípios: São Paulo. Available online: https://cidades.ibge.gov.br/brasil/sp/sao-paulo/pesquisa/38/47001?tipo=ranking&indicador=46997 (accessed on 1 September 2022).
20. Brazilian Institute of Geography and Statistics (IBGE). Produto Interno Bruto Dos Municípios: PIB por Município. Available online: https://www.ibge.gov.br/estatisticas/economicas/contas-nacionais/9088-produto-interno-bruto-dos-municipios.html?t=o-que-e&c=3550308 (accessed on 1 September 2022).
21. IAG-USP. Estação Meteorológica Do IAG/USP. Available online: http://estacao.iag.usp.br/seasons/index.php# (accessed on 30 March 2020).
22. De Almeida Albuquerque, T.T.; Andrade, M.F.; Ynoue, R.Y. Characterization of Atmospheric Aerosols in the City of São Paulo, Brazil: Comparisons between Polluted and Unpolluted Periods. *Environ. Monit. Assess.* **2012**, *184*, 969–984. [CrossRef]
23. Environmental Company of the State of São Paulo (CETESB). *Qualidade Do Ar No Estado de São Paulo 2020*; Companhia Ambiental do Estado de São Paulo: São Paulo, Brasil, 2021.

24. Brazilian Institute of Geography and Statistics (IBGE). Frota de Veículos. Available online: https://cidades.ibge.gov.br/brasil/sp/pesquisa/22/28120?localidade1=355030 (accessed on 1 September 2022).
25. Squizzato, R.; Nogueira, T.; Martins, L.D.; Martins, J.A.; Astolfo, R.; Machado, C.B.; Andrade, M.F.; Freitas, E.D. Beyond Megacities: Tracking Air Pollution from Urban Areas and Biomass Burning in Brazil. *Clim. Atmos. Sci.* **2021**, *4*, 17. [CrossRef]
26. Carslaw, D.C.; Ropkins, K. Openair—An R Package for Air Quality Data Analysis. *Environ. Model. Softw.* **2012**, *27–28*, 52–61. [CrossRef]
27. R Core Team. *R: A Language and Environment for Statistical Computing*; R Core Team: Vienna, Austria, 2021.
28. World Health Organization (WHO). *AirQ+: Key Features*; WHO Regional Office for Europe: Copenhagen, Denmark, 2016.
29. Brazilian Ministry of Health Informações de Saúde. TABNET. Available online: https://datasus.saude.gov.br/informacoes-de-saude-tabnet/ (accessed on 15 September 2022).
30. World Health Organization (WHO). *Air Quality Guidelines: Global Update 2005*; WHO Regional Office for Europe: Copenhagen, Denmark, 2006.
31. Fallahizadeh, S.; Kermani, M.; Esrafili, A.; Asadgol, Z.; Gholami, M. The Effects of Meteorological Parameters on PM10: Health Impacts Assessment Using AirQ+ Model and Prediction by an Artificial Neural Network (ANN). *Urban Clim.* **2021**, *38*, 100905. [CrossRef]
32. Hadei, M.; Nazari, S.S.H.; Eslami, A.; Khosravi, A.; Yarahmadi, M.; Naghdali, Z.; Shahsavani, A. Distribution and Number of Ischemic Heart Disease (IHD) and Stroke Deaths Due to Chronic Exposure to PM2.5 in 10 Cities of Iran (2013–2015); An AirQ+ Modelling. *J. Air Pollut. Health* **2017**, *2*, 129–136.
33. Carvalho, V.S.B.; Freitas, E.D.; Martins, L.D.; Martins, J.A.; Mazzoli, C.R.; Andrade, M.F. Air Quality Status and Trends over the Metropolitan Area of São Paulo, Brazil as a Result of Emission Control Policies. *Environ. Sci. Policy* **2015**, *47*, 68–79. [CrossRef]
34. Brazilian National Environmental Council Resolução CONAMA N° 491, de 19 de novembro de 2018. *Dispõe sobre Padrões de Qualidade do Ar*; Diário Oficial da União: Brasília, Brasil, 2018; pp. 155–156.
35. Brazilian National Environmental Council Resolução CONAMA N° 18, de 6 de maio de 1986. *Dispõe sobre a Criação Do Programa de Controle de Poluição do Ar Por Veículos Automotores—PROCONVE*; Diário Oficial da União: Brasília, Brazil, 1986; pp. 8792–8795.
36. Brazililian National Environmental Council Resolução CONAMA N° 297, de 26 de fevereiro de 2002. *Estabelece Os Limites Para Emissões de Gases Poluentes por Ciclomotores, Motociclos e Veículos Similares Novos*; Diário Oficial da União: Brasília, Brazil, 2002; pp. 86–88.
37. Seinfeld, J.H.; Pandis, S.N. *Atmospheric Chemistry and Physics: From Air Pollution to Climate Change*, 3rd ed.; John Wiley & Sons: Hoboken, NJ, USA, 2016.
38. Massambani, O.; Andrade, M.F. Seasonal Behavior of Tropospheric Ozone in the Sao Paulo (Brazil) Metropolitan Area. *Atmos. Environ.* **1994**, *28*, 3165–3169. [CrossRef]
39. Andrade, M.F.; Fornaro, A.; Freitas, E.D.; Mazzoli, C.R.; Martins, L.D.; Boian, C.; Oliveira, M.G.L.; Peres, J.; Carbone, S.; Alvalá, P.; et al. Ozone Sounding in the Metropolitan Area of São Paulo, Brazil: Wet and Dry Season Campaigns of 2006. *Atmos. Environ.* **2012**, *61*, 627–640. [CrossRef]
40. Traffic Engineering Company of São Paulo (CET). Especial COVID-19. Available online: http://www.cetsp.com.br/consultas/especial-covid-19.aspx (accessed on 20 July 2022).
41. Alvim, D.S.; Gatti, L.V.; Corrêa, S.M.; Chiquetto, J.B.; Santos, G.M.; De Souza Rossatti, C.; Pretto, A.; Rozante, J.R.; Figueroa, S.N.; Pendharkar, J.; et al. Determining VOCs Reactivity for Ozone Forming Potential in the Megacity of São Paulo. *Aerosol Air Qual. Res.* **2018**, *18*, 2460–2474. [CrossRef]
42. Schuch, D.; Freitas, E.D.; Espinosa, S.I.; Martins, L.D.; Carvalho, V.S.B.; Ramin, B.F.; Silva, J.S.; Martins, J.A.; Andrade, M.F. A Two Decades Study on Ozone Variability and Trend over the Main Urban Areas of the São Paulo State, Brazil. *Environ. Sci. Pollut. Res.* **2019**, *26*, 31699–31716. [CrossRef] [PubMed]
43. World Health Organization (WHO). *Health Risks of Ozone from Long-Range Transboundary Air Pollution*; WHO Regional Office for Europe: Copenhagen, Denmark, 2008.
44. Foster, W.M.; Freed, A.N. Regional Clearance of Solute from Peripheral Airway Epithelia: Recovery after Sublobar Exposure to Ozone. *J. Appl. Physiol.* **1999**, *86*, 641–646. [CrossRef] [PubMed]
45. Gryparis, A.; Forsberg, B.; Katsouyanni, K.; Analitis, A.; Touloumi, G.; Schwartz, J.; Samoli, E.; Medina, S.; Anderson, H.R.; Niciu, E.M.; et al. Acute Effects of Ozone on Mortality from the "Air Pollution and Health: A European Approach" Project. *Am. J. Respir. Crit. Care Med.* **2004**, *170*, 1080–1087. [CrossRef] [PubMed]
46. Chen, R.; Yin, P.; Meng, X.; Wang, L.; Liu, C.; Niu, Y.; Lin, Z.; Liu, Y.; Liu, J.; Qi, J.; et al. Associations between Ambient Nitrogen Dioxide and Daily Cause-Specific Mortality: Evidence from 272 Chinese Cities. *Epidemiology* **2018**, *29*, 482–489. [CrossRef]
47. He, M.Z.; Kinney, P.L.; Li, T.; Chen, C.; Sun, Q.; Ban, J.; Wang, J.; Liu, S.; Goldsmith, J.; Kioumourtzoglou, M.-A. Short- and Intermediate-Term Exposure to $NO_2$ and Mortality: A Multi-County Analysis in China. *Environ. Pollut.* **2020**, *261*, 114165. [CrossRef]
48. Hvidtfeldt, U.A.; Sørensen, M.; Geels, C.; Ketzel, M.; Khan, J.; Tjønneland, A.; Overvad, K.; Brandt, J.; Raaschou-Nielsen, O. Long-Term Residential Exposure to PM2.5, PM10, Black Carbon, $NO_2$, and Ozone and Mortality in a Danish Cohort. *Environ. Int.* **2019**, *123*, 265–272. [CrossRef]
49. Zhao, S.; Liu, S.; Sun, Y.; Liu, Y.; Beazley, R.; Hou, X. Assessing $NO_2$-Related Health Effects by Non-Linear and Linear Methods on a National Level. *Sci. Total Environ.* **2020**, *744*, 140909. [CrossRef]

50. Kliengchuay, W.; Srimanus, W.; Srimanus, R.; Kiangkoo, N.; Moonsri, K.; Niampradit, S.; Suwanmanee, S.; Tantrakarnapa, K. The Association of Meteorological Parameters and AirQ+ Health Risk Assessment of PM2.5 in Ratchaburi Province, Thailand. *Sci. Rep.* **2022**, *12*, 12971. [CrossRef]
51. Mirzaei, A.; Tahriri, H.; Khorsandi, B. Comparison between AirQ+ and BenMAP-CE in Estimating the Health Benefits of PM2.5 Reduction. *Air Qual. Atmos. Health* **2021**, *14*, 807–815. [CrossRef]
52. Naghan, D.J.; Neisi, A.; Goudarzi, G.; Dastoorpoor, M.; Fadaei, A.; Angali, K.A. Estimation of the Effects PM2.5, $NO_2$, $O_3$ Pollutants on the Health of Shahrekord Residents Based on AirQ+ Software during (2012–2018). *Toxicol. Rep.* **2022**, *9*, 842–847. [CrossRef]
53. Rovira, J.; Domingo, J.L.; Schuhmacher, M. Air Quality, Health Impacts and Burden of Disease Due to Air Pollution (PM10, PM2.5, $NO_2$ and $O_3$): Application of AirQ+ Model to the Camp de Tarragona County (Catalonia, Spain). *Sci. Total Environ.* **2020**, *703*, 135538. [CrossRef]
54. Sacks, J.D.; Fann, N.; Gumy, S.; Kim, I.; Ruggeri, G.; Mudu, P. Quantifying the Public Health Benefits of Reducing Air Pollution: Critically Assessing the Features and Capabilities of WHO's AirQ+ and U.S. EPA's Environmental Benefits Mapping and Analysis Program—Community Edition (BenMAP—CE). *Atmosphere* **2020**, *11*, 516. [CrossRef]
55. Brazilian Ministry of Health. Painel Coronavírus. Available online: https://covid.saude.gov.br/ (accessed on 10 April 2023).

International Journal of
*Environmental Research and Public Health*

MDPI

*Article*

# Evaluation of the Relationship between Fractional Exhaled Nitric Oxide (FeNO) with Indoor $PM_{10}$, $PM_{2.5}$ and $NO_2$ in Suburban and Urban Schools

**Khairul Nizam Mohd Isa [1,2], Juliana Jalaludin [1,*], Saliza Mohd Elias [1], Norlen Mohamed [3], Jamal Hisham Hashim [4] and Zailina Hashim [1]**

[1] Department of Environmental and Occupational Health, Faculty of Medicine and Health Sciences, Universiti Putra Malaysia, UPM, Serdang 43400, Selangor, Malaysia; khairulnizamm@unikl.edu.my (K.N.M.I.); saliza_me@upm.edu.my (S.M.E.); drzhashim@gmail.com (Z.H.)
[2] Environmental Health Research Cluster (EHRc), Environmental Healthcare Section, Institute of Medical Science Technology, Universiti Kuala Lumpur, Kajang 43000, Selangor, Malaysia
[3] Environmental Health Unit, Level 2, E3, Disease Control Division, Ministry of Health, Putrajaya 62590, Wilayah Persekutuan Putrajaya, Malaysia; norlen.mohamed@moh.gov.my
[4] Department of Health Sciences, Faculty of Engineering and Life Science, Universiti Selangor, Shah Alam Campus, Seksyen 7, Shah Alam 40000, Selangor, Malaysia; jamalhas@hotmail.com
* Correspondence: juliana@upm.edu.my; Tel.: +603-97692397



**Citation:** Mohd Isa, K.N.; Jalaludin, J.; Mohd Elias, S.; Mohamed, N.; Hashim, J.H.; Hashim, Z. Evaluation of the Relationship between Fractional Exhaled Nitric Oxide (FeNO) with Indoor $PM_{10}$, $PM_{2.5}$ and $NO_2$ in Suburban and Urban Schools. *Int. J. Environ. Res. Public Health* **2022**, *19*, 4580. https://doi.org/10.3390/ijerph19084580

Academic Editors: Nuno Canha, Marta Almeida and Evangelia Diapouli

Received: 29 December 2021
Accepted: 3 March 2022
Published: 11 April 2022



**Abstract:** Numerous epidemiological studies have evaluated the association of fractional exhaled nitric oxide (FeNO) and indoor air pollutants, but limited information available of the risks between schools located in suburban and urban areas. We therefore investigated the association of FeNO levels with indoor particulate matter ($PM_{10}$ and $PM_{2.5}$), and nitrogen dioxide ($NO_2$) exposure in suburban and urban school areas. A comparative cross-sectional study was undertaken among secondary school students in eight schools located in the suburban and urban areas in the district of Hulu Langat, Selangor, Malaysia. A total of 470 school children (aged 14 years old) were randomly selected, their FeNO levels were measured, and allergic skin prick tests were conducted. The $PM_{2.5}$, $PM_{10}$, $NO_2$, and carbon dioxide ($CO_2$), temperature, and relative humidity were measured inside the classrooms. We found that the median of FeNO in the school children from urban areas (22.0 ppb, IQR = 32.0) were slightly higher as compared to the suburban group (19.5 ppb, IQR = 24.0). After adjustment of potential confounders, the two-level hierarchical multiple logistic regression models showed that the concentrations of $PM_{2.5}$ were significantly associated with elevated of FeNO (>20 ppb) in school children from suburban (OR = 1.42, 95% CI = 1.17–1.72) and urban (OR = 1.30, 95% CI = 1.10–1.91) areas. Despite the concentrations of $NO_2$ being below the local and international recommendation guidelines, $NO_2$ was found to be significantly associated with the elevated FeNO levels among school children from suburban areas (OR = 1.11, 95% CI = 1.06–1.17). The findings of this study support the evidence of indoor pollutants in the school micro-environment associated with FeNO levels among school children from suburban and urban areas.

**Keywords:** FeNO; school children; urban; suburban; indoor pollutants

## 1. Introduction

Many health-related assessments on the indoor air pollutants in the school settings have been conducted. The evaluations of risk assessment on school children are particularly critical, because schools represent a distinctive and vital micro-environment. Consistent results supporting those higher risks of respiratory allergic diseases were significantly associated with poor indoor air quality setting [1,2]. Asthma is the most common illness of childhood, affecting 339 million and resulting in the deaths of 13,909 children globally in 2016 [3]. Additionally, Achakulwisut et al. [4] estimated that 4.0 million of new paediatric asthma cases at global levels could be linked to annual $NO_2$ pollution and 64.0% of cases occur in urban centres.

On another note, the concentrations of particulate matter (PM), black carbon, nitrogen dioxide ($NO_2$) and ozone ($O_3$) have always shown significant variation between urban (cities and megacities) and nonurban (suburban, rural, remote) areas [5]. For example, studies have shown that the concentration of $O_3$ is usually higher in suburban areas, while the concentration of $NO_2$ is generally opposite to $O_3$, which depends heavily on the road traffic [6,7]. Generally, air pollution in the urban environment can be associated with anthropogenic sources as well as motor vehicle exhaust and industrial emissions [8]. Additionally, the greater density of road traffic and buildings in urban areas may intensify the generation of urban heat island and urban pollution island phenomena than in nonurban areas [9]. These variations can have substantial impacts on vulnerable groups, including children and the elderly. Furthermore, previous research works have focused on urban areas with high pollution and emission levels. There is little data establishing the differences between exposure in urban and suburban school environments. Moreover, the exposure effects of air pollutants exposure in suburban areas cannot be overlooked. Hence, further evidence from epidemiological studies is needed to obtain more comprehensive knowledge about the impacts of the major air pollutants ($PM_{10}$, $PM_{2.5}$, $NO_2$) in school micro-environments in both, urban and suburban areas.

In recent years, the evaluation of airway inflammation which is associated with air pollutants had been made very simplistically and noninvasively with direct measurement of fractional exhaled nitric oxide concentration in exhaled breath (FeNO) using chemiluminescence analyser [10]. Nitric oxide (NO) is the most widely used biomarker and found to be directly correlated with the severity of airway inflammation [11]. A growing body of evidence shows that short term exposure to the major air pollutants, including $PM_{10}$, $PM_{2.5}$, $NO_2$ and $O_3$ are linked to changes in FeNO levels [12–14]. Nevertheless, the results are inconsistent. For instance, Gong et al. [12] and Prapamontol et al. [15] have demonstrated that ultrafine particles (UFP) and $PM_{10}$, respectively, measured in urban school environments, were negatively associated with FeNO levels, due to other independent factors including timing of exposure and prevailing weather conditions [16]. Moreover, due to lack of research in analysing the comparison of effects of these major pollutants on school children between urban and suburban areas, it is impossible to further characterise the strategies for air pollution controls by the local authorities.

For these reasons, we conducted the present study to investigate the association of FeNO with indoor $PM_{2.5}$, $PM_{10}$ and $NO_2$ exposure in urban and suburban school areas. This is critical in order to better understand the significant variations of indoor pollutants exposure that may contribute to respiratory allergic diseases among school children.

## 2. Materials and Methods

### 2.1. Study Population

This comparative cross-sectional study was conducted among school children aged 14 years old from secondary schools in the Hulu Langat district, Selangor, Malaysia. The Hulu Langat district features an urban sprawl from rapid urbanisation of Kuala Lumpur and Putrajaya, where there are massive construction development projects for the industrial and property estates [17]. A stratified random sampling design was followed for selection of schools. The school areas were classified as urban and suburban based on the locale classification of ecological measures by the Ministry of Education, Malaysia. The number of schools selected was defined based on the sample size of school children in relation to respiratory symptoms [18]. A total of eight schools were used to give satisfactory confidence. Thus, four single session schools (afternoon) were randomly selected from each school area. Sample size calculation was adjusted for stratification sampling using a design effect of 1.1 and 0.02 was considered as an anticipated value for inter-cluster correlation (ICC) based on a comparable study conducted in children [19]. A total of 470 school children were recruited for this study. They were randomly selected from four classrooms in each school.

In the selection process, school children who have been attending the same school since January 2017 (or more than 18 months) and obtained written consent from parents or legal guardians with the addition of their own assent were included. On the other hand, school children with concomitant heart diseases and severe asthma conditions were excluded. We also excluded school children who had incomplete data and their respective residential address outside of school area. The clinical assessment and indoor air monitoring were carried out at the same time from August until November 2018 and in early February 2019. This study was approved by the Ethics Committee for Research Involving Human Subjects Universiti Putra Malaysia (JKEUPM) (JKEUPM-2018-189).

### 2.2. Clinical Assessment

Information on demographic characteristics, doctor's diagnosed asthma, current asthma medication and any asthma attack during the last 12 months were collected by self-administrative questionnaire. The questionnaire was adapted from the International Study of Asthma and Allergies in Childhood (ISAAC), the European Community Respiratory Health Survey (ECRHS) and previous studies [20,21]. This information was verified during face-to-face interviews and telephone calls with the children's respective guardians.

Airway inflammation was assessed by measuring the fractional exhaled nitric oxide (FeNO) using chemiluminescence analyser (NIOX VERO, Circassia, Sweden) with a detection limit and accuracy of 5–300 ppb and ±5 ppb, respectively. This analyser has visual and audio signals guide the school children to achieve the desired expiratory flow of 50 mL/s in six to ten seconds. Samples of exhaled breath were taken in accordance with the standard and as recommended by the manufacturer [22]. A single-breath technique was used, and this procedure was repeated at least twice to obtain an average value. School children were instructed to avoid eating and drinking for at least an hour before the FeNO assessment. To exclude errors related to the time of sampling, all the FeNO assessment was performed in the afternoon between 1.00 PM and 6.00 PM by trained enumerators.

All school children with guardian consent and own assent underwent skin prick test with five common allergens purchased from ALK-Abelló, (Madrid, Spain): *Dermatophagoides pteronyssinus, Dermatophagoides farina, Cladosporium herbarum, Alternaria alternate,* and *Felis domesticus,* also Histamine (10 mg/mL) and glycerol-saline were used as the positive and negative controls, respectively. The procedures of skin prick test were carried out by trained medical assistants and in accordance with the Australasian Society of Clinical Immunology and Allergy guidelines. The reaction was measured after 15 min by measuring the wheal diameter. The allergen's wheal diameter of 3 mm was considered as a positive control. Atopy was defined as a significant positive skin prick test to at least one of the applied allergens [23].

### 2.3. Assessment of School Environment

Four classrooms in each school were randomly selected and inspected for signs of dampness or mould growth. Indoor temperature (°C), relative humidity (%), $CO_2$ (ppm) were monitored in the classrooms by using Q-TrakTM IAQ monitor (Model 7565 TSI Incorporated, Shoreview, MN, USA) with the average log interval values over one minute. The accuracy of this device on temperature, relative humidity, and $CO_2$ are ±0.6 °C, ±3%, and ±50 ppm, respectively. The concentrations of $PM_{10}$ and $PM_{2.5}$ (μg/m$^3$) were measured using two separate units of Dust-Trak monitor (Model 8532 TSI Incorporated, Shoreview, MN, USA) at a sampling rate of 1.7 L/min and a detection limit of 0.001–150 mg/m$^3$. All of these samplers were always placed one metre above floor level and one metre away from the school children in the centre of the classrooms.

In each school, a total of four hours of measurements were collected for all of these parameters during the learning session between 1.00 PM and 6.00 PM to give more realistic exposure estimation and has been previously described [14,24–26]. For measurement of $NO_2$ (μg/m$^3$), the IVL diffusion samplers (IVL, Goteborg, Sweden) were used with the limit of detection (LOD) of 0.5 μg/m$^3$ and 10.0% of measurement uncertainty [27]. This passive diffuser sampler was used to determine the average concentration of $NO_2$ in the air for a week.

### 2.4. Statistical Analysis

Descriptive test analysis was performed by the Statistical Package for Social Science (SPSS) 25.0. We used log transformed data of indoor parameters to improve normality in the regression analysis. The chi-squared test was used to compare the school children's characteristics with respect to school areas, while statistical comparisons for FeNO levels and indoor pollutant parameters were made using the Mann–Whitney test. The two-level hierarchical multiple logistic regression (school and school children) was performed using the Stata/MP 15.1 (StataCorp LLC, College Station, TX, USA) to evaluate the association between categorised FeNO levels (normal vs. elevated level; >20 ppb) as the dependent variable and indoor pollutant parameters, controlling for gender, atopy, doctor-diagnosed asthma, weight, height, smoking status, parental asthma/allergy and family member smoking status [28,29]. A FeNO level above 20 ppb was considered as an elevated level, as recommended by the American Thoracic Society (ATS) guidelines for children [30]. We analysed the association models for each school area separately. Additionally, we incorporated the sampling weights [31] with the formula recommended by Foy [32] in the multivariate analysis stage to compensate for unequal probability of selection at classroom and school children levels. All tests were 2-tailed, and a $p$-value of less than 0.05 was considered significant.

## 3. Results

### 3.1. Personal Characteristic and FeNO Levels

Table 1 summarises the personal characteristics of school children involved in this study. Two hundred (42.6%) of the school children were from suburban and 270 (57.4%) from urban school areas, of which 61.3% were female. The overall prevalence of doctor-diagnosed asthma was 10.6%. Moreover, a total of 57.7% of school children had been sensitised to at least one of the allergens tested. However, the prevalence of doctor-diagnosed asthma and atopy did not significantly differ by the school areas. A total of 56.1% of the school children from urban areas were more likely to be exposed to secondhand tobacco smoke (SHS) at home. The prevalence of allergy and/or asthma was slightly higher in the parents from urban (52.3%) than suburban (47.7%) areas.

The level of FeNO was nearly the same among school children in suburban (median = 19.5, IQR = 24.0) and urban areas (median = 22.0, IQR = 32.0) (Figure 1). Nevertheless, comparison analysis revealed that the FeNO levels were significantly different by gender among school children from urban school areas ($p < 0.05$). Additionally, there were statistically significant higher levels of FeNO in doctor-diagnosed asthma and atopy school children from both urban and suburban school areas ($p < 0.001$) (Table 2). In particular, 71.4% and 72.4% of school children with doctor-diagnosed asthma conditions from suburban and urban areas, respectively, were significantly recorded with elevated FeNO levels ($p < 0.05$). Similarly, a higher proportion of school children with elevated FeNO levels were observed among atopic groups from both areas ($p < 0.001$).

**Table 1.** Characteristics of school children in suburban and urban school areas.

| Characteristics | Overall ($n$ = 470) | Suburban ($n$ = 200) | Urban ($n$ = 270) | $p$-Value |
|---|---|---|---|---|
| Gender | | | | |
| Male | 182 (38.7) | 66 (36.3) | 116 (63.7) | 0.028 * |
| Female | 288 (61.3) | 134 (46.5) | 154 (53.5) | |
| Ethnicity | | | | |
| Malay | 408 (86.8) | 195 (47.8) | 213 (52.2) | <0.001 ** |
| Non-Malay | 62 (13.2) | 5 (8.1) | 57 (91.9) | |

**Table 1.** *Cont.*

| Characteristics | Overall ($n$ = 470) | Suburban ($n$ = 200) | Urban ($n$ = 270) | $p$-Value |
|---|---|---|---|---|
| Doctor-diagnosed asthma | | | | |
| Yes | 50 (10.6) | 21 (42.0) | 29 (58.0) | 0.933 |
| No | 420 (89.4) | 179 (42.6) | 241 (57.4) | |
| Atopic | | | | |
| Yes | 271 (57.7) | 110 (40.6) | 161 (59.4) | 0.315 |
| No | 199 (42.3) | 90 (45.2) | 109 (54.8) | |
| Parental allergy/asthma | | | | |
| Yes | 155 (33.0) | 74 (47.7) | 81 (52.3) | 0.110 |
| No | 315 (67.0) | 126 (40.0) | 189 (60.0) | |
| Smoking | | | | |
| Yes | 30 (6.4) | 9 (30.0) | 21 (70.0) | 0.151 |
| No | 440 (93.6) | 191 (43.4) | 249 (56.6) | |
| Parental/sibling smoking | | | | |
| Yes | 285 (60.6) | 125 (43.9) | 160 (56.1) | 0.477 |
| No | 185 (39.4) | 75 (40.5) | 110 (59.5) | |

* $p < 0.05$; ** $p < 0.001$.

**Table 2.** Difference of FeNO levels (ppb) and prevalence of elevated FeNO levels (>20 ppb) between school children from suburban and urban areas by their characteristics.

| Characteristics | Overall (N = 470) Median (IQR) | Suburban Normal $n$ (%) | Suburban Elevated $n$ (%) | $p$-Value | Urban Normal $n$ (%) | Urban Elevated $n$ (%) | $p$-Value |
|---|---|---|---|---|---|---|---|
| Gender | | | | | | | |
| Male | 26 (31) | 32 (48.5) | 34 (51.5) | 0.549 | 42 (36.2) | 74 (63.8) | 0.004 * |
| Female | 19 (26) | 71 (53.0) | 63 (47.0) | | 83 (53.9) | 71 (46.1) | |
| Ethnicity | | | | | | | |
| Malay | 21 (28) | 99 (50.8) | 96 (49.2) | 0.402 | 99 (46.5) | 114 (53.5) | 0.907 |
| Non-Malay | 21 (25) | 4 (80.0) | 1 (20.0) | | 26 (45.6) | 31 (54.4) | |
| Doctor-diagnosed asthma | | | | | | | |
| Yes | 56 (63) | 6 (28.6) | 15 (71.4) | 0.026 * | 8 (27.6) | 21 (72.4) | 0.032 * |
| No | 20 (23) | 97 (54.2) | 82 (45.8) | | 117 (48.5) | 124 (51.5) | |
| Atopic | | | | | | | |
| Yes | 32 (38) | 38 (34.5) | 72 (65.5) | <0.001 ** | 50 (31.1) | 111 (68.9) | <0.001 ** |
| No | 16 (12) | 65 (72.2) | 25 (27.8) | | 75 (68.8) | 34 (31.2) | |
| Parental allergy/asthma | | | | | | | |
| Yes | 22 (30) | 38 (51.4) | 36 (48.6) | 0.974 | 34 (42.0) | 47 (58.0) | 0.351 |
| No | 21 (27) | 65 (51.6) | 61 (48.4) | | 91 (48.1) | 98 (51.9) | |
| Smoking | | | | | | | |
| Yes | 19.5 (40) | 4 (44.4) | 5 (55.6) | 0.927 | 12 (57.1) | 9 (42.9) | 0.299 |
| No | 21 (27) | 99 (51.8) | 92 (48.2) | | 113 (45.4) | 136 (54.6) | |
| Parental/sibling smoking | | | | | | | |
| Yes | 21 (28) | 61 (48.8) | 64 (51.2) | 0.324 | 78 (48.8) | 82 (51.2) | 0.329 |
| No | 22 (29) | 42 (56.0) | 33 (44.0) | | 47 (42.7) | 63 (57.3) | |

* $p < 0.05$; ** $p < 0.001$.

| School areas | Mean (SD) | Median (IQR) | z | p-Value |
|---|---|---|---|---|
| Suburban | 29.2 (22.3) | 19.5 (24) | −0.734 | 0.463 |
| Urban | 31.8 (23.9) | 22.0 (32) | | |

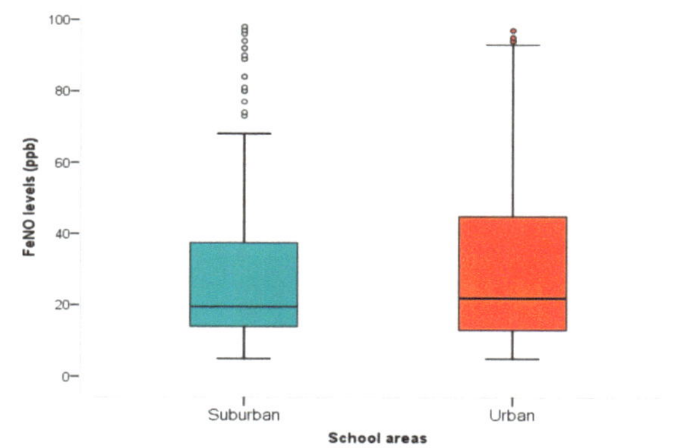


**Figure 1.** Summary of FeNO levels in school children by school areas, suburban and urban. The *p*-value refers to the comparison analysis using Mann–Whitney test. Box plots showing median values (presented by a horizontal line inside the box) and percentiles ranges (10th, 25th, 75th and 90th) of FeNO levels in school children by school areas. Circles represent outliers.

### *3.2. Classroom Inspection and Indoor Environmental Parameters*

Generally, all classrooms were equipped with three ceiling fans, naturally ventilated and designed with glass jalousie window panes on both sides of the wall. The floor surface was finished with concrete. None of the classrooms had signs of dampness or mould growth. About 25 classrooms were occupied with plastic tables and the rest of the classrooms were wooden tables. Only four classrooms used wooden chairs, while the others used plastic chairs. There were bookshelves, whiteboard, and soft boards in every classroom.

From the comparison analysis, there were significant differences ($p < 0.001$) in all indoor parameters observed between schools located in urban and suburban areas, except the concentration of $CO_2$ ($p > 0.05$). Generally, the temperature level, the concentrations of $NO_2$, $PM_{10}$ and $PM_{2.5}$ recorded in schools located in urban areas were moderately higher than in suburban areas (Table 3).

**Table 3.** Comparison of indoor air pollutants between suburban and urban areas.

| Parameter | Suburban Median (IQR) | Urban Median (IQR) | p-Value | Reference |
|---|---|---|---|---|
| | $n = 16$ | $n = 16$ | | |
| Temperature (°C) | 27.0 (1.0) | 29.0 (2.0) | <0.001 ** | 23–26 [a] |
| Relative humidity (%) | 80.4 (7.5) | 74.6 (9.5) | <0.001 ** | 40–70 [a] |
| $CO_2$ (ppm) | 456.0 (27.0) | 452.0 (33.0) | 0.068 | <1000 [a,b,c,d] |
| $NO_2$ (μg/m$^3$) | 20.0 (29.0) | 32.0 (5.0) | <0.001 ** | 200 [b], 100 ppb [c], 75 [d] |
| $PM_{2.5}$ (μg/m$^3$) | 21.9 (2.1) | 24.3 (2.5) | <0.001 ** | 25 [b], 35 [c], 50 [d] |
| $PM_{10}$ (μg/m$^3$) | 36.7 (2.7) | 41.0 (7.3) | <0.001 ** | 50 [b], 150 [c], 120 [d] |

N = 32. IQR = interquartile range. ** $p < 0.001$. [a] Industrial Code of Practice on Indoor Air Quality (ICOP-IAQ) 2010. [b] World Health Organization (WHO) guideline. [c] The National Ambient Air Quality Standard by USEPA. [d] The new Malaysian Ambient Air Quality Standard 2018 Interim Target-2.

### 3.3. Association of Indoor Pollutants with FeNO Levels

The results of two-level hierarchical multiple logistic regression models for school children from suburban and urban school areas are shown in Table 4. After controlling gender, atopy, doctor-diagnosed asthma, weight, height, parental asthma/allergy and smoking status, we found only the concentrations of $PM_{2.5}$ were significantly associated (OR = 1.30, 95% CI = 1.10–1.91) with elevated of FeNO levels in school children from urban areas ($p < 0.05$). In the second model, the concentration of $NO_2$ was found to be significantly associated with the elevated FeNO levels with an odd 1.11 (95% CI = 1.06–1.17) compared with the normal FeNO group in school children from suburban areas. Notably, the same model also shows that the concentration of $PM_{2.5}$ was positively associated with the elevation of FeNO levels with the odd 1.42 (95% CI = 1.17–1.72).

**Table 4.** Association of elevated FeNO levels (>20 ppb) in school children from suburban and urban areas with indoor air pollutants.

| Parameter | Suburban ($n$ = 200) | | Urban ($n$ = 270) | |
|---|---|---|---|---|
| | OR | 95% CI | OR | 95% CI |
| Temperature (°C) | 0.84 | (0.31–2.27) | 0.99 | (0.75–1.32) |
| Relative humidity (%) | 0.97 | (0.87–1.08) | 0.94 | (0.89–0.96) |
| $NO_2$ (µg/m$^3$) | 1.11 | (1.06–1.17) ** | 1.02 | (0.97–1.06) |
| $CO_2$ (ppm) | 1.02 | (0.99–1.05) | 1.00 | (0.98–1.03) |
| $PM_{2.5}$ (µg/m$^3$) | 1.42 | (1.17–1.72) * | 1.30 | (1.10–1.91) * |
| $PM_{10}$ (µg/m$^3$) | 0.89 | (0.75–1.08) | 1.09 | (0.98–1.20) |

* $p < 0.05$; ** $p < 0.001$. CI = confidence interval. OR calculated for 10 µg/m$^3$ increase in concentration of $NO_2$, $PM_{2.5}$, $PM_{10}$. OR calculated for 100 ppm increase in concentration of $CO_2$. OR (OR = odds ratio) was calculated by two-level hierarchical multiple logistic regression.

## 4. Discussion

This comparative cross-sectional study explored the association of FeNO levels with indoor air pollutants exposure among 470 school children in urban and suburban school areas. Further, this study demonstrated significant associations of indoor $PM_{2.5}$ and $NO_2$ concentrations with FeNO levels among school children in both school areas. To the best of our knowledge, our study is one of very few studies that provide epidemiological evidence for the link between FeNO levels and indoor pollutants measured in school's micro-environment in suburban and urban areas. Our findings also add to the scant evidence about the adverse effects of $PM_{2.5}$ and $NO_2$ in school micro-environment on airway inflammation in Southeast Asia.

In this current study, we found that the prevalence of doctor-diagnosed asthma and atopy was 10.6% and 57.7%, respectively. Nevertheless, the latest prevalence of doctor-diagnosed asthma among children aged 13 to 14 years old reported from local studies in Terengganu (Malaysia) and Penang (Malaysia) were 8.4% [21] and 10.3% [33], respectively. Compared to other studies in Southeast Asia, the prevalence of doctor-diagnosed asthma in Bangkok (Thailand), Singapore, and Surabaya (Indonesia) was 8.8% [34], 10.0% [35] and 6.8% [36], respectively. Therefore, there seems to be an indication that asthma prevalence is on the rise in Malaysia. Overall, in most countries, an increased prevalence of asthma has been documented compared to the past century [37]. Moreover, according to Sembajwe et al. [38], the prevalence of doctor-diagnosed asthma across the world regions was reflected by the national incomes. Likewise, the prevalence of atopy identified by previous studies also shows lower percentages than this current study. Previous studies in Terengganu (Malaysia), Surabaya (Indonesia) and South Korea found that the prevalence of atopy among similar age group were 40.3% [21], 29.0% [36] and 27.3% [39], respectively.

Furthermore, we observed that the prevalence of doctor-diagnosed asthma and atopy were more pronounced in school children from urban areas than suburban areas. These findings were consistent with those of other studies conducted among Chinese children [40] and Korean children [41]. This also accords with the extensive body of evidence on the role of atopy as a major risk factor for asthma, rhinitis and eczema in children [42].

With regard to the comparison analysis between urban and suburban areas, results from this current study shows that the FeNO levels were not statistically different between these groups. This result is supported by the previous comparative studies conducted in Selangor (Malaysia; urban/suburban), Terengganu, (Malaysia; urban/rural) and Bilthoven (Netherlands; urban/suburban) [43–45]. Thus, this indicates that school children from urban and suburban areas have a normal FeNO range, 20–35 ppb, as recommended by the American Thoracic Society (ATS) [30]. However, further comparison analysis provided additional evidence that the proportion of elevated FeNO levels were significantly higher in school children who had diagnosed asthma and atopy from both areas, suburban and urban, that could reflect a higher degree of airway inflammation. These results are consistent with the other studies and reviews reported [29,46,47]. A possible explanation for this might be that urban environmental irritants stimulate the development of allergic sensitisation [41].

Overall, we found that the concentrations of $PM_{10}$, $PM_{2.5}$, $NO_2$ and $CO_2$ measured inside the classrooms were below the guideline limits set by the WHO guidelines [3], the National Ambient Air Quality Standard by USEPA [48], the new Malaysian Ambient Air Quality Standard 2018 Interim Target-2 [49] and the Industrial Code of Practice on Indoor Air Quality (ICOP-IAQ) 2010 [50]. This was due to inflow of outdoor air through the jalousie window panes on both sides of the wall and adequate ceiling fans which could have made the natural ventilation effective [51,52].

In the multivariate analysis, we found that the $PM_{2.5}$ exposures were positively associated with elevated FeNO levels among school children from both areas, suburban and urban. In accordance with previous results, Shang et al. [53] reported that the ambient $PM_{2.5}$ measured in urban areas has strong association with FeNO levels. Another epidemiological study conducted in Taiyuan City, China also reported that only the concentration of $PM_{2.5}$ measured inside the classroom was significantly associated with FeNO levels [54]. These findings were consistent with the review articles by Qibin et al. [55] and Chen et al. [56]. They indicated that $PM_{2.5}$ can directly produce a significant number of free oxygen radicals, which lead to activation of inducible NOS (iNOS) expression and prolonged release of high amounts of NO in the airways. This has been confirmed by Long et al. [57] in their in vitro study, reported recently.

Another important finding in this current study was the significant association of the increased $NO_2$ concentration with elevated FeNO levels among school children from suburban areas. This finding support the results in the previous studies conducted by Olaniyan et al. [58] and Kamaruddin et al. [11], who also indicated that concentrations of $NO_2$ were positively associated with a three-fold and five-fold increased risk of elevated FeNO levels among school children aged 9–11 years old in the Western Cape Province of South Africa and Terengganu, Malaysia, respectively. Nevertheless, we fail to find a significant association between elevated FeNO levels with concentration of $NO_2$ in urban school areas, which is in agreement with those obtained by Gaffin et al. [59] and Carlsen et al. [60]. This result may be explained by the fact that differences in study design or characteristics of school children (history of asthma, smoking status) or different methods of assessing the FeNO levels [61]. The underlying mechanisms of $NO_2$ influences the FeNO levels are unclear [62]. However, there is some evidence reported that the FeNO may be modulated by DNA methylation which is involved in the arginase–nitric oxide synthase pathway [16,56,62]. Interestingly, a longitudinal study conducted by Zhang et al. [54] and Jiang et al. [62] reported that short-term exposure to $NO_2$ and $PM_{2.5}$ significantly decreased the NOS2A and increased the ARG2 methylation. Therefore, their findings provide insights to enhance our understanding of the $NO_2$ pathophysiology mechanisms, for which further studies are therefore suggested.

Some limitations of this study should be noted. First, the exposure assessment was collected over an average of one week for $NO_2$, while the other indoor pollutants were collected with an average of a four-hour time frame in each school, which might have introduced exposure measurement errors in evaluating the multiple-pollutants effects. Nevertheless, we attempted to specify almost 80% of the exposure time frame during the school hours. Despite this, we found compelling evidence that linked the $PM_{2.5}$ and $NO_2$ exposure to FeNO levels. Second, the measurement of air pollutants parameters could be improved by incorporating outdoor air quality monitoring and at-home indoor settings, which potentially contribute to the respiratory outcomes. However, the measurements for indoor and outdoor parameters are expected to be constant throughout the year. This is supported by several studies conducted across Peninsular Malaysia [63,64]. Third, the other possible confounding factors in the home environment such as dampness/mould, furry pets, environmental tobacco smoke (ETS), residential materials and redecoration activities were not comprehensively addressed in this current study. Furthermore, it should also be noted that the questionnaire used was based on self-reporting and may have introduced recall bias. To address this limitation, a face-to-face interview session was conducted following the completion of the questionnaire and telephone call, with their respective parents, to verify the self-reporting information. Another limitation of this study is that the classification of atopy was based on a small number of allergens, which may have underestimated the prevalence of atopy. Finally, through the nature of the study design, the cross-sectional study design utilised here preludes establishing causation inferences.

## 5. Conclusions

In summary, the exposure levels of indoor pollutants in the school environment in this current study are in accordance with the WHO guidelines, the National Ambient Air Quality Standard by USEPA, the ICOP-IAQ 2010 and the new Malaysian Ambient Air Quality Standard 2018 Interim Target-2. Moreover, this study suggested that there is an independent relationship between $PM_{2.5}$ and $NO_2$, although less than in the existing guidelines, that adversely affects the FeNO levels in school children from suburban and urban areas. Therefore, the results from this current study have provided additional evidence to reinforce the effects of indoor pollutants in both school environments on the respiratory outcomes among school children. This is an important finding and indicates that further intervention studies are needed to identify the most effective mechanisms to comprehensively reduce the risks of indoor pollutants exposure in school micro-environment settings. Taking into consideration the school buildings in Malaysia, which are designed with a natural ventilation system and often situated nearby heavy traffic roads, there is also a need to formulate policies to control air pollution in urban and suburban areas, and to strengthen the prevention measures. For example, the implementation of strategies such as locating buildings away from high density traffic roads and creation of asthma-friendly school programs are recommended.

**Author Contributions:** Conceptualization, K.N.M.I., Z.H. and J.H.H.; methodology, K.N.M.I., Z.H., J.J., J.H.H. and S.M.E.; software, K.N.M.I.; validation, K.N.M.I., Z.H. and J.J.; formal analysis, K.N.M.I.; investigation, K.N.M.I. and Z.H.; resources, Z.H. and S.M.E.; data curation, K.N.M.I. and Z.H.; writing—original draft preparation, K.N.M.I.; writing—review and editing, Z.H. and J.J.; visualization, K.N.M.I.; supervision, Z.H. and J.J.; project administration, K.N.M.I., Z.H. and N.M.; funding acquisition, Z.H. and S.M.E. All authors have read and agreed to the published version of the manuscript.

**Funding:** This research was funded in part by grants from Universiti Putra Malaysia (UPM) (High Impact Putra Grant—VOT 9598000 and Putra Grant-Putra Graduate Initiative—GP-IPS/2018/9648600).

**Institutional Review Board Statement:** The study was conducted in accordance with the Declaration of Helsinki, and approved by the Ethics Committee for Research Involving Human Subjects Universiti Putra Malaysia (JKEUPM) (JKEUPM-2018-189).

**Informed Consent Statement:** Informed consent was obtained from all subjects involved in the study.

**Data Availability Statement:** The data presented in this study are available on request from the first author. The data are not publicly available due to privacy.

**Acknowledgments:** The researchers are grateful to all school students who had participated in this study.

**Conflicts of Interest:** The authors declare no conflict of interest.

## References

1. Ha, J.; Lee, S.W.; Yon, D.K. Ten-Year Trends and Prevalence of Asthma, Allergic Rhinitis, and Atopic Dermatitis among the Korean Population, 2008–2017. *Clin. Exp. Pediatr.* **2020**, *63*, 278–283. [CrossRef]
2. Madureira, J.; Paciencia, I.; Rufo, J.; Ramos, E.; Barros, H.; Teixeira, J.P.; de Oliveira Fernandesa, E. Indoor Air Quality in Schools and Its Relationship with Children's Respiratory Symptoms. *Atmos. Environ.* **2015**, *118*, 145–156. [CrossRef]
3. WHO. *World Health Statistics 2018: Monitoring Health for the SDGs, Sustainable Development Goals*; World Health Organization: Geneva, Switzerland, 2018.
4. Achakulwisut, P.; Brauer, M.; Hystad, P.; Anenberg, S.C. Global, National, and Urban Burdens of Paediatric Asthma Incidence Attributable to Ambient $NO_2$ Pollution: Estimates from Global Datasets. *Lancet Planet. Health* **2019**, *3*, e166–e178. [CrossRef]
5. Al Mamun, A.; Cheng, I.; Zhang, L.; Dabek-Zlotorzynska, E.; Charland, J.-P. Overview of Size Distribution, Concentration, and Dry Deposition of Airborne Particulate Elements Measured Worldwide. *Environ. Rev.* **2020**, *28*, 77–88. [CrossRef]
6. Tong, L.; Zhang, H.; Yu, J.; He, M.; Xu, N.; Zhang, J.; Qian, F.; Feng, J.; Xiao, H. Characteristics of Surface Ozone and Nitrogen Oxides at Urban, Suburban and Rural Sites in Ningbo, China. *Atmos. Res.* **2017**, *187*, 57–68. [CrossRef]
7. Huang, D.; Li, Q.; Wang, X.; Li, G.; Sun, L.; He, B.; Zhang, L.; Zhang, C. Characteristics and Trends of Ambient Ozone and Nitrogen Oxides at Urban, Suburban, and Rural Sites from 2011 to 2017 in Shenzhen, China. *Sustainability* **2018**, *10*, 4530. [CrossRef]
8. Gulia, S.; Nagendra, S.M.S.; Khare, M.; Khanna, I. Urban Air Quality Management—A Review. *Atmos. Pollut. Res.* **2015**, *6*, 286–304. [CrossRef]
9. Ulpiani, G. On the Linkage between Urban Heat Island and Urban Pollution Island: Three-Decade Literature Review towards a Conceptual Framework. *Sci. Total Environ.* **2021**, *751*, 141727. [CrossRef]
10. Karrasch, S.; Linde, K.; Rücker, G.; Sommer, H.; Karsch-Völk, M.; Kleijnen, J.; Jörres, R.A.; Schneider, A. Accuracy of FENO for Diagnosing Asthma: A Systematic Review. *Thorax* **2017**, *72*, 109–116. [CrossRef] [PubMed]
11. Kamaruddin, A.S.; Jalaludin, J.; Hamedon, T.R.; Hisamuddin, N.H. FeNO as a Biomarker for Airway Inflammation Due to Exposure to Air Pollutants among School Children Nearby Industrial Areas in Terengganu. *Pertanika J. Sci. Technol.* **2019**, *27*, 589–600.
12. Gong, J.; Zhu, T.; Hu, M.; Wu, Z.; Zhang, J.J. Different Metrics (Number, Surface Area, and Volume Concentration) of Urban Particles with Varying Sizes in Relation to Fractional Exhaled Nitric Oxide (FeNO). *J. Thorac. Dis.* **2019**, *11*, 1714–1726. [CrossRef] [PubMed]
13. Shi, J.; Chen, R.; Yang, C.; Lin, Z.; Cai, J.; Xia, Y.; Wang, C.; Li, H.; Johnson, N.; Xu, X.; et al. Association between Fine Particulate Matter Chemical Constituents and Airway Inflammation: A Panel Study among Healthy Adults in China. *Environ. Res.* **2016**, *150*, 264–268. [CrossRef] [PubMed]
14. Mohd Isa, K.N.; Hashim, Z.; Jalaludin, J.; Than, L.T.L.; Hashim, J.H. The Effects of Indoor Pollutants Exposure on Allergy and Lung Inflammation: An Activation State of Neutrophils and Eosinophils in Sputum. *Int. J. Environ. Res. Public Health* **2020**, *17*, 5413. [CrossRef]
15. Prapamontol, T.; Norbäck, D.; Thongjan, N.; Suwannarin, N.; Somsunun, K.; Ponsawansong, P.; Khuanpan, T.; Kawichai, S.; Naksen, W. Fractional Exhaled Nitric Oxide (FeNO) in Students in Northern Thailand: Associations with Respiratory Symptoms, Diagnosed Allergy and the Home Environment. *J. Asthma* **2021**, 1–9. [CrossRef] [PubMed]
16. Zhang, Q.; Wang, W.; Niu, Y.; Xia, Y.; Lei, X.; Huo, J. The Effects of Fine Particulate Matter Constituents on Exhaled Nitric Oxide and DNA Methylation in the Arginase–nitric Oxide Synthase Pathway. *Environ. Int.* **2019**, *131*, 105019. [CrossRef]
17. Bakeri, A.; Ramli, Z.; Choy, E.A.; Awang, A. Rising Property Price: The Effects And The Preparations of the Malay People In The Suburbs. *Malaysian J. Soc. Adm.* **2020**, *14*, 39–59.
18. Asrul, S.; Juliana, J. Indoor Air Quality and Its Association with Respiratory Health among Preschool Children in Urban and Suburban Area. *Malaysian J. Public Health Med.* **2017**, *1*, 78–88.
19. Liaw, S.-T.; Sulaiman, N.D.; Barton, C.A.; Chondros, P.; Harris, C.A.; Sawyer, S.; Dharmage, S.C. Improve General Practitioners Asthma Management and Knowledge: A Cluster Randomised Trial in the Australian Setting. *BMC Fam. Pract.* **2008**, *9*, 22. [CrossRef]
20. Cai, G.; Hashim, J.H.; Hashim, Z.; Ali, F.; Bloom, E.; Larsson, L.; Lampa, E.; Norbäck, D. Fungal DNA, Allergens, Mycotoxins and Associations with Asthmatic Symptoms among Pupils in Schools from Johor Bahru, Malaysia. *Pediatr. Allergy Immunol.* **2011**, *22*, 290–297. [CrossRef]
21. Ma'pol, A.; Hashim, J.H.; Norbäck, D.; Weislander, G.; Hashim, Z.; Isa, Z.M. FeNO Level and Allergy Status among School Children in Terengganu, Malaysia. *J. Asthma* **2019**, *57*, 842–849. [CrossRef]
22. Circassia. *NIOX VERO Airway Inflammation Monitor: User Manual*; Circassia: Uppsala, Sweden, 2020.
23. ASCIA (Australasian Society of Clinical Immunology and Allergy). Skin Prick Testing for the Diagnosis of Allergic Disease—A Manual for Practitioners. Available online: https://www.allergy.org.au/images/stories/pospapers/ASCIA_SPT_Manual_March_2016.pdf (accessed on 14 January 2018).

24. Kamaruddin, A.S.; Jalaludin, J.; Hamedon, T. Exposure to Industrial Air Pollutants and Respiratory Health School and Home Exposure among Primary School Children in Kemaman, Terengganu. *Int. J. Appl. Chem.* **2016**, *12*, 45–50.
25. Suhaimi, N.F.; Jalaludin, J.; Bakar, S.A. Cysteinyl Leukotrienes as Biomarkers of Effect in Linking Exposure to Air Pollutants and Respiratory Inflammation among School Children. *Ann. Trop. Med. Public Health* **2017**, *10*, 423–431.
26. Pallares, S.; Gomez, E.; Martínez, A.; Jordan, M.M. The Relationship between Indoor and Outdoor Levels of $PM_{10}$ and Its Chemical Composition at Schools in a Coastal Region in Spain. *Heliyon* **2019**, *5*, e02270. [CrossRef]
27. Foldvary, V.; Beko, G.; Langer, S.; Arrhenius, K.; Petras, D. Effect of Energy Renovation on Indoor Air Quality in Multifamily Residential Buildings in Slovakia. *Build. Environ.* **2017**, *122*, 363–372. [CrossRef]
28. Horvath, I.; Hunt, J.; Barnes, P.J. Exhaled Breath Condensate: Methodological Recommendations and Unresolved Questions. *Eur. Respir. J.* **2005**, *26*, 523–548. [CrossRef] [PubMed]
29. Pignatti, P.; Visca, D.; Loukides, S.; Märtson, A.; Alffenaar, J.C.; Battista, G.; Spanevello, A. A Snapshot of Exhaled Nitric Oxide and Asthma Characteristics: Experience from High to Low Income Countries. *Pulmonology* **2020**, *28*, 44–58. [CrossRef] [PubMed]
30. Dweik, R.A.; Boggs, P.B.; Erzurum, S.C.; Irvin, C.G.; Leigh, M.W.; Lundberg, J.O.; Olin, A.; Plummer, A.L.; Taylor, D.R. American Thoracic Society Documents An Official ATS Clinical Practice Guideline: Interpretation of Exhaled Nitric Oxide Levels (FENO) for Clinical Applications. *Am. J. Respir. Crit. Care Med.* **2011**, *184*, 602–615. [CrossRef]
31. Foy, P. Calculation of Sampling Weights. In *Third International Mathematics and Science Study Technical Report, Volume II: Implementation and Analysis—Primary and Middle School Years*; Martin, M.O., Kelly, D.L., Eds.; TIMSS International Study Center: Chestnut Hill, MA, USA, 1997; pp. 71–79.
32. Hahs-Vaughn, D.L. A Primer for Using and Understanding Weights With National Datasets. *J. Exp. Educ.* **2005**, *73*, 221–248. [CrossRef]
33. Norbäck, D.; Hashim, J.H.; Hashim, Z.; Cai, G.-H.; Sooria, V.; Ismail, S.A.; Wieslander, G. Respiratory Symptoms and Fractional Exhaled Nitric Oxide (FeNO) among Students in Penang, Malaysia in Relation to Sign of Dampness at School and Funfal DNA in School Dust. *Sci. Total Environ.* **2017**, *577*, 148–154. [CrossRef]
34. Chinratanapisit, S.; Suratannon, N.; Pacharn, P.; Sritipsukho, P.; Vichyanond, P. Prevalence and Severity of Asthma, Rhinoconjunctivitis and Eczema in Children from the Bangkok Area: The Global Asthma Network (GAN) Phase I. *Asian Pacific J. Allergy Immunol.* **2019**, *37*, 226–231. [CrossRef]
35. Goh, S.H.; Chong, K.W.; Chiang, W.C.; Goh, A.; Loh, W. Outcome of Drug Provocation Testing in Children with Suspected Beta-Lactam Hypersensitivity. *Asia Pac. Allergy* **2021**, *11*, e3. [CrossRef] [PubMed]
36. Soegiarto, G.; Abdullah, M.S.; Damayanti, L.A.; Suseno, A.; Effendi, C. The Prevalence of Allergic Diseases in School Children of Metropolitan City in Indonesia Shows a Similar Pattern to That of Developed Countries. *Asia Pac. Allergy* **2019**, *9*, e17. [CrossRef] [PubMed]
37. Lundbäck, B.; Backman, H.; Lötvall, J.; Rönmark, E. Is Asthma Prevalence Still Increasing? *Expert Rev. Respir. Med.* **2016**, *10*, 39–51. [CrossRef] [PubMed]
38. Sembajwe, G.; Cifuentes, M.; Tak, S.W.; Kriebel, D.; Gore, R.; Punnett, L. National Income, Self-Reported Wheezing and Asthma Diagnosis from the World Health Survey. *Eur. Respir. J.* **2010**, *35*, 279–286. [CrossRef] [PubMed]
39. Park, Y.M.; Lee, S.; Kim, W.K.; Han, M.Y.; Kim, J.; Chae, Y.; Hahm, M.; Lee, K.; Kwon, H.; Park, K.S.; et al. Original Article Risk Factors of Atopic Dermatitis in Korean Schoolchildren: 2010 International Study of Asthma and Allergies in Childhood. *Asian Pacific J. Allergy Immunol.* **2016**, *34*, 65–72. [CrossRef]
40. Norbäck, D.; Lu, C.; Wang, J.; Zhang, Y.; Li, B.; Zhao, Z.; Huang, C.; Zhang, X.; Qian, H.; Sun, Y.; et al. Asthma and Rhinitis among Chinese Children-Indoor and Outdoor Air Pollution and Indicators of Socioeconomic Status (SES). *Environ. Int.* **2018**, *115*, 1–8. [CrossRef]
41. Kim, J.; Hahm, M.-I.; Lee, S.-Y.; Kim, W.K.; Chae, Y.; Park, Y.M.; Han, M.Y.; Lee, K.-J.; Kwon, H.-J.; Jung, J.-A.; et al. Sensitization to Aeroallergens in Korean Children: A Population-Based Study in 2010. *J. Korean Med. Sci.* **2011**, *26*, 1165–1172. [CrossRef]
42. Di Cicco, M.; D'Elios, S.; Peroni, D.G.; Comberiati, P. The Role of Atopy in Asthma Development and Persistence. *Curr. Opin. Allergy Clin. Immunol.* **2020**, *20*, 131–137. [CrossRef]
43. Ma'pol, A.; Hashim, J.H.; Norbäck, D.; Weislander, G.; Hashim, Z.; Isa, Z.M. Prevalence of Asthma and Level of Fractional Exhaled Nitrogen Oxide among Malaysian School Children. *BMC Public Health* **2014**, *14* (Suppl. S1), O27. [CrossRef]
44. Steerenberg, P.A.; Nierkens, S.; Fischer, P.H.; Van Loveren, H.; Opperhuizen, A.; Vos, J.G.; van Amsterdam, J.G.C. Traffic-Related Air Pollution Affects Peak Expiratory Flow, Exhaled Nitric Oxide, and Inflammatory Nasal Markers. *Arch. Environ. Health Int. J.* **2001**, *56*, 167–174. [CrossRef]
45. Zainudin, M.A.; Jalaludin, J.; Sopian, N.A. Indoor Air Quality (IAQ) in Preschools and Its Association with Respiratory Inflammation among Pre-Schoolers. *Malaysian J. Med. Health Sci.* **2019**, *15*, 12–18.
46. Zhao, Z.; Huang, C.; Zhang, X.; Xu, F.; Kan, H.; Song, W.; Wieslander, G.; Norback, D. Fractional Exhaled Nitric Oxide in Chinese Children with Asthma and Allergies e A Two-City Study. *Respir. Med.* **2013**, *107*, 161–171. [CrossRef] [PubMed]
47. Mpairwe, H.; Namutebi, M.; Nkurunungi, G.; Tumwesige, P.; Nambuya, I.; Mukasa, M.; Onen, C.; Nnaluwooza, M.; Apule, B.; Katongole, T.; et al. Risk Factors for Asthma among Schoolchildren Who Participated in a Case-Control Study in Urban Uganda. *Elife* **2019**, *8*, e49496. [CrossRef]
48. Environmental Protection Agency. NAAQS Table. Available online: https://www.epa.gov/criteria-air-pollutants/naaqs-table (accessed on 12 May 2019).

49. Department of Environment Malaysia. New Malaysia Ambient Air Quality Standard. Available online: https://www.doe.gov.my/portalv1/en/info-umum/english-air-quality-trend/108 (accessed on 12 May 2019).
50. Department of Occupational Safety and Health. *Industry Code of Practice on Indoor Air Quality 2010*; Department of Occupational Safety and Health: Putrajaya, Malaysia, 2010.
51. Bayoumi, M. Improving Indoor Air Quality in Classrooms via Wind-Induced Natural Ventilation. *Model. Simul. Eng.* **2021**, *2021*, 6668031. [CrossRef]
52. Vassella, C.C.; Koch, J.; Henzi, A.; Jordan, A.; Waeber, R.; Iannaccone, R.; Charri, R. From Spontaneous to Strategic Natural Window Ventilation: Improving Indoor Air Quality in Swiss Schools. *Int. J. Hyg. Environ. Health* **2021**, *234*, 113746. [CrossRef]
53. Shang, J.; Zhang, Y.; Schauer, J.J.; Tian, J.; Hua, J.; Han, T.; Fang, D.; An, J. Associations between Source-Resolved PM 2.5 and Airway Inflammation at Urban and Rural Locations in Beijing. *Environ. Int.* **2020**, *139*, 105635. [CrossRef]
54. Zhang, X.; Fan, Q.; Bai, X.; Li, T.; Zhao, Z.; Fan, X.; Norback, D. Levels of Fractional Exhaled Nitric Oxide in Children in Relation to Air Pollution in Chinese Day Care Centres. *Int. J. Tuberc. Lung Dis.* **2018**, *22*, 813–819. [CrossRef]
55. Qibin, L.; Yacan, L.; Minli, J.; Meixi, Z.; Chengye, L.; Yuping, L.; Chang, C. The Impact of $PM_{2.5}$ on Lung Function in Adults with Asthma. *Int. J. Tuberc. Lung Dis.* **2020**, *24*, 570–576. [CrossRef]
56. Chen, X.; Liu, F.; Niu, Z.; Mao, S.; Tang, H.; Li, N.; Chen, G.; Liu, S.; Lu, Y.; Xiang, H. The Association between Short-Term Exposure to Ambient Air Pollution and Fractional Exhaled Nitric Oxide Level: A Systematic Review and Meta-Analysis of Panel Studies. *Environ. Pollut.* **2020**, *265*, 114833. [CrossRef]
57. Long, M.-H.; Zhu, X.; Wang, Q.; Chen, Y.; Gan, X.; Li, F.; Fu, W.-L.; Xing, W.-W.; Xu, D.-Q.; Xu, D.-G. $PM_{2.5}$ Exposure Induces Vascular Dysfunction via NO Generated by INOS in Lung of ApoE-/-Mouse. *Int. J. Biol. Sci.* **2020**, *16*, 49–60. [CrossRef]
58. Olaniyan, T.; Jeebhay, M.; Röösli, M.; Naidoo, R.N.; Künzli, N.; de Hoogh, K.; Saucy, A.; Badpa, M.; Baatjies, R.; Parker, B.; et al. The Association between Ambient $NO_2$ and $PM_{2.5}$ with the Respiratory Health of School Children Residing in Informal Settlements: A Prospective Cohort Study. *Environ. Res.* **2020**, *186*, 109606. [CrossRef] [PubMed]
59. Gaffin, J.M.; Hauptman, M.; Petty, C.R.; Sheehan, W.J.; Lai, P.S.; Wolfson, J.M.; Gold, D.R.; Coull, B.A. Nitrogen Dioxide Exposure in School Classrooms of Inner-City Children with Asthma. *J. Allergy Clin. Immunol.* **2018**, *141*, 2249–2255.e2. [CrossRef] [PubMed]
60. Carlsen, H.K.; Boman, P.; Bjor, B.; Olin, A.-C.; Forsberg, B. Coarse Fraction Particle Matter and Exhaled Nitric Oxide in Non-Asthmatic Children. *Int. J. Environ. Res. Public Health* **2016**, *13*, 621. [CrossRef] [PubMed]
61. Delfino, R.J.; Staimer, N.; Tjoa, T.; Gillen, D.L.; Schauer, J.J.; Shafer, M.M. Airway Inflammation and Oxidative Potential of Air Pollutant Particles in a Pediatric Asthma Panel. *J. Expo. Sci. Environ. Epidemiol.* **2013**, *23*, 466–473. [CrossRef]
62. Jiang, Y.; Niu, Y.; Xia, Y.; Liu, C.; Lin, Z.; Wang, W.; Ge, Y.; Lei, X.; Wang, C.; Cai, J.; et al. Effects of Personal Nitrogen Dioxide Exposure on Airway Inflammation and Lung Function. *Environ. Res.* **2019**, *177*, 108620. [CrossRef]
63. Abidin, E.Z.; Semple, S.; Rasdi, I.; Ismail, S.N.S.; Ayres, J.G. The Relationship between Air Pollution and Asthma in Malaysian Schoolchildren. *Air Qual. Atmos. Health* **2014**, *7*, 421–432. [CrossRef]
64. Norbäck, D.; Hisham, J.; Hashim, Z.; Ali, F. Volatile Organic Compounds (VOC), Formaldehyde and Nitrogen Dioxide ($NO_2$) in Schools in Johor Bahru, Malaysia: Associations with Rhinitis, Ocular, Throat and Dermal Symptoms, Headache and Fatigue. *Sci. Total Environ.* **2017**, *592*, 153–160. [CrossRef]

International Journal of
*Environmental Research and Public Health*

MDPI

*Article*

# The Influence of Environmental Polycyclic Aromatic Hydrocarbons (PAHs) Exposure on DNA Damage among School Children in Urban Traffic Area, Malaysia

**Nur Hazirah Hisamuddin [1], Juliana Jalaludin [1,*], Suhaili Abu Bakar [2] and Mohd Talib Latif [3]**

[1] Department of Environmental and Occupational Health, Faculty of Medicine and Health Sciences, Universiti Putra Malaysia (UPM), Serdang 43400, Malaysia; gs49787@student.upm.edu.my
[2] Department of Biomedical Sciences, Faculty of Medicine and Health Sciences, Universiti Putra Malaysia (UPM), Serdang 43400, Malaysia; suhaili_ab@upm.edu.my
[3] Department of Earth Sciences and Environment, Faculty of Science and Technology, Universiti Kebangsaan Malaysia, Bangi 43600, Malaysia; talib@ukm.edu.my
* Correspondence: juliana@upm.edu.my

**Abstract:** This study aimed to investigate the association between particulate PAHs exposure and DNA damage in Malaysian schoolchildren in heavy traffic (HT) and low traffic (LT) areas. PAH samples at eight schools were collected using a low volume sampler for 24 h and quantified using Gas Chromatography-Mass Spectrometry. Two hundred and twenty-eight buccal cells of children were assessed for DNA damage using Comet Assay. Monte-Carlo simulation was performed to determine incremental lifetime cancer risk (ILCR) and to check the uncertainty and sensitivity of the estimated risk. Total PAH concentrations in the schools in HT area were higher than LT area ranging from 4.4 to 5.76 ng $m^{-3}$ and 1.36 to 3.79 ng $m^{-3}$, respectively. The source diagnostic ratio showed that PAHs in the HT area is pyrogenic, mainly from diesel emission. The 95th percentile of the ILCR for children in HT and LT area were $2.80 \times 10^{-7}$ and $1.43 \times 10^{-7}$, respectively. The degree of DNA damage was significantly more severe in children in the HT group compared to LT group. This study shows that total indoor PAH exposure was the most significant factor that influenced the DNA damage among children. Further investigation of the relationship between PAH exposure and genomic integrity in children is required to shed additional light on potential health risks.

**Keywords:** polycyclic aromatic hydrocarbons (PAHs); children; urban traffic area; DNA damage



**Citation:** Hisamuddin, N.H.; Jalaludin, J.; Abu Bakar, S.; Latif, M.T. The Influence of Environmental Polycyclic Aromatic Hydrocarbons (PAHs) Exposure on DNA Damage among School Children in Urban Traffic Area, Malaysia. *Int. J. Environ. Res. Public Health* **2022**, *19*, 2193. https://doi.org/10.3390/ijerph19042193

Academic Editors: Nuno Canha, Marta Almeida and Evangelia Diapouli

Received: 30 December 2021
Accepted: 12 February 2022
Published: 15 February 2022



## 1. Introduction

Urbanisation and economic growth have been strongly associated with increased transportation demand and the number of road vehicles within cities [1,2]. Pollutants from traffic emission or known as traffic related air pollution (TRAP), were the major contributor to air pollution in Malaysia, particularly in urban areas. Particulate matter (PM) is considered one of the most harmful components of ambient pollutants to humans [3]. Recent studies in the city of Kuala Lumpur, Malaysia demonstrated that PM pollution in this area is greatly contributed by vehicle emissions [4–8].

Polycyclic aromatic hydrocarbons (PAHs) are one of the most important organic groups bound to particulate matter in terms of health risk. PAHs are ubiquitous, semi-volatile and persistent organic pollutants that are formed as by-products from the incomplete combustion of organic materials [9–11]. Due to their resistance to degradation processes, especially when bound to particles, they are transported over long distances and could be identified even in remote areas [12]. Several studies have shown that motor vehicle emissions contribute to PAHs pollution in urban areas [13–17] especially high-molecular weight particulate PAHs, which are primarily present in $PM_{2.5}$ [18]. Several studies conducted in the Klang Valley also indicated the contribution of motor vehicle emissions to PAHs concentration [10,11,17–19].

The impact of PAHs on the environment and public health has sparked widespread concern due to their mutagenic and carcinogenic properties. Exposure to genotoxic substances such as PAHs can cause oxidative stress, which could lead to DNA damage and disturbances in DNA replication. Changes in DNA replication may cause mutation and lead to carcinogenic effects [20,21]. Evidence from previous studies suggests that environmental PAHs exposure may lead to increased DNA damage in children attending schools within 500 m of busy roads [22,23]. Children have a longer lifespan and are more likely to develop chronic diseases that can be triggered by early exposure. Chronic diseases and cancer caused by environmental toxicants are thought to evolve in phases that take years or even decades to develop from their initiation to clinical manifestation. Carcinogenic and toxic exposures in early childhood seems more likely to lead to disease than similar exposures later in life [24,25].

Previous studies that highlighted traffic emission as the primary contributor of PAHs in the Klang Valley [10,11,18,19] give the concern to assess the chronic health effects of inhalation PAHs among the susceptible population living in this area, especially children. In order to estimate the health risks from the exposure to PAHs, it is also essential to quantify the resultant changes that are effected through the application of biomarkers associated with the toxicants [23]. Educational buildings such as primary schools are among the most important buildings where children typically spend up to 1/3 of their time [26]. Therefore, understanding exposure to health concerned pollutants in these locations has become a priority for the scientific community [27–30]. To date, there is limited information on PAH exposure in school settings in Malaysia. Since children attending schools near congested urban areas are constantly exposed to air pollutants, this study is carried out to fill the knowledge gap on the association between PAHs from traffic emission and the chronic health effect concerning DNA damage among Malaysian children in the urban city of the Klang Valley.

## 2. Materials and Methods

### 2.1. Study Location

Klang Valley was selected as the location for this study because it is known as fast growing area and the most populated area in Malaysia. The cities in the Klang Valley region have a well-developed road network, with approximately 379,146 vehicles travelling around daily [31]. The school selection was based on proximity to high and low traffic areas. High traffic area is defined as areas within 500 m on either side of highways with an average daily traffic (ADT) volume of $\geq$18,000 vehicles, or within 100 m on either side of major roads with an ADT volume of $\geq$15,000 vehicles [32]. Children from primary schools in Kuala Lumpur and Gombak, were categorised as high traffic (HT) group. The schools in HT were designated H1, H2, H3 and H4. Meanwhile, low traffic is defined as area located 5 km away from nearby highways, major roads and industrial sites. The children of primary schools in Hulu Langat, were categorised as the low traffic (LT) group, with the schools on LT designated as L1, L2, L3, and L4. The map of study area is shown in Figure 1.

### 2.2. Study Population

This study includes a total of 228 school children studying at the selected primary schools in Klang Valley, specifically age between 7 to 11 years old who met the inclusion and exclusion criteria. Both genders, which are female and male, were selected among the Malay population to prevent gender bias of collected data. Only Malaysian citizens and Malay ethnicity were recruited to homogenize the samples. These criteria have been decided to control genetic differential factors, which could affect the final outcomes. Children who had a known history of medical problems were excluded in this study. This criterion has been decided because it has the potential to impair children's physiological function and increase their susceptibility to pollutants [22]. Besides, children who had radiotherapy or chemotherapy in the previous 12 months or X-rays in the last three months were excluded. Any radiation exposure can affect DNA structure by inducing DNA strand breaks and thus affecting the significance of the results of this study [33]. Children who had a family history of cancer also were excluded. This criterion has been decided because he or she have higher risk of getting cancer [34].

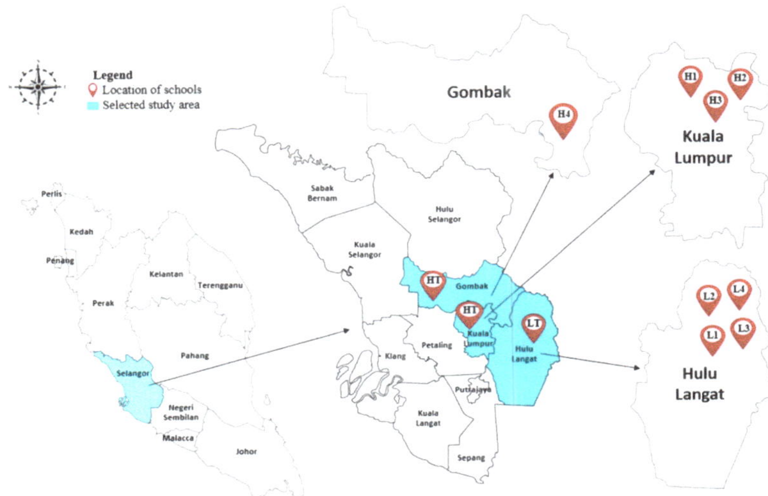


**Figure 1.** Locations of the selected primary schools. Primary schools in Kuala Lumpur and Gombak, were categorised as high traffic (HT) group with the schools in HT were designated H1, H2, H3 and H4. Primary schools in Hulu Langat, were categorised as the low traffic (LT) group, with the schools on LT designated as L1, L2, L3, and L4.

### 2.3. Questionnaires

A total of 280 questionnaires and consent forms were distributed to eight schools and 228 respondents completed the questionnaire and gave consent for biological sampling, representing 81.4% of response rate. A validated questionnaire from American Thoracic Society (ATS-DLD-78-C) and International Study of Asthma and Allergies in Childhood (ISAAC) translated from English to Malay (Supplementary S1) were distributed to parents or guardians of children.

### 2.4. $PM_{2.5}$ Sampling

Gravimetric sampling of $PM_{2.5}$ were conducted using a low volume sampler, MiniVol Air Sampler (Airmetrics, Springfield, OR, USA; model 4.2). The filter paper used was 47 mm quartz microfiber filter papers (Whatman, Maidstone, Kent, UK; catalogue no. 1851-047). PAH samples were extracted from the filter paper. MiniVol Air Sampler was set up on the ground with its stand to sample air at a rate of 5 L/min, indoor and outdoor for 24 h. The air sampler was located approximately 1.0 m above the floor and placed at the back of the selected classroom. For outdoor monitoring, the sampling was conducted in a safe area, which is near the main entrance gate of the school or guard posts. A total of 64 samples and 8 blank samples (one for each school) were collected during the sampling campaign.

### 2.5. PAHs Extraction

The filter paper was cut into small pieces (approximately 1 cm × 1 cm) in a 50 mL glass bottle. 0.1 mL of 1.5 ppm PAHs surrogate internal standards; anthracene-$d_{10}$, p-terphenyl-$d_{14}$, benz[a]anthracene-$d_{12}$, and perylene-$d_{14}$ (SUPELCO, St. Louis, MO, USA) were spiked into all samples for recovery assessment. 10 mL of dichloromethane (DCM) and n-hexane with ratio 5 mL:5 mL was added into the glass bottle. Next, the mixture was sonicated using a bath sonicator (Elma, Singen, Germany) for 30 min (2 min run and 1 min rest ×

15 cycles). This procedure was repeated three times and the extracts were combined. Then, the extract was concentrated to approximately 0.2 mL under a gentle blow of nitrogen gas ($N_2$). Silica SPE cartridges (LiChrolut RP-18 1000 mg 6 mL, Merck, Germany; catalogue no. 1021220001) were used for clean-up process and pre-concentration of samples. The final step of SPE was elution by DCM:n-hexane (3.5 mL:6.5 mL). The eluent was further reduced to 0.1 mL via a gentle stream of $N_2$ gas before transferring to a 2 mL autosampler vial containing a vial insert.

The PAHs were quantified using gas chromatography–mass spectrometry (GC-MS) instrument (Agilent Technologies, Santa Clara, CA, USA; model 6890N/5975), fitted with HP-5MS capillary column (30 m × 250 μm). Helium was used as a carrier gas with a flow rate of 1.0 mL/min. The sample was injected at 200 °C using the splitless mode. The temperature of the GC column was programmed as follows: initial 40 °C, followed by a temperature increase to 150 °C (8 °C per min), and an increment of 4 °C per min to 310 °C for a 6 min hold. Mass spectrometry was acquired using the electron ionization (E.I.) mode.

The concentrations of 16 United States Environmental Protection Agency (US EPA) priority PAHs were determined in this study including 3-rings PAHs; acenaphthene (ACP), acenaphthylene (ACY), anthracene (ANT), fluorene (FLR), phenanthrene (PHE), 4-rings PAHs; fluoranthene (FLT), pyrene (PYR), benzo(a)anthracene (BaA), chrysene (CYR), 5-rings PAHs; benzo(k)fluoranthene (BkF), Benzo[b]fluoranthene (BbF), benzo(a)pyrene (BaP), dibenzo(a,h)anthracene (DhA) and 6-rings PAHs; indeno(1,2,3-cd)pyrene (IcP) and benzo(ghi)perylene (BgP); except for 2-rings PAHs, naphthalene (NAP) owing to its high volatilities. The GC-MS instrument was calibrated with standard mixtures of PAHs (PAH Calibration Mix, SUPELCO, USA; catalogue no. SA_CRM47940). Five points (0.2, 0.5, 1.0, 1.5, 3.0 ppm) of standard PAHs mixtures were analysed for the establishment of the calibration curves. The correlation coefficients ($R^2$) for linear regressions of the calibration curves were >0.99 in all cases. For every sample, procedural blanks were run to ensure that there were no significant background interferences. Blank filters were extracted and analysed using the same method with the actual samples. The limit of detection (LOD) of each PAHs were calculated based on five independent measurements of blank samples and its standard deviation. The recovery efficiency for the internal standards ranged from 79–113%. The LOD for individual PAHs compounds ranged from 0.01 to 0.17 ng $m^{-3}$.

### 2.6. Health Risk Assessment

BaP-equivalent concentration (BaPeq), also known as toxicity equivalent concentration (TEQ), was used to evaluate the toxicity of PAHs. To calculate TEQ, the reference toxic equivalent factor (TEFs) of PAHs with respect to BaP were multiplied with the concentration of PAHs species [35] as shown on the following equation:

$$\begin{aligned} TEQ = 0.001\,(ACY + ACE + FLR + PHE + FLT + PYR) + 0.01\,(ANT + BgP + CYR) \\ +0.1(BaA + BkF + BbF + IND) + BaP + DhA \end{aligned} \tag{1}$$

The carcinogenic risk of PAHs by respiratory exposure was estimated using the incremental lifetime cancer risk (ILCR) model [36,37]. The equation is as follows:

$$ILCR = \frac{C \times \left(CSF \times \sqrt[3]{\frac{BW}{70}}\right) \times IR \times ED \times EF}{BW \times AT} \times cf \tag{2}$$

where:

| | | |
|---|---|---|
| C (ng $m^{-3}$) | = | TEQ |
| CSF (3.85 mg $kg^{-1}$ $day^{-1}$) | = | Inhalation cancer slope factor of BaP |
| BW (kg) | = | Body weight (kg) |
| IR (12 $m^3$ $day^{-1}$); | = | Inhalation rate |
| ED (6 years) | = | Exposure duration |
| EF (250 day $year^{-1}$) | = | Exposure frequency |
| AT (70 years × 365 days) | = | Averaging time of carcinogenic PAHs exposure [38] |
| cf | = | Conversion factor |

To reduce the uncertainties in risk assessment models, Monte-Carlo simulation using Crystal Ball software version 11.1.2.4 (Oracle Corp., Austin, TX, USA) was applied for probability and sensitivity analysis, where parameters were expressed by a series of data based on probability distribution. The results of sensitivity analysis were shown as rank correlation coefficients and the higher coefficient represents the most contributor in the uncertainty of calculated risk. Table 1 shows the variable distribution types applied in the Monte Carlo simulation for both schools in HT and LT areas.

**Table 1.** The variable distribution types applied in the Monte Carlo simulation for both schools in HT and LT areas.

| Parameter | Distribution Mode | HT Schools | LT Schools |
|---|---|---|---|
| TEQ (ng $m^{-3}$) | Logistic * | 2.32 | 1.02 |
| Inhalation rate ($m^3$ $day^{-1}$) | Constant | 12 | 12 |
| Exposure frequency (day $year^{-1}$) | Constant | 250 | 250 |
| Exposure duration (year) | Constant | 6 | 6 |
| Averaging time (days) | Constant | 25, 500 | 25, 500 |
| Body weight (kg) | Log normal * | 33.13, 14.54 | 28.33, 10.86 |
| Cancer slope factor (mg $kg^{-1}$ $day^{-1}$) | Constant | 3.85 | 3.85 |

* Logistic data are represented as arithmetic mean, log normal represented as LN (arithmetic mean and standard deviation).

### 2.7. Comet Assay

Buccal epithelial cells of children were collected for the analysis of DNA damage using Comet assay. The assay was conducted following the protocol in Comet assay Kit (Trevigen, Gaithersburg, MD, USA; catalogue no. 4253-096-K). Firstly, the buccal cells suspension were washed with 1x PBS, calcium and magnesium free (Thermo Fisher Scientific, Waltham, MA, USA; catalogue no. 70011-044) and centrifuged at 2500 rpm for 1 min. 10 µL of cells were combined with 75 µL of low melting agarose and immediately pipetted onto the comet slide (Trevigen, Gaithersburg, MD, USA; catalogue no. 4250-200-03). The slides were placed flat in the dark at a 4° C chiller. After 30 min, the slides were immersed in a pre-chilled lysis solution (Trevigen, Gaithersburg, MD, USA; catalogue no. 4250-050-01) and left for 60 min in 4 °C chiller. Next, the excess buffer was discarded from the slides. The slides were immersed in the freshly prepared alkaline unwinding solution for 60 min at room temperature. The slides were removed from the alkaline solution and were placed horizontal onto the electrophoresis slide tray. Then, alkaline electrophoresis solution was poured in the tray until the level covered over the slides. Electrophoresis was run for 20 min at 21 V constant voltage. The slides were washed twice in deionized water for 5 min and then washed once in 70% ethanol (R&M Chemicals, Birmingham, UK; CAS no. 64-17-5) for 10 min at room temperature. After drying, the slides were stained with 50 µL SYBR Green (Invitrogen, Carlsbad, CA, USA; catalogue no. S7563). The comet images were captured using fluorescence microscope (Olympus, Tokyo, Japan; model BX51) under 20× magnification. The images were analysed using OpenComet software version 1.3 (https://www.cometbio.org, accessed on 24 September 2021). The extent of DNA damage was reported as tail moment (the product of %T and tail length).

### 2.8. Statistical Analysis

The data collected were analysed using Statistical Package for Social Science (SPSS) Version 25 (IBM, Armonk, NY, USA). The data were checked for missing values and outliers. Data normality of continuous variables were determined based on Shapiro-Wilks. The statistical tests include univariate, bivariate, and multivariate analysis to prove the relationship between PAHs and DNA damage. Univariate analysis included descriptive statistics for sociodemographic data, PAHs concentration, and tail moment (descriptor for DNA damage). Bivariate analysis involved comparison of means or medians of the variables studied, particularly for continuous data. The mean for tail moment was log10-transformed to get normal-distributed data prior to statistical analysis. Meanwhile, multiple linear regression was performed to assess the main variables that influence the DNA damage among the respondents.

### 2.9. Quality Control

The pre-test of the questionnaire was conducted on at least 10% of the total respondents, which was among 28 respondents. The Cronbach's $\alpha$ value was 0.812 which considered acceptable ($\alpha \geq 0.7$). The buccal samples were kept at 4 °C in an icebox with ice packs until being transported to laboratory. The comet cells were examined in a zig-zag pattern under fluorescence microscope to prevent over counting or repetition of cells. Filter papers were pre-baked at 500 °C for 4 h using muffle furnace (Carbolite, Sheffield, UK; model CWF11/23). To avoid photodegradation, the filter papers were wrapped up in aluminium foil in sealed bags. No plasticware was used for PAHs analysis to avoid any cross-contamination from other sources. All glassware used during the procedure were cleaned, rinsed with n-hexane and rinsed with distilled water and acetone (R&M Chemicals, Birmingham, UK) prior to being baked in a furnace at 500 °C for 4 h, to volatilize and remove any organic contaminants.

## 3. Results and Discussion

### 3.1. Distributions of PAHs Species at Schools

Figure 2 shows the concentrations of total PAHs ($\sum$PAHs) in indoor and outdoor environment at each school. The distributions of PAHs species in indoor and outdoor $PM_{2.5}$ samples were shown in Supplementary S2. As portrayed in Figure 2, the trend of indoor $\sum$PAHs concentration for HT schools are as follows H1 (5.58 ± 4.72 ng $m^{-3}$) > H3 (4.86 ± 3.17 ng $m^{-3}$) > H4 (4.65 ± 1.39 ng $m^{-3}$) > H2 (4.19 ± 0.91 ng $m^{-3}$). Meanwhile for LT schools, L2 demonstrated the highest indoor $\sum$PAHs concentration (3.69 ± 3.19 ng $m^{-3}$) followed by L1 (3.48 ± 3.14 ng $m^{-3}$), L4 (2.26 ± 1.32 ng $m^{-3}$) and L3 (1.25 ± 0.83 ng $m^{-3}$). The results clearly showed that a higher concentration of outdoor $\sum$PAHs was seen in HT schools compared to LT schools with the highest concentration recorded at school H1 (5.76 ± 2.20 ng $m^{-3}$), closely followed by school H3 (5.69 ± 3.22 ng $m^{-3}$), school H4 (4.96 ± 1.96 ng $m^{-3}$) and school H2 (4.40 ± 1.79 ng $m^{-3}$). On the other hand, the schools in LT group portrayed low PAHs concentration, especially school L3 (1.36 ± 0.69 ng $m^{-3}$) and L4 (2.63 ± 1.96 ng $m^{-3}$). H1 is located less than 50 m away from the main road and 160 m from highways. Meanwhile, L3 recorded the lowest $\sum$PAHs concentrations, with 1.25 ± 0.83 ng $m^{-3}$ for indoor and 1.36 ± 0.69 ng $m^{-3}$ for outdoor. The low $\sum$PAHs obtained were probably due to less traffic densities in that area. Moreover, the school environment surrounded by forest may act as a natural filter that contributes to removing air pollutants, thus lowering the concentration of total $\sum$PAHs [39]. Trees can help minimize air pollution by absorbing the particulate matter via stoma uptake [40,41].

In this study, BkF was the dominant species presence in particulate PAHs samples for all HT schools with concentration range from 0.64 to 1.84 ng $m^{-3}$ for indoor and 0.71 to 1.04 ng $m^{-3}$ for outdoor. The result is consistent with study by Suradi et al. [18] which detected BkF out of 13 measured PAHs species, as the dominant species at Kuala Lumpur City Hall (DBKL) with the mean concentration 0.42 ng $m^{-3}$. On the other hand, ACP was the highest species found in the indoor samples of school L1 and L2 with the concentration of 0.91 ng $m^{-3}$ and 0.84 ng $m^{-3}$, respectively.

It was found that the range of the concentration of indoor $\sum$PAHs obtained in this study was quite similar with study by Ismail et al. [39] conducted in three primary schools in Kuala Lumpur, Malaysia with the concentration ranged from 1.6 to 8.0 ng $m^{-3}$ for indoor PAHs. BgP was the dominant species found in the study, which indicate the vehicular emission. Another study conducted by Sopian et al. [42] in school environment in Terengganu, Malaysia reported a higher concentration than the present study with concentration ranged from 4.21 to 63.22 ng $m^{-3}$ and 5.93 to 67.72 ng $m^{-3}$ for indoor and outdoor, respectively. Furthermore, the present study reported a much lower PAHs concentration than other studies dealing with the school environment in China [37,43], Portugal [26] and Lithuania [44].

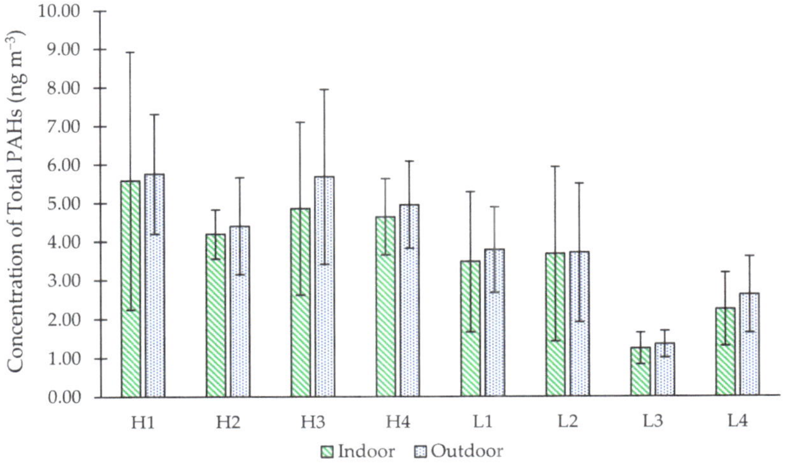


**Figure 2.** The concentrations of total PAHs in indoor and outdoor environment at each school.

An urban middle school in Beijing China recorded 6 times higher total PAHs concentration than the present study with the concentration of 29.8 ng m$^{-3}$ and 33.7 ng m$^{-3}$ for indoor and outdoor, respectively [37]. They had identified BbF, CYR, IcP, Flu, BgP and BaP as the most prominent species and might be emitted from coal combustion, residential cooking and traffic exhausts. Wang et al. [43] found a higher level of indoor and outdoor PAHs in school located in commercial and residential area of Xian, China with concentrations of 79.9 and 92.00 ng m$^{-3}$, respectively. BbF, IcP, BgP, CYR and BkF were the most abundant species found in that study, which contributed by biomass burning, vehicle emissions, and coal combustion activity. PAHs exposure in primary urban schools in Portugal recorded a higher total PAHs concentration range from 2.8 to 54 ng m$^{-3}$ for indoor and 7.1 to 48 ng m$^{-3}$ for outdoor which are highly dominated by DhA, Acy and BjF species [26]. Meanwhile, Krugly et al. [44] reported the concentrations of particulate and gaseous PAHs in indoor and outdoor air from five Lithuanian urban primary schools with concentration range from 20.3 to 131.1 ng m$^{-3}$ and 40.7 to 121.2 ng m$^{-3}$ for indoor and outdoor, respectively. In that study, naphthalene appeared be the most abundant PAH species in all sampling sites. Motor vehicle emissions and fuel combustion for heating purposes were the main sources of PAHs in the study.

### 3.2. Indoor and Outdoor PAHs

Figure 3 shows the indoor to outdoor (I/O) ratio of PAHs concentrations. The I/O ratio can be used to explain the relationship between indoor and outdoor pollution states [37]. If the I/O ratio were >1, indoor sources are stronger than outdoor sources. On the other hand, if it were <1, the indoor sources are weaker [45]. This study shows that all schools had I/O ratios below 1, ranging from 0.86 to 0.99, indicating that the indoor sources were relatively weak. The results of I/O ratio show that among eight schools, school L2 recorded the highest ratio (0.99). This situation can be explained by a higher penetration of outdoor particles into indoor classrooms resulting in higher I/O ratios at L2. This phenomenon may be due to the indoor and outdoor air exchange due to the natural ventilation system, which refers to the opening of windows and doors in this school. In addition, the windows in school L2 were open most of the time, which increases the airflow into the classrooms. The pathway of pollutants from outdoor air to indoor spaces depends on ventilation and infiltration. Prevailing wind provides natural ventilation when doors and windows are open [46]. Infiltration can also occur through cracks and leaks in the building, which can be significant for a poorly sealed building [47]. Because of these mechanisms, outdoor air pollutants can enter indoor spaces and can either be diluted or accumulated depending on ventilation conditions [47].

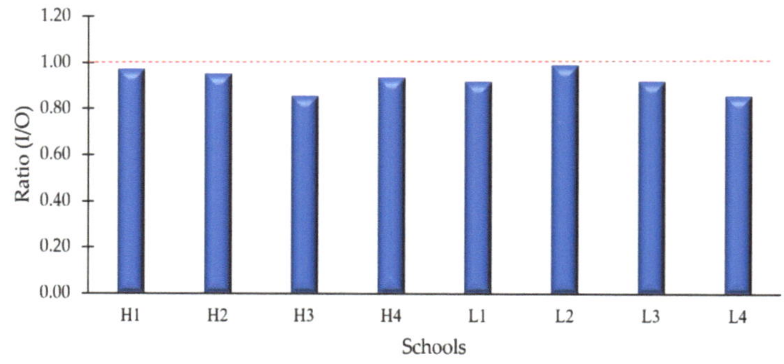


**Figure 3.** Indoor to outdoor (I/O) ratios of PAHs concentration in each school.

Previous work has conducted studies on the I/O ratio in school environments. Wang et al. [43] reviewed more than 16 studies from different schools around the world and found that the I/O ratio of PAHs ranged from 0.43 to 0.93. A study in a middle school in Beijing, China reported almost similar I/O ratio with the present study, with a value of 0.98 [37]. According to Long and Sarnat [48], an indoor source is present when the I/O ratio is greater than 1.15. The interpretation of the I/O ratios in the present study supports the idea that indoor PAH concentrations are mainly influenced by the ambient atmosphere. Analysis of the diagnostic ratios of individual PAHs may provide further insight into the origin of PAHs.

### *3.3. Distribution of PAHs Based on Number of Rings*

The percentage distribution of outdoor and indoor PAHs based on number of rings was shown in Supplementary S3. In general, the distribution of individual PAHs was mainly dominated by high molecular weight (HMW) PAHs structured by four to six rings (FLT, PYR, BaA, CYR, BaP, BbF, BkF, BgP, DhA, IcP), compared to low molecular weight (LMW) PAHs with three rings (ACY, ACP, FLR, PHE, ANT). However, school L1, L2, and L4 demonstrated a contra finding as the LMW conquered the total PAHs concentration compared to HMW. The indoor and outdoor PAHs at school L1, L2 and L4 had higher contribution of three aromatic ring PAHs, accounting for more than 50% of the total measured PAHs. The 4-ring PAHs in all schools accounted for 10 to 20% of the total PAHs. The outdoor HT school, H3 had the highest fraction of 4 ring (20%). The 5 and 6 ring PAHs were vastly abundant (ranges between 51 to 79%) in all schools except for three schools L1, L2 and L4. The three mentioned schools had 32.62% (L1), 27.24% (L2) and 37.68% (L3) of HMW PAHs for the outdoor samples. Furthermore, the indoor samples of 5 and 6 ring PAHs were highly detected in school H3 with a percentage of 78.88% of the total measured PAHs.

In urban areas, pyrogenic sources are the main source of PAHs, especially HMW-PAHs, which are mainly present in $PM_{2.5}$. [18]. As suggested by Yunker et al. [49], HMW PAHs are more likely to be portioned in $PM_{2.5}$ compared to LMW PAHs. Based on the finding, a higher distribution of HMW PAHs was seen in all HT schools which consistent with previous studies in urban traffic areas in the Klang Valley [10,11,19,36,50]. HMW PAHs that consist of more than 5 ring are an indicator of traffic emission [51]. HMW PAHs are usually formed during the process involving high temperature such as fuel combustion [39].

Meanwhile, the formation of LMW PAHs is associated with low-temperature combustion, such as wood-burning [52].

### 3.4. Source Diagnostic Ratio

The source diagnostic ratio was used to identify possible PAH sources in each school. The application of diagnostic ratios involves comparing ratio between specific pairs of PAHs compounds with the same molar mass and similar physicochemical properties [52]. Bivariate plots of selected PAHs based on source diagnostic ratio are shown in Figure 4. The ratio ANT/(ANT + PHE) was indicative of an anthropogenic source of PAHs emissions, with values below 0.1 indicating a petrogenic source. In contrast, any value above 0.1 indicates a pyrogenic source [53]. In this study, the ratio ANT/(ANT + PHE) had a value greater than 0.1, indicating a strong contribution from the pyrogenic source. In addition, the ratio of IcP/(IcP + BgP) was greater than 0.2 for all samples, indicating a contribution from a pyrogenic source such as fossil fuel combustion, grass, wood, or coal.

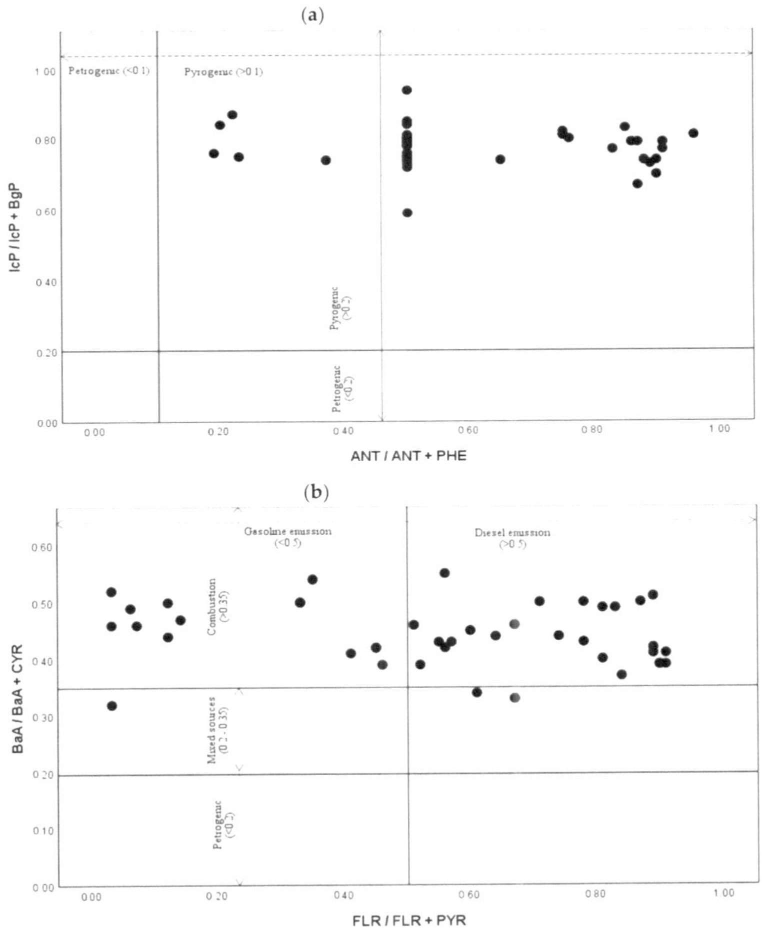


**Figure 4.** Result of source diagnostic ratio method. (**a**) IcP/IcP + BgP and ANT/ANT + PHE (**b**) BaA/BaA + CYR and FLR/FLR + PYR.

The BaA/(BaA + CYR) ratio is reported as an indicator for petrogenic (unburned petroleum) sources when <0.2 and combustion sources when >0.35. A ratio between 0.2 and 0.35 indicates that it is from mixed sources [49]. Most of the samples had the ratios of BaA/(BaA + CYR) higher than 0.35 suggesting that the combustion activity is a primary source. Akyüz and Çabuk [54] also suggested that a ratio greater than 0.35 denotes vehicular emissions. Two important anthropogenic sources, namely gasoline and diesel emissions, contribute to traffic-related air pollution. The FLR/(FLR + PYR) ratio can identify diesel and gasoline vehicle sources with the higher ratio (>0.5) of FLR/(FLR + PYR) indicate a diesel emission. In contrast, a lower ratio (<0.5) indicates a gasoline emission [55,56]. Majority of schools in this study were impacted by diesel emission, except for H3 and L3. Interestingly, 100% of samples from H2 and H4 were originated from diesel emission, which could be contributed by the pass-by of heavy-duty vehicles. Meanwhile, L3 had the highest number of samples (n = 6, 75%) originated from the exhausts of gasoline engine vehicles. This study's findings suggested that PAH sources mainly from vehicular emission and other pyrogenic contributions such as grass, wood, and coal combustion. A similar finding was found in a previous study that reported PAHs compound in Kuala Lumpur may not only originated from urban traffic combustion, but also contributed by coal, grass, and wood burning activities [18].

### 3.5. Health Risk Assessment

The obtained values for ΣTEQ–PAHs in outdoor air ranged from 0.67 to 2.77 ng $m^{-3}$ with an overall average of 1.64 ng $m^{-3}$ (Supplementary S4). Meanwhile, the obtained values for ΣTEQ–PAHs in indoor air ranged from 0.86 to 2.87 ng $m^{-3}$ with an overall average of 1.62 ng $m^{-3}$ (Supplementary S5). Generally, the TEQ values reported in this study, except for school L3 and L4, has exceeded the maximum permissible risk level of 1 ng $m^{-3}$ of BaP as set by European Guidelines. Oliveira et al. [28] reported that TEQ values were mainly higher in Asian schools (range: 4.70–49.4 ng $m^{-3}$) compared to European school environments (range: 0.04–29.8 ng $m^{-3}$).

DhA congener exhibited the highest carcinogenic potency of the PAHs in all schools, and this was most likely due its high toxicity factor (TEF value of 5). On average, DhA contributed up to 80% and 76% of outdoor and indoor ΣTEQ–PAHs, respectively. In contrast, BaP, the most known and studied carcinogenic PAHs, was the second-highest contributor for ΣTEQ–PAHs, accounting for 13.5% at outdoor and 17.5% at indoor air. The result is consistent with several previous studies [43,57,58], which reported DhA and BaP as the most influential components in the TEQ values.

The children in the HT group had a significantly higher risk of cancer than the children in the LT group. It can be arranged in ascending order: L3 ($5.41 \times 10^{-8}$) < L4 ($6.18 \times 10^{-8}$) < L2 ($9.39 \times 10^{-8}$) < L1 ($9.43 \times 10^{-8}$) < H2 ($1.04 \times 10^{-7}$) < H4 ($1.27 \times 10^{-7}$) < H3 ($1.66 \times 10^{-7}$) < H1 ($1.99 \times 10^{-7}$). Monte Carlo simulations were performed to make an accurate cancer risk estimation of PAH inhalation, as shown in Figure 5. The mean value of the probability cancer risk for the HT and LT groups were $1.54 \times 10^{-7}$ and $7.59 \times 10^{-8}$, respectively, and both means show similarity to the actual ILCR generated in the SPSS software ($1.59 \times 10^{-7}$ and $7.58 \times 10^{-8}$). The 95th percentiles of the ILCR calculated the risk for the HT and LT populations were $2.80 \times 10^{-7}$ and $1.43 \times 10^{-7}$, respectively. These ILCR values indicate that the daily inhalation dose of PAHs and cancer risk for children in the HT group is lower than the acceptable levels of $10^{-6}$ to $10^{-4}$ as proposed by the USEPA (2011). The estimated carcinogen risk for the HT group was higher than previous studies in Kuala Lumpur with a calculated ILCR of $2.64 \times 10^{-8}$ and $5.51 \times 10^{-8}$ in 2019 and 2021, respectively [11,18].

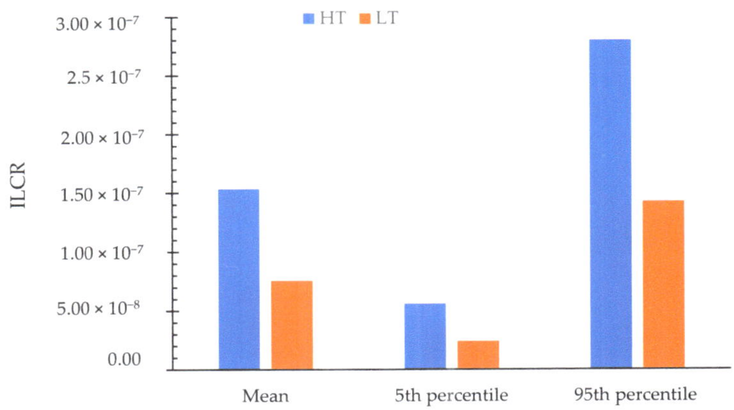


**Figure 5.** Probabilistic distribution of ILCR for children in HT and LT groups.

A sensitivity analysis was performed to find the parameter with the greatest impact on the overall risk outcome. In this study, the TEQ value and body weight (BW) were the most influential parameters for the carcinogenic risk due to PAH inhalation. As shown in Figure 6, the PAHs concentration in the TEQ value has a greater impact on the ILCR estimation in HT and LT group, with a correlation coefficient of 0.58 and 0.71, respectively. In contrast, an inverse relationship appeared between BW and estimated carcinogenic risk for HT and LT group (correlation coefficient: −0.72 and −0.59) which similarly found by several previous studies [42,59–61]. A negative correlation value indicates that the increase in predictor is associated with a decrease in ILCR prediction. The results of the sensitivity analysis suggest that the values and probability distributions of TEQ and BW should be correctly determined to increase the accuracy of the estimated risks.

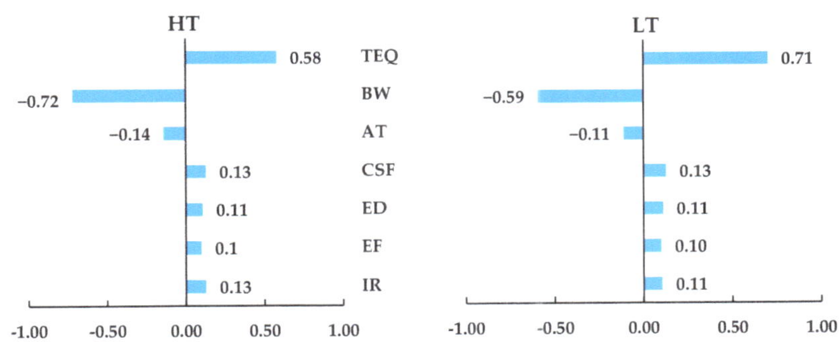


**Figure 6.** Sensitivity analysis of carcinogenic risk for children in HT and LT groups. TEQ = Toxic equivalent concentration; BW = Body weight; AT = Averaging time of carcinogenic PAHs exposure; CSF = Inhalation cancer slope factor; ED = exposure duration; EF = Exposure frequency; IR = Inhalation rate.

### *3.6. Individual Factors on DNA Damage*

The result shows that the tail moment (parameter of DNA damage) of the HT group ($3.13 \pm 0.53$) were significantly higher than the value recorded in the LT group ($2.80 \pm 0.81$) with $p < 0.05$. The individual factors on tail moment were stratified and compared using independent t-test and one-way ANOVA, as shown in Table 2. Considering all children together, the present study revealed that children aged 10–11 years old had significantly higher tail moment compared to children aged 7 to 9 years old (3.03 ± 0.62 vs. 2.84 ± 0.71). Several studies have reported an increased DNA damage by older age as assessed by the comet assay [62,63]; meanwhile there are studies that showed no effect of age on the extent of DNA damage [64,65].

**Table 2.** Comparison of Tail Moment according to a different individual factor.

| Variables | HT (N = 113) | | LT (N = 115) | | All Children (N = 228) | |
|---|---|---|---|---|---|---|
| | GM ± SD | p-Value ‡ | GM ± SD | p-Value ‡ | GM ± SD | p-Value ‡ |
| Age | | | | | | |
| 7–9 | 3.15 ± 0.50 | 0.914 | 2.74 ± 0.74 | 0.302 | 2.84 ± 0.71 | 0.040 * |
| 10–11 | 3.13 ± 0.54 | | 2.88 ± 0.71 | | 3.03 ± 0.62 | |
| Gender | | | | | | |
| Boy | 3.09 ± 0.53 | 0.310 | 2.86 ± 0.66 | 0.556 | 2.99 ± 0.60 | 0.773 |
| Girl | 3.19 ± 0.54 | | 2.78 ± 0.77 | | 2.96 ± 0.71 | |
| BMI categories | | | | | | |
| Underweight | 3.20 ± 0.59 | 0.338 | 2.98 ± 0.62 | 0.421 | 3.10 ± 0.60 | 0.138 |
| Normal | 3.19 ± 0.45 | | 2.87 ± 0.64 | | 3.02 ± 0.58 | |
| Overweight | 2.84 ± 0.61 | | 2.53 ± 1.13 | | 2.68 ± 0.88 | |
| Obese | 3.07 ± 0.63 | | 2.69 ± 0.85 | | 2.88 ± 0.76 | |
| House distance from main road | | | | | | |
| <500 m | 3.13 ± 0.53 | 0.846 | 2.85 ± 0.71 | 0.243 | 2.99 ± 0.64 | 0.465 |
| ≥500 m | 3.16 ± 0.58 | | 2.63 ± 0.82 | | 2.90 ± 0.75 | |
| House distance from highway | | | | | | |
| <500 m | 3.16 ± 0.49 | 0.504 | 2.59 ± 0.44 | 0.393 | 3.11 ± 0.51 | 0.028 * |
| ≥500 m | 3.09 ± 0.59 | | 2.83 ± 0.74 | | 2.91 ± 0.71 | |
| Mode of transportation to school | | | | | | |
| Active mode | 3.04 ± 0.59 | 0.509 | 2.79 ± 0.98 | 0.928 | 2.95 ± 0.73 | 0.899 |
| Motorized mode | 3.15 ± 0.53 | | 2.82 ± 0.71 | | 2.98 ± 0.65 | |
| Grilled food | | | | | | |
| Yes | 3.14 ± 0.54 | 0.990 | 2.94 ± 0.66 | 0.387 | 3.03 ± 0.61 | 0.554 |
| No | 3.13 ± 0.54 | | 2.79 ± 0.74 | | 2.96 ± 0.67 | |
| Supplement intake | | | | | | |
| Yes | 3.12 ± 0.49 | 0.750 | 2.76 ± 0.69 | 0.466 | 2.96 ± 0.61 | 0.763 |
| No | 3.15 ± 0.59 | | 2.86 ± 0.75 | | 2.99 ± 0.70 | |
| Fruit consumption | | | | | | |
| Yes | 3.14 ± 0.54 | 0.617 | 2.85 ± 0.72 | 0.274 | 3.00 ± 0.65 | 0.123 |
| No | 3.06 ± 0.47 | | 2.66 ± 0.73 | | 2.81 ± 0.66 | |
| ETS exposure | | | | | | |
| Yes | 3.16 ± 0.52 | 0.672 | 2.87 ± 0.58 | 0.516 | 3.01 ± 0.57 | 0.495 |
| No | 3.12 ± 0.55 | | 2.78 ± 0.80 | | 2.95 ± 0.70 | |

‡ ANOVA and Independent Sample $t$-test were computed based on log-transformed DNA damage; * Significant at $p < 0.05$.

Gender, age, and BMI are among the important demographic parameters that need to be emphasized in epidemiological studies because of their influence on genotoxic effects [66]. In this study, no significant difference in the tail moment were found for gender. Gajski et al. [65], in a study of healthy children living in an urban area in Croatia, found a significantly higher mean value of comet assay parameters in female children compared to male children. In the present study, no significant difference in tail moment measurement was found for BMI categories, which is in agreement with study by Sopian et al. [42]. It is well documented that being overweight and obese are associated with an increased DNA damage [67]. Gandhi [68] reported that DNA damage was almost five folds higher in the young obese subjects when DNA migration (strand breaks) was compared to the healthy control. The study speculated that increased oxidative stress and depletion of antioxidant in obese subjects resulted in increased genetic damage.

The effects of tobacco smoking on DNA damage have been widely investigated because tobacco contains carcinogenic and genotoxic substances. Therefore, smoking is also considered as a confounder [69]. In the present study, children who exposed to environmental tobacco smoke (ETS) had a higher tail moment (3.01 ± 0.57) than those children who did not exposed to ETS (2.95 ± 0.70); however, no significant difference detected from the mean comparison. Studies by Zalata et al. [70] and Beyoglu et al. [71] demonstrated that exposure to ETS among children was positively proven can increase DNA damage. School-aged children are particularly more exposed to secondhand smoke compared with other age groups. School-aged children spend a lot of time at home and stay close to their parents, suggesting that living with parents who smoke may be a strong predictor of increased exposure to substances contained in cigarettes [72].

Interestingly, the results showed that children living less than 500 m from main road and highway had a longer tail moment than the children living more than 500 m from main road and highway. However, the differences were statistically significant only for house distance from highway ($p$ = 0.028) but not with house distance from main road ($p$ = 0.465). Meanwhile, mode of transportation to school, did not exhibit any significant difference in tail moment measurement. In this study, grilled food consumers had a higher tail moment than children who consume less frequently (3.03 ± 0.61 vs. 2.96 ± 0.67, $p$ = 0.554). Children who less frequently consume supplement also were observed had a slightly higher tail moment (2.99 ± 0.70) than children who frequently take supplement (2.96 ± 0.61). Fruit consumption, however, showed inverse findings, as children who frequently eat fruit had a higher tail moment than those who less frequently eat fruit (3.00 ± 0.65 vs. 2.81 ± 0.66, $p$ = 0.123). High consumption of vegetables, fruits and juices rich in antioxidant vitamins and phytophenols has been shown to be positively associated with low levels of endogenous DNA strand breaks and oxidised DNA bases, and protective against ex vivo generation of DNA damage [69,73].

### 3.7. Factors That Influence DNA Damage

Simple linear regression was applied to investigate the relationship between DNA damage (tail moment) with possible predictor factors of DNA damage among children. Tail moment was used as a dependent variable in the linear regression. Supplementary S6 shows that the total particulate PAHs and carcinogenic PAHs in indoor and outdoor environment had significant influence on the occurrence of DNA damage in buccal epithelial cells of children in this study. In addition, the analysis also revealed a significant association between house distance from highway and DNA damage. Other individual factors did not exhibit significant association in the regression model.

The significant variables were further computed into MLR analysis using a stepwise method to determine the best predictor of the dependent variable. It was revealed that total indoor PAH exposure was the most significant factor that influenced the DNA damage among children, as portrayed in equation below:

$$Log\ tail\ moment = 2.734 + 0.063\ (Exposure\ to\ total\ indoor\ PAHs) \tag{3}$$

As shown in Table 3, the model predicted every unit increase of exposure to total indoor PAHs will lead to increment of log tail moment by 0.063. For the model, the beta value was significant at the 0.05 level. VIF value was <5, which suggested that there was no problem with multicollinearity. There was a significant direct linear relationship between total indoor PAHs with DNA damage ($p < 0.05$). Total indoor PAHs explained 4.4% of the variance in tail moment, adjusted $R^2$ = 0.044, F (1, 226) = 11.54, $p < 0.05$.

The present study strongly suggests that DNA damage is significantly affected by indoor particulate PAHs after controlling all possible confounding factors such as age, gender, BMI, house distance from main road and highway, mode of transportation to school, dietary habit and ETS exposure. Since all classrooms in this study has natural ventilation system, the present study provide evidence that children could be subjected to outdoor PAHs that infiltrate the schools' indoor environment via open windows and doors, in which poses a higher risk of genotoxicity. In addition, the indoor-outdoor PAHs ratio in all schools ranged from 0.86 to 0.99, indicating a high penetration of outdoor PAHs into indoor classrooms.

**Table 3.** Factor that influences DNA damage among children after controlling all confounders.

| Variable | *B* (95% CI) | β | *p*-Value | Adjusted $R^2$ |
|---|---|---|---|---|
| Constant | 2.734 (2.572, 2.897) | | <0.001 * | 0.044 |
| Indoor tPAHs | 0.063 (0.026, 0.100) | 0.220 | 0.001 * | |

* Significant at $p < 0.05$; Method: Stepwise.

The present study is in line with those of Sopian et al. [42], who discovered severe DNA damage in children living near a petrochemical plant in Terengganu, which exposed them to high amounts of particulate PAHs. In that study, the total indoor PAHs and open burning were the significant factor that influence the tail moment (adjusted $R^2 = 0.127$). In other words, the interaction of indoor PAH emissions and open burning significantly explained a 12.7% variation of the tail moment. In another study by Jasso-Pineda et al. [27], a significant DNA damage was found in a group of children living in a family that utilised biomass combustion. The study revealed a significant correlation between internal PAHs exposure (1-OHP) and DNA damage (r = 0.65, $p < 0.01$). Similarly, Sanchez Guerra et al. [74] also found positive associations between internal PAHs exposure (1-OHP) and DNA damage among Mexican children living near petrochemical industries. In addition, the results of this study are supported by a study by Ismail et al. [75], which demonstrated a high risk of DNA damage and respiratory symptoms in children who attended school near busy roads in Selangor.

A fact that could be considered a limitation of the present study was other toxic pollutants emitted from motor vehicles were not investigated in this study. DNA damage could also be induced by numerous environmental pollutants besides PAHs (i.e., Benzene and 1,3-butadiene) [76,77]. Future research should be conducted on quantification of other carcinogenic pollutants such as toxic gases (benzene, toluene, ethylbenzene, xylenes) to evaluate comprehensively the combine effects of exposure to urban air pollutants on genotoxic effects in children.

## 4. Conclusions

The findings provided evidence that children living near busy roads are more likely to be exposed to environmental PAHs and have a higher risk of genotoxicity than children living in low traffic areas. This study has successfully reduced the knowledge gap on relationship between urban traffic pollution and genotoxicity in children in the Southeast Asia region. The results not only contribute to the understanding of the levels, distribution, and sources of PAHs in educational settings, but also shed light on the governance of children's living environments and well-being, especially in urban areas. Although damage to genetic material is ubiquitous and inevitable for organisms, effective DNA repair systems can protect against these negative effects and maintain genetic stability. However, if the rate of DNA damage exceeds the capacity of the cell to repair it, detrimental biological consequences such as genotoxic damage and carcinogenesis will occur. Further study on the link between PAH exposure and genomic integrity in children should be investigated to shed additional light on potential health risks. This study recommends increasing the distance between future school sites and busy roads to reduce children's exposure to PAHs from traffic. Although the location of schools should be further away from busy roads, it is not easy to relocate existing schools, especially in the Klang Valley where the population is growing but there is less and less land available. Therefore, traffic density around existing schools should be reduced.

**Supplementary Materials:** The following supporting information can be downloaded at: https://www.mdpi.com/article/10.3390/ijerph19042193/s1, Supplementary S1: Questionnaire; Supplementary S2: Distributions of PAHs species in indoor and outdoor $PM_{2.5}$ samples; Supplementary S3: Percentage distribution of outdoor and indoor PAHs based on number of rings; Supplementary S4: TEQ of PAHs compounds in outdoor air of studied primary schools; Supplementary S5: TEQ of PAHs compounds in indoor air of studied primary schools; Supplementary S6: Predictor factors of DNA damage (Tail moment) among children.

**Author Contributions:** N.H.H.: writing—original draft and investigation. J.J.: writing—review and editing, conceptualisation and supervision. S.A.B.: validation and supervision. M.T.L.: writing—review and editing, methodology and supervision. All authors have read and agreed to the published version of the manuscript.

**Funding:** This project has received funding from Univerisiti Putra Malaysia under Impact Putra Grant (Project Code: UPM/800-3/3/1/GPB/2018/9659700).

**Institutional Review Board Statement:** The study was conducted according to the guidelines of the Declaration of Helsinki and approved by the ethics committee for research involving human subjects, Universiti Putra Malaysia (Reference no.: JKEUPM-2018-324).

**Informed Consent Statement:** Informed consent was obtained from all subjects involved in the study.

**Data Availability Statement:** Not applicable.

**Acknowledgments:** The authors gratefully acknowledge the generous assistance and support from the Environmental Health Lab, Faculty of Medicine and Health Sciences, Universiti Putra Malaysia, and Laboratory 2146, Faculty of Science and Technology, Universiti Kebangsaan Malaysia. Deepest appreciation to Anas Ahmad Jamhari and Haris Hafizal Abd Hamid for their technical assistance.

**Conflicts of Interest:** The authors declare no conflict of interest.

## References

1. Nagpure, A.S.; Gurjar, B.; Kumar, V.; Kumar, P. Estimation of exhaust and non-exhaust gaseous, particulate matter and air toxics emissions from on-road vehicles in Delhi. *Atmospheric Environ.* **2016**, *127*, 118–124. [CrossRef]
2. Azhari, A.; Latif, M.T.; Mohamed, A.F. Road traffic as an air pollutant contributor within an industrial park environment. *Atmospheric Pollut. Res.* **2018**, *9*, 680–687. [CrossRef]
3. Hamanaka, R.B.; Mutlu, G.M. Particulate Matter Air Pollution: Effects on the Cardiovascular System. *Front. Endocrinol.* **2018**, *9*, 680. [CrossRef]
4. Azhari, A.; Halim, N.; Mohtar, A.; Aiyub, K.; Latif, M.; Ketzel, M. Evaluation and prediction of $PM_{10}$ and $PM_{2.5}$ from road source emissions in Kuala Lumpur City Centre. *Sustainability* **2021**, *13*, 5402. [CrossRef]
5. Khan, F.; Latif, M.T.; Juneng, L.; Amil, N.; Nadzir, M.S.M.; Hoque, H.M.S. Physicochemical factors and sources of particulate matter at residential urban environment in Kuala Lumpur. *J. Air Waste Manag. Assoc.* **2015**, *65*, 958–969. [CrossRef] [PubMed]
6. Suhaimi, N.; Jalaludin, J.; Abu Bakar, S. The influence of Traffic-Related Air Pollution (TRAP) in primary schools and residential proximity to traffic sources on Histone H3 level in selected Malaysian children. *Int. J. Environ. Res. Public Heal.* **2021**, *18*, 7995. [CrossRef] [PubMed]
7. Fadzir, N.A.S.; Jalaludin, J. Indoor particulate matter 2.5 ($PM_{2.5}$) and lung function among children living near busy road in Cheras, Kuala Lumpur. *Health Environ. J.* **2013**, *4*, 1–19.
8. Hisamuddin, N.H.; Jalaludin, J.; Yusof, A.N.; Tualeka, A.R. Genotoxic effects of exposure to urban traffic related air pollutants on children in Klang Valley, Malaysia. *Aerosol Air Qual. Res.* **2020**, *20*, 2614–2623. [CrossRef]
9. Al-Daghri, N.M.; Alokail, M.S.; Abd-Alrahman, S.H.; Draz, H.M.; Yakout, S.M.; Clerici, M. Polycyclic aromatic hydrocarbon exposure and pediatric asthma in children: A case–control study. *Environ. Health* **2013**, *12*, 1. [CrossRef]
10. Khan, M.F.; Latif, M.T.; Lim, C.H.; Amil, N.; Jaafar, S.A.; Dominick, D.; Mohd Nadzir, M.S.; Sahani, M.; Tahir, N.M. Seasonal effect and source apportionment of polycyclic aromatic hydrocarbons in $PM_{2.5}$. *Atmos. Environ.* **2015**, *106*, 178–190. [CrossRef]
11. Sulong, N.A.; Latif, M.T.; Sahani, M.; Khan, M.F.; Fadzil, M.F.; Mohd Tahir, N.; Mohamad, N.; Sakai, N.; Fujii, Y.; Othman, M.; et al. Distribution, sources and potential health risks of polycyclic aromatic hydrocarbons (PAHs) in $PM_{2.5}$ collected during different monsoon seasons and haze episode in Kuala Lumpur. *Chemosphere* **2019**, *219*, 1–14. [CrossRef] [PubMed]
12. Liu, Y.; Yu, Y.; Liu, M.; Ge, R.; Li, S.; Liu, X.; Dong, W.; Qadeer, A. Characterization and source identification of PM2.5-bound polycyclic aromatic hydrocarbons (PAHs) in different seasons from Shanghai, China. *Sci. Total Environ.* **2018**, *644*, 725–735. [CrossRef] [PubMed]
13. Bahry, P.S.; Zakaria, M.P.; Bin Abdullah, A.M.; Abdullah, D.K.; Sakari, M.; Chandru, K.; Shahbazi, A. Forensic characterization of Polycyclic Aromatic Hydrocarbons and Hopanes in aerosols from Peninsular Malaysia. *Environ. Forensics* **2009**, *10*, 240–252. [CrossRef]

14. Caricchia, A.M.; Chiavarini, S.; Pezza, M. Polycyclic aromatic hydrocarbons in the urban atmospheric particulate matter in the city of Naples (Italy). *Atmospheric Environ.* **1999**, *33*, 3731–3738. [CrossRef]
15. Okuda, T.; Kumata, H.; Zakaria, M.P.; Naraoka, H.; Ishiwatari, R.; Takada, H. Source identification of Malaysian atmospheric polycyclic aromatic hydrocarbons nearby forest fires using molecular and isotopic compositions. *Atmospheric Environ.* **2002**, *36*, 611–618. [CrossRef]
16. Omar, N.Y.M.; Mon, T.C.; Rahman, N.A.; Bin Abas, M.R. Distributions and health risks of polycyclic aromatic hydrocarbons (PAHs) in atmospheric aerosols of Kuala Lumpur, Malaysia. *Sci. Total Environ.* **2006**, *369*, 76–81. [CrossRef]
17. Omar, N.Y.; Bin Abas, M.; Ketuly, K.A.; Tahir, N.M. Concentrations of PAHs in atmospheric particles (PM-10) and roadside soil particles collected in Kuala Lumpur, Malaysia. *Atmospheric Environ.* **2002**, *36*, 247–254. [CrossRef]
18. Suradi, H.; Khan, F.; Sairi, N.; Rahim, H.; Yusoff, S.; Fujii, Y.; Qin, K.; Bari, A.; Othman, M.; Latif, M. Ambient levels, emission sources and health effect of $PM_{2.5}$-Bound carbonaceous particles and Polycyclic Aromatic Hydrocarbons in the city of Kuala Lumpur, Malaysia. *Atmosphere* **2021**, *12*, 549. [CrossRef]
19. Jamhari, A.A.; Sahani, M.; Latif, M.T.; Chan, K.M.; Seng, H.; Khan, M.F.; Mohd Tahir, N. Concentration and source identification of polycyclic aromatic hydrocarbons (PAHs) in PM10 of urban, industrial and semi-urban areas in Malaysia. *Atmos. Environ.* **2014**, *86*, 16–27. [CrossRef]
20. Sopian, N.A.; Jalaludin, J. The Application of Biomarker in Determining Genotoxic Potential of Polyaromatic Hydrocarbon Exposure among children. *Ann. Trop. Med. Public Heal.* **2017**, *10*, 533.
21. Hisamuddin, N.H.; Jalaludin, J. Children's exposure to polycyclic aromatic hydrocarbon (PAHs): A review on urinary 1-hydroxypyrene and associated health effects. *Rev. Environ. Health* **2021**. [CrossRef] [PubMed]
22. Ruchirawat, M.; Settachan, D.; Navasumrit, P.; Tuntawiroon, J.; Autrup, H. Assessment of potential cancer risk in children exposed to urban air pollution in Bangkok, Thailand. *Toxicol. Lett.* **2007**, *168*, 200–209. [CrossRef] [PubMed]
23. Tuntawiroon, J.; Mahidol, C.; Navasumrit, P.; Autrup, H.; Ruchirawat, M. Increased health risk in Bangkok children exposed to polycyclic aromatic hydrocarbons from traffic-related sources. *Carcinogenesis* **2006**, *28*, 816–822. [CrossRef]
24. Mielżyńska, D.; Siwinska, E.; Kapka, L.; Szyfter, K.; Knudsen, L.E.; Merlo, D.F. The influence of environmental exposure to complex mixtures including PAHs and lead on genotoxic effects in children living in Upper Silesia, Poland. *Mutagenesis* **2006**, *21*, 295–304. [CrossRef] [PubMed]
25. Suk, W.A.; Murray, K.; Avakian, M.D. Environmental hazards to children's health in the modern world. *Mutat. Res. Mutat. Res.* **2003**, *544*, 235–242. [CrossRef] [PubMed]
26. Oliveira, M.; Slezakova, K.; Delerue-Matos, C.; Pereira, M.D.C.; Morais, S. Assessment of exposure to polycyclic aromatic hydrocarbons in preschool children: Levels and impact of preschool indoor air on excretion of main urinary monohydroxyl metabolites. *J. Hazard. Mater.* **2017**, *322*, 357–369. [CrossRef]
27. Jasso-Pineda, Y.; Barriga, F.D.; Yáñez-Estrada, L.; Pérez-Vázquez, F.J.; Pérez-Maldonado, I.N. DNA damage in Mexican children living in high-risk contaminated scenarios. *Sci. Total Environ.* **2015**, *518–519*, 38–48. [CrossRef]
28. Ochoa-Martinez, A.C.; Orta-Garcia, S.T.; Rico-Escobar, E.M.; Carrizales-Yañez, L.; Del Campo, J.D.M.; Pruneda-Alvarez, L.G.; Ruiz-Vera, T.; Gonzalez-Palomo, A.K.; Piña-Lopez, I.G.; Torres-Dosal, A.; et al. Exposure assessment to environmental chemicals in children from Ciudad Juarez, Chihuahua, Mexico. *Arch. Environ. Contam. Toxicol.* **2016**, *70*, 657–670. [CrossRef]
29. Oliveira, M.; Slezakova, K.; Delerue-Matos, C.; Pereira, M.C.; Morais, S. Children environmental exposure to particulate matter and polycyclic aromatic hydrocarbons and biomonitoring in school environments: A review on indoor and outdoor exposure levels, major sources and health impacts. *Environ. Int.* **2019**, *124*, 180–204. [CrossRef]
30. Junaidi, E.S.; Jalaludin, J.; Tualeka, A.R. A Review on the Exposure to Benzene among children in schools, preschools and daycare centres. *Asian J. Atmospheric Environ.* **2019**, *13*, 151–160. [CrossRef]
31. Ministry of Transport. Transport Statistics Malaysia 2019. Available online: https://www.mot.gov.my/en/StatistikTahunanPengangkutan/TransportStatisticsMalaysia2019.pdf (accessed on 30 June 2021).
32. Brauer, M.; Reynolds, C.; Hystad, P. Traffic-related air pollution and health in Canada. *Can. Med. Assoc. J.* **2013**, *185*, 1557–1558. [CrossRef]
33. Borrego-Soto, G.; Ortiz-Lopez, R.; Rojas-Martinez, A. Ionizing radiation-induced DNA injury and damage detection in patients with breast cancer. *Genet. Mol. Biol.* **2015**, *38*, 420–432. [CrossRef] [PubMed]
34. Heath, J.A.; Smibert, E.; Algar, E.M.; Dite, G.; Hopper, J.L. Cancer risks for relatives of children with cancer. *J. Cancer Epidemiol.* **2014**, *2014*, 806076. [CrossRef]
35. Nisbet, I.C.; Lagoy, P.K. Toxic equivalency factors (TEFs) for polycyclic aromatic hydrocarbons (PAHs). *Regul. Toxicol. Pharmacol.* **1992**, *16*, 290–300. [CrossRef]
36. Othman, M.; Latif, M.T.; Jamhari, A.A.; Abd Hamid, H.H.; Uning, R.; Khan, M.F.; Mohd Nadzir, M.S.; Sahani, M.; Abdul Wahab, M.I.; Chan, K.M. Spatial distribution of fine and coarse particulate matter during a southwest monsoon in Peninsular Malaysia. *Chemosphere* **2021**, *262*, 127767. [CrossRef]
37. Ouyang, R.; Yang, S.; Xu, L. Analysis and risk assessment of $PM_{2.5}$-Bound PAHs in a comparison of indoor and outdoor environments in a middle school: A Case Study in Beijing, China. *Atmosphere* **2020**, *11*, 904. [CrossRef]
38. US EPA. Exposure Factors Handbook Chapter 6. Available online: https://www.epa.gov/expobox/exposure-factors-handbook-chapter-6 (accessed on 8 September 2021).

39. Ismail, M.F.; Fadzil, M.F.; Tahir, N.M.; Latif, M.T.; Mohamad, N. Preliminary assessment of the distribution of $PM_{2.5}$-bound Polycyclic Aromatic Hydrocarbons in primary school environments in Kuala Lumpur. *Univ. Malays. Teren. J. Undergrad. Res.* **2019**, *1*, 51–58. [CrossRef]
40. Letter, C.; Jäger, G. Simulating the potential of trees to reduce particulate matter pollution in urban areas throughout the year. *Environ. Dev. Sustain.* **2019**, *22*, 4311–4321. [CrossRef]
41. Terzaghi, E.; De Nicola, F.; Cerabolini, B.E.; Posada-Baquero, R.; Ortega-Calvo, J.-J.; Di Guardo, A. Role of photo- and biodegradation of two PAHs on leaves: Modelling the impact on air quality ecosystem services provided by urban trees. *Sci. Total Environ.* **2020**, *739*, 139893. [CrossRef]
42. Sopian, N.; Jalaludin, J.; Abu Bakar, S.; Hamedon, T.; Latif, M. Exposure to particulate PAHs on potential genotoxicity and cancer risk among school children living near the Petrochemical Industry. *Int. J. Environ. Res. Public Health* **2021**, *18*, 2575. [CrossRef]
43. Wang, J.; Xu, H.; Guinot, B.; Li, L.; Ho, S.S.H.; Liu, S.; Li, X.; Cao, J. Concentrations, sources and health effects of parent, oxygenated- and nitrated- polycyclic aromatic hydrocarbons (PAHs) in middle-school air in Xi'an, China. *Atmospheric Res.* **2017**, *192*, 1–10. [CrossRef]
44. Krugly, E.; Martuzevicius, D.; Sidaraviciute, R.; Ciuzas, D.; Prasauskas, T.; Kauneliene, V.; Stasiulaitiene, I.; Kliucininkas, L. Characterization of particulate and vapor phase polycyclic aromatic hydrocarbons in indoor and outdoor air of primary schools. *Atmospheric Environ.* **2014**, *82*, 298–306. [CrossRef]
45. Crist, K.C.; Liu, B.; Kim, M.; Deshpande, S.R.; John, K. Characterization of fine particulate matter in Ohio: Indoor, outdoor, and personal exposures. *Environ. Res.* **2008**, *106*, 62–71. [CrossRef]
46. Srithawirat, T.; Latif, M.T.; Sulaiman, F.R. Indoor $PM_{10}$ and its heavy metal composition at a roadside residential environment, Phitsanulok, Thailand. *Atmósfera* **2016**, *29*, 311–322. [CrossRef]
47. Leung, D.Y.C. Outdoor-indoor air pollution in urban environment: Challenges and opportunity. *Front. Environ. Sci.* **2015**, *2*, 69. [CrossRef]
48. Long, C.M.; Sarnat, J.A. Indoor-Outdoor relationships and infiltration behavior of elemental components of outdoor $PM_{2.5}$ for Boston-Area homes. *Aerosol Sci. Technol.* **2004**, *38*, 91–104. [CrossRef]
49. Yunker, M.B.; Macdonald, R.W.; Vingarzan, R.; Mitchell, R.D.; Goyette, D.; Sylvestre, S. PAHs in the Fraser River basin: A critical appraisal of PAH ratios as indicators of PAH source and composition. *Org. Geochem.* **2002**, *33*, 489–515. [CrossRef]
50. Khan, F.; Hwa, S.W.; Hou, L.C.; Mustaffa, N.I.H.; Amil, N.; Mohamad, N.; Sahani, M.; Jaafar, S.A.; Nadzir, M.S.M.; Latif, M.T. Influences of inorganic and polycyclic aromatic hydrocarbons on the sources of PM2.5 in the Southeast Asian urban sites. *Air Qual. Atmosphere Health* **2017**, *10*, 999–1013. [CrossRef]
51. Venkataraman, C.; Lyons, J.M.; Friedlander, S.K. Size distributions of Polycyclic Aromatic Hydrocarbons and elemental carbon. 1. Sampling, Measurement Methods, and Source Characterization. *Environ. Sci. Technol.* **1994**, *28*, 555–562. [CrossRef]
52. Tobiszewski, M.; Namiesnik, J. PAH diagnostic ratios for the identification of pollution emission sources. *Environ. Pollut.* **2012**, *162*, 110–119. [CrossRef]
53. Pies, C.; Hoffmann, B.; Petrowsky, J.; Yang, Y.; Ternes, T.A.; Hofmann, T. Characterization and source identification of polycyclic aromatic hydrocarbons (PAHs) in river bank soils. *Chemosphere* **2008**, *72*, 1594–1601. [CrossRef] [PubMed]
54. Akyüz, M.; Çabuk, H. Gas–particle partitioning and seasonal variation of polycyclic aromatic hydrocarbons in the atmosphere of Zonguldak, Turkey. *Sci. Total Environ.* **2010**, *408*, 5550–5558. [CrossRef] [PubMed]
55. Ravindra, K.; Bencs, L.; Wauters, E.; de Hoog, J.; Deutsch, F.; Roekens, E.; Bleux, N.; Berghmans, P.; Van Grieken, R. Seasonal and site-specific variation in vapour and aerosol phase PAHs over Flanders (Belgium) and their relation with anthropogenic activities. *Atmospheric Environ.* **2006**, *40*, 771–785. [CrossRef]
56. Fang, G.-C.; Wu, Y.-S.; Chen, M.-H.; Ho, T.-T.; Huang, S.-H.; Rau, J.-Y. Polycyclic aromatic hydrocarbons study in Taichung, Taiwan, during 2002?2003. *Atmospheric Environ.* **2004**, *38*, 3385–3391. [CrossRef]
57. Oliveira, M.; Slezakova, K.; Madureira, J.; De Oliveira Fernandes, E.; Delerue-Matos, C.; Morais, S.; do Carmo Pereira, M. Polycyclic aromatic hydrocarbons in primary school environments: Levels and potential risks. *Sci. Total Environ.* **2017**, *575*, 1156–1167. [CrossRef]
58. Slezakova, K.; Oliveira, M.; Madureira, J.; de Oliveira Fernandes, E.; Delerue-matos, C.; Morais, S.; do Carmo Pereira, M. Polycyclic aromatic hydrocarbons (PAH) in Portuguese educational settings: A comparison between preschools and elementary schools. *J. Toxicol. Environ. Health Part A* **2017**, *80*, 630–640. [CrossRef]
59. Fandi, N.F.M.; Jalaludin, J.; Latif, M.T.; Hamid, H.H.A.; Awang, M.F. BTEX exposure assessment and inhalation health risks to traffic policemen in the Klang Valley region, Malaysia. *Aerosol Air Qual. Res.* **2020**, *20*, 1922–1937. [CrossRef]
60. Miri, M.; Alahabadi, A.; Ehrampoush, M.H.; Ghaffari, H.R.; Sakhvidi, M.J.Z.; Eskandari, M.; Rad, A.; Lotfi, M.H.; Sheikhha, M.H. Environmental determinants of polycyclic aromatic hydrocarbons exposure at home, at kindergartens and during a commute. *Environ. Int.* **2018**, *118*, 266–273. [CrossRef]
61. Tarafdar, A.; Oh, M.-J.; Nguyen-Phuong, Q.; Kwon, J.-H. Profiling and potential cancer risk assessment on children exposed to PAHs in playground dust/soil: A comparative study on poured rubber surfaced and classical soil playgrounds in Seoul. *Environ. Geochem. Health* **2019**, *42*, 1691–1704. [CrossRef]
62. Milić, M.; Ceppi, M.; Bruzzone, M.; Azqueta, A.; Brunborg, G.; Godschalk, R.; Koppen, G.; Langie, S.; Møller, P.; Teixeira, J.P.; et al. The hCOMET project: International database comparison of results with the comet assay in human biomonitoring. Baseline frequency of DNA damage and effect of main confounders. *Mutat. Res. Mutat. Res.* **2021**, *787*, 108371. [CrossRef]

63. Piperakis, S.M.; Kontogianni, K.; Karanastasi, G.; Iakovidou-Kritsi, Z.; Piperakis, M.M. The use of comet assay in measuring DNA damage and repair efficiency in child, adult, and old age populations. *Cell Biol. Toxicol.* **2007**, *25*, 65–71. [CrossRef]
64. Locken-Castilla, A.; Pacheco-Pantoja, E.L.; Rodríguez-Brito, F.; May-Kim, S.; López-Rivas, V.; Ceballos-Cruz, A. Smoking index, lifestyle factors, and genomic instability assessed by single-cell gel electrophoresis: A cross-sectional study in subjects from Yucatan, Mexico. *Clin. Epigenetics* **2019**, *11*, 150. [CrossRef]
65. Gajski, G.; Gerić, M.; Oreščanin, V.; Garaj-Vrhovac, V. Cytogenetic status of healthy children assessed with the alkaline comet assay and the cytokinesis-block micronucleus cytome assay. *Mutat. Res. Toxicol. Environ. Mutagen.* **2013**, *750*, 55–62. [CrossRef]
66. How, V.; Hashim, Z.; Ismail, P.; Said, S.M.; Omar, D.; Tamrin, S.B.M. Exploring Cancer Development in Adulthood: Cholinesterase Depression and Genotoxic Effect From Chronic Exposure to Organophosphate Pesticides Among Rural Farm Children. *J. Agromedicine* **2014**, *19*, 35–43. [CrossRef] [PubMed]
67. Setayesh, T.; Nersesyan, A.; Mišík, M.; Ferk, F.; Langie, S.; Andrade, V.M.; Haslberger, A.; Knasmüller, S. Impact of obesity and overweight on DNA stability: Few facts and many hypotheses. *Mutat. Res. Rev. Mutat. Res.* **2018**, *777*, 64–91. [CrossRef] [PubMed]
68. Gandhi, G.; Kaur, G. Assessment of DNA damage in obese individuals. *Res. J. Biol.* **2012**, *2*, 37–44.
69. Azqueta, A.; Ladeira, C.; Giovannelli, L.; Boutet-Robinet, E.; Bonassi, S.; Neri, M.; Gajski, G.; Duthie, S.; Del Bo', C.; Riso, P.; et al. Application of the comet assay in human biomonitoring: An hCOMET perspective. *Mutat. Res. Mutat. Res.* **2019**, *783*, 108288. [CrossRef] [PubMed]
70. Zalata, A.; Yahia, S.; Elbakary, A.; Elsheikha, H. Increased DNA damage in children caused by passive smoking as assessed by comet assay and oxidative stress. *Mutat. Res. Toxicol. Environ. Mutagen.* **2007**, *629*, 140–147. [CrossRef]
71. Beyoglu, D.; Ozkozaci, T.; Akici, N.; Omurtag, G.Z.; Akici, A.; Ceran, O.; Sardas, S. Assessment of DNA damage in children exposed to indoor tobacco smoke. *Int. J. Hyg. Environ. Health* **2010**, *213*, 40–43. [CrossRef] [PubMed]
72. Jeong, C.-H.; Salehi, S.; Wu, J.; North, M.L.; Kim, J.S.; Chow, C.-W.; Evans, G.J. Indoor measurements of air pollutants in residential houses in urban and suburban areas: Indoor versus ambient concentrations. *Sci. Total Environ.* **2019**, *693*, 133446. [CrossRef]
73. Dusinska, M.; Collins, A. The comet assay in human biomonitoring: Gene-environment interactions. *Mutagenesis* **2008**, *23*, 191–205. [CrossRef] [PubMed]
74. Sánchez-Guerra, M.; Pelallo-Martínez, N.; Barriga, F.D.; Rothenberg, S.J.; Cadena, L.H.; Faugeron, S.; Oropeza-Hernández, L.F.; Guaderrama-Díaz, M.; Quintanilla-Vega, B. Environmental polycyclic aromatic hydrocarbon (PAH) exposure and DNA damage in Mexican children. *Mutat. Res. Toxicol. Environ. Mutagen.* **2012**, *742*, 66–71. [CrossRef] [PubMed]
75. Ismail, I.N.; Jalaludin, J.; Abu Bakar, S.; Hisamuddin, N.H.; Suhaimi, N.F. Association of raffic-Related Air Pollution (TRAP) with DNA damage and respiratory health symptoms among primary school children in Selangor. *Asian J. Atmospheric Environ.* **2019**, *13*, 106–116. [CrossRef]
76. Ruchirawat, M.; Navasumrit, P.; Settachan, D. Exposure to benzene in various susceptible populations: Co-exposures to 1,3-butadiene and PAHs and implications for carcinogenic risk. *Chem. Interactions* **2010**, *184*, 67–76. [CrossRef]
77. Awang, M.F.; Jalaludin, J.; Abu Bakar, S.; Latif, M.T.; Fandi, N.F.M.; Hamid, H.H.A. Assessment of micronucleus frequency and respiratory health symptoms among traffic policemen exposed to BTEX and $PM_{2.5}$ in Klang Valley, Malaysia. *J. Teknol.* **2020**, *82*. [CrossRef]

International Journal of
*Environmental Research and Public Health*

MDPI

*Review*

# Synergistic or Antagonistic Health Effects of Long- and Short-Term Exposure to Ambient $NO_2$ and $PM_{2.5}$: A Review

**Anna Mainka * and Magdalena Żak**

Department of Air Protection, Silesian University of Technology, 22B Konarskiego St., 44-100 Gliwice, Poland
* Correspondence: anna.mainka@polsl.pl; Tel.: +48-322371060

**Abstract:** Studies on adverse health effects associated with air pollution mostly focus on individual pollutants. However, the air is a complex medium, and thus epidemiological studies face many challenges and limitations in the multipollutant approach. $NO_2$ and $PM_{2.5}$ have been selected as both originating from combustion processes and are considered to be the main pollutants associated with traffic; moreover, both elicit oxidative stress responses. An answer to the question of whether synergistic or antagonistic health effects of combined pollutants are demonstrated by pollutants monitored in ambient air is not explicit. Among the analyzed studies, only a few revealed statistical significance. Exposure to a single pollutant ($PM_{2.5}$ or $NO_2$) was mostly associated with a small increase in non-accidental mortality (HR:1.01–1.03). $PM_{2.5}$ increase of <10 μg/m$^3$ adjusted for $NO_2$ as well as $NO_2$ adjusted for $PM_{2.5}$ resulted in a slightly lower health risk than a single pollutant. In the case of cardiovascular heart disease, mortality evoked by exposure to $PM_{2.5}$ or $NO_2$ adjusted for $NO_2$ and $PM_{2.5}$, respectively, revealed an antagonistic effect on health risk compared to the single pollutant. Both short- and long-term exposure to $PM_{2.5}$ or $NO_2$ adjusted for $NO_2$ and $PM_{2.5}$, respectively, revealed a synergistic effect appearing as higher mortality from respiratory diseases.

**Keywords:** air pollutants; $NO_2$; $PM_{2.5}$; multi-pollutant approach; mortality; morbidity



**Citation:** Mainka, A.; Żak, M. Synergistic or Antagonistic Health Effects of Long- and Short-Term Exposure to Ambient $NO_2$ and $PM_{2.5}$: A Review. *Int. J. Environ. Res. Public Health* **2022**, *19*, 14079. https://doi.org/10.3390/ijerph192114079

Academic Editors: Nuno Canha, Marta Almeida and Evangelia Diapouli

Received: 30 September 2022
Accepted: 26 October 2022
Published: 28 October 2022



## 1. Introduction

Recently, there has been a significant increase in publications on air pollution and human health [1]. As an example, Sun and Zhu [2] in a scoping review analyzed 361 articles covering studies on both short- and long-term health consequences of exposure to various ambient air pollutants. Among ambient air pollutants most commonly investigated are particulate matter (PM) with diameters < 10 μm and <2.5 μm ($PM_{2.5}$, $PM_{10}$), as well as nitrogen dioxide ($NO_2$). To a lesser extent, studies included general air pollution of gases such as ozone ($O_3$), sulfur dioxide ($SO_2$), and carbon monoxide (CO), as well as hazardous air pollutants (HAPs) such as metals and volatile organic compounds (VOC) [3–16]. Health effects are analyzed in a wide range. Dominski et al. [1] ranked the most frequently studied health effects, starting with (1) respiratory diseases including asthma, respiratory infections, respiratory disorders, and chronic obstructive pulmonary disease (COPD). Following this were (2) cardiovascular diseases including hypertension, heart rate variability, heart attack, cardiopulmonary disease, ischemic heart disease, blood coagulation, deep vein thrombosis, and stroke; (3) other diseases such as DNA methylation changes, neurobehavioral disfunctions, inflammatory disease, skin disease, and disability; (4) pregnancy and children characterized by such parameters as birth weight, infant death, infantile eczema, preterm birth, fertility, pregnancy-induced hypertension, and other diseases; (5) mental disorders: Alzheimer's disease, Parkinson's disease, depression and stress, annoyance, autism spectrum disorder (ASD), and reduced cognitive function; (6) all causes, when there are no specific causes of disease, including morbidity, hospital admissions, outpatient visits, emergency room visits, and mortality; (7) chronic diseases: diabetes and chronic respiratory diseases; (8) general health outcomes; and (9) cancer, including bladder cancer, brain tumor, breast cancer, liver cancer, lung cancer, and unspecified cancer.

Thus far, we have learned much about the relationships between air pollution and health. In last year's updated World Health Organization (WHO) air quality guidelines [17], the limit values were significantly decreased (Table 1). We know that the majority of health effects present a positive correlation to air pollution levels and that the harmful outcomes are evident. The plausibility of these associations is supported by results from experimental exposures to humans, animals, and cells [18]. For some pollutants, the results are ambiguous, there is more than one interpretation, or only limited evidence is available for harmful health effects. Moreover, for some pollutants, especially those characterized by levels below the reference concentration, no health effects are to be expected [19].

There are significant statistical associations between concentrations of individual pollutants and population health outcomes. A single-pollutant approach to air pollution management and research was partly motivated by the Clean Air Act [20], which identifies six criteria air pollutants (CO, tropospheric $O_3$, lead—Pb, nitrogen oxides ($NO_x$), PM, and $SO_2$), that are the motivation of air quality regulations based on monitored air concentrations. WHO air quality guidelines [17] point out that among priorities for policy-relevant scientific questions such as "how, why and for whom do the health effects of air pollution exist?" there has been a call to move from a single-pollutant to a multi-pollutant approach. An amendment from the single-pollutant approach requires a shift in air pollution health research to provide a sound basis for multi-pollutant air quality management.

**Table 1.** EU limit values and WHO Air Quality Guidelines (AQG) for considered air pollutants.

| Air Pollutant | EU Limit Values [21] | WHO AQG [17] | |
|---|---|---|---|
| | Annual [a] | Annual | Daily |
| $PM_{2.5}$, μg/m$^3$ | 25 | 5 | 15 |
| $NO_2$, μg/m$^3$ | 40 | 10 | 25 |

n.a.—not applicable; [a] daily values are not available.

WHO guidelines [17] point out that studies of multi-pollutant exposures including the examination of additive, synergistic (greater than additive), or antagonistic (less than additive) effects play an important role. The Environmental Protection Agency USA provided definitions for terms that describe various types of toxicologic interactions, including forms of additivity, antagonism, inhibition, masking, potentiation, synergism, and other toxicologic phenomena [18,22], defined below:

(1) Additivity is when the "effect" of the combination is estimated by the sum of the exposure levels or the effects of the individual chemicals. The effect may refer to the measured response or the incidence of adversely affected animals. The sum may be a weighted sum (the sum of component doses scaled by their toxic potency relative to the index chemical) or a conditional sum (when the toxic response from the chemical mixture is equal to the sum of independent component responses).
(2) Antagonism is when the effect of the combination is less than that suggested by the component's toxic effects. Antagonism must be defined in the context of the definition of "no interaction", which is usually dose or response addition.
(3) Inhibition is when one substance does not have a toxic effect on a certain organ system, but when added to a toxic chemical, it makes the latter less toxic.
(4) Masking is when the compounds produce opposite or functionally competing effects at the same site or sites so that the effects produced by the combination are less than suggested by the component's toxic effects.
(5) Potentiation is when one substance does not have a toxic effect on a certain organ or system, but when added to a toxic chemical, it makes the latter more toxic.
(6) Synergism is when the effect of the combination is greater than that suggested by the component's toxic effects. Synergism must be defined in the context of the definition

of "no interaction", when neither compound by itself produces an effect, and no effect is seen when they are administered together.

(7) Confounding is another cause of concern in studies of air pollution. If not adequately addressed, cofounding may either increase the apparel effect of air pollution. Confounding occurs when a confounder, a risk predictor for the outcome of concern, co-varies with the air pollutant being investigated.

A fundamental aspect of the multi-pollutant approach to air pollution epidemiology is that the air pollutant concept is being replaced by a common source perspective. To assess the multi-index analysis of ecological risks from the source perspective, a geographic information system (GIS) and remote sensing (RS) can be used. Among indicators reflecting environmental risks, the pressure from urban expansion, land use, and degradation as well as cropland proportion can be included [23]. One of the sources identified as being responsible for adverse health effects is transport. For automobile traffic, the characteristic pollutants are $NO_2$ and $PM_{2.5}$. Both are derived from combustion processes and increase oxidative stress [24]. $NO_2$ is often treated as a surrogate for PM [25,26]; on the other hand, 14% to 27% of the measured secondary $PM_{2.5}$ is generated from $NO_x$ set of reactions [27,28]. Nevertheless, $NO_2$ or $PM_{2.5}$ alone are not sufficient to fully characterize the toxicity of the atmospheric mixture or to fully explain the risk of mortality and morbidity associated with exposure to ambient air pollution [6].

Other factors, besides ambient air quality, that determine the length and quality of life is a healthy lifestyle including both dietary habits [29,30] and physical activity [31].

This paper aims to synthesize worldwide evidence on the health effects of short-term and long-term exposure to both $NO_2$ and $PM_{2.5}$ on all-cause and/or cause-specific mortality and morbidity. As most of the studies focus on single pollutants, we aim to present the knowledge gaps in the multi-pollutant approach.

Our intention is to provide an illustrative overview of the available health research performed with respect to the concurrent $NO_2$ and $PM_{2.5}$ exposure and should be considered rather in a narrative context, not as a systematic review. To identify publications reporting results from studies on the impact of $NO_2$ and $PM_{2.5}$ on health effects, we conducted a broad search of the online databases, particularly with articles published during the last 10 years. Both research articles and literature reviews on the relationship between simultaneous $NO_2$ and $PM_{2.5}$ exposure and physical and/or mental health effects were included. Inclusion criteria were (1) cohort studies; (2) long-term exposure metrics, i.e., annual or multi-year averages; (3) mean daily or monthly exposure metrics (short-term exposure); (4) cross-sectional, case–control studies. A good example is the systematic review published by Mills et al. [32], presenting the results of 60 articles that provided estimates of both (1) $NO_2$ as a single-pollutant and (2) a two-pollutant model including $NO_2$ adjusted for PM and to a lesser extent PM meta-analyses for $NO_2$ adjusted. However, within a study [32], $PM_{10}$ was the most used metric (67%). Orellano et al. [10] presented evidence on the physical health effects of short-term exposure to $PM_{10}$ and $PM_{2.5}$, $NO_2$, and ozone ($O_3$) on all-cause mortality, and $PM_{10}$ and $PM_{2.5}$ on cardiovascular, respiratory, and cerebrovascular mortality. A highly valuable work covering both the physical and mental health effects with respect to $NO_x$ exposure available risk ratios was included in the review of Shaw and Heyst [33].

The results presented here do not represent an exhaustive list of the literature on physical and/or mental health, but rather offer an illustration of findings in this area, particularly studies presenting statistic parameters: risk, odds, or hazard ratios. It is worth underlining that risk ratios (RR), odds ratios (OR), and hazard ratios (HR) are three common, but often misused, statistical measures, especially in clinical research, or are misunderstood in their interpretation of a study's results [34].

Risk ratio (RR) is also known as relative risk. The definition from the Dictionary of Environmental Health [35] states that RR is an expression of the occurrence (such as either a percentage or ratio) of the particular phenomenon in an exposed population compared to the appearance of the same phenomenon in a population that has not been exposed. A RR of 1 means that the probability or risk of an event occurring in either population when compared to the other is even, and therefore both groups have the same amount of risk. If the risk is doubled, the RR is 2, and if it is halved, it is 0.5.

Odds ratio (OR) is a statistical technique used in epidemiological case–control studies derived by dividing the odds of an event happening in the exposed group by the odds of the same event happening in the control group [35]. OR provides a measure of the strength of association between two variables on a scale, with 1 being no association, above 1 being a positive association, and below 1 being a negative association. While risk reports the number of events of interest in relation to the total number of trials, odds report the number of events of interest in relation to the number of events not of interest. Stated differently, it reports the number of events to nonevents. For example, the risk of flipping a coin to be heads is 1:2 or 50%; the odds of flipping a coin to be heads is 1:1, as there is one desired outcome (event), and one undesired outcome (nonevent). Misreporting of the OR as the RR, then, can often exaggerate data. It is important to remember that OR is a relative measure just as RR, and thus sometimes a large OR can correspond with a small difference between odds [36].

Hazard ratio (HR) is a similar but distinct measure. It concerns rates of change as a comparison of two hazards. It can present how quickly two survivorship curves diverge through comparison of the slopes of the curves. An HR of 1 indicates no divergence—within both curves, the likelihood of the event was equally likely at any given time. An HR not equal to 1 indicates that two events are not occurring at an equal rate, and the risk of an individual in one group is different than the risk of an individual in another at any given time interval. An important aspect of HRs is the proportional hazard assumption. To report a singular hazard ratio, it must be assumed that the two hazard rates are constant [36]. Regardless of the value of RR/OR/HR, the interpretation should only be made after determining whether the result provides statistically significant evidence towards a conclusion (verified by a $p$-value < 0.05 or 95% confidence interval). Keeping these principles and the RR/OR/HR framework in mind minimizes bias and prevents erroneous conclusions being drawn from the results of a published study on different samples. Table 2 shows the characteristics of RR/OR/HR ratios along with correct and incorrect use.

**Table 2.** Risk ratio (RR) vs. odds ratio (OR) vs. hazard ratio (HR) [36].

| Parameter | RR | OR | HR |
|---|---|---|---|
| Aim | Determination of relationships in risk potential based on some variable. | Identify the relationship between two variables. | Determine how one group changes relative to the other. |
| Usage | Informs how an intervention changes risk. | Informs whether there is a relationship between the intervention and the risk; estimates how this relationship occurs. | Informs how intervention changes the rate at which an event is experienced. |
| Limitations | Only applicable if the study design is representative of the population. Cannot be used for case–control studies. | It can generally be used anywhere but is not always a useful statistic on its own. It exaggerates the risks. | In order to be typically useful, the rate of change within the two groups should be relatively consistent. |
| Timeline | Static. Summarizes an overall study. | Static. Summarizes an overall study. | Dynamic. Provides information about the way a study progresses over time. |

**Table 2.** *Cont.*

| Parameter | RR | OR | HR |
| --- | --- | --- | --- |
| Correct | Most useful and often preferred statistic. Most intuitive and easiest to understand. | Can be more widely used. Can sometimes approximate relative risk in rare cases. Does not require a random sampling. Useful for logistic regression and case–control studies. | Allows a member of association in outcomes in survivorship curves. |
| Incorrect | Must assume causal direction (change in an independent variable changes the outcome variable). Only effective and useful with randomized something. | Less intuitive and can seem to exaggerate data. Only shows correlation, not causation. | Requires proportional hazards assumption (all data groups must show a roughly linear relationship between number of events and time). Should be reported with median time-to-event. |

## 2. Considerations and Discussion

Air pollution has an impact on various health effects. The simplest division includes mortality and morbidity. Mortality reflects the reduction of life expectancy, while morbidity relates to illness occurrence. Mortality is the most studied health endpoint in association with air pollution [37]. The analysis includes all-cause and cause-specific mortality. Other terms used for this indicator are premature death, additional mortality, and death postponed. In all of these metrics, the health effect is expressed by the number of deaths [38]. Morbidity indicator estimates changes in new or existing diseases in a target population [39]. Low air pollution levels can impact on the health of susceptible and sensitive groups, especially with already fragile immune systems (e.g., immunodeficiency, asthma, malnutrition, old age, infancy) [40].

Studies showed a relationship between air pollutants and adverse health effects that appear after short-term (acute) or long-term (chronic) exposure [41]. Short-term exposure to air pollutants is closely related to COPD, cough, shortness of breath, wheezing, asthma, respiratory disease, and high rates of hospitalization (a measurement of morbidity). The long-term effects connected with exposure to air pollution are chronic asthma, pulmonary insufficiency, cardiovascular diseases, and cardiovascular mortality. According to a Swedish cohort study [42], diabetes seems to be induced after long-term air pollution exposure. Moreover, air pollution appears to have various malign health effects in early human life, such as respiratory, cardiovascular, mental, and perinatal disorders, leading to infant mortality or chronic disease in adult age [43]. Exposure to pollutants in utero increases the risk of neuro-developmental delay. Childhood exposure has been inversely associated with neuro-developmental outcomes in younger children, as well as with academic achievement and neurocognitive performance in older children. In older adults, air pollution is associated with accelerated cognitive decline [44].

There is also emerging evidence that air pollutants may adversely affect cognitive development [45], cognitive performance [15,44–47], and stress [48] and psychological well-being [49,50]. Air pollution impairs verbal tests, and the effect becomes stronger as people age. Cognitive decline or impairment are risk factors for Alzheimer's disease and other forms of dementia, especially for elderly persons [47,51]. The wide-ranging impacts of air pollutants on brain health and functioning may support the hypothesis of an association with clinically relevant mental health outcomes [9].

Acute and long-term effects are partly interrelated; however, long-term effects are not the sum of short-term effects [26]. The effects of long-term exposure are much greater than those observed for short-term exposure, suggesting that effects are not just due to exacerbations, but maybe also could be due to the progression of underlying diseases. On the other hand, the effects of continued short-term exposure may contribute to the initiation or exacerbation of the chronic disease, and those affected by the acute exposures may reflect a distinct, susceptible subgroup with underlying or existing disease or unrecognized vulnerability [26].

### 2.1. Health Effects of $PM_{2.5}$

PM is a widespread air pollutant, consisting of a mixture of solid and liquid particles suspended in the air, covering a wide range of sizes and chemical compositions [52]. PM is considered so hazardous that in 2016, the WHO and the International Agency for Research on Cancer (IARC) classified ambient PM as Group 1—compounds carcinogenic to humans. The authors of the IARC monograph [53] underline that PM exposure from different sources features mutagenic and carcinogenic effects in people. Unfortunately, the number of air monitoring ground stations is limited, and the spatial distribution is discontinuous, and thus to obtain a fine-grained spatiotemporal distribution of $PM_{2.5}$, a retrieval model can be used [54]. In an estimation of future global mortality from changes in air pollution, Silva et al. [55] predicted 55,600 (−34,300 to 164,000) deaths in 2030 and 215,000 (−76,100 to 595,000) in 2100 due to $PM_{2.5}$ worldwide (countering by 16% the global decrease in $PM_{2.5}$-related mortality.

Fine particles with a diameter < 2.5 μm ($PM_{2.5}$) are considered one of the leading environmental health risk factors due to the potential penetration of particles deeper into the lungs [56]. Total lung deposition of $PM_{2.5}$ particles is about 60% for ultrafine particles and 20% for fine particles [57]. In adults at rest, the nasal deposition of $PM_{2.5}$ is about 20% and increases to 30–40% during exercise. Lower values (about 10–20%) are reached in children aged 5–15 years [58]. Once deposited in the lung, most particles are removed through several clearance mechanisms. Insoluble particles deposited on ciliated airways are generally removed from the respiratory tract by mucociliary activity within 24–48 h [57]. Several $PM_{2.5}$ can also be absorbed into the bloodstream through alveolar capillaries, causing lung and systemic inflammation [59–61]. The inflammatory reactions of the lungs and bronchi are due to oxidative stress produced by increased levels of reactive oxygen species (ROS) that responsible for many of the cardiovascular and respiratory health effects [62]. Among the mechanisms related to cardiovascular outcomes, the most influential is the release of pro-oxidative and proinflammatory mediators from the lungs into the circulation and autonomic nervous system imbalance [63].

#### 2.1.1. Short-Term Exposure to $PM_{2.5}$

Short-term exposure to $PM_{2.5}$ evidences a positive relationship with hospital admissions or emergency room visits for cardiovascular outcomes or respiratory effects, increasing from a 0.5 to 3.4% per 10 $\mu g/m^3$ rise in $PM_{2.5}$ levels [37]. Significant associations between short-term (averaging time over 2 or 24 h) $PM_{2.5}$ concentrations and myocardial infarction [27,64] cardiac arrhythmias as well as ischemic stroke [65] were observed. Studies on specific cardiovascular diseases indicate that ischemic heart disease, and congestive heart failure may be influential for the observed associations. Although estimates from studies of cerebrovascular diseases are less precise and consistent, ischemic diseases are more strongly associated with $PM_{2.5}$ compared to hemorrhagic stroke. The available evidence suggests that cerebrovascular effects occur at short lags (approximately 1 day), while effects at longer legs are rarely evaluated. Cardiovascular hospital admissions are especially reported in areas with concentrations ranging from 7 to 18 $\mu g/m^3$ [22,27]. Toxicological studies associated with $PM_{2.5}$ exposure show reduced myocardial blood flow during ischemia and altered vascular reactivity, providing myocardial ischemia. In addition to ischemia, plausible biological mechanisms include increased right ventricular pressure and decreased myocardial contractility in the association between $PM_{2.5}$ and congestive heart failure. Additionally, systemic inflammation and oxidative stress are cited [55,66,67].

In a meta-analysis of 33 studies from China addressing the short-term effects of air pollution, a 10 μg/m$^3$ increase in $PM_{2.5}$ was associated with a 0.38% increase in total mortality [68]. Han et al. [69], on the basis of the measurements conducted from 2015 to 2019 in 296 cities across China, estimated that long-term exposure to $PM_{2.5}$ levels exceeding current WHO guidelines (5 μg/m$^3$) was associated with 17% average all-cause mortality. Data for China are relevant to analysis performed, for example, in Poland, as PM concentrations in China are far higher than the Western European or North American averages but comparable to results from Poland [68]. The estimated short-term effects of daily $PM_{2.5}$ concentrations on mortality were even more profound than in Western Europe or China. In the short-term exposure to $PM_{2.5}$, the number of emergency hospitalizations for pneumonia increased by 1.7% per 10 μg/m$^3$ of $PM_{2.5}$ [70]. In contrast, the increase in the number of deaths for Tricity and Warsaw (Poland) were 2.1% and 2.6% per 10 μg/m$^3$ $PM_{2.5}$, respectively [71]. Several lines of evidence suggest that $PM_{2.5}$ promotes and exacerbates allergic disease, which often underlies asthma [72]. A burst of reactive oxygen species induced by $PM_{2.5}$ was found in the neutrophils of asthmatic patients [73]. Braithwaite et al. [9] in a systematic review included 11 studies on short-term (<6 months) associations between eligible mental health outcomes and $PM_{2.5}$.

#### 2.1.2. Long-Term Exposure to $PM_{2.5}$

Longer-term effects of $PM_{2.5}$ exposure could be greater than the immediate ones [71]. PM2.5 is recognized as a key air pollutant significantly related to premature mortality attributed to cardiovascular diseases and lung cancer [74,75]. The results of the European Study of Cohorts for Air Pollution Effects (ESCAPE) showed higher overall mortality due to long-term $PM_{2.5}$ exposure, with statistically significant associations also reported for individuals exposed to $PM_{2.5}$ concentrations below the European threshold of 25 μg/m$^3$ [76]. Data from many European cohorts found an association between long-term exposure to $PM_{2.5}$ and lung and kidney cancer [77–83]. At concentrations exceeding the threshold value, pediatric $PM_{2.5}$ exposures deliver health interventions before the development of obesity and identify and mitigate environmental factors influencing obesity and Alzheimer's disease. Braithwaite et al. [9], through the analysis of nine articles, support the hypothesis of an association between long-term ($\geq$6 months) $PM_{2.5}$ exposure and depression, as well as anxiety. Calderón-Garcidueñas et al. [84] pointed to diffuse neuroinflammation, damage to the neurovascular unit, and the production of autoantibodies to neural and tight-junction proteins as the worrisome findings in children chronically exposed to concentrations above $PM_{2.5}$ current standards, potentially constituting significant risk factors for the development of Alzheimer's disease later in life. The results of the REVIHAAP (Review of Evidence on Health Aspects of Air Pollution) project [26] point to additional systemic health effects beyond the respiratory and cardiovascular systems—for example, effects on the central nervous system, the progression of Alzheimer's and Parkinson's diseases, developmental outcomes in children, and reproductive health outcomes such as low birth weight.

#### 2.1.3. $PM_{2.5}$ Health Effect Mechanism

There are different mechanisms of effect considering both short- and long-term exposure. The health effects of $PM_{2.5}$, as well as the mechanisms underlying these effects, were investigated by Feng et al. [85] in 132 articles published from 2005 to 2015. The central mechanisms of harmful effects of $PM_{2.5}$ are oxidative stress (intracellular), mutagenicity/genotoxicity, and inflammation [85]. Initially, across the absorption of $PM_{2.5}$ by the targeting cells, toxic effects ensue due to the release of organic chemicals from the pollutant, which is metabolically activated by enzyme systems. Oxidative stress through free radicals and activation of inflammatory cells is an important mechanism [86]. Furthermore, impairment of the antioxidant system is an adverse effect and could be a mechanism for damage. Some organic extracts from $PM_{2.5}$, such as polycyclic aromatic hydrocarbons (PAHs), are responsible for mutagenicity, including DNA damage responses, promoting changes in the biochemistry and physiology of cells. However, one of the main mechanisms of almost all the adverse health effects of $PM_{2.5}$ is inflammation. Systemic inflammation seen through bio-markers (e.g., C-reactive protein) is a consequence induced by $PM_{2.5}$. Moreover, several studies have shown a decrease in immune function caused by $PM_{2.5}$ [85]. The modifications in physiological and biochemical functions, as well as damage to some tissues and organs, are factors that can lead to the development of several negative consequences, including cardiovascular diseases [85].

### 2.2. Health Effects of $NO_2$

$NO_2$ has low water solubility (0.037 $cm^3/cm^3$ $H_2O$ at 35 °C), and therefore a large fraction of inhaled $NO_2$ could be deposited in the peripheral airways. Inhaled $NO_2$, or a component or reaction product of $NO_2$, could subsequently be delivered via tissue absorption and transfer across the blood–gas interface to the blood; therefore, systemic effects are possible [87]. It is absorbed along the entire respiratory tract, but exposure studies indicate that the major target site for the action of $NO_2$ is the terminal bronchioles. Despite laboratory, clinical, and epidemiological research, the health effects of $NO_2$ exposure on humans are not well understood. The toxicological evidence suggests that increased susceptibility to infection; functional deficits from effects on airways; and deterioration of the status of persons with chronic respiratory conditions, including asthmatics, are of potential concern [88].

#### 2.2.1. Short-Term Exposure to $NO_2$

In the short-term exposure, a 0.3% increase in total daily mortality was reported per 10 $\mu g/m^3$ $NO_2$ increase [88]. The strongest short-term effects of $NO_2$ exposition are respiratory hospital admissions (all ages), and the following are all-cause mortality (all ages). Adebayo-Ojo et al. [4] point to a significant increase in hospital admissions for respiratory disease per lower increase in $NO_2$ concentration (interquartile range IQR of 7.3 $\mu g/m^3$). A positive association among all ages was 2.3; however, among children under 15 years old, the estimate increased to 3.1%. Cardiovascular hospital admissions are also included; however, a sensitivity analysis showed that in the case of cardiovascular effects, PM is more indicative. The study in Poland [71] showed that the acute health outcomes associated with $NO_2$ exposure might be worse than anticipated [76].

Tsai and Yang [89] found a significant association between hospital admissions for pneumonia and $PM_{2.5}$ in Taipei, Taiwan. Polish findings highlight the prominent impact of $NO_2$ on pneumonia-related hospitalizations. Pneumonia is an important cause of death, especially in the sensitive groups of young children and the elderly. In 2017, worldwide, 15% of all deaths of children under 5 were caused by pneumonia [90]. In patients with asthma, $NO_2$ potentiates bronchial responsiveness [91] and triggers both allergen-dependent [92] and -independent [93] eosinophilic inflammation. At concentrations $\geq 2$ ppm, $NO_2$ disrupts the tracheobronchial epithelial monolayer [94] and modifies the severity of viral infections [95]. In the Polish study [71], the greatest immediate effect on pneumonia-related hospitalizations in all investigated agglomerations had $NO_2$, and a $NO_2$ increase of 10 $\mu g/m^3$ was associated with up to an 11% increase in the number of daily pneumonia-related hospitalizations.

#### 2.2.2. Long-Term Exposure to $NO_2$

In the ESCAPE project [96], the long-term effect of $NO_2$ on natural-cause mortality was not shown. On the other hand, Faustini et al. [96], in a meta-analysis of 19 studies, demonstrated that an annual increase in $NO_2$ of 10 $\mu g/m^3$ was associated with a 4.1% increase in natural mortality. The magnitude of $NO_2$ effects on mortality was similar to that of $PM_{2.5}$ [96]. Two large cohort studies [97,98] examined the relationship between lung function growth in children and long-term exposure to $NO_2$. Although in California [98] the mean $NO_2$ concentrations were rather low (from 7.5 to 71.4 $\mu g/m^3$), and in Mexico City [97] were rather high (51–80 $\mu g/m^3$), both demonstrated deficits in lung function growth in children associated with $NO_2$ exposition. In Norway and Sweden, exposure to

$NO_2$ from traffic was associated with decreased lung function in children 9–10 years old in Oslo [99] and 8 years old in Stockholm [100].

#### 2.2.3. $NO_2$ Health Effect Mechanism

The main mechanism of $NO_2$ toxicity has been suggested to involve lipid peroxidation in cell membranes and various actions of free radicals on structural and functional molecules. The concentrations over 0.2 ppm produce the above-mentioned adverse effects in humans, while concentrations higher than 2.0 ppm affect the cytotoxic T lymphocytes (CD8+ T cells) and natural killer cells (NK cells) that produce our immune response [41].

Associations between $NO_2$ pollution and mortality [71,101], as well as the exacerbation of asthma [102], enhance the allergic response to inhaled allergens [103,104], manifesting as respiratory viral infection and its associated inflammation. Spannhake et al. [95] investigated the interactive effects of human rhinovirus type 16 (RV16) and the $NO_2$ on markers of proinflammatory activity in human bronchial and nasal epithelial cells. Dose–response experiments for $NO_2$ indicated that 3 h exposure to concentrations from 1.0 to 3.0 ppm induced cytokine release from bronchial epithelial cells in a dose-dependent manner with a minimal effect on cell viability. Additionally, the authors of [95] analyzed the interaction between asthmatic symptoms and viral infection within their common epithelial cell targets in the upper and lower respiratory tracks. The results demonstrated that the expression and release of IL-8 (interleukin 8, a pro-inflammatory cytokine that has a role in neutrophil activation and has been identified within the pathogenesis and progression of the disease [105]) in response to combined infection and $NO_2$ exposure was significantly higher than the sum of release from the cells that underwent only infection or only $NO_2$ exposure.

$NO_2$ damages the lung and proteins vital to its function. The mechanism includes exposure of $\alpha$-1-proteinase inhibitor ($\alpha$1PI) to $NO_2$, which resulted in a 50% loss of immunoreactivity with either monoclonal or polyclonal antibodies in an enzyme-linked immunosorbent assay at molar ratios of NO:$\alpha$12PI of 100:1 and greater. Additionally, the results of parallel O-phthalaldehyde and bicinchoninic acid protein assays as well as amino acid analysis on control and $NO_2$-exposed $\alpha$1PI suggested a reactivity of $NO_2$ with lysine residues [106].

### 2.3. *$PM_{2.5}$ and $NO_2$*

The epidemiological evidence has consistently shown that the $NO_2$ associations do not reflect adverse effects of $NO_2$ itself but rather the health effects of other air pollutants, mainly PM or other components of the complex mixture of traffic-related air pollutants [32]. Predominantly, this is due to the strong correlations between $NO_2$ and other combustion-derived air pollutants, especially PM. Many studies [7,14,26,107–114] underline a positive correlation between $NO_2$ and $PM_{2.5}$. In some studies, $NO_2$ has been proposed as a surrogate for PM [115,116]; in others, the confounding effects of $PM_{2.5}$ are also underlined [88,117]. Based on 7 years of air pollutant measurements in Iran, analyzed by the AirQ+ model, Naghan et al. [101] observed that the short-term effects of $PM_{2.5}$ on health were greater than those of $NO_2$. However, the long-term effects of $NO_2$ were greater than $PM_{2.5}$.

A number of studies confirm the relationship between multiple pollutants. For example, in 17 Chinese cities (characterized by average $NO_2$ levels of 26–67 $\mu g/m^3$), each short-term 10 $\mu g/m^3$ increase in $NO_2$ corresponded to a 1.63% increase in mortality [118]. The association stayed significant when adjusted for PM [118]. In comparison with Chinese findings [118], research performed in Poland [71] showed an increase in daily mortality of 1.7% in Cracow and of 3.9% in the Tricity per 10 $\mu g/m^3$ $NO_2$ increase; however, the effect of $NO_2$ was independent of other air pollutants including $PM_{2.5}$. In a Dutch birth cohort, $NO_2$ and $PM_{2.5}$ were highly correlated (0.93), and a regression model confirmed the association with some outcomes of asthma and allergy during the first 4 years of life [119]. Interestingly, $NO_2$ is likely a potential modifier for the association between $PM_{2.5}$ and the risk of inhaled allergies because the risk estimates for $PM_{2.5}$ in higher $NO_2$ ($\geq$42.0, $\mu g/m^3$) are statistically different from those in lower $NO_2$ levels ($p$ = 0.046) [120]. However only a few articles have investigated the relationship between $PM_{2.5}$ and $NO_2$, and our analysis identified only eight publications with specific data on RR, OR, or HR (Table 3). In our review, only three articles [16,114,121] covered concentration ranges exceeding EU guidelines (Table 1) for both pollutants.

**Table 3.** Risk ratio (RR), odds ratio (OR), and hazard ratio (HR), and health effects linked to $PM_{2.5}$ and $NO_2$ exposure.

| Health Outcomes | Exposure Duration | OR/HR/RR | Mean (Min.-Max.) Concentration μg/m³ | | References |
|---|---|---|---|---|---|
| | | | $PM_{2.5}$ | $NO_2$ | |
| Non-Accidental Cause Mortality | | | | | |
| Non-accidental causes | Chronic | HR: 1.01 (95% CI 0.99–1.02) 10 μg/m³ $PM_{2.5}$ adjusted for $NO_2$ HR: 1.02 (95% CI 1.01–1.03) 10 μg/m³ $NO_2$ adjusted for $PM_{2.5}$ | 23.0 (7.2–32.1) | 43.6 (13.0–75.2) | [110] |
| | Chronic | HR: 1.02 (95% CI 1.00–1.04) 2.8 μg/m³ $PM_{2.5}$ (IQR) adjusted for $NO_2$ HR: 1.03 (95% CI 1.01–1.05) 12.5 μg/m³ $NO_2$ (IQR) adjusted for $PM_{2.5}$ | 5.9 (0.4–17.2) | 8.6 (0.0–69.1) | [122] |
| Natural cause | Chronic | HR: 1.06 (95% CI 0.98–1.15) 5 μg/m³ $PM_{2.5}$ adjusted for $NO_2$ HR: 1.01 (95% CI 0.97–1.05) 10 μg/m³ $NO_2$ adjusted for $PM_{2.5}$ | 6.6–31 | 5.2–59.8 | [76] |
| All-cause mortality | Chronic | RR: 1.015 (95% CI 0.980–1.050) 5.03 μg/m³ $PM_{2.5}$ (IQR) adjusted for $NO_2$ RR: 1.015 (95% CI 0.098–1.050) 7.74 μg/m³ $NO_2$ (IQR) adjusted for $PM_{2.5}$ | 14.1 (4.3–25.1) | 12.3 (3.0–21.9) | [123] |
| | Chronic | OR: 1.023 (95% CI 0.814–1.279) 35.6 μg/m³ $PM_{2.5}$ (IQR) adjusted for $NO_2$ OR: 1.457 (95% CI 1.076–2.152) 16.9 μg/m³ $NO_2$ (IQR) adjusted for $PM_{2.5}$ | 80.7 (37.0–142.4) | 49.7 (33.6–68.1) | [16] |
| | Short term (1 h-days) | RR: 1.0004 (95% CI 0.9926–1.0082) 10 μg/m³ $PM_{2.5}$ adjusted for $NO_2$ | 5.7–176.7 | 18.4–99.2 (24-h average) 40.0–161.2 (1 h max.) | [10] |
| Cardiovascular Disease Mortality | | | | | |
| Cardiovascular | Chronic | RR: 1.043 (95% CI 0.989–1.101) 5.03 μg/m³ $PM_{2.5}$ (IQR) adjusted for $NO_2$ RR: 1.030 (95% CI 0.987–1.075) 7.74 μg/m³ $NO_2$ (IQR) adjusted for $PM_{2.5}$ | 14.1 (4.3–25.1) | 12.3 (3.0–21.9) | [123] |
| | Short term (1 h-days) | RR: 1.0092 (95% CI 0.9945–1.0241) 10 μg/m³ $PM_{2.5}$ adjusted for $NO_2$ | 5.7–176.7 | 18.4–99.2 (24-h average) 40.0–161.2 (1 h max.) | [10] |
| Congenital heart diseases | Chronic | OR: 1.267 (95% CI 0.643–2.404) 35.6 μg/m³ $PM_{2.5}$ (IQR) adjusted for $NO_2$ OR: 1.667 (95% CI 1.011–2.738) 16.9 μg/m³ $NO_2$ (IQR) adjusted for $PM_{2.5}$ | 80.7 (37.0–142.4) | 49.7 (33.6–68.1) | [16] |
| Ischemic heart disease | Chronic | RR: 1.090 (95% CI 1.015–1.170) 5.03 μg/m³ $PM_{2.5}$ (IQR) adjusted for $NO_2$ RR: 1.029 (95% CI 0.972–1.090) 7.74 μg/m³ $NO_2$ (IQR) adjusted for $PM_{2.5}$ | 14.1 (4.3–25.1) | 12.3 (3.0–21.9) | [123] |
| Out-of-hospital cardiac arrest | Chronic | OR: 1.07 (95% CI 1.03–1.11) 10 μg/m³ $PM_{2.5}$ adjusted for $NO_2$ | 76.0 (5.0–476.0) | 51.7 (7.8–136.2) | [114] |
| Respiratory Disease Mortality | | | | | |
| Respiratory | Short term (1h-days) | RR: 1.0135 (95% CI 1.0008–1.0263) 10 μg/m³ $PM_{2.5}$ adjusted for $NO_2$ | 5.7–176.7 | 18.4–99.2 (24-h average) 40.0–161.2 (1 h max.) | [10] |
| | Chronic | RR: 1.064 (95% CI 0.954–1.185) 5.03 μg/m³ $PM_{2.5}$ (IQR) adjusted for $NO_2$ RR: 0.973 (95% CI 0.891–1.063) 7.74 μg/m³ $NO_2$ (IQR) adjusted for $PM_{2.5}$ | 14.1 (4.3–25.1) | 12.3 (3.0–21.9) | [123] |

**Table 3.** *Cont.*

| Health Outcomes | Exposure Duration | OR/HR/RR | Mean (Min.-Max.) Concentration μg/m$^3$ | | References |
|---|---|---|---|---|---|
| | | | $PM_{2.5}$ | $NO_2$ | |
| Pneumonia-related | Chronic | OR: 0.961 (95% CI 0.754–1.145) 35.6 μg/m$^3$ $PM_{2.5}$ (IQR) adjusted for $NO_2$ OR: 1.781 (95% CI 1.011–2.738) 16.9 μg/m$^3$ $NO_2$ (IQR) adjusted for $PM_{2.5}$ | 80.7 (37.0–142.4) | 49.7 (33.6–68.1) | [16] |
| Lung cancer | Chronic | RR: 0.985 (95% CI 0.867–1.119) 5.03 μg/m$^3$ $PM_{2.5}$ (IQR) adjusted for $NO_2$ RR: 1.118 (95% CI 1.010–1.236) 7.74 μg/m$^3$ $NO_2$ (IQR) adjusted for $PM_{2.5}$ | 14.1 (4.3–25.1) | 12.3 (3.0–21.9) | [123] |
| Cerebrovascular Disease Mortality | | | | | |
| Stroke | Chronic | RR: 1.019 (95% CI 0.934–1.112) 5.03 μg/m$^3$ $PM_{2.5}$ (IQR) adjusted for $NO_2$ RR: 1.070 (95% CI 0.998–1.147) 7.74 μg/m$^3$ $NO_2$ (IQR) adjusted for $PM_{2.5}$ | 14.1 (4.3–25.1) | 12.3 (3.0–21.9) | [123] |
| Chronic Diseases | | | | | |
| Chronic kidney diseases | Chronic | HR: 1.43 (95% CI 0.98–2.09) 10 μg/m$^3$ $PM_{2.5}$ adjusted for $NO_2$ HR: 1.05 (95% CI 0.97–1.14) 10 μg/m$^3$ $NO_2$ adjusted for $PM_{2.5}$ | 26.6 | 44.8 | [121] |

IQR—interquartile range.

The health outcomes have been grouped among non-accidental mortality [10,16,76,110,122,123], cardiovascular disease mortality [10,16,114,123], respiratory disease mortality [10,16,123], and additionally stroke [123] and chronic kidney diseases [121]. Tables 4 and 5 present the comparison of RR, OR, and HR between $NO_2$ (Table 4) and $PM_{2.5}$ (Table 5) adjusted for $PM_{2.5}$ and $NO_2$, respectively. Generally, the highest risk of mortality is connected with congenital heart diseases (cardiovascular mortality) and pneumonia (pneumonia), both reported in [16]. Wang et al. [16] analyzed the associations between ambient air pollution and the number of deaths among children under 5 years old. The authors found that the associations between pollutants and infant deaths in Beijing (between January 2014 and September 2014) were more pronounced than those in children aged 1–5 years at lag1–lag2 and lag01–lag02. In the case of $NO_2$ in the multi-pollutant analysis, the OR was higher at all-cause mortality and pneumonia-related mortality than for a single pollutant. However, in the case of mortality caused by congenital heart diseases, the adjustment of $PM_{2.5}$ brought an antagonistic effect in comparison to $NO_2$ alone. An antagonistic relationship was also found for $PM_{2.5}$ exposure adjusted for $NO_2$, for all-cause mortality, congenital heart diseases, and pneumonia-related mortality; however, PM2.5 exposure was not statistically significant. As presented in Tables 3–5, for congenital heart disease, OR were 1.267 (95% CI 0.643–2.404) for $PM_{2.5}$ adjusted for $NO_2$ and 1.667 (95% CI 1.011–2.738) for $NO_2$ adjusted for $PM_{2.5}$, while for single pollutants were 1.155 (95% CI 0.953–1.390) and 1.383 (95% CI 1.113–1.718) for $PM_{2.5}$ and $NO_2$, respectively. Since the confidence interval (CI) included the value 1, we cannot be sure that the influence of the $PM_{2.5}$ increase was statistically significant. Statistically significant positive associations were observed for $NO_2$. Moreover, the pneumonia-related associations of $NO_2$ were stronger than those observed for overall death.

Jerret et al. [123] in California cohort studies including 73,711 subjects used land use regression (LUR) models to evaluate the association between air pollution, including $PM_{2.5}$ and $NO_2$ and several causes of death, such as cardiovascular disease, ischemic heart disease, stroke, respiratory disease, and lung cancer. In models that included both $PM_{2.5}$ and $NO_2$, the $PM_{2.5}$ associations with mortality from all causes were reduced to about half the size of those in the single-pollutant models, and the estimates became insignificant. The increase in $NO_2$ and $PM_{2.5}$ concentration on cardiovascular disease mortality revealed the highest risk among the analyzed results. $PM_{2.5}$ exposition caused a significantly higher risk of death due to congenital heart diseases and ischemic heart diseases, but the effects of $PM_{2.5}$ were attenuated with $NO_2$. $PM_{2.5}$ had elevated but insignificant risk estimates for respiratory deaths, whereas neither of the other pollutants was associated with respiratory mortality. For lung cancer, $NO_2$ consistently elevated risks in two-pollutant models.

**Table 4.** Risk ratio (RR), odds ratio (OR), and hazard ratio (HR) with corresponding health effects linked to $NO_2$ exposure alone and adjusted for $PM_{2.5}$.

| Health Outcomes | Exposure Duration | $NO_2$ Adjusted for $PM_{2.5}$ | | | $NO_2$ | | | Synergistic (+) Antagonistic (−) No Significant Difference (0) | Reference |
|---|---|---|---|---|---|---|---|---|---|
| | | HR | OR | RR | HR | OR | RR | | |
| Non-Accidental Cause Mortality | | | | | | | | | |
| Non-accidental causes | Chronic | **1.02** | | | **1.01** | | | + | [110] |
| | Chronic | **1.03** | | | **1.03** | | | 0 | [122] |
| Natural cause | Chronic | 1.01 | | | 1.01 | | | 0 | [76] |
| All-cause mortality | Chronic | | | 1.025 | | | **1.031** | − | [123] |
| | Chronic | | **1.457** | | | **1.383** | | + | [16] |
| Cardiovascular Disease Mortality | | | | | | | | | |
| Cardiovascular | Chronic | | | 1.03 | | | **1.048** | − | [123] |
| Congenital heart diseases | Chronic | | **1.667** | | | **2.103** | | − − | [16] |
| Ischemic heart disease | Chronic | | | 1.029 | | | **1.066** | − | [123] |
| Respiratory Disease Mortality | | | | | | | | | |
| Respiratory | Chronic | | | 0.973 | | | 0.999 | − | [123] |
| Pneumonia-related | Chronic | | **1.781** | | | **1.74** | | + | [16] |
| Lung cancer | Chronic | | | 1.118 | | | 1.111 | + | [123] |
| Cerebrovascular Disease Mortality | | | | | | | | | |
| Stroke | Chronic | | | 1.07 | | | **1.078** | − | [123] |
| Chronic Diseases | | | | | | | | | |
| Chronic kidney diseases | Chronic | 1.05 | | | **1.07** | | | − | [121] |

$p > 0.05$ no significant association. Statistically significant risks values are in bold.

The influence of the increase in $NO_2$ and $PM_{2.5}$ concentration on non-accidental mortality was featured as 1–3% of health risk increase [110,122]. The adjustment for $PM_{2.5}$ in $NO_2$ increase pointing towards small synergy or no influence was observed, while adjustment for $NO_2$ in $PM_{2.5}$ increase revealed a small antagonistic effect. Cesaroni et al. [110] analyzed cause-specific mortality of adults in Rome in accordance with two GIS indicators of traffic exposure (distance to heavy traffic roads with >10,000 vehicles per day, and traffic intensity in 150 m); however, these associations were evaluated separately for $NO_2$ and $PM_{2.5}$. In the two-pollutant model, long-term exposure to both $NO_2$ and $PM_{2.5}$ were analyzed only in the aspect of non-accidental mortality with HR: 1.02 (95% CI 1.01–1.03) per 10 $\mu g/m^3$ of $NO_2$ adjusted for $PM_{2.5}$, and HR: 1.01 (95% CI 0.99–1.02) per 10 $\mu g/m^3$ of $PM_{2.5}$ adjusted for $NO_2$.

The single pollutant estimations point to similar non-accidental mortality risk for $PM_{2.5}$ and $NO_2$, with HRs of 1.01 (95% CI 1.00–1.02) and 1.01 (95% CI 1.00–1.01), respectively. In the two-pollutant model, the associations were similar to single-pollutant risks; however, $PM_{2.5}$ adjusted for $NO_2$ were not statistically significant. Christidis et al. [122] analyzed the data from 1981 to 2016 in a cohort of Canadian communities exposed to low concentrations of $PM_{2.5}$ (average 5.9 $\mu g/m^3$). The unadjusted model had a HR: 0.96 (95% CI 0.92–1.00), which increased to 1.11 (95% CI 1.04–1.18) when adjusted by the socio-economic, behavioral, and contextual covariates. The authors evaluated the impact of individual-level behavioral risk factors on the $PM_{2.5}$ mortality association relationship. The inclusion of behavioral covariates in a model including socioeconomic and ecological covariates lowered the $PM_{2.5}$ hazard ratio by 2% (from 1.13 to 1.11). This modest change in the hazard ratio can indicate that the behavioral covariates were being adequately controlled for by the socio-economic and ecological covariates in the established relationship between $PM_{2.5}$ exposure and non-accidental mortality. In the two-pollutant model, the estimated effect of an increase in $NO_2$ on mortality was independent from the $PM_{2.5}$ adjustment, while the HR associated with an increase in $PM_{2.5}$ concentration in the adjustment for $NO_2$ was only slightly lower than the HR of a single pollutant (HR: 1.03, 95% CI 1.01–1.05, and HR: 1.03, 95% CI 1.02–1.05 for $PM_{2.5}$ and $NO_2$, respectively).

**Table 5.** Risk ratio (RR), odds ratio (OR), and hazard ratio (HR) with corresponding health effects linked to $PM_{2.5}$ exposure alone and adjusted for $NO_2$.

| Health Outcomes | Exposure Duration | $PM_{2.5}$ Adjusted for $NO_2$ | | | $PM_{2.5}$ | | | Synergistic (+) Antagonistic (−) No Significant Difference (0) | Reference |
|---|---|---|---|---|---|---|---|---|---|
| | | HR | OR | RR | HR | OR | RR | | |
| Non-Accidental Cause Mortality | | | | | | | | | |
| Non-accidental causes | Chronic | 1.01 | | | **1.01** | | | 0 | [110] |
| | Chronic | **1.02** | | | **1.03** | | | − | [122] |
| Natural cause | Chronic | 1.06 | | | **1.07** | | | − | [76] |
| All-cause mortality | Chronic | | | 1.015 | | | **1.032** | − | [123] |
| | Chronic | | 1.023 | | | 1.155 | | − | [16] |
| | Short term (1 h-days) | | | 1.00 | | | **1.0065** | − | [10] |
| Cardiovascular Disease Mortality | | | | | | | | | |
| Cardiovascular | Chronic | | | 1.043 | | | **1.064** | − | [123] |
| | Short term (1 h-days) | | | 1.009 | | | **1.0092** | 0 | [10] |
| Congenital heart diseases | Chronic | | 1.267 | | | **1.653** | | − − | [16] |
| Ischemic heart disease | Chronic | | | **1.09** | | | **1.111** | − | [123] |
| Out-of-hospital cardiac arrest | Chronic | | **1.07** | | | 1.07 | | 0 | [114] |
| Respiratory Disease Mortality | | | | | | | | | |
| Respiratory | Short term (1 h-days) | | | **1.014** | | | **1.0073** | + | [10] |
| | Chronic | | | 1.064 | | | **1.046** | + | [123] |
| Pneumonia-related | Chronic | | 0.961 | | | 1.171 | | − − | [16] |
| Lung cancer | Chronic | | | 0.985 | | | 1.062 | − | [123] |
| Cerebrovascular Disease Mortality | | | | | | | | | |
| Stroke | Chronic | | | 1.019 | | | 1.065 | − | [123] |
| Chronic Diseases | | | | | | | | | |
| Chronic kidney diseases | Chronic | 1.43 | | | **1.53** | | | − − | [121] |

$p > 0.05$ non-significant association. Statistically significant risks values are in bold.

In the article on all-cause mortality by Beelen et al. [76], the authors analyzed the data from 22 European cohorts recruited mainly in the 1990s. Exposure to $PM_{2.5}$ and $NO_2$ as separate pollutants resulted in HR: 1.07 (95% CI 1.02–1.13) and 1.01 (95% CI 0.99–1.03), respectively, for a 5 $\mu g/m^3$ increase in $PM_{2.5}$ concentration and 10 $\mu g/m^3$ for $NO_2$. However, the health risk of the increase in $NO_2$ concentration was not statistically significant (CI includes value 1). In the two-pollutant model, the health effect estimates either for $PM_{2.5}$ adjusted for $NO_2$ (HR: 1.06, 95% CI 0.98–1.15) or for $NO_2$ adjusted for $PM_{2.5}$ (HR: 1.01, 95% CI 0.97–1.05) did not differ from the single-pollutant model.

The development of chronic kidney disease and confirmed long-term exposure to ambient $PM_{2.5}$ and $NO_2$ was examined by Guo et al. [121]. Every 10 $\mu g/m^3$ increase in $PM_{2.5}$ or $NO_2$ concentrations was associated with a higher risk of developing chronic kidney disease. The increase in $PM_{2.5}$ concentration resulted in a higher risk of developing chronic kidney disease than the increase in $NO_2$ concentration. HR for PM2.5 alone was 1.53 (95% CI 1.07–2.2), while for $NO_2$, the risk was eightfold lower (HR: 1.07, 95% CI 1.00–1.15). The adjustment for $PM_{2.5}$ or $NO_2$ appeared to be antagonistic in comparison to single-pollutant risk. However, the results were not statistically significant.

In one study [114], the authors analyzed the increase in out-of-hospital cardiac arrest and calculated the ORs per 10 $\mu g/m^3$ increase in $NO_2$ levels. Furthermore, they statistically compared the concentrations on control days with the daily average concentrations on the onset day (Lag 0) and on days 1–5 before onset (Lags 1, 2, 3, 4, and 5). In a multi-pollutant model including $NO_2$ adjusted for $PM_{2.5}$, the estimated effect was comparable to a single-pollutant risk. For $PM_{2.5}$, OR was 1.07 (95% CI 1.04–1.10), and for $NO_2$, it was 1.05 (95% CI 0.98–1.11); however, the calculated OR for $PM_{2.5}$ adjustment was not statistically significant.

Only one study included short-term exposition. Orellano [10], in an extensive review including 196 articles, reported evidence of a positive association between short-term (1 h do 1 day) exposure to air pollutants and all-cause mortality. In the case of single pollutants, the relative risk levels were 1.0065 (95% CI 1.0044–1.0086) and 1.072 (95% CI 1.0059–1.0085) for $PM_{2.5}$ and $NO_2$, respectively. As can be seen (Tables 3–5), the association values were higher for $NO_2$ adjustment in comparison to a single-pollutant model in the case of respiratory diseases and were lower in the case of all-cause mortality; however, in the later, the statistical significance was irrespective.

## 3. Conclusions

A literature review of the health effects of the co-interactions of the pollutants under consideration ($PM_{2.5}$ and $NO_2$) showed that it depended on the type of pollutant, the cause of death/disease, and the magnitude of the increase in its concentration.

The increased risk of non-accidental mortality was observed with higher $NO_2$ and $PM_{2.5}$ concentrations. $NO_2$ was significantly associated with mortality when adjusted for $PM_{2.5}$; however, the estimated effect of $PM_{2.5}$ was no longer significant. With a small increase in concentration, the risk for all-cause mortality was raised from 1 to 3%. $NO_2$ exposure adjusted for $PM_{2.5}$ at a high concentration of ambient $PM_{2.5}$ was particularly influential, revealing significant synergy between both pollutants. $PM_{2.5}$ was associated more with cardiovascular disease mortality, whereas $NO_2$ exposure was associated more with respiratory mortality. Regardless of which pollutant was major or adjusted, in cardiovascular mortality, the adjustment resulted in a rather antagonistic effect, while in respiratory disease mortality, particularly by pneumonia and lung cancer, $NO_2$ exposure adjusted for $PM_{2.5}$ revealed a synergistic effect, and conversely, $PM_{2.5}$ exposure adjusted for $NO_2$ brought about an antagonistic effect.

The high correlations between pollutants were associated with serious limitations in the use of multi-pollutant models. Because of this, the health effects of many pollutants are still not fully known. To better understand the health effects of many pollutants, more research needs to be conducted on biological mechanisms, preferably grouping pollutants according to their mode of action.

**Author Contributions:** Conceptualization A.M. and MŻ.; investigation A.M. and M.Ż.; data curation, M.Ż.; writing—original draft preparation, A.M.; writing—review and editing A.M. and M.Ż.; visualization, A.M. and M.Ż.; funding acquisition A.M. and M.Ż. All authors have read and agreed to the published version of the manuscript.

**Funding:** Research funded by the statutory research by the Faculty of Energy and Environmental Engineering, Silesian University of Technology.

**Conflicts of Interest:** The authors declare no conflict of interest.

## References

1. Dominski, F.H.; Lorenzetti Branco, J.H.; Buonanno, G.; Stabile, L.; Gameiro da Silva, M.; Andrade, A. Effects of Air Pollution on Health: A Mapping Review of Systematic Reviews and Meta-Analyses. *Environ. Res.* **2021**, *201*, 111487. [CrossRef] [PubMed]
2. Sun, Z.; Zhu, D. Exposure to Outdoor Air Pollution and Its Human-Related Health Outcomes: An Evidence Gap Map. *BMJ Open* **2019**, *9*, e031312. [CrossRef] [PubMed]
3. Adebayo-Ojo, T.C.; Wichmann, J.; Arowosegbe, O.O.; Probst-Hensch, N.; Schindler, C.; Künzli, N. Short-Term Effects of PM10, $NO_2$, $SO_2$ and $O_3$ on Cardio-Respiratory Mortality in Cape Town, South Africa, 2006–2015. *Int. J. Environ. Res. Public Health* **2022**, *19*, 8078. [CrossRef] [PubMed]
4. Adebayo-Ojo, T.C.; Wichmann, J.; Arowosegbe, O.O.; Probst-Hensch, N.; Schindler, C.; Künzli, N. Short-Term Joint Effects of $PM_{10}$, $NO_2$ and $SO_2$ on Cardio-Respiratory Disease Hospital Admissions in Cape Town, South Africa. *Int. J. Environ. Res. Public Health* **2022**, *19*, 495. [CrossRef]
5. Chen, T.M.; Gokhale, J.; Shofer, S.; Kuschner, W.G. Outdoor Air Pollution: Nitrogen Dioxide, Sulfur Dioxide, and Carbon Monoxide Health Effects. *Am. J. Med. Sci.* **2007**, *333*, 249–256. [CrossRef]
6. Crouse, D.L.; Peters, P.A.; Hystad, P.; Brook, J.R.; van Donkelaar, A.; Martin, R.V.; Villeneuve, P.J.; Jerrett, M.; Goldberg, M.S.; Arden Pope, C.; et al. Ambient $PM_{2.5}$, $O_3$, and $NO_2$ Exposures and Associations with Mortality over 16 Years of Follow-up in the Canadian Census Health and Environment Cohort (CanCHEC). *Environ. Health Perspect.* **2015**, *123*, 1180–1186. [CrossRef]
7. Riches, N.O.; Gouripeddi, R.; Payan-Medina, A.; Facelli, J.C. K-Means Cluster Analysis of Cooperative Effects of CO, $NO_2$, $O_3$, $PM_{2.5}$, $PM_{10}$, and $SO_2$ on Incidence of Type 2 Diabetes Mellitus in the US. *Environ. Res.* **2022**, *212*, 113259. [CrossRef]
8. Jo, S.; Kim, Y.-J.; Park, K.W.; Hwang, Y.S.; Lee, S.H.; Kim, B.J.; Chung, S.J. Association of $NO_2$ and Other Air Pollution Exposures With the Risk of Parkinson Disease. *JAMA Neurol.* **2021**, *78*, 800–808. [CrossRef]
9. Braithwaite, I.; Zhang, S.; Kirkbride, J.B.; Osborn, D.P.J.; Hayes, J.F. Air Pollution (Particulate Matter) Exposure and Associations with Depression, Anxiety, Bipolar, Psychosis and Suicide Risk: A Systematic Review and Meta-Analysis. *Environ. Health Perspect.* **2019**, *127*, 126002. [CrossRef]
10. Orellano, P.; Reynoso, J.; Quaranta, N.; Bardach, A.; Ciapponi, A. Short-Term Exposure to Particulate Matter ($PM_{10}$ and $PM_{2.5}$), Nitrogen Dioxide ($NO_2$), and Ozone ($O_3$) and All-Cause and Cause-Specific Mortality: Systematic Review and Meta-Analysis. *Environ. Int.* **2020**, *142*, 105876. [CrossRef]
11. Li, S.; Wang, H.; Hu, H.; Wu, Z.; Chen, K.; Mao, Z. Effect of Ambient Air Pollution on Premature SGA in Changzhou City, 2013-2016: A Retrospective Study. *BMC Public Health* **2019**, *19*, 705. [CrossRef] [PubMed]
12. Zhao, N.; Smargiassi, A.; Hatzopoulou, M.; Colmegna, I.; Hudson, M.; Fritzler, M.J.; Awadalla, P.; Bernatsky, S. Long-Term Exposure to a Mixture of Industrial $SO_2$, $NO_2$, and $PM_{2.5}$ and Anti-Citrullinated Protein Antibody Positivity. *Environ. Health* **2020**, *19*, 86. [CrossRef] [PubMed]
13. Litchfield, I.J.; Ayres, J.G.; Jaakkola, J.J.K.; Mohammed, N.I. Is Ambient Air Pollution Associated with Onset of Sudden Infant Death Syndrome: A Case-Crossover Study in the UK. *BMJ Open* **2018**, *8*, e018341. [PubMed]
14. Ma, T.; Yazdi, M.D.; Schwartz, J.; Réquia, W.J.; Di, Q.; Wei, Y.; Chang, H.H.; Vaccarino, V.; Liu, P.; Shi, L. Long-Term Air Pollution Exposure and Incident Stroke in American Older Adults: A National Cohort Study. *Glob. Epidemiol.* **2022**, *4*, 100073. [CrossRef]
15. Hu, X.; Nie, Z.; Ou, Y.; Qian, Z.; McMillin, S.E.; Aaron, H.E.; Zhou, Y.; Dong, G.; Dong, H. Air Quality Improvement and Cognitive Function Benefit: Insight from Clean Air Action in China. *Environ Res* **2022**, *214*, 114200. [CrossRef]
16. Wang, J.; Cao, H.; Sun, D.; Qi, Z.; Guo, C.; Peng, W.; Sun, Y.; Xie, Y.; Liu, X.; Li, B.; et al. Associations between Ambient Air Pollution and Mortality from All Causes, Pneumonia, and Congenital Heart Diseases among Children Aged under 5 Years in Beijing, China: A Population-Based Time Series Study. *Environ. Res.* **2019**, *176*, 108531. [CrossRef]
17. WHO. Global Air Quality Guidelines: Particulate Matter (PM2.5 and PM10), Ozone, Nitrogen Dioxide, Sulfur Dioxide and Carbon Monoxide. Available online: https://apps.who.int/iris/handle/10665/345329 (accessed on 28 September 2022).
18. Mauderly, J.L.; Burnett, R.T.; Castillejos, M.; Özkaynak, H.; Samet, J.M.; Stieb, D.M.; Vedal, S.; Wyzga, R.E. Is the Air Pollution Health Research Community Prepared to Support a Multipollutant Air Quality Management Framework? *Inhal. Toxicol.* **2010**, *22*, 1–19. [CrossRef]
19. De Leeuw, F.; Horálek, J. *Quantifying the Health Impacts of Ambient Air Pollution: Methodology and Input Data*; ETC/ACM: Bilthoven, The Netherlands, 2016.
20. Regulatory and Guidance Information by Topic: Air | US EPA. Available online: https://www.epa.gov/regulatory-information-topic/regulatory-and-guidance-information-topic-air#criteriapollutants (accessed on 9 September 2022).

21. Union, P. *Directive 2008/50/EC of the European Parliament and of the Council of 21 May 2008 on Ambient Air Quality and Cleaner Air for Europe*; Official Journal of the European Union: Aberdeen, UK, 2008.
22. US EPA. *Supplementary Guidance for Conducting Health Risk Assessment of Chemical Mixtures*; EPA: Washington, DC, USA, 2000.
23. Gao, L.; Tao, F.; Liu, R.; Wang, Z.; Leng, H.; Zhou, T. Multi-Scenario Simulation and Ecological Risk Analysis of Land Use Based on the PLUS Model: A Case Study of Nanjing. *Sustain. Cities Soc.* **2022**, *85*, 104055. [CrossRef]
24. Huang, Y.C.T.; Rappold, A.G.; Graff, D.W.; Ghio, A.J.; Devlin, R.B. Synergistic Effects of Exposure to Concentrated Ambient Fine Pollution Particles and Nitrogen Dioxide in Humans. *Inhal. Toxicol.* **2012**, *24*, 790–797. [CrossRef]
25. Ashmore, M.R.; Dimitroulopoulou, C. Personal Exposure of Children to Air Pollution. *Atmos. Environ.* **2009**, *43*, 128–141. [CrossRef]
26. WHO. *Review of Evidence on Health Aspects of Air Pollution—REVIHAAP Project*; World Health Organization: Geneva, Switzerland, 2013.
27. Zhao, Z.; Lv, S.; Zhang, Y.; Zhao, Q.; Shen, L.; Xu, S.; Yu, J.; Hou, J.; Jin, C. Characteristics and Source Apportionment of $PM_{2.5}$ in Jiaxing, China. *Environ. Sci. Pollut. Res. Int.* **2019**, *26*, 7497–7511. [CrossRef] [PubMed]
28. Cheng, M.; Tang, G.; Lv, B.; Li, X.; Wu, X.; Wang, Y.; Wang, Y. Source Apportionment of $PM_{2.5}$ and Visibility in Jinan, China. *J. Environ. Sci.* **2021**, *102*, 207–215. [CrossRef]
29. Scicchitano, P.; Cameli, M.; Maiello, M.; Modesti, P.A.; Muiesan, M.L.; Novo, S.; Palmiero, P.; Saba, P.S.; Pedrinelli, R.; Ciccone, M.M. Nutraceuticals and Dyslipidaemia: Beyond the Common Therapeutics. *J. Funct. Foods* **2014**, *6*, 11–32. [CrossRef]
30. Afshin, A.; Sur, P.J.; Fay, K.A.; Cornaby, L.; Ferrara, G.; Salama, J.S.; Mullany, E.C.; Abate, K.H.; Abbafati, C.; Abebe, Z.; et al. Health Effects of Dietary Risks in 195 Countries, 1990–2017: A Systematic Analysis for the Global Burden of Disease Study 2017. *Lancet* **2019**, *393*, 1958–1972. [CrossRef]
31. Zhao, M.; Veeranki, S.P.; Magnussen, C.G.; Xi, B. Recommended Physical Activity and All Cause and Cause Specific Mortality in US Adults: Prospective Cohort Study. *BMJ* **2020**, *370*, 2031. [CrossRef] [PubMed]
32. Mills, I.C.; Atkinson, R.W.; Anderson, H.R.; Maynard, R.L.; Strachan, D.P. Distinguishing the Associations between Daily Mortality and Hospital Admissions and Nitrogen Dioxide from Those of Particulate Matter: A Systematic Review and Meta-Analysis. *BMJ Open* **2016**, *6*, e010751. [CrossRef]
33. Shaw, S.; van Heyst, B. An Evaluation of Risk Ratios on Physical and Mental Health Correlations Due to Increases in Ambient Nitrogen Oxide (NOx) Concentrations. *Atmosphere* **2022**, *13*, 967. [CrossRef]
34. Simon, S.D. Understanding the Odds Ratio and the Relative Risk. *J. Androl.* **2001**, *22*, 533–536.
35. Worthington, D. *Dictionary of Environmental Health*; Routledge, Taylor and Francis Group: New York, NY, USA, 2020; ISBN 9780367578572.
36. George, A.; Stead, T.S.; Ganti, L. What's the Risk: Differentiating Risk Ratios, Odds Ratios, and Hazard Ratios? *Cureus* **2020**, *12*, 6–13. [CrossRef]
37. Costa, S.; Ferreira, J.; Silveira, C.; Costa, C.; Lopes, D.; Relvas, H.; Borrego, C.; Roebeling, P.; Miranda, A.I.; Paulo Teixeira, J. Integrating Health on Air Quality Assessment-Review Report on Health Risks of Two Major European Outdoor Air Pollutants: PM and $NO_2$. *J. Toxicol. Environ. Health Part B* **2014**, *17*, 307–340. [CrossRef]
38. Krewski, D.; Burnett, R.; Jerrett, M.; Pope, C.A.; Rainham, D.; Calle, E.; Thurston, G.; Thun, M. Mortality and Long-Term Exposure to Ambient Air Pollution: Ongoing Analyses Based on the American Cancer Society Cohort. *J. Toxicol. Environ. Health Part A* **2006**, *68*, 1093–1109. [CrossRef] [PubMed]
39. Viegi, G.; Maio, S.; Pistelli, F.; Baldacci, S.; Carrozzi, L. Epidemiology of Chronic Obstructive Pulmonary Disease: Health Effects of Air Pollution. *Respirology* **2006**, *11*, 523–532. [CrossRef] [PubMed]
40. Utell, M.J.; Frank, R. *Susceptibility to Inhaled Pollutants*; ASTM: Balimore, MD, USA, 1989.
41. Manisalidis, I.; Stavropoulou, E.; Stavropoulos, A.; Bezirtzoglou, E. Environmental and Health Impacts of Air Pollution: A Review. *Front. Public Health* **2020**, *8*, 14. [CrossRef] [PubMed]
42. Eze, I.C.; Schaffner, E.; Fischer, E.; Schikowski, T.; Adam, M.; Imboden, M.; Tsai, M.; Carballo, D.; von Eckardstein, A.; Künzli, N.; et al. Long-Term Air Pollution Exposure and Diabetes in a Population-Based Swiss Cohort. *Environ. Int.* **2014**, *70*, 95–105. [CrossRef] [PubMed]
43. Kelishadi, R.; Poursafa, P. Air Pollution and Non-Respiratory Health Hazards for Children. *Arch. Med. Sci.* **2010**, *6*, 483–495. [CrossRef]
44. Clifford, A.; Lang, L.; Chen, R.; Anstey, K.J.; Seaton, A. Exposure to Air Pollution and Cognitive Functioning across the Life Course—A Systematic Literature Review. *Environ. Res.* **2016**, *147*, 383–398. [CrossRef]
45. Zhang, X.; Chen, X.; Zhang, X. The Impact of Exposure to Air Pollution on Cognitive Performance. *Proc. Natl. Acad. Sci. USA* **2018**, *115*, 9193–9197. [CrossRef]
46. Freire, C.; Ramos, R.; Puertas, R.; Lopez-Espinosa, M.J.; Julvez, J.; Aguilera, I.; Cruz, F.; Fernandez, M.F.; Sunyer, J.; Olea, N. Association of Traffic-Related Air Pollution with Cognitive Development in Children. *J. Epidemiol. Community Health* **2010**, *64*, 223–228. [CrossRef]
47. Wang, J.N.; Wang, Q.; Li, T.T.; Shi, X.M. Association between Air Pollution and Cognitive Function in the Elderly. *Zhonghua Yu Fang Yi Xue Za Zhi* **2017**, *51*, 364–368. [CrossRef]

48. Mehta, A.J.; Kubzansky, L.D.; Coull, B.A.; Kloog, I.; Koutrakis, P.; Sparrow, D.; Spiro, A.; Vokonas, P.; Schwartz, J. Associations between Air Pollution and Perceived Stress: The Veterans Administration Normative Aging Study. *Environ. Health* **2015**, *14*, 10. [CrossRef]
49. Lim, Y.H.; Kim, H.; Kim, J.H.; Bae, S.; Park, H.Y.; Hong, Y.C. Air Pollution and Symptoms of Depression in Elderly Adults. *Environ. Health Perspect.* **2012**, *120*, 1023–1028. [CrossRef] [PubMed]
50. Qiu, X.; Danesh-Yazdi, M.; Weisskopf, M.; Kosheleva, A.; Spiro, A.S.; Wang, C.; Coull, B.A.; Koutrakis, P.; Schwartz, J.D. Associations between Air Pollution and Psychiatric Symptoms in The Normative Aging Study. *Environ. Res. Lett.* **2022**, *17*, 034004. [CrossRef] [PubMed]
51. Chen, H.; Kwong, J.C.; Copes, R.; Hystad, P.; van Donkelaar, A.; Tu, K.; Brook, J.R.; Goldberg, M.S.; Martin, R.V.; Murray, B.J.; et al. Exposure to Ambient Air Pollution and the Incidence of Dementia: A Population-Based Cohort Study. *Environ. Int.* **2017**, *108*, 271–277. [CrossRef] [PubMed]
52. Ghio, A.J.; Carraway, M.S.; Madden, M.C. Composition of Air Pollution Particles and Oxidative Stress in Cells, Tissues, and Living Systems. *J. Toxicol. Environ. Health B* **2012**, *15*, 1–21. [CrossRef] [PubMed]
53. IARC. *Monographs on the Evaluation of Carcinogenic Risks to Humans*; International Agency for Research on Cancer: Lyon, France, 2010; Volume 96.
54. Ma, P.; Tao, F.; Gao, L.; Leng, S.; Yang, K.; Zhou, T. Retrieval of Fine-Grained $PM_{2.5}$ Spatiotemporal Resolution Based on Multiple Machine Learning Models. *Remote Sens.* **2022**, *14*, 599. [CrossRef]
55. Silva, R.A.; West, J.J.; Lamarque, J.F.; Shindell, D.T.; Collins, W.J.; Faluvegi, G.; Folberth, G.A.; Horowitz, L.W.; Nagashima, T.; Naik, V.; et al. Future Global Mortality from Changes in Air Pollution Attributable to Climate Change. *Nat. Clim. Chang.* **2017**, *7*, 647–651. [CrossRef] [PubMed]
56. Fiordelisi, A.; Piscitelli, P.; Trimarco, B.; Coscioni, E.; Iaccarino, G.; Sorriento, D. The Mechanisms of Air Pollution and Particulate Matter in Cardiovascular Diseases. *Heart Fail. Rev.* **2017**, *22*, 337–347. [CrossRef]
57. Schlesinger, R.B. The Interaction of Inhaled Toxicants with Respiratory Tract Clearance Mechanisms. *Crit. Rev. Toxicol.* **2008**, *20*, 257–286. [CrossRef]
58. Becquemin, M.H.; Swift, D.L.; Bouchikhi, A.; Roy, M.; Teillac, A. Particle Deposition and Resistance in the Noses of Adults and Children. *Eur. Respir. J.* **1991**, *4*, 694–702.
59. Brook, R.D.; Franklin, B.; Cascio, W.; Hong, Y.; Howard, G.; Lipsett, M.; Luepker, R.; Mittleman, M.; Samet, J.; Smith, S.C.; et al. Air Pollution and Cardiovascular Disease: A Statement for Healthcare Professionals from the Expert Panel on Population and Prevention Science of the American Heart Association. *Circulation* **2004**, *109*, 2655–2671. [CrossRef]
60. Miller, M.R.; Shaw, C.A.; Langrish, J.P. From Particles to Patients: Oxidative Stress and the Cardiovascular Effects of Air Pollution. *Future Cardiol.* **2012**, *8*, 577–602. [CrossRef]
61. Ni, L.; Chuang, C.C.; Zuo, L. Fine Particulate Matter in Acute Exacerbation of COPD. *Front. Physiol.* **2015**, *6*, 294. [CrossRef] [PubMed]
62. Esworthy, R. *Air Quality: EPA's 2013 Changes to the Particulate Matter (PM) Standard*; Library of Congress: Washington, DC, USA; Congressional Research Service: Washington, DC, USA, 2013.
63. Brook, R.D. Cardiovascular Effects of Air Pollution. *Clin. Sci.* **2008**, *115*, 175–187. [CrossRef] [PubMed]
64. Chang, C.C.; Kuo, C.C.; Liou, S.H.; Yang, C.Y. Fine Particulate Air Pollution and Hospital Admissions for Myocardial Infarction in a Subtropical City: Taipei, Taiwan. *J. Toxicol. Environ. Health Part A* **2013**, *76*, 440–448. [CrossRef] [PubMed]
65. Chiu, H.F.; Tsai, S.S.; Weng, H.H.; Yang, C.Y. Short-Term Effects of Fine Particulate Air Pollution on Emergency Room Visits for Cardiac Arrhythmias: A Case-Crossover Study in Taipei. *J. Toxicol. Environ. Health Part A* **2013**, *76*, 614–623. [CrossRef]
66. Brocato, J.; Sun, H.; Shamy, M.; Kluz, T.; Alghamdi, M.A.; Khoder, M.I.; Chen, L.C.; Costa, M. Particulate Matter From Saudi Arabia Induces Genes Involved in Inflammation, Metabolic Syndrome and Atherosclerosis. *J. Toxicol. Environ. Health Part A* **2014**, *77*, 751–766. [CrossRef]
67. Brook, R.D.; Rajagopalan, S.; Pope, C.A.; Brook, J.R.; Bhatnagar, A.; Diez-Roux, A.V.; Holguin, F.; Hong, Y.; Luepker, R.V.; Mittleman, M.A.; et al. Particulate Matter Air Pollution and Cardiovascular Disease: An Update to the Scientific Statement from the American Heart Association. *Circulation* **2010**, *121*, 2331–2378. [CrossRef]
68. Shang, Y.; Sun, Z.; Cao, J.; Wang, X.; Zhong, L.; Bi, X.; Li, H.; Liu, W.; Zhu, T.; Huang, W. Systematic Review of Chinese Studies of Short-Term Exposure to Air Pollution and Daily Mortality. *Environ. Int.* **2013**, *54*, 100–111. [CrossRef]
69. Han, C.; Xu, R.; Ye, T.; Xie, Y.; Zhao, Y.; Liu, H.; Yu, W.; Zhang, Y.; Li, S.; Zhang, Z.; et al. Mortality Burden Due to Long-Term Exposure to Ambient PM2.5 above the New WHO Air Quality Guideline Based on 296 Cities in China. *Environ. Int.* **2022**, *166*, 107331. [CrossRef]
70. Qiu, H.; Tian, L.W.; Pun, V.C.; Ho, K.F.; Wong, T.W.; Yu, I.T.S. Coarse Particulate Matter Associated with Increased Risk of Emergency Hospital Admissions for Pneumonia in Hong Kong. *Thorax* **2014**, *69*, 1027–1033. [CrossRef]
71. Dąbrowiecki, P.; Badyda, A.; Chciałowski, A.; Czechowski, P.O.; Wrotek, A. Influence of Selected Air Pollutants on Mortality and Pneumonia Burden in Three Polish Cities over the Years 2011–2018. *J. Clin. Med.* **2022**, *11*, 3084. [CrossRef] [PubMed]
72. Wu, J.-Z.; Ge, D.-D.; Zhou, L.-F.; Hou, L.-Y.; Zhou, Y.; Li, Q.-Y. Effects of Particulate Matter on Allergic Respiratory Diseases. *Chronic. Dis. Transl. Med.* **2018**, *4*, 95–102. [CrossRef] [PubMed]

73. Deng, X.; Rui, W.; Zhang, F.; Ding, W. $PM_{2.5}$ Induces Nrf2-Mediated Defense Mechanisms against Oxidative Stress by Activating PIK3/AKT Signaling Pathway in Human Lung Alveolar Epithelial A549 Cells. *Cell Biol. Toxicol.* **2013**, *29*, 143–157. [CrossRef] [PubMed]
74. Cohen, A.J.; Brauer, M.; Burnett, R.; Anderson, H.R.; Frostad, J.; Estep, K.; Balakrishnan, K.; Brunekreef, B.; Dandona, L.; Dandona, R.; et al. Estimates and 25-Year Trends of the Global Burden of Disease Attributable to Ambient Air Pollution: An Analysis of Data from the Global Burden of Diseases Study 2015. *Lancet* **2017**, *389*, 1907–1918. [CrossRef]
75. Health Effects Institute. *State of Global Air 2020*; Health Effects Institute: Boston, MA, USA, 2020.
76. Beelen, R.; Raaschou-Nielsen, O.; Stafoggia, M.; Andersen, Z.J.; Weinmayr, G.; Hoffmann, B.; Wolf, K.; Samoli, E.; Fischer, P.; Nieuwenhuijsen, M.; et al. Effects of Long-Term Exposure to Air Pollution on Natural-Cause Mortality: An Analysis of 22 European Cohorts within the Multicentre ESCAPE Project. *Lancet* **2014**, *383*, 785–795. [CrossRef]
77. Raaschou-Nielsen, O.; Pedersen, M.; Stafoggia, M.; Weinmayr, G.; Andersen, Z.J.; Galassi, C.; Sommar, J.; Forsberg, B.; Olsson, D.; Oftedal, B.; et al. Outdoor Air Pollution and Risk for Kidney Parenchyma Cancer in 14 European Cohorts. *Int. J. Cancer* **2017**, *140*, 1528–1537. [CrossRef]
78. Pyo, J.-S.; Kim, N.Y.; Kang, D.-W. Impacts of Outdoor Particulate Matter Exposure on the Incidence of Lung Cancer and Mortality. *Medicina* **2022**, *58*, 1159. [CrossRef]
79. Raaschou-Nielsen, O.; Andersen, Z.J.; Beelen, R.; Samoli, E.; Stafoggia, M.; Weinmayr, G.; Hoffmann, B.; Fischer, P.; Nieuwenhuijsen, M.J.; Brunekreef, B.; et al. Air Pollution and Lung Cancer Incidence in 17 European Cohorts: Prospective Analyses from the European Study of Cohorts for Air Pollution Effects (ESCAPE). *Lancet Oncol.* **2013**, *14*, 813–822. [CrossRef]
80. Hamra, G.B.; Guha, N.; Cohen, A.; Laden, F.; Raaschou-Nielsen, O.; Samet, J.M.; Vineis, P.; Forastiere, F.; Saldiva, P.; Yorifuji, T.; et al. Outdoor Particulate Matter Exposure and Lung Cancer: A Systematic Review and Meta-Analysis. *Environ. Health Perspect.* **2014**, *122*, 906–911. [CrossRef]
81. Yorifuji, T.; Kashima, S. Air Pollution: Another Cause of Lung Cancer. *Lancet Oncol.* **2013**, *14*, 788–789. [CrossRef]
82. Sax, S.N.; Zu, K.; Goodman, J.E. Air Pollution and Lung Cancer in Europe. *Lancet Oncol.* **2013**, *14*, e439–e440. [CrossRef]
83. Oh, J.; Ye, S.; Kang, D.H.; Ha, E. Association between Exposure to Fine Particulate Matter and Kidney Function: Results from the Korea National Health and Nutrition Examination Survey. *Environ. Res.* **2022**, *212*, 113080. [CrossRef] [PubMed]
84. Calderón-Garcidueñas, L.; Leray, E.; Heydarpour, P.; Torres-Jardón, R.; Reis, J. Air Pollution, a Rising Environmental Risk Factor for Cognition, Neuroinflammation and Neurodegeneration: The Clinical Impact on Children and Beyond. *Rev. Neurol.* **2016**, *172*, 69–80. [CrossRef] [PubMed]
85. Feng, S.; Gao, D.; Liao, F.; Zhou, F.; Wang, X. The Health Effects of Ambient PM2.5 and Potential Mechanisms. *Ecotoxicol. Environ. Saf.* **2016**, *128*, 67–74. [CrossRef] [PubMed]
86. Xing, Y.F.; Xu, Y.H.; Shi, M.H.; Lian, Y.X. The Impact of PM2.5 on the Human Respiratory System. *J. Thorac. Dis.* **2016**, *8*, E69–E74. [CrossRef]
87. Solomon, C.; Christian, D.; Chen, L.; Welch, B.; Kleinman, M.; Dunham, E.; Erle, D.; Balmes, J. Effect of Serial-Day Exposure to Nitrogen Dioxide on Airway and Blood Leukocytes and Lymphocyte Subsets. *Eur. Respir. J.* **2000**, *15*, 922–928. [CrossRef]
88. Samoli, E.; Aga, E.; Touloumi, G.; Nisiotis, K.; Forsberg, B.; Lefranc, A.; Pekkanen, J.; Wojtyniak, B.; Schindler, C.; Niclu, E.; et al. Short-Term Effects of Nitrogen Dioxide on Mortality: An Analysis within the APHEA Project. *Eur. Respir. J.* **2006**, *27*, 1129–1137. [CrossRef]
89. Tsai, S.S.; Yang, C.Y. Fine Particulate Air Pollution and Hospital Admissions for Pneumonia in a Subtropical City: Taipei, Taiwan. *J. Toxicol. Environ. Health A* **2014**, *77*, 192–201. [CrossRef]
90. World Health Organization. Pneumonia. Available online: https://www.who.int/health-topics/pneumonia#tab=tab_1 (accessed on 28 September 2022).
91. Brown, J.S. Nitrogen Dioxide Exposure and Airway Responsiveness in Individuals with Asthma. *Inhal. Toxicol.* **2015**, *27*, 1–14. [CrossRef]
92. Barck, C.; Lundahl, J.; Halldén, G.; Bylin, G. Brief Exposures to $NO_2$ Augment the Allergic Inflammation in Asthmatics. *Environ. Res.* **2005**, *97*, 58–66. [CrossRef]
93. Ezratty, V.; Guillossou, G.; Neukirch, C.; Dehoux, M.; Koscielny, S.; Bonay, M.; Cabanes, P.A.; Samet, J.M.; Mure, P.; Ropert, L.; et al. Repeated Nitrogen Dioxide Exposures and Eosinophilic Airway Inflammation in Asthmatics: A Randomized Crossover Study. *Environ. Health Perspect.* **2014**, *122*, 850–855. [CrossRef] [PubMed]
94. Robison, T.W.; Kwang-Jin, K. Dual Effect of Nitrogen Dioxide on Barrier Properties of Guinea Pig Tracheobronchial Epithelial Monolayers Cultured in an Air Interface. *J. Toxicol. Environ. Health* **2009**, *44*, 57–71. [CrossRef] [PubMed]
95. William Spannhake, E.; Reddy, S.P.M.; Jacoby, D.B.; Yu, X.Y.; Saatian, B.; Tian, J. Synergism between Rhinovirus Infection and Oxidant Pollutant Exposure Enhances Airway Epithelial Cell Cytokine Production. *Environ. Health Perspect.* **2002**, *110*, 665–670. [CrossRef]
96. Faustini, A.; Rapp, R.; Forastiere, F. Nitrogen Dioxide and Mortality: Review and Meta-Analysis of Long-Term Studies. *Eur. Respir. J.* **2014**, *44*, 744–753. [CrossRef]
97. Rojas-Martinez, R.; Perez-Padilla, R.; Olaiz-Fernandez, G.; Mendoza-Alvarado, L.; Moreno-Macias, H.; Fortoul, T.; McDonnell, W.; Loomis, D.; Romieu, I. Lung Function Growth in Children with Long-Term Exposure to Air Pollutants in Mexico City. *Am. J. Respir. Crit. Care Med.* **2007**, *176*, 377–384. [CrossRef] [PubMed]

98. Gauderman, W.J.; Vora, H.; McConnell, R.; Berhane, K.; Gilliland, F.; Thomas, D.; Lurmann, F.; Avol, E.; Kunzli, N.; Jerrett, M.; et al. Effect of Exposure to Traffic on Lung Development from 10 to 18 Years of Age: A Cohort Study. *Lancet* **2007**, *369*, 571–577. [CrossRef]
99. Oftedal, B.; Brunekreef, B.; Nystad, W.; Madsen, C.; Walker, S.E.; Nafstad, P. Residential Outdoor Air Pollution and Lung Function in Schoolchildren. *Epidemiology* **2008**, *19*, 129–137. [CrossRef] [PubMed]
100. Schultz, E.S.; Gruzieva, O.; Bellander, T.; Bottai, M.; Hallberg, J.; Kull, I.; Svartengren, M.; Melén, E.; Pershagen, G. Traffic-Related Air Pollution and Lung Function in Children at 8 Years of Age: A Birth Cohort Study. *Am. J. Respir. Crit. Care Med.* **2012**, *186*, 1286–1291. [CrossRef]
101. Naghan, D.J.; Neisi, A.; Goudarzi, G.; Dastoorpoor, M.; Fadaei, A.; Angali, K.A. Estimation of the Effects $PM_{2.5}$, $NO_2$, $O_3$ Pollutants on the Health of Shahrekord Residents Based on AirQ+ Software during (2012–2018). *Toxicol. Rep.* **2022**, *9*, 842–847. [CrossRef]
102. Kelly, F.J.; Fussell, J.C. Air Pollution and Airway Disease. *Clin. Exp. Allergy* **2011**, *41*, 1059–1071. [CrossRef]
103. Tunnicliffe, W.S.; Burge, P.S.; Ayres, J.G. Effect of Domestic Concentrations of Nitrogen Dioxide on Airway Responses to Inhaled Allergen in Asthmatic Patients. *Lancet* **1994**, *344*, 1733–1736. [CrossRef]
104. Strand, V.; Svartengren, M.; Rak, S.; Barck, C.; Bylin, G. Repeated Exposure to an Ambient Level of $NO_2$ Enhances Asthmatic Response to a Nonsymptomatic Allergen Dose. *Eur. Respir. J.* **1998**, *12*, 6–12. [CrossRef] [PubMed]
105. Cesta, M.C.; Zippoli, M.; Marsiglia, C.; Gavioli, E.M.; Mantelli, F.; Allegretti, M.; Balk, R.A. The Role of Interleukin-8 in Lung Inflammation and Injury: Implications for the Management of COVID-19 and Hyperinflammatory Acute Respiratory Distress Syndrome. *Front. Pharmacol.* **2022**, *12*, 808797. [CrossRef] [PubMed]
106. Hood, D.B.; Gettins, P.; Johnson, D.A. Nitrogen Dioxide Reactivity with Proteins: Effects on Activity and Immunoreactivity with $\alpha$-1-Proteinase Inhibitor and Implications for NO2-Mediated Peptide Degradation. *Arch. Biochem. Biophys.* **1993**, *304*, 17–26. [CrossRef]
107. Dominici, F.; Peng, R.D.; Barr, C.D.; Bell, M.L. Protecting Human Health from Air Pollution: Shifting from a Single-Pollutant to a Multipollutant Approach. *Epidemiology* **2010**, *21*, 187–194. [CrossRef]
108. Siddika, N.; Rantala, A.K.; Antikainen, H.; Balogun, H.; Amegah, A.K.; Ryti, N.R.I.; Kukkonen, J.; Sofiev, M.; Jaakkola, M.S.; Jaakkola, J.J.K. Short-Term Prenatal Exposure to Ambient Air Pollution and Risk of Preterm Birth—A Population-Based Cohort Study in Finland. *Environ. Res.* **2020**, *184*, 109290. [CrossRef]
109. Szyszkowicz, M. An Approach to Represent a Combined Exposure to Air Pollution. *Int. J. Occup. Med. Environ. Health* **2015**, *28*, 823–830. [CrossRef]
110. Cesaroni, G.; Badaloni, C.; Gariazzo, C.; Stafoggia, M.; Sozzi, R.; Davoli, M.; Forastiere, F. Long-Term Exposure to Urban Air Pollution and Mortality in a Cohort of More than a Million Adults in Rome. *Environ. Health Perspect.* **2013**, *121*, 324–331. [CrossRef]
111. Gass, K.; Klein, M.; Sarnat, S.E.; Winquist, A.; Darrow, L.A.; Flanders, W.D.; Chang, H.H.; Mulholland, J.A.; Tolbert, P.E.; Strickland, M.J. Associations between Ambient Air Pollutant Mixtures and Pediatric Asthma Emergency Department Visits in Three Cities: A Classification and Regression Tree Approach. *Environ. Health* **2015**, *14*, 58. [CrossRef]
112. Evangelopoulos, D.; Katsouyanni, K.; Keogh, R.H.; Samoli, E.; Schwartz, J.; Barratt, B.; Zhang, H.; Walton, H. $PM_{2.5}$ and $NO_2$ Exposure Errors Using Proxy Measures, Including Derived Personal Exposure from Outdoor Sources: A Systematic Review and Meta-Analysis. *Environ. Int.* **2020**, *137*, 105500. [CrossRef]
113. Public Health England. *The Effects of Long-Term Exposure to Ambient Air Pollution on Cardiovascular Morbidity: Mechanistic Evidence. A Report by the Committee on the Medical Effects of Air Pollutants*; Public Health England: London, UK, 2018.
114. Xia, R.; Zhou, G.; Zhu, T.; Li, X.; Wang, G. Ambient Air Pollution and Out-of-Hospital Cardiac Arrest in Beijing, China. *Int. J. Environ. Res. Public Health* **2017**, *14*, 423. [CrossRef] [PubMed]
115. Seaton, A.; Dennekamp, M. Hypothesis: Ill Health Associated with Low Concentrations of Nitrogen Dioxide—An Effect of Ultrafine Particles? *Thorax* **2003**, *58*, 1012–1015. [CrossRef] [PubMed]
116. Brook, J.R.; Burnett, R.T.; Dann, T.F.; Cakmak, S.; Goldberg, M.S.; Fan, X.; Wheeler, A.J. Further Interpretation of the Acute Effect of Nitrogen Dioxide Observed in Canadian Time-Series Studies. *J. Expo. Sci. Environ. Epidemiol.* **2007**, *17*, S36–S44. [CrossRef] [PubMed]
117. Sarnat, J.A.; Schwartz, J.; Catalano, P.J.; Suh, H.H. Gaseous Pollutants in Particulate Matter Epidemiology: Confounders or Surrogates? *Environ. Health Perspect.* **2001**, *109*, 1053–1061. [CrossRef]
118. Chen, R.; Samoli, E.; Wong, C.M.; Huang, W.; Wang, Z.; Chen, B.; Kan, H. Associations between Short-Term Exposure to Nitrogen Dioxide and Mortality in 17 Chinese Cities: The China Air Pollution and Health Effects Study (CAPES). *Environ. Int.* **2012**, *45*, 32–38. [CrossRef] [PubMed]
119. Brauer, M.; Hoek, G.; Smit, H.A.; de Jongste, J.C.; Gerritsen, J.; Postma, D.S.; Kerkhof, M.; Brunekreef, B. Air Pollution and Development of Asthma, Allergy and Infections in a Birth Cohort. *Eur. Respir. J.* **2007**, *29*, 879–888. [CrossRef]
120. Hou, X.; Huang, H.; Hu, H.; Wang, D.; Sun, B.; Zhang, X.D. Short-Term Exposure to Ambient Air Pollution and Hospital Visits for IgE-Mediated Allergy: A Time-Stratified Case-Crossover Study in Southern China from 2012 to 2019. *eClinicalMedicine* **2021**, *37*, 100949. [CrossRef]
121. Guo, C.; Chang, L.Y.; Wei, X.; Lin, C.; Zeng, Y.; Yu, Z.; Tam, T.; Lau, A.K.H.; Huang, B.; Lao, X.Q. Multi-Pollutant Air Pollution and Renal Health in Asian Children and Adolescents: An 18-Year Longitudinal Study. *Environ. Res.* **2022**, *214*, 114144. [CrossRef]

122. Christidis, T.; Erickson, A.C.; Pappin, A.J.; Crouse, D.L.; Pinault, L.L.; Weichenthal, S.A.; Brook, J.R.; van Donkelaar, A.; Hystad, P.; Martin, R.V.; et al. Low Concentrations of Fine Particle Air Pollution and Mortality in the Canadian Community Health Survey Cohort. *Environ. Health* **2019**, *18*, 84. [CrossRef]
123. Jerrett, M.; Burnett, R.T.; Beckerman, B.S.; Turner, M.C.; Krewski, D.; Thurston, G.; Martin, R.V.; van Donkelaar, A.; Hughes, E.; Shi, Y.; et al. Spatial Analysis of Air Pollution and Mortality in California. *Am. J. Respir. Crit. Care Med.* **2013**, *188*, 593–599. [CrossRef]

MDPI AG
Grosspeteranlage 5
4052 Basel
Switzerland
Tel.: +41 61 683 77 34

*International Journal of Environmental Research and Public Health* Editorial Office
E-mail: ijerph@mdpi.com
www.mdpi.com/journal/ijerph

www.ingramcontent.com/pod-product-compliance
Lightning Source LLC
Chambersburg PA
CBHW041422210926
51850CB00013B/3329
* 9 7 8 3 7 2 5 8 2 3 3 4 5 *